# Mass Spectrometry
# A Tutorial

# NATO ASI Series

**Advanced Science Institutes Series**

*A Series presenting the results of activities sponsored by the NATO Science Committee, which aims at the dissemination of advanced scientific and technological knowledge, with a view to strengthening links between scientific communities.*

The Series is published by an international board of publishers in conjunction with the NATO Scientific Affairs Division

| | |
|---|---|
| **A Life Sciences**<br>**B Physics** | Plenum Publishing Corporation<br>London and New York |
| **C Mathematical and Physical Sciences**<br>**D Behavioural and Social Sciences**<br>**E Applied Sciences** | Kluwer Academic Publishers<br>Dordrecht, Boston and London |
| **F Computer and Systems Sciences**<br>**G Ecological Sciences**<br>**H Cell Biology**<br>**I Global Environmental Change** | Springer-Verlag<br>Berlin, Heidelberg, New York, London, Paris and Tokyo |

**NATO-PCO-DATA BASE**

The electronic index to the NATO ASI Series provides full bibliographical references (with keywords and/or abstracts) to more than 30000 contributions from international scientists published in all sections of the NATO ASI Series.
Access to the NATO-PCO-DATA BASE is possible in two ways:

– via online FILE 128 (NATO-PCO-DATA BASE) hosted by ESRIN, Via Galileo Galilei, I-00044 Frascati, Italy.

– via CD-ROM "NATO-PCO-DATA BASE" with user-friendly retrieval software in English, French and German (© WTV GmbH and DATAWARE Technologies Inc. 1989).

The CD-ROM can be ordered through any member of the Board of Publishers or through NATO-PCO, Overijse, Belgium.

**Series C: Mathematical and Physical Sciences - Vol. 353**

# Mass Spectrometry in the Biological Sciences: A Tutorial

edited by

Michael L. Gross
Midwest Center for Mass Spectrometry,
Department of Chemistry,
University of Nebraska-Lincoln,
Lincoln, Nebraska, U.S.A.

Springer Science+Business Media, B.V.

Proceedings of the NATO Advanced Study Institute on
Mass Spectrometry in the Molecular Sciences
Cetraro, Italy
June 17–29, 1990

**Library of Congress Cataloging-in-Publication Data**

NATO Advanced Study Institute on "Mass Spectrometry in the Molecular Sciences" (1990 : Cetraro, Italy)
Mass spectrometry in the biological sciences : a tutorial / edited by Michael L. Gross.
p. cm. -- (NATO ASI series. Series C, Mathematical and physical sciences ; v. 353)
"Proceedings of the NATO Advanced Study Institute on 'Mass Spectrometry in the Molecular Sciences,' held in Cetraro, Italy, June 17-129, 1990."
"Published in cooperation with NATO Scientific Affairs Division."
Includes index.
ISBN 978-94-010-5157-6 ISBN 978-94-011-2618-2 (eBook)
DOI 10.1007/978-94-011-2618-2
1. Mass spectrometry--Congresses. 2. Biomolecules--Analysis--Congresses. I. Gross, Michael L. II. Title. III. Series.
QP519.9.M3N37 1990
574.19'285--dc20 91-40542

ISBN 978-94-010-5157-6

# CONTENTS

# PREFACE

The developments in mass spectrometry over the past fifteen years have been impressive in their implications in bioanalytical chemistry. The achievements begin with the inventions of Cf-252 Plasma Desorption Mass Spectrometry by Macfarlane and Fourier Transform Mass Spectrometry by Comisarow and Marshall in the mid 1970s. The former showed the feasibility of producing large gas-phase ions from large biomolecules whereas the latter enhanced the capabilities for ion trapping especially in analytical mass spectrometry.

A major achievement was the development by Barber of Fast Atom Bombardment (FAB) mass spectrometry, an advance that heralded a new era in biological mass spectrometry. Contemporary and routine instruments such as magnetic sectors and quadrupoles were rapidly adapted to FAB, and nearly the entire universe of small molecules became amenable to study by mass spectrometry.

The introduction of FAB also paved the way for improvement of instrument capability. For example, the upper mass limit of magnet sector mass spectrometers was increased by nearly an order of magnitude by the instrument manufacturers. Furthermore, the technique of tandem mass spectrometry (MS/MS) was given new meaning because important structural information for biomolecules could now be produced for ions introduced by FAB into the tandem instrument. The evolution of MS/MS continues today with the development of ion traps, time-of-flight, and sector instruments equipped with array detection.

A second wave of the revolution began in the late 1980s with the development of electrospray ionization by Fenn and matrix-assisted laser desorption by Hillenkamp. These remarkable methods moved molecular weight determination into the range of a few hundred thousand. In addition, the matrix-assisted laser desorption provided even more driving force for the renaissance of time-of-flight mass spectrometry. The quadrupole ion trap was also reborn, and, borrowing ideas from FTMS, rapidly showed potential in biological molecule determination.

Improvements in separation science and new interfaces for liquid separation methods and mass spectrometers pushed forward new ideas such as thermospray, electrospray, and continuous flow FAB. Less dramatic but continuing growth in the understanding of ion chemistry and the introduction of new methods such as neutralization-reionization mass spectrometry and ion traps promise new applications in the biological sciences.

These spectacular developments in the science of mass spectrometry formed the backdrop for the NATO Advanced Study Institute (ASI) on the subject of "Mass Spectrometry in the Molecular Sciences". The institute was held at the Hotel San Michele just outside of the small village of Cetraro in Southern Italy, from June 17-29, 1990. From a perch overlooking the Mediterranean, 101 scientists discussed, among other subjects, the developments outlined above.

Credit for the idea for the ASI belongs to Professor G. Sindona (University of Calabria) and his colleagues Professors A. Malorni and G. Marino at the University of Naples. Local arrangements, invitations to European participants and help with the design of the scientific program were also handled by Professor Sindona.

Special thanks go to the NATO scientific committee and the Italian Research Council (CNR), who provided most of the funds for the institute, and to the University of Calabria, Finnigan MAT (Italy) and VG (Fisons) Instruments (Italy), who contributed supplementary funding. Travel grants for five young U.S. scientists were generously provided by the U.S. National Science Foundation. We are also grateful to the "Comunitá Montana" of Paola and the Tourist Board of Calabria Region for sponsoring social activities.

The editor acknowledges the special efforts by the seventeen other senior lecturers: P. Arpino (France), A. Burlingame (USA), R. Caprioli (USA), C. Costello (USA), P. Derrick (UK), J. Fenn (USA), S. Gaskell (USA), S. Hammerum (Denmark), F. Hillenkamp (Germany), A. Malorni (Italy), G. Marino (Italy), H. Morris (UK), P. Roepstorff (Denmark), N. Russo (Italy), H. Schwarz (Germany), G. Sindona (Italy), and H. Wollnik (Germany). Most have generously contributed to this book.

The senior lecturers, in discussions with the editor, decided that yet another compilation of short research/review articles was not needed in mass spectrometry. In an effort to provide a more useful record of the Advanced Study Institute, members of this group promoted the idea of a tutorial/review. Thus, the book that follows is intended to be principally a tutorial on the current status of biological mass spectrometry. It is divided into three sections. The first section "Instrumentation for Mass Analysis and Detection" is designed to introduce and to review the basic principles of the major methods for conducting mass analysis and ion detection in mass spectrometry.

Section II "Methods in Mass Spectrometry" treats the various approaches that are useful for studying biological molecules. Included are the newer methods of ionization and sample introduction, as well as neutralization-reionization MS and molecular orbital theory.

Section III, "Applications to Biomolecules", comprises a set of chapters on how the various methods and instruments can be used to solve structural problems in peptide and protein, carbohydrate, nucleoside and nucleotide, and lipid biochemistry.

A purpose of the book, in addition to providing a record of the meeting, is to serve as an introduction to mass spectrometry to newcomers to the field as well as to graduate students. The contributors and the editor hope that this tutorial may find use for some years as a learning instrument in graduate classrooms and seminars.

The editor wishes to acknowledge Kathy Schinstock and Lorraine Moon for their expert help in preparing the book.

Michael L. Gross
Lincoln, NE (USA)
September, 1991

# CONTRIBUTING AUTHORS


| | |
|---|---|
| I. J. Amster | Department of Chemistry, University of Georgia, Athens, Georgia 30206 USA |
| P. J. Arpino | Institut National Agronomique, Laboratoire de Chimie Analytique, 16, rue Claude Bernard, 75231 Paris 05, France |
| K. D. Ballard | Center for Experimental Therapeutics, Baylor College of Medicine, Houston, TX 77030, USA |
| A. L. Burlingame | Department of Pharmaceutical Chemistry, Mass Spectrometry Facility, University of California, San Francisco, CA 94143-0446, USA |
| K. A. Caldwell | Midwest Center for Mass Spectrometry, Department of Chemistry, University of Nebraska-Lincoln, Lincoln, NE 68588-0362, USA |
| R. M. Caprioli | The Analytical Chemistry Center and the Department of Biochemistry and Molecular Biology, The University of Texas Medical School, P. O. Box 20708, Houston, TX 77225, USA |
| R. L. Cerny | Midwest Center for Mass Spectrometry, Department of Chemistry, University of Nebraska-Lincoln, Lincoln, NE 68588-0362, USA |
| C. E. Costello | Mass Spectrometry Facility, Department of Chemistry, Massachusetts Institute of Technology, Cambridge, MA 02139, USA |
| F. De Angelis | University of l'Aquila, Department of Chemistry, Via Assergi, 4, I-67100 l'Aquila, Italy |
| P. J. Derrick | Institute of Mass Spectrometry, University of Warwick, Coventry CV4 7AL, United Kingdom |
| R. C. Dorey | Department of Chemistry, Wichita State University, Wichita, KS 67208, USA |
| D. F. Dyckes | Department of Chemistry, University of Houston, Houston, TX 77204-5641, USA |
| H. Ehring | Institute für Medizinische Physik der Universität Münster, Robert-Koch-Str. 31 D-4400 Münster, Germany |
| J. B. Fenn | Department of Chemical Engineering, Yale University, New Haven, CT USA |
| P. Ferranti | Servizio di Spettrometria di Massa, CNR-Universitá di Napoli, Via Pansini, 5, 80131 Napoli, Italy |
| S. W. Fink | Department of Chemistry, University of Houston, Houston, TX 77204-5641, USA |

R. B. Freas — Department of Chemistry, University of Houston, Houston, TX 77204-5641, USA

S. J. Gaskell — Center for Experimental Therapeutics, Baylor College of Medicine, Houston, TX 77030, USA

P. W. Geno — Southwest Research Institute, 6220 Culebra Rd., P. O. Drawer 28510, San Antonio, TX 78228-0510 USA

G. S. Gorman — Department of Chemistry, University of Georgia, Athens, Georgia 30206 USA

M. L. Gross — Midwest Center for Mass Spectrometry, Department of Chemistry, University of Nebraska-Lincoln, Lincoln, NE 68588-0362, USA

U. Grüner — II. Physikalisches Institut der Universität Giessen, 6300 Giessen, Germany

A. Hassenbürger — Institute für Medizinische Physik der Universität Münster, Robert-Koch-Str. 31 D-4400 Münster, Germany

F. Hillenkamp — Institute für Medizinische Physik der Universität Münster, Robert-Koch-Str. 31 D-4400 Münster, Germany

C. L. Holliman — Midwest Center for Mass Spectrometry, Department of Chemistry, University of Nebraska-Lincoln, Lincoln, NE 68588-0362, USA

C. B. Jacoby — Midwest Center for Mass Spectrometry, Department of Chemistry, University of Nebrasaka-Lincoln, Lincoln, NE 68588-0362, USA

G. Li — II. Physikalisches Institut der Universität Giessen, 6300 Giessen, Germany

A. Malorni — Servizio di Spettrometria di Massa, CNR-Universitá di Napoli, Via Pansini, 5, 80131 Napoli, Italy

M. Mann — Department of Molecular Biology, Odense University, DK-5230 Odense, Denmark

G. Marino — Dipartimento di Chimica Organica e Biologica, Universitá di Napoli, 80134 Napoli, Italy

M. A. McLean — Department of Chemistry, University of Houston, Houston, TX 77204-5641, USA

W. T. Moore — The Analytical Chemistry Center and the Department of Biochemistry and Molecular Biology, The University of Texas Medical School, P. O. Box 20708, Houston, TX 77225, USA

A. Overberg — Institute für Medizinische Physik der Universität Münster, Robert-Koch-Str. 31 D-4400 Münster, Germany

M. J. Polce — Department of Chemistry, The University of Akron, Akron, Ohio 44325-3601, USA

| | |
|---|---|
| P. Pucci | Servizio di Spettrometria di Massa, CNR-Universitá di Napoli, Via Pansini, 5, 80131 Napoli, Italy |
| P. Roepstorff | Department of Molecular Biology, Odense University, DK-5230 Odense M Denmark |
| N. Russo | Dipartimento di Chimica, Universitá della Calabria, I-87030 Arcavacata di Rende (CS), Italy |
| C. A. Settineri | Department of Pharmaceutical Chemistry, Mass Spectrometry Facility, University of California, San Francisco, CA 94143-0446, USA |
| S. Shen | Department of Chemical Engineering, Yale University, New Haven, CT USA |
| R. Siciliano | Servizio di Spettrometria di Massa, CNR-Universitá di Napoli, Via Pansini, 5, 80131 Napoli, Italy |
| G. Sindona | Dipartimento di Chimica, Universitá della Calabria, I-87030 Arcavacata di Rende (CS), Italy |
| J. P. Speir | Department of Chemistry, University of Georgia, Athens, Georgia 30206 USA |
| K. Stachowiak | Department of Chemistry, University of Houston, Houston, TX 77204-5641, USA |
| S. Suwanrumpha | Department of Chemistry, University of Houston, Houston, TX 77204-5641, USA |
| M. Toscano | Dipartimento di Chimica, Universitá della Calabria, I-87030 Arcavacata di Rende (CS), Italy |
| J. R. Trainor | Institute of Mass Spectrometry, University of Warwick, Coventry CV4 7AL, United Kingdom |
| R. B. van Breemen | Department of Chemistry, North Carolina State University, Raleigh, North Carolina 27695-8204, USA |
| R. Wait | Division of Pathology, Public Health Laboratory Service Centre for Applied Microbiology and Research, Porton Down, Salisbury, United Kingdom |
| C. Wesdemiotis | Department of Chemistry, The University of Akron, Akron, Ohio 44325-3601, USA |
| C. Wilder | Department of Chemistry, University of Houston, Houston, TX 77204-5641, USA |
| H. Wollnik | II. Physikalisches Institut der Universität Giessen, 6300 Giessen, Germany |
| V. H. Wysocki | Department of Chemistry, Box 2006, Virginia Commonwealth University, Richmond, VA 23225-2006, USA |

# PARTICIPANTS IN THE INSTITUTE

## CANADA


J. H. Banoub
Science Branch,
Fisheries and Oceans Canada
P. O. Box 5667
St. John's, Newfoundland A1C 5X1
Canada

J. R. Jocelyn Paré
Head, Analytical Programs
R. R. T. E. C.
Ottawa, Ontario K1A OH3
Canada


## DENMARK


Gustav Bojesen
Department of Chemistry
Odense Universitet
Campusvej 55
DK-5230 Odense
Denmark

Søren Bøwadt
Department of Chemistry
Odense Universitet
Campusvej 55
DK-5230 Odense
Denmark

L Steen Hammerum
Dept. of General and Organic Chemistry
The H. C. Orsted Institute
University of Copenhagen
DK-2100 Copenhagen
Denmark

Peter Højrup
Department of Molecular Biology
Odense Universitet
Campusvej 55
DK-5230 Odense
Denmark

Ulla Justesen
Department of Chemistry
Odense Universitet
Campusvej 55
DK-5230 Odense
Denmark

Klaus Klarskov
Department of Molecular Biology
Odense Universitet
Campusvej 55
KD-5230 Odense
Denmark

Gunhild Khristiansen
Department of Chemistry
Odense Universitet
Campusvej 55
DK-5230 Odense
Denmark

Matthias Mann
Department of Molecular Biology
Odense Universitet
Campusvej 55
DK-5230 Odense
Denmark

L/A Peter Roepstorff
Department of Molecular Biology
Odense Universitet
Campusvej 55
DK-5230 Odense
Denmark

Gert Talbo
Department of Molecular Biology
Odense Universitet
Campusvej 55
DK-5230 Odense
Denmark


L = Lecturer
A = Principal Author

## FRANCE

L/A Patrick Arpino
Ist. National Agronomique
Laboratoire de Chimie Analytique
16, Rue Claude-Bernard
F-75231 Paris-Cedex
France

Isabelle Gentais
Institut National Agronomique
16, Rue Claude-Bernard
F-75231 Paris-Cedex
France

Thierry Blasco
Laboratoire de Chimie Orgtanique et Structurale
Université Pierre et Marie Curie
4, Place Jussieu
F-75231 Paris Cedex
France

Klaus D. Haegele
Department of Clinical Research
Merrel Dow Res. Institute
2, Rue de Stockholm
F-67000 Strasbourg
France

## GERMANY

Hanno Ehring
Institut fuer Medizinishce Physik
Westf. Wilhelms Universitate
Huefferstrasse 68
D-4400 Muenster
Germany

L/A Franz Hillenkamp
Inst. fuer Medizinische Physik
Universitaet Munster
Heufferstrasse 68
Muenster
Germany

Andreas Overberg
Institut fuer Medizinische Physik
Westf. Wilhelms Universitate
Huefferstrasse 68
D-4400 Muenster
Germany

L/A Hermann Wollnik
Physikalisches Institut
Justus-Liebig Universitaet
D-6300 Giessen
Germany

Anna Hassenbuerger
Institut fuer Medizinische Physik
Westf. Wilhelms Universitaet
Huefferstrasse 68
D-4400 Muenster
Germany

Per Nielsen
Department of Chemistry
Universitaet Konstanz
Postfach 5560
D-7750 Konstanz
Germany

L Helmut Schwarz
Inst. fuer Organische Chemie
Technische Universitaet Berlin
Strasse des 17 Juni, 135
D-1000 Berlin 12
Germany

## ITALY

Maria Vittoria Barbarulo
Dip. di Chimica
Universitá "La Sapienza"
P. le Aldo Moro, 5
I-00185 Roma
Italy

Francesca Buiarelli
Dip. Chimica Analitica
Universitá "La Sapienza"
P. le Aldo Moro, 5
I-00185 Roma
Italy

Tiziano Bullani
Finnigan MAT
Milano
Italy

Alessandro Cracco
Dip. di Pediatria
Universitá di Padova
Via Giustiniani, 6
I-35100 Padova
Italy

Enrico Davoli
Instituto "Mario Negri"
Via Eritrea, 62
I-20157 Milano
Italy

Giuseppe Elli
VG Instruments
Viale dell'assunta 101
I-20063 Cernusco sul Naviglio (MI)

Domenico Garozzo
CNR, Ist. Per la Chimica e la
Tecnologia dei Materiali Polimerici
Viale Andrea Doria, 6
I-95125 Catania
Italy

Bruno Malavasi
Ist. Scienze Farmacologiche
Universitá di Milano
Via Balzaretti, 9
I-20133 Milano
Italy

A/L Gennaro Marino
Dip. Chimica Organica e Biologica
Universitá di Napoli
Via Mezzocannone 16
I-80134 Napoli
Italy

Gerardo Palla
Ist. Chimica Organica
Universitá di Parma
I-43100 Parma
Italy

Giampiero Carreca
Policlinico Militare
Via Saint Bon, 7
I 20147 Milano
Italy

Giovanna Danza
Dip. Fisiopatologia Clinica
Universitá di Firenze
Via Pieraccini, 6
I-50100 Firenze
Italy

A Francesco De Angelis
Dip. di Chimica
Universitá di L'Aquila
Via Assergi 4
I-67100 L'Aquila
Italy

Pasquale Ferrante
Serv. di Spettrometria di Massa
II Facoltá di Medicina
Via S. Pansini, 5
I-80131 Napoli
Italy

Francesco Greco
Dip. di Chimica
Universitá della Calabria
I-87030 Arcavacata Di Rende (CS)
Italy

L/A Antonio Malorni
Serv. di Spettrometria di Massa
II Facoltá di Medicina
Via S. Pansini, 5
I-80131 Napoli
Italy

Paolo Mascaro
Dip. di Chimica
Universitá della Calabria
I-87030 Arcavacata di Rende (CS)
Italy

Giuseppe Pieraccini
Dip. di Farmacologia
Universitá di Firenze
Viale Morgagni 65
I-50134 Firenze
Italy

Piero Pucci
Instituto di Chimica
Universitá della Basilicata
I-85100 Potenza
Italy

L/A Nino Russo
Dip. di Chimica
Universitá della Calabria
I-87030 Arcavacata di Rende (CS)
Italy

Eligio Sebastiani
VG Instruments
Viale dell'assunta 101
I-20063 Cernusco sul Naviglio (MI)
Italy

Luca Simonotti
Finnigan MAT
Milano
Italy

Francesca Zappacosta
Serv. di Spettrometria di Massa
II Facoltá di Medicina
via S. Pansini, 5
I-80131 Napoli
Italy

Margherita Ruoppolo
Serv. di Spettrometria di Massa
II Facoltá di Medicina
via S. Pansini, 5
I-80131 Napoli
Italy

Marina Scandola
Dept. of Chemistry
Glaxo S. p. A.
Via Fleming, 2
I-37100 Verona
Italy

Rosa Siciliano
Serv. di Spettrometria di Massa
II Facoltá di Medicina
via S. Pansini, 5
I-80131 Napoli
Italy

L/A Giovanni Sindona
Dip. di Chimica
Universitá della Calabria
I-87030 Arcavacata di Rende (CS)
Italy

**THE NETHERLANDS**

H. P. Benschop
TNO Prins Maurits Laboratory
Chem. & Technological Res. Institue
Lange Kleiweg 137
NL 2288 GJ Rijswijk ZH
The Netherlands

**PORTUGAL**

Graca Santana-Marques
Department of Chemistry
University of Aveiro
P-3800 Averio
Portugal

**POLAND**

Witold Danikiewicz
Polish Academy of Sciences
Institute of Organic Chemistry
ul. Kasprzka 44
01-224 Warszawa
Poland

## SPAIN

Joaquin Abian
Depto. Neuroquimica
CID-CSIC
Jordi Giona 18-26
E-08034 Barcelona
Spain

Carlos Celma
Depto. de Quimica Organica
Facultad de Quimica
Universidad de Barcelona
Marti i Franquest 1-11
E-08028 Barcelona
Spain

Cinta Prote
Depto. de Quimica Ambiental
CID-CSIC
Jordi Giona 18-26
E-08034 Barcelona
Spain

Pilar Fernandez Roman
Env. Chemistry Dept.
CID-CSIC
Jordi Giona 18-26
E-08034 Barcelona
Spain

## TURKEY

Nalan Akcakaya
Department of Chemistry
Istanbul Technical University
Maslak, Instanbul
Turkey

Sevim Bilgic
Anadolu Universitesi
Fen Edebiyat Fakultesi
Kimya Bolumu
Eskisehir, Turkey

Sevim Dalkara
Pharmaceutical Chem. Dept.
Faculty of Pharmacy
Hacettepe University
Ankara, Turkey

Leyla Kalaycioglu
Department of Biochemistry
Veterinary Faculty
Selcuk University
Konya, Turkey

Semih Otles
Faculty of Engineering
Department of Food Engineering
University of EGE
Bornova, Izmir
Turkey

Tamerkan Ozgen
Dept. of Chemistry
Faculty of Science
EGE University
Bornova/Izmir
Turkey

Cihangir Tanyeli
Middle East Technical Univ.
Department of Chemistry
0653 Ankara Turkey

## UNITED KINGDOM

Parastoo Azadi
Department of Biochemistry
Imperial College
Imperial College Road
London SW7 2AZ
United Kingdom

Caroline D. Bradley
Department of Chemistry
Warwick Inst. of Mass Spectrometry
University of Warwick
Coventry CV4 7AL
United Kingdom

Anna Chan
Department of Biochemistry
Imperial College
Imperial College Road
London SW7 2AZ
United Kingdom

Dominic Chan
Department of Chemistry
Warwick Inst. of Mass Spectrometry
University of Warwick
Coventry CV4 7AL
United Kingdom

Alex Colburn
Department of Chemistry
Warwick Inst. of Mass Spectrometry
University of Warwick
Coventry CV4 7AL
United Kingdom

James I. Langridge
Biochemistry Res. Group
University College of Swansea
Swansea SA2 8PP
United Kingdom

G. A. Mills
South Laboratory & Pathology Block
Southampton General Hospital
Tremona Road
Southampton SO9 4XY
United Kindgom

Maria Panico
Department of Biochemistry
Imperial Colege
Imperial College Road
London SW7 2AZ
United Kingdom

A Robin Wait
P.H.L.S., Centre for Applied Microbiology and Research
Division of Pathology
Porton Down
Salisbury, Wiltshire SP4 0JG
United Kingdom

L/A Peter Derrick
Department of Chemistry
Warwick Inst. of Mass Spectrometry
University of Warwick
Coventry CV4 7AL
United Kingdom

John Letcher
VG Instruments
Floats Road, Wythenshawe
Manchester M23 9LE
United Kingdom

L Howard R. Morris
Department of Biochemistry
Imperial College
Imperial College Road
London SW7 2AZ
United Kingdom

Andrew Reason
Department of Biochemistry
Imperial College
Imperial College Road
London SW7 2AZ
United Kingdom

## UNITED STATES

A I. Jonathan Amster
Dept. of Chemistry
University of Georgia
Athens, Georgia 30602 (USA)

Daniel D. Bombick
Wright State University
Dayton, OH (USA)

L/A A. L. Burlingame
Department of Chemistry
University of California
San Francisco, CA 94143 (USA)

L/A Richard M. Caprioli
Analytical Chemistry Center
The University of Texas
Houston, TX 77030 (USA)

Kevin Ballard
Center for Experimental Therapeutics
Baylor College of Medicine
One Baylor Plaza, Room 826E
Houston, TX 77030 (USA)

James G. Boyle
Department of Chemical Engineering
Yale University
New Haven, CT 06520 (USA)

Kenny Caldwell
Department of Chemistry
University of Nebraska
Lincoln, NE 68588-0304 (USA)

Ronald L. Cerny
Midwest Center for Mass Spectrometry
Dept. of Chemistry
Univeristy of Nebraska
Lincoln, NE 68588-0362 (USA)

L/A Catherine Costello
Department of Chemistry
Massachusetts Institute of Technology
Cambridge, MA 02139 (USA)

A R. Cameron Dorey
Dept. of Chemistry
University of Central Arkansas
Conway, AR 72032 (USA)

L/A John Fenn
Department of Chemical Engineering
Yale University
New Haven, CT 06520 (USA)

A Royal B. Freas
Dept. of Chemistry
University of Houston
Houston, TX 77209-5641 (USA)

Douglas A. Gage
MSU-NIH Mass Spectrometry Facility
Dept. of Biochemistry
Michigan State University
East Lansing, MI 48824-1319 (USA)

L/A Simon Gaskell
Center for Experimental Therapeutics
Baylor College of Medicine
One Baylor Plaza, Room 826E
Houston, TX 77030 (USA)

A Paul W. Geno
Southwest Research Institute
6220 Culebra Road
Post Office Drawer 28510
San Antonio, TX 78228-0510

L/A Michael L. Gross
Department of Chemistry
University of Nebraska
Lincoln, NE 68588-0304 (USA)

Ronald C. Orlando
Structural Biochemistry Center
Department of Chemistry
Univ. of Maryland-Baltimore County
Baltimore, MD 21228 (USA)

Damon Papac
Dept. of Pharmacology
Medical University of South Carolina
Charleston, SC 29425 (USA)

Christine Settineri
Dept. of Pharamaceutical Chemistry
University of California
521 Parrnasus Street, Rm C18
San Francisco, CA 94143-0446 (USA)

Elisabeth A. Stemmler
Dept. of Chemistry
Bowdoin College
Brunswick, ME 04011 (USA)

A Richard van Breemen
Department of Chemistry
North Carolina State University
Raleigh, NC 27695 (USA)

A Chrys Wesdemiotis
Department of Chemistry
Univeristy of Akron
Akron, OH 44325-3601 (USA)

A Vicki Wysocki
Department of Chemistry
Box 2006
Virginia Commonwealth University
Richmond, Virginia 23225-2006

Joseph Zaia
Department of Chemistry
Massachusetts Institute of Technology
Cambridge, MA 02139 (USA)

# Part I
# Instrumentation for Mass Analysis and Detection

# SECTORS AND TANDEM SECTORS

J. R. TRAINOR AND P. J. DERRICK
*Institute of Mass Spectrometry*
*University of Warwick*
*Coventry CV4 7AL*
*United Kingdom*

ABSTRACT. The ion optics of sector and tandem sector mass spectrometers is a complex concern, integral to all their applications. This article provides an introduction to instrumental aspects and ion optics of sector mass spectrometers. The $\alpha^2$ aberration is derived as an example of the imperfections in first order ion optic theory, which affect the all-important instrumental qualities of sensitivity, resolution and precision in mass determinations. Some other aberrations are discussed in light of modern methods for correction or compensation for those imperfections. Configurations of tandem sector instruments are introduced and discussed in the context of the requirements of different experiments.

## 1. Introduction

In this review article we cover some instrumental aspects of sector-field mass spectrometers. It is important to establish at the outset that, while the focus of the current paper is on the magnetic and electric sectors of the spectrometer, the operation of the high-performance mass spectrometer depends on additional slits, ion guides, lenses, the detector and critically on the ion source. The electric and magnetic fields can be combined in many ways including arrangements such as in the Wien filter, but in every case the overall effect of the combination of these fields is to act as either a mass-to-charge ratio or as a velocity filter (or both simultaneously). This emerges from the mathematics of the two most important equations in the ion physics relating to one of the simplest combinations, where an accelerating electric field precedes a magnetic field.

$$Bzq = mv/r \tag{1}$$

$$zqV = mv^2/2 \tag{2}$$

$B$ is the homogeneous magnetic induction normal to the velocity $v$ of the ion, $m$ is the mass of the ion, $q$ is the elementary charge, $z$ is the number and sign of the elementary charges, $V$ is the the accelerating potential experienced by the ion prior to entering the magnetic field and $r$ is the radius of the ion orbit under the influence of the magnetic field.

Either $m$ or $v$ can be eliminated from equations (1) and (2) giving:

$$m/z = B^2r^2q/2V \tag{3}$$

$$v = V/2Br \tag{4}$$



*M. L. Gross (ed.), Mass Spectrometry in the Biological Sciences: A Tutorial*, 3–27.

An interdependence of mass and velocity is evident. This principle has been recognised in the practice of mass spectrometry since Thomson's paraboloids [1]. In his arrangement of electric and magnetic fields, aligned parallel and superimposed, information on the mass and velocity distribution is simultaneously available from the images left on photographic film. To obtain good sensitivity and resolution, even from the most carefully designed and constructed sectors, the generation of a beam of minimal angular divergence and minimal energy spread is an integral part of the problem.

## 2. Sector optics

There are many similarities between light and ion optics such that much understanding of ion optics can be gained from a knowledge of light optics. The analogy, however, is far from formally complete. The physical forms of the lenses are different. Light-optic lenses are hard, solid, well-defined volumes in which the refractive faces can be precisely shaped. The fields which cause refraction in ion-optic lenses change their refractive index continuously. The shapes of these fields are restricted by the Laplace equation. Ion optic lenses are rather intangible, without sharp boundaries to the fields of which they are comprised. There are many further limitations to the analogy including the fact that electrons and ions carry charge which introduces a problem in that electrostatic repulsive forces exist between particles in the beam. A more complete set of differences can be read in any of many established text books [2].

Not all light optic lenses have formal equivalents in ion optics, in particular few ion optic lenses are diverging at all and this point is worth highlighting. Simple homogeneous magnetic sectors for example are prisms, yet these also cause direction refocussing. The electrostatic immersion lens and electrostatic quadrupole lens may be diverging; the latter is highly astigmatic producing a line focus because it is always simultaneously convergent and divergent in its two perpendicular planes of action. The thin convex (converging) light optic lens has a formal analogy in ion optics. Parallel rays of light from an object at infinity are brought to a real (positive) focal point by a convex lens. (As a note of digression, we add that as the thickness of the lens increases, the cardinal points of the lens separate. It can then be noticed that the principle plane of the light optic lenses are on the same side as their respective foci, whereas in the electrostatic ion optic lens the principle planes are *crossed over*. Crossing-over of principle planes is a feature of ion optic lenses.) Further inspection of the simple convex ion optic lens reveals that both positively and negatively charged particles alike are focussed by the lens (however by different trajectories within the action of the field). To illustrate this point an einzel lens is considered in Figure 1. The potentials on this electrostatic lens cannot be adjusted to produce a virtual focus from rays of either charge. A simple concave lens in light optics produces a virtual focus from parallel object rays, by diverging the rays as they pass through the refractive medium: optical extrapolation of these rays produces the virtual image. <u>There is no discrete ion optic lens which is formally analogous to such a simple concave light optic lens</u>. By discrete we mean that the object and image rays begin and end in regions of the same electrostatic potential (refractive index) which are separated and distinct from the lens. A virtual focus, however, can be formed from a simple electrostatic ion optic lens, if the object lies within the lens as is the case with an immersion lens (see Figure 2). This point impinges on the scan laws of sector tandem mass spectrometry when a floated collision cell is employed in experiments.

Consideration of texts covering electron optics and the modern ray-tracing techniques will show that ion-optic design has never been so easy to implement as it is nowadays. There are now powerful computer programs to take the weight of the intense theoretical

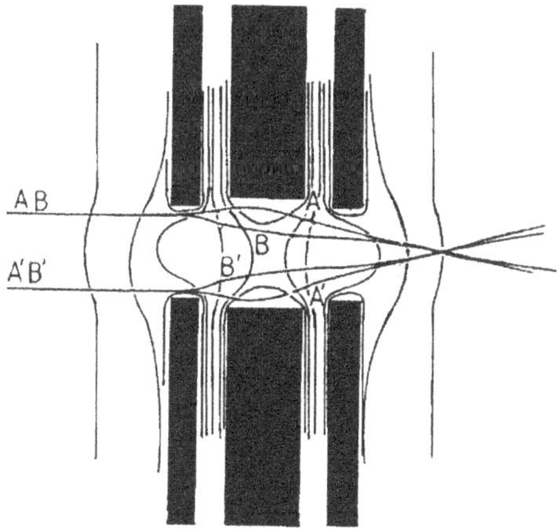


Figure 1. An Einzel lens showing two positive ion and two negative ion trajectories. The lens is convergent for charged particles of all energies. In this example the outer electrodes are at +200 V and the inner electrode at -200 V. The trajectories A and A' are for a 16 Da ion carrying one + charge at total energy (kinetic + potential) of 350 eV. The B, B' trajectories represent the trajectories of a 16 Da ion carrying one - charge with a total energy of 350 eV.

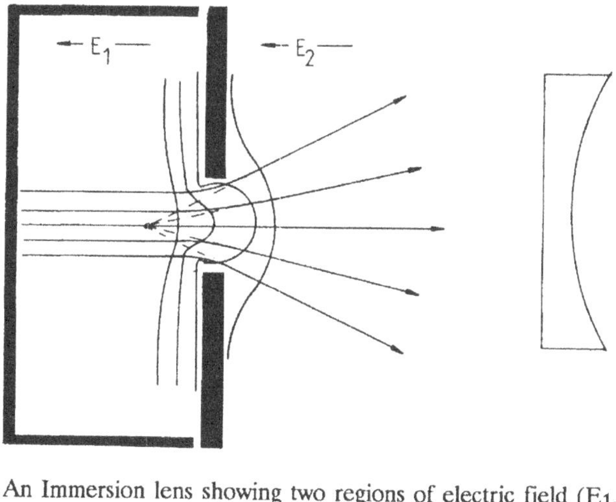


Figure 2. An Immersion lens showing two regions of electric field ($E_1$, $E_2$). $E_1 >$ $E_2$ and the beam of ions is diverged forming a virtual focus inside region $E_1$. An equivalent ion optic lens is shown on the right although strictly speaking the flat glass region should extend backwards toward infinity.

calculations, thus allowing the designer to make and test bold designs quickly and accurately. This capability will become more evident and more widely applicable with advancements in computer technology. The legacy of electron physics is the fundamental theory which is equally applicable to the study of ions.

Mass spectroscopists took advantage of the empirical observation that the trajectories of a beam of ions of given physical attributes (viz., one mass and energy), emerging from a point source under the influence of either a homogeneous magnetic, or radial, homogeneous electric field, reconverge in a plane at points determined by the magnitude of their mass and energy. In the case of the magnetic field this plane is inclined at $180^o$ from the plane of entry into the field. In the case of an electric field direction refocussing occurs along a line in the plane $127^o\ 17'$ from the plane of entry into the field. Note that the electric sector disperses ions of the same mass travelling at different speeds in direction foci along a line in that plane. Early mass spectroscopes such as those of Dempster [3] and Bainbridge [4] were based on $180^o$ magnetic analyzers where the detector or detector slit was placed at the edge of the analyzer where direction refocussing occurs.

We follow the notation of Hintenberger and König [5] in reference to mass spectroscopes as a term encompassing: a) mass spectrometers with point foci (and usually electronic detection); b) mass spectrographs with focal planes (with array or photographic plate detectors). We have also reserved the term 'sector' for the context of analyzers having a wedge like geometry which form a focus at a point in a remote, field free region. The introduction of 'sectors' was a genuine landmark in the development of modern mass spectrometers.

Nier [6] introduced the use of magnetic sector analyzers into mass spectrometry in 1940, after Stephens [7] and Barber [8] showed that the reconvergence of ion beams discussed above was a specific case of a more general refocussing property belonging to all geometries of magnetic analyzers. The introduction of true sectors into mass spectrometry fuelled the study and treatment of aberrations in the search for better performance from the new instruments. Aberrations are the terms given to the non-ideal behaviour of lenses. By this we mean that these are corrections to the oversimplified mathematics required to simulate the operation of real lenses. Contrary to the implication of the word, there is nothing wrong with the physics of the lens. Aberrations could be described as *corrections to the imperfect Gaussian theory of the operation of real lenses.*

## 2.1. DERIVATION OF THE DISTORTION-TERM ($\alpha^2$) OF A SIMPLE MAGNETIC SECTOR

We follow Stephens [6] in his considerations of non-ideal behaviour in a magnetic lens and apply this to ions.

Ions are considered to pass into the field of action of a wedge-shaped magnetic sector with perfect field-boundaries (i.e., an imaginary magnetic sector with no stray field). A diagram of the system is shown in Figure 3. To clarify Stephens' method, we also add dummy variables **l** and **m**. We consider rays of ions emerging from **A**, entering the magnetic sector with angles $\pm\,\alpha$ about the instrument axis **APWB**.

The position of **Q**,**T**,**V** and **U** are defined in the two-dimensional co-ordinate system as $(x_1,y_1)$, $(x_2,y_2)$, $(x_3,y_3)$ and $(x_4,y_4)$. With the origin fixed at a point **A**, we can show that the positions **Q**,**T**,**V** and **U** are transferred on to the following coordinates.

$$\mathbf{A} = (x,y) = (0,0) \tag{5}$$

$$\mathbf{Q} = (x_1,y_1) = \mathbf{l}\cos(\theta+\alpha),\ \mathbf{l}\sin(\theta+\alpha) \tag{6}$$

because

$$\mathbf{l} = a\cos\theta/\cos\alpha \tag{7}$$

we may substitute into the above equation to obtain

$$\mathbf{Q} = (x_1,y_1) = a\cos\theta\cos(\theta + \alpha)/\cos\alpha,\ \ a\cos\theta\sin(\theta + \alpha)/\cos\alpha. \tag{8}$$

Similarly

$$\mathbf{T} = (x_2,y_2) = x_1 + \mathbf{m}\sin(\theta + \alpha),\ \ y_1 - \mathbf{m}\cos(\theta + \alpha) \tag{9}$$

but $\mathbf{m} = r = a\sin\theta$ and when substituted into the above gives

$$\mathbf{T} = (x_2,y_2) = x_1 + a\sin\theta\sin(\theta + \alpha),\ \ y_1 - a\sin\theta\cos(\theta + \alpha). \tag{10}$$

And the co-ordinates of **Q** can be fed into these equations. [Before continuing we define quantities $e = r\cos(\omega)$, $f = r\sin(\omega)$].

$$\mathbf{V} = (x_3,y_3) = x_2 + e,\ \ (x_3 - a)/\tan\varphi \tag{11}$$

$$\mathbf{V} = (x_3,y_3) = x_2 + e,\ \ (x_3 - a)\cot\varphi \tag{12}$$

Therefore, $x_3$ is required to solve $y_3$.

Writing:

$$e^2 = r^2 - f^2 \text{ (Pythagoras) and} \tag{13}$$

$$x_3^2 = x_2^2 + 2x_2e + e^2 \tag{14}$$

we obtain

$$x_3^2 = 2x_2x_3 - x_2^2 + r^2 - f^2. \tag{15}$$

By substitution and rearrangement, c can be written from the equation for $y_3$, as:

$$f = (x_3 - a)\cot\varphi - y_2 \tag{16}$$

or by squaring as:

$$f^2 + 2cy_2 + y_2^2 = (x_3^2 - 2x_3a + a^2)\cot^2\varphi. \tag{17}$$

Therefore,

$$f^2 = x_3^2\cot^2\varphi - 2x_3a\cot^2\varphi + a^2\cot^2\varphi - 2y_2x_3\cot\varphi + 2y_2a\cot\varphi + y_2^2. \tag{18}$$

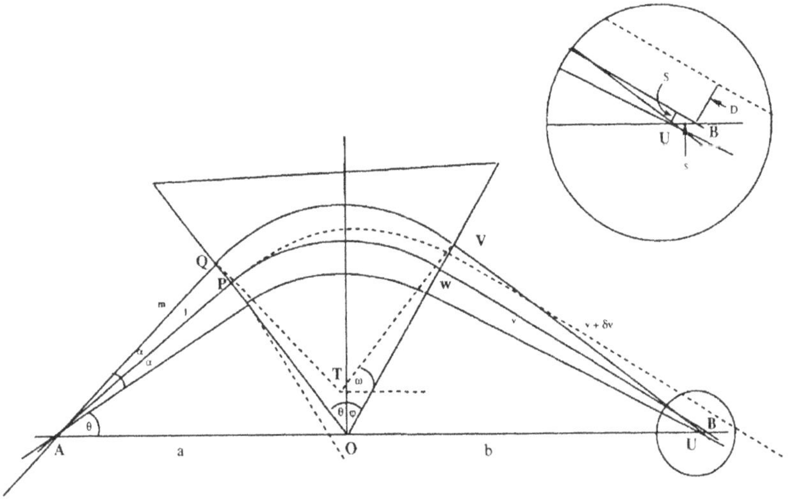


Figure 3. A magnetic sector analyzer. The ray APWB is the axial ray of ions with velocity v. The dotted ray has the same direction of propagation along the axis of the sector, but has a different velocity v + δv. AQVU represents an ion of velocity |v| emerging from A and directed along a line having an angle +α from the axis. The fourth ray is directed towards the sector in a line inclined at -α to the axis. This ray is refocussed at U along the trajectory shown. The inset shows an expansion of the region close to the direction focus.

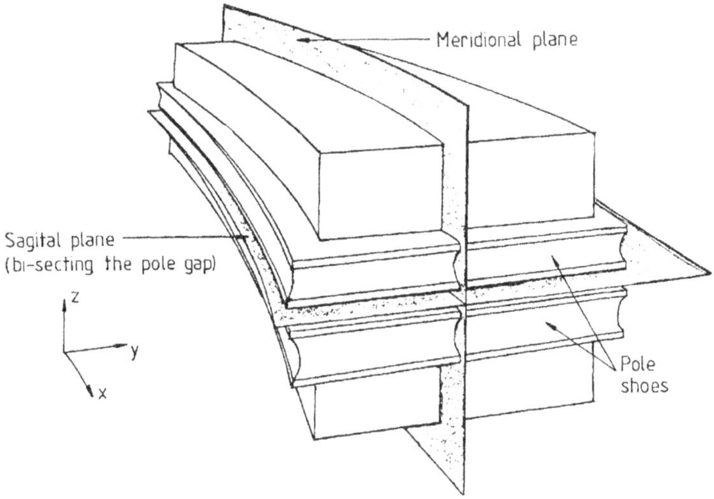


Figure 4. A 3-D diagram of a magnetic sector showing the perpendicular sagital and meridional planes.

Again by substitution, rearrangement, and expansion, we obtain

$$x_3^2 \csc^2\varphi = 2x_3\{x_2 + a\cot^2\varphi + y_2\cot\varphi\} - x_2^2 + r^2 - a^2\cot^2\varphi - 2y_2 a\cot\varphi - y_2^2. \tag{19}$$

By using the method of *completing the square* , this equation can be solved for $x_3$. Note that only the positive root of the equation is meaningful here.

$$U = (x_4, y_4) = x_3 + y_3\tan\omega,\ 0 \tag{20}$$

$$U = (x_4, y_4) = x_3 + y_3\ \{(y_3 - y_2)/(x_3 - x_2)\},\ 0 \tag{21}$$

The above reduces to the following co-ordinates:

$$x_1 = [a\cos\theta\cos(\theta + \alpha)]/\cos\alpha \tag{22a}$$

$$y_1 = [a\cos\theta\sin(\theta + \alpha)]/\cos\alpha \tag{22b}$$

$$x_2 = a\ \{(1/\cos\alpha)\cos\theta\cos(\theta + \alpha) + \sin\theta\sin(\theta + \alpha)\} \tag{23a}$$

$$y_2 = a\ \{(1/\cos\alpha)\cos\theta\sin(\theta + \alpha) - \cos\theta\sin(\theta + \alpha)\} \tag{23b}$$

$$x_3 = \{-n + (n^2 - 4mc)^{0.5}\}/2m \tag{24a}$$

$$y_3 = (x_3 - a)\cot\alpha \tag{24b}$$

$$x_4 = x_3 + y_3(y_3 - y_2)/(x_3 - x_2) \tag{25a}$$

$$y_4 = 0 \tag{25b}$$

Where $m = \csc^2\varphi$,

$n = -2(x_2 + y_2\cot\varphi + a\cot^2\varphi)$, and

$c = (x_2^2 + y_2^2 + 2ay_2\cot\varphi + a^2\cot^2\varphi - a^2\sin^2\theta)$,

The object of the mathematics is to derive the analytical solutions for S and D from Figure 3.

$$s = (a+b) - x_4 \tag{26}$$

$$S = a\ (\sin\varphi + \sin\theta) - \{\sin\varphi[x_3(x_3 - x_2) - y_3(y_3 - y_2)]/(x_3 - x_2)\} \tag{27}$$

We proceed by expanding functions of $\alpha$, one-by-one into Taylor series. These are truncated for terms of higher order than $\alpha^2$ and substituted into equation (27). The mathematics is laborious; each co-ordinate above must be separately treated and re-expanded where necessary, until an expansion of no more than three terms is formed.

$$S = (a\alpha^2/2)[\sin^2\theta/\sin\varphi + \sin^2\varphi/\sin\theta]. \tag{28}$$

When $\varphi = \theta$, this further reduces to:

$$S = (a\alpha^2/2)\sin\theta. \tag{29}$$

D is determined from an analogous derivation using R extended to $R + \Delta R$, following from an increase in velocity of $\Delta v$ in v. The terms R and v are related by the ion physics (i.e., $RH(q/m) = v$), and all powers of $\Delta R > 1$ are discarded in approximating the expansion. We obtain:

$$D = a(\sin\theta/\sin\varphi)(\Delta v/v)(\sin\theta + \sin\varphi) \tag{30}$$

The mathematics, as illustrated by this relatively simple case, would soon become impossible for any detailed calculation taking account of higher-order effects. These more complex problems are solved much more readily by using phase space [9] or the method of transfer matrices. The committed pupils of ion optics are pointed to the excellent text of Wollnik [10], who deals with the method of transfer matrices and other points in much greater detail than is desirable in this overview.

### 2.2. OTHER ABERRATIONS IN MAGNETIC SECTORS

Angular aberrations, as described in Figure 3, in the (x,y) plane of the mass spectrometer are described as $\alpha$ aberrations. Those angular aberrations, which caused by the deviations in the perpendicular (x,z) plane along the axis of the ion path, are known as $\beta$ aberrations. Simple mass spectrometers do not correct this divergence in the $\beta$ plane; however, the design of modern commercial mass spectrometers is directed towards the minimization of $\alpha$ and $\beta$ aberrations. Note that $\beta$ is also used in mass spectrometry by some authors to denote energy; and this difference in nomenclature has caused confusion in the past.

Another seriously deleterious effect on the performance of a mass spectrometer is the $\beta$ or out-of-plane dispersion, which can limit the sensitivity of an instrument. This is a simple first order dispersion, which if uncorrected simply causes the ion beam to spread out along the axis of the mass spectrometer.

There are a total of 10 aberrations to the first order focussing of magnetic sectors arising from the second order terms (ignored in the first order approximation) which describe specific image defects [11]. Of the 10 aberrations, six are sagital and four meridional. A visual definition of these planes is given in Figure 4. Distortion is the first of the six sagital aberrations (derived Section 2.1). If uncorrected, these aberrations compel the operator to work at lower performance with the slits opened wider than the theoretical optimum in order to obtain sensitivity. Resolution is sacrificed by this attempt to obtain satisfactory sensitivity, and, moreover, the peak shape becomes asymmetric and variable, changing with the opening and closing of the slits. Measurements requiring peak centroids such as are needed for accurate mass determination and energy release measurements suffer if the higher-order aberrations are not minimized.

## 2.3. MAGNET DESIGN AND CURATIVE METHODS FOR MAGNETIC IMAGE ABERRATIONS

Blake [12,13] describes the considerations required for the design and construction of simple, homogeneous electromagnets employing DC electrical supplies. Basic electromechanical equations relevant to design are discussed, and materials available at that time for the fabrication of the magnet core are compared. Magnets used in mass spectrometers must have a field spatially confined to between the pole pieces (i.e., the fringe fields should be minimized). Following Blake's criteria, 'C' shaped magnets with the windings coiled around the poles are most suitable of the three methods of winding that are discussed. Modern magnet materials and magnet designs are removed by a generation from Blake's considerations.

Many inventive methods have been described to correct for the image aberrations in a magnetic sector. The electric sector of a double-focussing instrument can be designed to compensate to some extent for some of the aberrations introduced by a non-ideal magnetic field. The different aberrations are mathematically separable, but not all mutually independent and inevitably occur simultaneously. The effect of each may vary and more than one corrective measure may be used in combination. Corrective methods tend to address the different aberrations as though they were independent, thus successful correction must minimize aggravation of the other aberrations which are present. Among the methods that have been applied are angling the detection or image plane to counter image plane rotation aberrations. The distortion term from the magnet that was derived in Section 2.1 tends to be the most serious second-order aberration. This has been tackled by substituting the homogeneous magnetic field with a carefully designed inhomogeneous field. Cottrell and Greathead [14] describe the use of such inhomogeneous magnetic fields. The magnet construction is complex with ridged, curved pole pieces and *on-board* field shims. An alternative method to address the $\alpha^2$ distortion is to shape, or angle the magnet poles in order to approximate a perfect field. The latter method was discussed by Hintenberger and König [5] and has influenced the design of commercial instruments. Another alternative method for correction of $\alpha^2$ aberration is the incorporation of a pre-sector magnetic field shim, which is used to trim the entrance trajectories of ions into the sector field.

A method that has proved popular with the instrument manufacturers for the minimization of stigmatic aberrations is the use of highly astigmatic lenses derived from electrostatic multipole combinations. The hexapole has been used in mass spectrometer optics. Matsuda and co-workers [15] have favoured the use of quadrupole doublets. The addition of quadrupole doublets to correct for the aberration of $\beta$ divergence has certain advantages of flexibility over the utilization of more-difficult-to-fabricate, toroidal electric sectors [16]. The foci in orthogonal planes can be independently varied and in combination allow a method of zooming the mass scale. This has been exploited in the use of an array detector to vary the dynamic range while increasing sensitivity [17].

## 2.4. ELECTRIC SECTORS

The most common type of electrostatic analyzers have a cylindrical geometry. The cylindrical sector is characterized by a circular arc along the instrument's axis, with parallel curved electrodes perpendicular to the $\alpha$-plane. Like a homogeneous magnetic sector, the simple cylindrical electric sector has no ability to focus in the $\beta$ plane. The action on the beam is, therefore, in the $\alpha$ plane, and the beam is separated according to ion-energy. Cylindrical sectors are used most frequently, because they are easy to manufacture. The first-order theory of the action of the electrostatic analyzer on ions was described by Boerboom [18]. The ions, within a small acceptance about the spectrometer axis, are caused to follow spherical orbits in an electric sector and are separated according to kinetic energy [19].

More complex sectors may be designed with toroidal electrodes [20]. These have the same spherical radius in the (x,y) plane, but also have a curvature in the (x,z) plane. Such analyzers in the limit would become hemispherical and focus equally well $\alpha$ and $\beta$ divergence. An ingenious example of an inhomogeneous hemispherical analyzer has been fabricated by Futrell and co-workers [21]. It is this focussing property in $\beta$ that can be used to compensate for some of the aberrations of the magnet as described in Section 2.3 above.

### 2.5. ELECTRIC SECTOR ABERRATIONS AND THEIR TREATMENT

The performance of an electric sector is affected by dirt, errors in fabrication and defects in the electronic supplies. The most important imperfections tend to arise from errors in fabrication or mounting of the plates, and from depletion of the fringe field at the electrode boundaries. This has a penetrative effect on the electric field deep between the electrodes.

Boerboom [22] considered the effects of imperfect, electrostatic-analyzer-electrode fabrication on the performance of the mass spectrometer. The paper deals with the effects of incorrectly aligned or twisted electrodes, and considers details such as the effect of surface finish on the performance. These effects lead inevitably to some blurring of the image and the formation of an oblique (twisted) image. The paper is largely a cautionary text, to inform designers on the tolerances that should be demanded for the fabrication and alignment of the electrodes.

The fringe field of an electrostatic analyzer can affect the homogeneity of the inner electric field. The effect lessens towards the beam axis deep between the electrodes. This effect can be addressed in a number of ways. Among these are the inclusion of field shims or alternatively Matsuda plates (an electrostatic device applied across the top and bottom of the sector), which are used to manipulate the inner electric field [15]. Field shims are electrodes attached along the rim of the electrode surfaces, which can be varied to increase the field close to the boundaries, thus correcting the inner deep electric field. Matsuda plates can produce the same linear correction or be used to mimic the field of a toroidal sector close to the the beam axis [16].

By varying the potentials on the Matsuda plates as a function of the applied electric sector electrode potential, the degree of curvature of the inner field can be tuned. The potential applied to the Matsuda plate to generate the closest approximation to the toroidal field depends on parameter such as the spacing of the sector electrodes. Matsuda and Fujita [16] showed an example of the potential contour map for a case where the Matsuda plate potential is 30% of the potential difference between the sector electrodes. Thus, the Matsuda plates can transform the focussing action of a cylindrical sector towards divergence in the $\beta$ plane. As mentioned, this effect can also be produced with additional multipole lenses.

## 3. Double focussing instruments

First consider reverse geometry, 'C'-shaped mass spectrometers. It is clear from equation (1) that in a homogeneous magnetic sector, the trajectories of ions of different momentum to charge ratios describe different arcs. First order spatial refocussing is achieved even in aberrated, simple homogeneous magnets for ions of given momentum and charge, emerging from a point in different directions (having angular divergence). In a more complex magnetic sector instrument that was manufactured to limit the effect of equation

(29) and to compensate for the other aberrations, second order angular focussing is achieved. For ions of constant mass the magnet disperses velocity [see equation (30)]; thus, the magnet turns an object point into a series of image points. For a continuous distribution of velocities, these points coalesce into a line built up from discrete foci: this is an example of a one to many mathematical mapping. The electric sector also disperses velocity equal in magnitude and opposite direction to that of the magnet, and it focusses the energy of the ions. Ions of a given mass may be dispersed along the line image of the magnetic sector (the new object for the electric sector). These ion trajectories are recombined into a single image point in the electric sector. For each mass considered, many object points are mapped onto one image point now focussed both in space and energy. A diagram to illustrate the double focussing principle is shown in Figure 5.

If the sectors are combined in the forward geometry 'C'-shape, the initial dispersion is made by the electric sector. It provides a one to many, object to image spatial focussing that can be observed between sectors. If the geometry is made 'S'-shaped, as in the Mattauch-Herzog mass spectrometer, the velocity dispersion of the sectors combines rather than cancels. Double focussing occurs simultaneously at an image plane after the magnetic (second) sector. Note that the Mattauch-Herzog type of instrument has no intermediate focus between sectors. Thus, the sectors in this type of mass spectrometer can be thought of in terms of a compound lens rather than as a combination of discrete lenses that is the case for the forward and reverse geometry instruments. Diagrams of two forward geometry instruments: a Nier-Johnson ('C'-shaped) and a Mattauch-Herzog ('S'-shaped) are shown in Figure 6.

In summary, therefore, the term double focussing is derived from the mass spectrometer's ability to focus two variable properties of the ion beam as opposed to simply having two foci. Double focussing mass spectrometers cause convergence of ion beams having different energies ($E \pm \delta E$, where $E$ is in thousands of volts and $\delta E$ is two or three orders of magnitude less) and angular divergence ($\pm\alpha$ where $\alpha$ is small). There are many examples of such mass spectrometers reported in the literature. The general features of these instruments are high resolution and sensitivity, combined with accurate mass calibration over a large mass range. The last condition is particularly important if large biological molecules are to be studied.

There are two principal geometries of choice for 'C'-shaped instruments. One is the Nier-Johnson arrangement with the electric sector preceding the magnet. The electric sector is used to disperse the disparate beam energies while providing spatial refocussing of like energies. Thus, there is a narrow band of energies passed, and this primes the beam for high resolution momentum analysis in the magnetic sector. The alternative is the reverse geometry spectrometer, which has the order of the magnetic and electric sectors swapped on the instrument axis. The ZAB-2F reported by Morgan et al. [23] is an important commercial instrument. The instrument also possesses high mass-resolution and high sensitivity, but, in addition, provides a method of energy analysis unavailable on forward-geometry spectrometers, facilitating experiments into the decomposition reactions of metastable ions. The technique of Mass-analyzed Ion Kinetic Energy Spectroscopy (MIKES) is a simple tandem mass spectrometry (MS/MS) experiment that can also be used to study spontaneous or induced decompositions occurring in the region following the magnetic sector (called the second field-free region). In the experiment, the magnet is parked at a magnet current selected to transmit single mass ions of interest. The electric sector potential is then scanned to detect ions decomposing in the second field free region of the mass spectrometer, thereby leading to both structural information for the parent ion and information concerning energy release information involved in breaking chemical bonds.

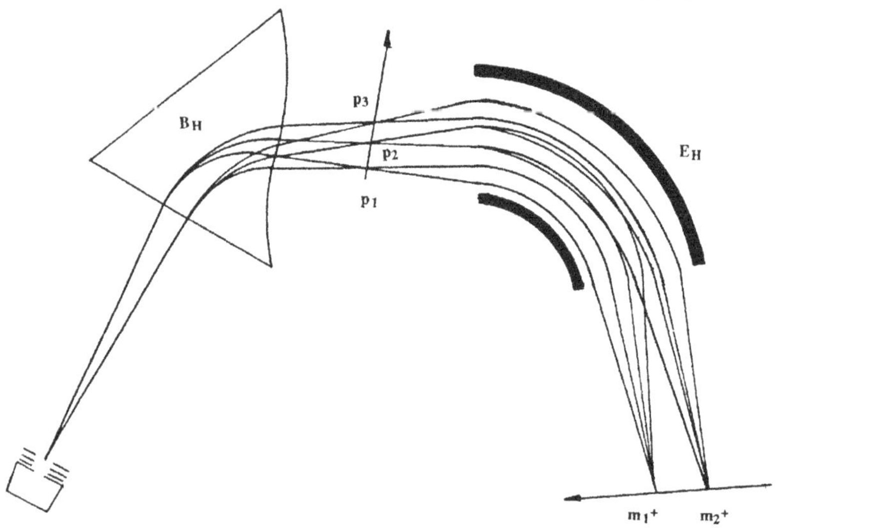


Figure 5. A 'C'-shaped, reverse-geometry mass spectrometer showing the double-focussing principle. Consider a point source generating a beam having angular divergence $\alpha$ and containing two particles $m_1^+$ and $m_2^+$. The ions in the beam are formed with one of two energies E or E + $\delta$E where the small difference in energy is expressed as $\delta$E. This beam is mapped on to an intermediate line image by the magnet. The line image is the object of the electric sector, which maps the object onto an image plane in two discrete, point foci. Each focus in this plane is a double-focus in $\alpha$ and $\delta$E. Note that the difference in masses of particles $m_1^+$ and $m_2^+$ is slight (< 1 Da) and $m_1 < m_2$. In this example there are a total of four speeds $|v_1|...|v_4|$, two speeds for each particle, forming a multitude of velocities propagating in all directions within $\alpha$. Using p to represent momentum, values are chosen to satisfy the following criteria:

$$m_1|v_1| < m_1|v_2| \quad = \quad m_2|v_3| < m_2|v_4|.$$
$$(p_1, E)\ (p_2, E + \delta E)\ \ (p_2, E)\ (p_3, E + \delta E)$$

The energy spread featured in the figure represents the energy spread $\delta$E in a real source arising from both the thermal energy of precursors and the potential drop across the element of space in which ions are made.

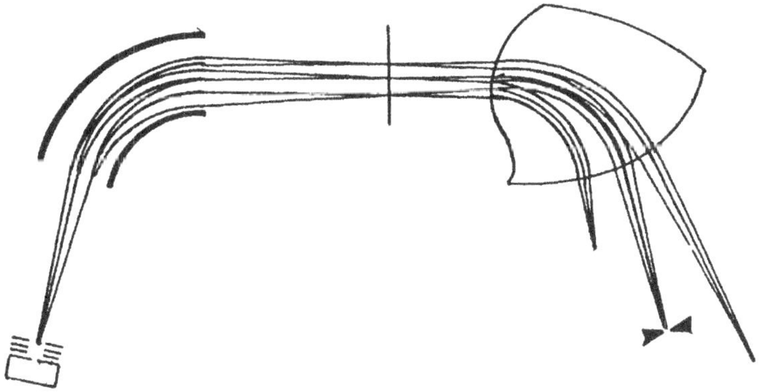

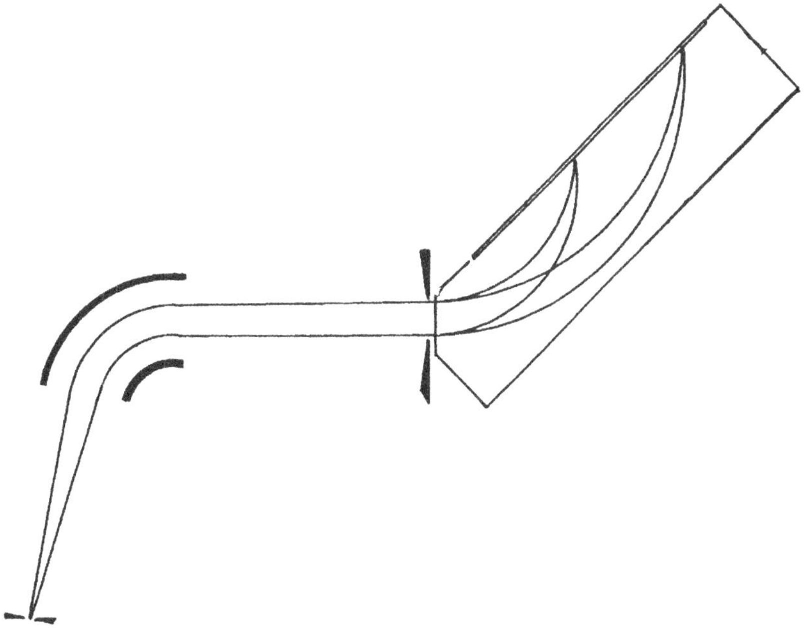

Figure 6. Top: a Nier-Johnson, forward geometry, 'C'-shaped, double-focussing mass spectrometer. Bottom: a Mattauch-Herzog, 'S'-shaped, forward geometry, double-focussing mass spectrometer. Note that the Mattauch-Herzog instrument has no intermediate foci, as such the sectors are like a compound lens.

Figures 7 and 8 show two mass spectrometers, the unusually large research instrument MMM built in Melbourne and now situated at Warwick University [24] and a large scale mass spectrometer (GEMMY) [25] built and situated at Osaka University. Comparison of these instruments will reveal some examples of the different methods used to bring about second-order double focussing. The MMM spectrometer has few optic lenses, being designed to use a magnet shim to correct the distortion, whereas the Osaka instrument incorporates electrostatic quadrupole lenses. The foci of the quadrupole lenses are used simultaneously for different effects in the meridional and sagital planes. The quadrupole is focussed in the $\alpha$ plane and defocussed in the $\beta$ plane. Both instruments contain Matsuda plates on the electric sectors and simultaneously seek to maximize sensitivity and resolution.

## 4. Tandem (4-Sector) mass spectrometry

The distinguishable arrangements of two pairs of objects can be obtained from the binomial coefficient X!/((X-Y)!Y!); here this is 4!/2!2! (= 6). Hence, two magnetic and two electric sectors can be arranged in six ways of which two configurations are 2-sector equivalents (EEBB and BBEE). For these latter configurations, the sectors can be arranged to cancel the first order aberrations, but are expensive 2-sector equivalents and can be disregarded in the context of this section. In addition to the four distinguishable arrangements of sectors, there are three joints or nodes connecting the chain of sectors each of which can be arranged independently in two ways (turning left or right): this gives a further $2^3$ arrangements. The combination of the above means that there are 32 possible different arrangements of instruments with different ion optic properties. The choice and evaluation of instruments is clearly a difficult matter and may lead to difficulties in standardizing results.

Figure 9 shows a schematic example of a commercial 4-sector mass spectrometer. These instruments are used for structural investigations. An ion of interest is selected with MS-1 and focussed into a collision cell. In the collision-cell, which is held at a relatively high pressure, there is a high probability that ions in the beam will experience a collision(s). Some of the energy (centre-of-mass) is converted into internal energy of the ion under investigation. Addition of sufficient energy, so suddenly, may activate the molecule into an energy state from which it can dissociate. Dissociations generally occur through weaker bonds located in kinetically accessible channels on a potential hypersurface. The number of channels increases with the number of constituent atoms in the ion and the fragments formed, tend to be chunks of the molecule which reflect the structural composition of the parent. Following the dissociation of a single charged parent ion, a charged portion and one or more neutrals is left. To a first approximation the ion retains the velocity of its parent; its mass and therefore energy, however have suddenly changed. The portfolio of charged fragments (daughter ions) from a parent ion can be detected by analysis in MS-2. In a tandem 4-sector instrument, high resolution can be obtained for both parent and daughter ions because MS-1 and MS-2 are each double focussing. The energy regime in which experiments can be performed is substantially higher than those of quadrupole-containing instruments. Experiments are limited to lower energies in quadrupoles because mass filtration can only occur if the electric field is oscillating rapidly compared to the residence time of ions in a device of limited size.

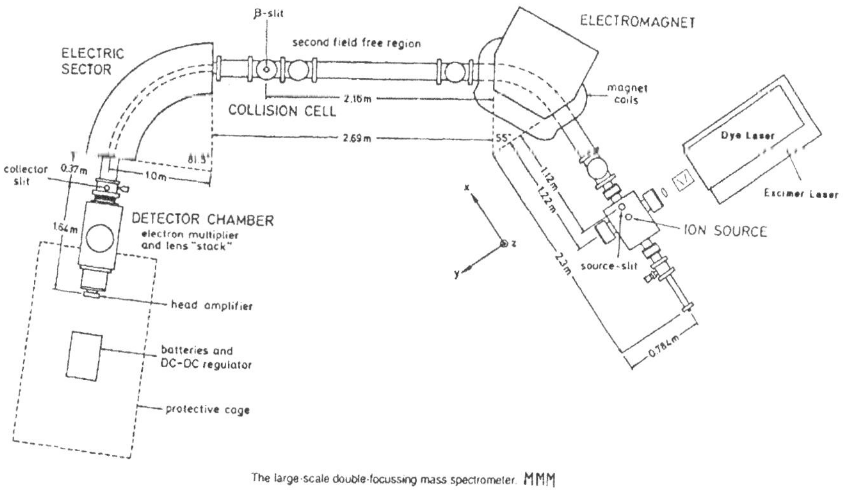


Figure 7. The MMM reverse geometry large scale mass spectrometer.

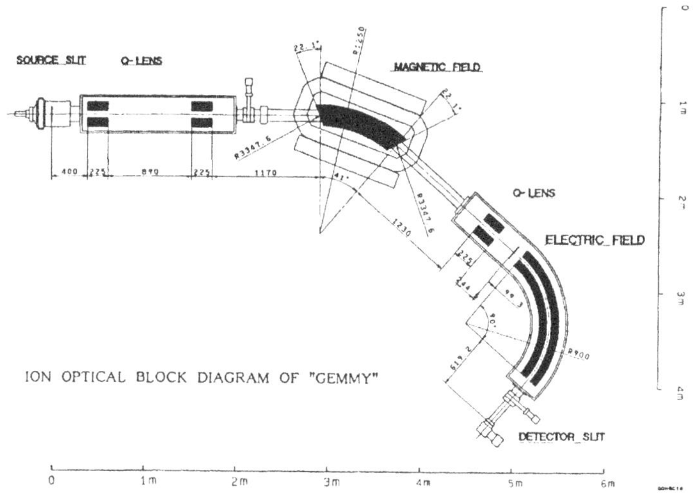


Figure 8. The GEMMY reverse geometry large scale mass spectrometer.

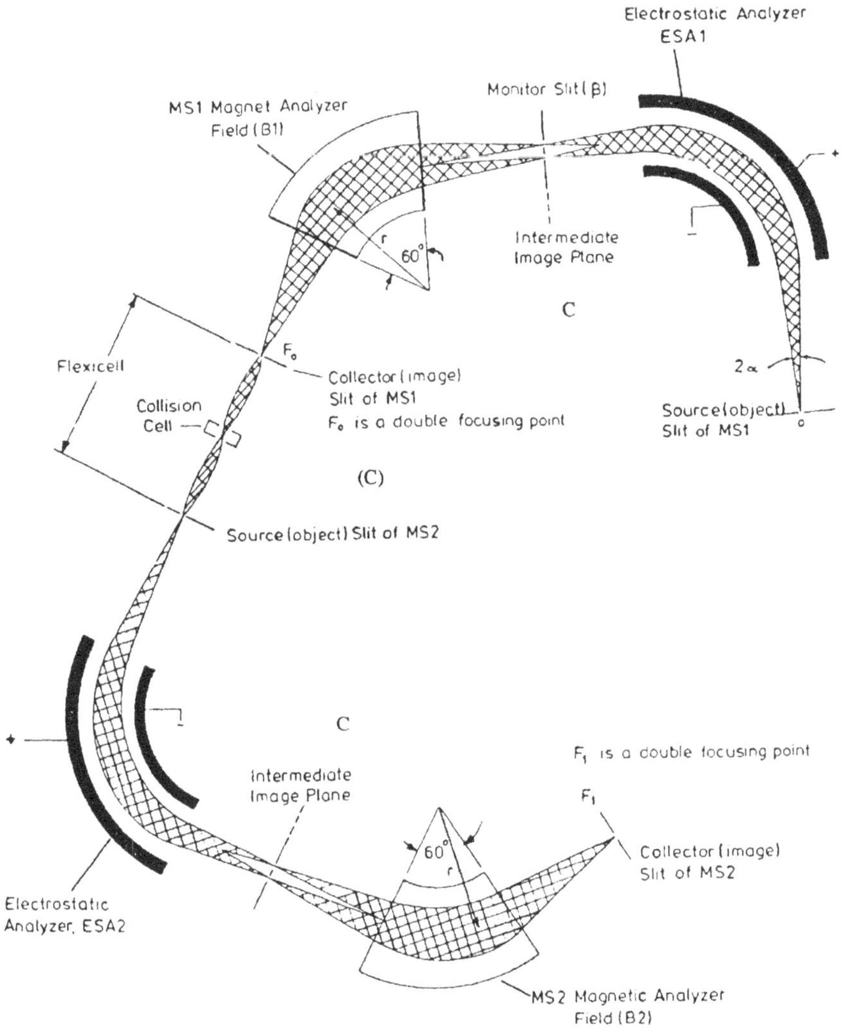


Figure 9. Kratos concept HH mass spectrometer. Note the C(C)C geometry of the sectors arranged in an EBEB geometry.

## 4.1. OPTIMAL CONFIGURATIONS

The new generation of 4-sector mass spectrometers may be operated in a number of different modes. The instrument, therefore, needs to be versatile to provide high mass resolution and collision-induced reactions for structural identification. To ensure that the instrument should be able to do both of these tasks, the 4-sector mass spectrometer must have additional ion optic stages. A super-high resolution mode of operation requires the foci of the second and third sectors to be coincident. On the other hand, incorporation of a collision cell requires those same sectors to focus on the entrance and exit slits of that collision cell. Beer's law, which is the basis of simple collision theory (SCT) [26], can be used to calculate the optimum length for the collision cell in the context of various design parameters. The fact that this cannot be infinitesimal constrains the second and third sectors to focus on different principal planes in contrast to the requirements for obtaining super-high resolution. Thus, if an instrument is to perform both tasks described above, it must incorporate additional optic devices to transpose one plane onto the other.

### 4.1.1. *Super-High Resolution*

Matsuo et al. [27] used the method of transfer matrices, a six parameter specification of the ion optic position vector similar to that used to describe phase space, to calculate the theoretical mass resolution of a JEOL HX110 and the tandem instrument HX110/HX110. The measured mass resolution broadly follows the theoretical resolution, which is based on the assumption that the mass dispersion can be summed for tandem sectors. Results show that an increase in the phenomenological *optical Q value* (the product of resolution and transmission) can be made and that super-high resolution is therefore possible with tandem 4-sector mass spectrometers. This application leads to double the resolution of the single mass spectrometer at a cost of just 10% attenuation in the total transmitted signal.

Matsuo and Matsuda [28] addressed the practical issues of super-high resolution in tandem (4-sector) instruments. The design of an instrument to verify the measurements made in Fourier transform mass spectrometers is discussed together with the theory of its operation. The discussion includes an analysis of the optics of eight configurations of tandem sector instruments (where the configuration of the first and second, and third and fourth sectors were fixed in a 'C'-shape). They conclude that optimum resolving power can be obtained only with tandem (4-sector) instruments capable of a super-high resolution mode (i.e., EBEB and BEBE configurations). For many applications the performance of modern high resolution mass spectrometers is sufficient, however the use of a super-high resolution mode may increase in popularity for instance to resolve structure within spectra of multiply charged ions.

### 4.1.2. *Collision-Activated Decomposition*

The design of the collision cell is almost as important as the ion source and sector analyzers. Much of the difference between the performance of different tandem 4-sector instruments will be traced to this. Direct comparisons of commercial instrumentation have been made under the identical experimental conditions, even with the same operator [29]. Although it appears that all of the data are present in comparative spectra, there are gross differences between the instruments in the relative transmission of high and low mass ions. It is clear that some sort of standardization in the design of collision cells is desirable if results are to be quantitative. Perhaps in the next decade, commercial instrumentation will be available with cells of a variety of standard designs, if a single standard is not adopted. The analogy could be drawn with the standardization of electron impact ion-sources in Nier-type designs with the use of 70 eV electron energies.

A popular technique has been developed to vary the collision energy in MS/MS, which benefits the collection of low-mass daughter ions. The collision cell is floated to a potential several kV above the relative instrument ground. This has effects on the ion optics and scanning of MS-2. The floated collision cell is the region in which daughter ions are formed. The ions experience a step function in their potential energy during formation which reflects the kinetic energy of the parent ion and the local electrostatic potential. Account of this feature must be taken in the linked scan of MS-2. The object which must be imaged onto the acceptance plane of MS-2 is part of an immersion lens discussed earlier and a series of lenses must be included to maintain a constant focal point. Thus, lenses following the collision cell must be scanned as part of the linked scan for daughter ions. A visual aid of this effect is shown in Figure 10.

## 5. Instruments and Instrumental methods

There have been a number of different 4-sector instruments reported. One of the first such instruments of a sound ion-optic design was reported by Futrell and Miller [30] in 1965. The apparatus was used to study the chemical-physics and reaction dynamics of simple processes up to initial reaction energies of 100 eV. This instrument was focussing in the second order, utilizing two back-to-back $60^{\circ}$ magnetic sectors to facilitate correction of second order aberrations in MS-1. MS-2 was a standard Mattauch-Herzog geometry mass spectrometer. White and Forman [31] built an instrument during the same era. The instrument was constructed in the C-geometry from four $90^{\circ}$ sectors in EBBE geometry: unusual in that, to our knowledge, it is the only instrument of this mixed geometry ever built. This instrument was constructed to study the physical properties of processes such as sputtering, and in the hope of measuring isotopic abundance ratios with an accuracy of $1{:}10^8$ at a resolution of $1{:}10^5$.

The implications for analytical mass spectrometry had not been clearly expounded at that time. It was more than a decade before McLafferty [32] suggested that this type of instrumentation could be used for chemical studies of sequential decompositions, metastable ions and collision-activated decomposition.

Tandem mass spectrometry for purposes of chemical analysis did not begin with the high performance 4-sector mass spectrometers. The first tandem mass spectrometry experiment was MIKES, performed on a reverse geometry, two sector mass spectrometer. A history of tandem mass spectrometry to 1983 is given by McLafferty [33]. Experiments have been progressively improved, following the success of the technique, in the search for higher quality spectra [34]. Initially a second electric sector was added to give greater resolution to the initial reactants. Unfortunately no improvement in the low resolution of the daughter scan is gained by this addition. Even though the quality of the information achieved in three sectors is improved, it was soon realized that a whole extra mass spectrometer would be required to gain the highest quality data. In recent years this extension of tandem MS has been exploited to the full, and a large number and variety of 4-sector mass spectrometers have been constructed.

An alternative method of making a tandem mass spectrometer is to couple a quadrupole mass filter to the front end of a sector mass spectrometer (QEB). For collisionally-activated decompositions (CAD) of small ions at low energy, it is sufficient to produce fragmentation in a low energy (eV) regime with unit mass separation of the parent. To a first approximation, acceleration for mass separation of the daughter ions is independent of the induced fragmentation because the energy loss and chemical energy release are both

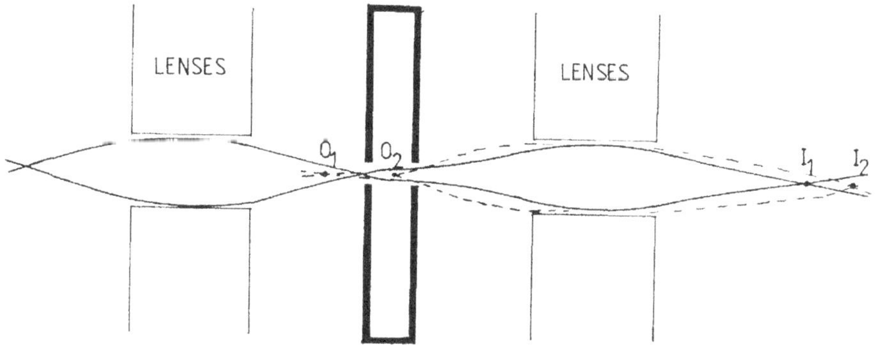


Figure 10. A schematic diagram of a hypothetical 4-sector collision cell. Collisions between the parent ion and one or more gas particles deposit sufficient internal energy into the parent ion to induce fragmentation and produce daughter ions. The approximation is made that the velocity of the parent and daughter ions are equal. Thus, the total energy of the daughter ion upon formation is the sum of its kinetic energy (determined by the daughter ion mass and the velocity *inherited* from the parent ion) and the potential energy at the point of its formation. The total energy, therefore, is a linear function of ion mass. Because the masses of daughters vary widely, the total energy also varies. The collision cell produces a different virtual object for each ion energy. The point at which this virtual object is brought to a real image varies with energy for a constant lens power. To avoid discrimination between different energies (masses) the image point must be kept constant.
Consider the diagram. For a daughter $m_1^+$, having energy $E_1$, a virtual object $O_1$ is formed which is refocussed at an image point $I_1$. The electric sector potential is tuned to analyze a narrow band of energies that are focussed on to its acceptance plane in which the point $I_1$ lies. Ions thus focussed are efficiently analyzed. For a second daughter $m_2^+$, having energy $E_2$, a virtual object $O_2$ is formed that has a natural refocus point $I_2$ at the lens power used to focus $O_1$ on to $I_1$. The electric sector's acceptance plane is fixed, thus if uncorrected, discrimination effects will be observed. The power of the compound lens can be adjusted to focus different objective positions on to the same point. This is achieved by scanning the lenses in the collision cell along with the electric sector, bringing ions of interest into the electric sector as they are required for analysis.

very much less than the energy due to acceleration in this regime; hence high resolution tandem mass spectrometry can be performed [35]. Data produced in this way provide complementary information to those from the three sector experiments, and the apparatus is relatively cheap to construct. The type of experiments that can be performed in this way are somewhat limited in terms of the selectivity that can be achieved in selecting the parent. These experiments become increasingly less informative as the mass range is significantly extended. The much more commonly used hybrid instruments employ quadrupoles after the sectors [e.g., BE(BE)QQ], to extend the range of sequential experiments which can be performed where unit mass resolution suffices. Hybrid instruments are discussed by Gaskell and Ballard in Chapter 2 of this volume.

Hill et al. [36] constructed a home-built tandem instrument at Lausanne comprising two forward Mattauch-Herzog geometry mass spectrometers (EBEB). The instrument has relatively poor transmission, due to problems in mechanical alignment. There are at least three companies manufacturing 4-sector (tandem) mass spectrometers. Instruments presently for sale include the JEOL HX110/HX110, Kratos Concept HH and the VG ZAB-4T. Each of these instruments has sufficient ion optic lenses to satisfy both criteria of optimal performance.

Ionization techniques applied in other areas of mass spectrometry are of course applicable to tandem sector studies. Martin and Biemann [37] investigated the differences that their instrument made to static FAB tandem mass spectrometry measurements. A 4-sector instrument normally has sufficiently high resolution to pick isotopically pure $^{12}C$ $(M + H)^+$ by using MS1. In combination with the high resolution, linked scans in MS-2, the six peptide analytes tested were shown to have far more distinct spectra in the 4-sector measurements. This separation is effective due to an instrumental deconvolution of the chemistry by the additional resolution. Similar improvements in daughter ion spectra captured in 4-sector mass spectra of nucleosides and nucleotides have been reported by other groups [38,39,40,41,42].

Instrumental improvements can be achieved by alternative means than producing ever more perfect ion optics. Cottrell and Evans [43] demonstrated the advantages of the introduction of a microchannel array detector in mass spectrometers. Addition of an array detector combines the benefits of photograhic detection mass spectrographs with modern, linear electronic ion multiplication. Ions of all masses are being constantly produced in the ion source. If all these ions could be used simultaneously, experiments would have far greater sensitivity and samples could be analyzed in less time. The array detector goes part way to satisfying this ideal. An array works using optics which image a portion or window of the mass range in an image plane (Figure 11). This image plane is segmented and ion amplification from each segment is independent. Thus, a digital array of space is made whereby resolution is limited by the definition of the segment compared to the magnification of the mass range imaged onto the plane. The best optical systems allow the mass window and its magnification to be varied to allow the relative benefits of resolution to be contrasted with the rate of data collection. When an array is incorporated in an instrument the analyzers are scanned in steps across the mass range. The steps should allow time to stabilize the magnetic field, because high inductance means that this cannot instantly hop from one field to the next. Faster scanning methods have been achieved with a relatively new instrument using an inhomogeneous electric sector [44]. In addition to stabilization, care must be taken to ensure good overlap between exposures so that a continuous spectrum can be constructed.

Biemann and co-workers [45, 46] obtained two orders of magnitude more sensitivity in daughter ion scans of the substance P $(M + H)^+$, by integrating an array into a 4-sector mass spectrometer. The incorporation of an array detector is an ion optic problem in its own right [17]. The principal focus must be a plane rather than a point and the inclusion

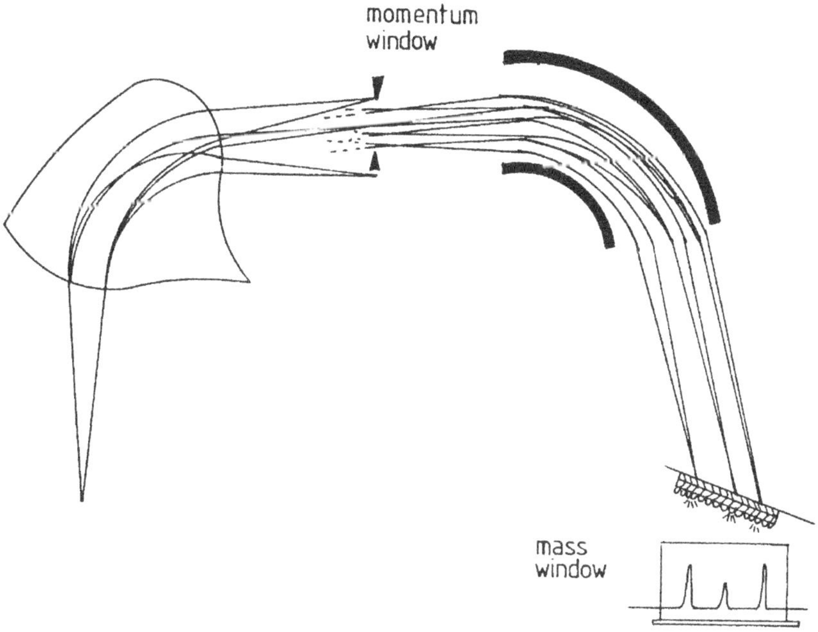


Figure 11. A schematic diagram mass spectrometer with an array detector. The magnet is stepped across the different momentum windows in turn. A broad bandwidth of energies is passed through the electric sector. These are focussed onto an image plane. In this plane different masses are detected with respect to their positions in the image plane simultaneously to produce a snapshot of a portion of the whole spectrum.

of a zooming mass scale is an advantage. Further discussions of the third order focussing characteristics can be read in an alternative paper [47].

Linked scans is a name for a group of mass spectrometer scanning methods that can produce some information not available in experiments such as MIKES. The development of linked scans has had an important effect on the development of mass spectrometers. In principle, there are no experimental data that can be captured on a reverse geometry instrument, that cannot be recovered on a forward geometry instrument. Much less information on the energy distribution is available than with MIKES. The technique of linked scans has become highly developed. Scan laws are described for instruments with floated collision cells which can generate improvements in sensitivity at low mass. The technique is dogged by some problems with coincidences giving rise to artifacts in the daughter ion scans, and some data are occasionally missed. In addition, the mass range is more limited than MIKES. Forward geometry mass spectrometers can at best, however, reproduce equivalent structural information to that obtained with MIKES.

Scan laws required to perform high-resolution linked-scan CAD experiments in tandem sector and forward geometry mass spectrometers are fraught with difficulty. Sheil et al. [48] report on induced dissociation of massive ions with 15 keV of kinetic energy, where the energy loss in the parent is in the order of hundreds of electron volts from collisions with atomic target gases. This result has marked implications for the techniques of tandem sector mass spectrometry. Linked scan spectra may miss peaks with such large shifts. A second problem of linked scans is the observation of artifacts arising from coincident signals that are unrelated. Such potential problems are discussed by Boyd [49] in his paper describing the scan laws for various sector instruments with floated collision cells. The mathematics of the scans described in this paper contain the more simple results for linked scans in the limit of placing the collision cell voltage at *instrument ground potential* (usually earth). The calibration and performance of such 4-sector scans has been reported for constant B/E scans of the second mass spectrometer (MS-2) [50]. The problems have largely been solved by the software development engineers; however, cross confirmation of linkage is advised to those performing linked scan experiments.

A BEEB mass spectrometer, in use [51], provides certain advantages in configuration for CAD studies. If a collision cell is incorporated in the second, third and fourth field free regions with detectors after sectors 2-4, then experiments can be performed at all levels of resolution. MIKES with MS-1 and MS-1E at higher resolution, with coupled linked scan experiments for the highest resolution. This arrangement ensures experimental integrity, especially if comparisons between the linked scan and MIKES data (e.g., to check for missing daughters or artifact peaks) are required. Experiments designed to investigate sequential mass spectrometry, known as $MS^n$ or *consecutive ion monitoring*, have been made on just such a BEEB instrument [52]. This is now a new experiment that cannot be done on apparatus with less sectors and trades resolution for the novelty of the experiment. The principal limitation of this experiment in studies of polymer and biopolymer molecule structure elucidation is the expectation that the signal would become attenuated by some four or five orders of magnitude (for $MS^4$).

Finally, instrumental methods can also be extended by development and modification of the other components of sector spectrometers such as the collision cell. A modified collision cell can extend the range of experiments to the studies of ionization-neutralization-reionization phenomena as demonstrated by Feng et al. [53] and is discussed by Wesdemiotis in this volume.

## 6. Conclusion

We have examined some of the basic theory and problems of electric and magnetic sectors and focussed on some recent developments in tandem mass spectrometry. The types of experiments that one *might* want to do with a tandem spectrometer must always be carefully considered before the instrument is procured or constructed No apparatus is versatile enough to allow all conceivable experiments to be performed optimally.

## 7. Acknowledgements

Thanks are extended to Professor Michael L. Gross and Ms Caroline D. Bradley for their comments and for criticizing the text.

## 8. References


11. Thomson J.J. 'Positive Rays (Canal rays)' (1907) *Phil. Mag.* **13**, 561-567.
2. Klemperer O. and Barnet M.E. (1971), Electron Optics 3rd Edition, Cambridge University Press. Grivet P. (1972), Electron Optics 2nd Edition, Pergammon Press. El-Kareh A.B. and El-Kareh J.C.J. (1970), Electron Optics **1,2**, Academic Press.
3. Dempster A.J. 'Positive ray analysis', (1918) *Phys. Rev.* **11**, 316-324.
4. Bainbridge K.T. 'Atomic masses and structure of Atomic nuclei', (1933) *J. Franklin Inst.* **215**, 509-534.
Bainbridge K.T. 'A mass spectrograph', (1932) *Phys. Rev.* **40**, 130-132.
5. Hintenberger H. and König L.A. 'Mass spectrometers and mass spectrographs corrected for image defects', (1959) in Waldron J. (Ed.), Advances in mass spectrometry, Pergammon Press, New York p. 16-35.
6. Nier A.O. 'A mass spectrometer for routine isotope-abundance measurements', (1940) *Rev. Sci. Instrum.* **11**, 212-216.
7. Stephens W.E. 'Magnetic Refocusing of Electron Paths', (1934) *Phys. Rev.* **45**, 513-518.
8. Barber N.F. 'Shape of an electron beam bent in a magnetic field', (1933) *Proc. Leeds Phil. Lit. Soc. Sci. Sec.* **2**, 427-434.
9. Boerboom A.J.H. and Laranjeira M.F. 'Application of Liouville's theorem in mass spectrometry', (1973) *Portugal Phys..* **8**, 245-260.
10. Charged Particle Optics Wollnik H. (1987), Charged particle Optics, Academic-Press.
11. Boerboom A.J.H. 'Theory of the aberrations in the focussing properties of a magnetic field with one plane of symmetry', (1959) *Appl. Sci. Res.* **B7**, 52-62. See also Boerboom A.J.H. 'Modern methods in ion optics' in Reed R.I. (ed.), Mass Spectrometry, Academic Press, London. p 73-79
12. Blake L.R. 'The design of electromagnets 1', (1953) *Electronic Engng.* **25**, 380-385.
13. Blake L.R. 'The design of electromagnets 2', (1953) *Electronic Engng.* **25**, 432-435.
14. Cottrell J.S. and Greathead R.J. 'Extending the mass range of a sector mass spectrometer', (1986) *Mass Spectrom. Rev.* **5**, 215-247.
15. Matsuda H. 'Electric field in cylinder condensers of finite length', (1976) *Int. J. Mass Spectrom. Ion Proc.* **22**, 95-102.

16. Matsuda H. and Fujita Y. 'Potential distribution in a cylinder condenser terminated by Matsuda Plates', (1975) *Int. J. Mass Spectrom. Ion Proc.* **16**, 395-404.
17. Itagaki Y., Ishihara M., Otsuka K., Kammei Y. and Kubata E. 'The role of a quadrupole doublet (QPD) and the zooming mass range capability on an array detector' JEOL Ltd, 1-2 Musashino 3-chrome Akishima, Tokyo 196, Japan
18. Boerboom A.J.H. 'Ion optics of fields with rotational symmetry and circular main orbit' , (1964) *J. Appl. Phys.* **35(6)**, 1691-1697.
19. Rumpf B.A. and Derrick P.J. 'Determination of translational energy release distributions through analysis of metastable peaks' (1988) *Int. J. Mass Spectrom. Ion Proc.* **82**, 239-257.
20. Matsuda H. 'Double focusing mass spectrometers of second order', (1976) Saunders J.H. and Wapstra A.H. (Ed.), Atomic masses and fundamental constants, **5**, 185-191.
21. Anderson S.G., Sohlberg K.W., Howard S.L., Shukla A.K. and Futrell J.H. 'A high transmission hemispherical energy analyzer for ion spectrometry' *J. Am. Chem. Soc. Mass Spectrom.* In press
22. Boerboom A.J.H. 'Image aberrations in an imperfectly shaped electrostatic analyser', (1976) *Int. J. Mass Spectrom. Ion Proc.* **22**, 259-271.
23. Morgan R.P., Beynon J.H., Bateman R.H., and Green B.N. 'The MM-ZAB-2F double focusing mass spectrometer and MIKE spectrometer', (1978) *Int. J. Mass Spectrom. Ion Phys.* **28**, 171-191.
24. Cullis P.M., Neumann G.M., Rogers D.E. and Derrick P.J. 'A new grand-scale mass spectrometer: preliminary measurements with field desorption under rapid heating' (1980) Adv. Mass Spectrom. **8B**, 1729-1738.
25. Matsuo T. 'High performance sector mass spectrometers: past and present', (1989) *Mass Spectrom Rev.* **8**, 203-236
26. Kim M.S. 'Collision induced dissociation', (1991) *Org. Mass Spectrom.*
27. Matsuo T., Ishihara M., Martin S.A. and Biemann K. 'Enhanced mass resolution without decrease of beam intensity in a four sector mass spectrometer', (1988) *Int. J. Mass Spectrom. Ion Proc.* **86**, 83-94.
28. Matsuo T. and Matsuda H. 'A super-high-resolution tandem mass spectrometer', (1989) *Int. J. Mass Spectrom. Ion Proc.* **91**, 27-40.
29. Bordas-Nagy J., Tandem Mass Spectometry Club (1991) and British Mass Spectrometry Society (1990).
30. Futrell J.H. and Miller C.D. 'Tandem mass spectrometer for study of ion-molecule reactions', (1966) *Rev. Sci. Instrum.* **37**, 1521-1526.
31. White F.A. and Forman L. 'Four stage mass spectrometer of large radius', (1967) *Rev. Sci. Instrum.* **38**, 355-359.
32. McLafferty F.W. (1977) in Jones C.E.R and Cramers C.A. (Ed.), Analytical Pyrolysis, Elsevier, Amsterdam.
33. McLafferty F.W.,Tandem Mass spectrometry, (1983), Wiley, New York.
34. Gross M.L., Chess E.K., Lyon P.A., Crow F.W., Evans S. and Tudge H. 'Triple analyser mass spectrometry for high resolution MS/MS studies' (1982) *Int J Mass Spectrom. Ion Proc.* **42**, 243-254
35. Glish G.L. and McLuckey S.A. 'High resolution detection of daughter ions with a hybrid mass spectrometer', (1986) *Anal. Chem.* **58(8)**, 1887-1889.
36. Hill N., Stahl D., and Bolton P. 'A BEBE four-sector mass spectrometer: new lease of life for two old instruments', (1990) *Int. J. Mass Spectrom. Ion Proc.* **95(3)**, 347-358.

37. Martin S.A. and Biemann K. 'A comparison of keV atom bombardment mass spectra of peptides obtained with a two-sector mass spectrometer with those from a four-sector tandem mass spectrometer', (1989) *Int. J. Mass Spectrom. Ion Proc.* **78**, 213-228.
38. Sakurai T., Matsuo T., Kusai A. and Nomjima K. 'Collisionally activated decomposition spectra of normal nuleosides and nucleotides using four sector tandem mass spectrometry', (1989) *Rapid Commun. Mass Spectrom.* **3(7)**, 212-216.
39. Jensen N., Lam K., Cody R.B. and Tamura J. 'Evidence for the distinction of cis and trans isomers of mono-unsaturated fatty acids by fast atom bombardment tandem mass spectrometric analysis', (1990) *Rapid Commun. Mass Spectrom.* **4(7)**, 239-241.
40. Domon B., Vath J.E. and Costello C.E. 'Analysis of derivatised ceramides and neutral glycosphingolipids by high performance tandem mass spectrometry', (1989) *Anal. Biochem.* **184(1)**, 151-164.
41. Takayama M., Fukai T., Nomura T. and Nojima K. 'Tandem mass spectrometric study on fragmentation in fast atom bombardment mass spectra of prenylated flavinoids', (1989) *Shitsuryo Bunseki* **37(4)**, 239-247.
42. Gerd G. and Rinehart K.L. Jr 'Primary structure of the hypertrehalosemic factor II from the corpus cardiacum of the Indian stick insect, Carausius morosus, determined by fast atom bombardment', (1987) *Biol. Chem. Hoppe-Seyler* **368(1)**, 67-75.
43. Cottrell J.S. and Evans S. 'Characteristics of a multichannel electro-optical detection system and its application to the analysis of large molecules by fast atom bombardment mass spectrometry', (1987) *Anal. Chem.* **59**, 1990-1994
44. Rollins K., Scrivens J.H., Jennings R.C.K., Morden W.E., Welby J.K. and Bateman R.H. 'The use of tandem mass spectrometry as a problem solving tool in the industrial environment' (1990) *Rapid Commun. Mass Spectrom.* **4**, 454-461
45. Hill J.A., Biller J.E., Martin S.A., Biemann K., Yoshidome K. and Sato K. 'Design considerations, calibration and applications of an array detector for a four-sector tandem mass spectrometer', (1989) *Int. J. Mass Spectrom. Ion Proc.* **92**, 211-230.
46. Hill J.A., Martin S.A., Biller J.E. and Biemann K. 'Use of a microchannel array detector in a four sector tandem mass spectrometer', (1988) *Biomed. Environ. Mass Spectrom.* **17(2)**, 147-151.
47. Matsuo T., Sakurai T. and Derrick P.J. 'Characterisation of the focal planes of a mass spectrometer using the method of transfer matrices', (1989) *Int. J. Mass Spectrom. Ion Proc.* **91**, 41-49.
48. Sheil M.M. and Derrick P.J. 'Tandem mass spectrometry of peptides: Relationship between translational energy loss and fragment ion mass', (1988) *Org. Mass Spectrom.* **13(5)**,429-435.
49. Boyd R.K. 'Scan laws for tandem mass spectrometry using a floated gas collision cell', (1987) *Int. J. Mass Spectrom Ion Proc.* **75**, 243-264.
50. Sato K., Asuda T., Ishihara M., Kunihiro F., Kammei Y., Kubota E., Costello C.E., Martin S.A., Scoble H.A. and Biemann K.. 'High performance mass spectrometry: Calibration and performance of linked scans of a four sector instrument', (1987) *Anal. Chem.* **59(13)**, 1652-1659.
51. Desiderio D.M. 'FAB-MS/MS study of two neuro peptides, dynorphins 1-7 and 1-13', (1986) *Int. J. Mass Spectrom. Ion Proc.* **74**, 217-233.
52. Tomer K.B., Guenat C.R. and Deterding L.J. 'Consecutive reaction monitoring in a four-sector mass spectrometer: $MS^4$ and one step beyond', (1988) *Anal. Chem.* **60(20)**, 2232-2236.
53. Feng R., Weisdemiotis C., Baldwin M.A. and McLafferty F.W. 'An improved tandem double-focusing mass spectrometer for ionization-reionization and collisional activation studies' (1988) *Int. J. Mass spectrom. Ion Proc.*, **86**, 95-107.

# HYBRID TANDEM MASS SPECTROMETRY

SIMON J. GASKELL and KEVIN D. BALLARD
*Center for Experimental Therapeutics*
*Baylor College of Medicine*
*Houston TX 77030*

ABSTRACT. The development of hybrid sector/quadrupole instruments for tandem mass spectrometry is described. The operational modes of these instruments are illustrated by MS/MS and MS/MS/MS analyses of peptides, xenobiotic metabolites and other compounds of biological importance. The versatility of operation (with respect to scan modes and the conditions utilized for collisionally activated decomposition) is a particular advantage of hybrid instruments.

## 1. Origins And Development Of Hybrid Instruments

A hybrid mass spectrometer can be defined as a serial connection of two or more ion analyzers, each of which is based on different separation principles. Using this definition, we may regard the familiar double-focussing instruments as representing one of the earliest hybrid designs. These instruments consist of a magnetic sector (B) which separates ions according to their momentum/charge ratio, and an electric sector (E), which separates according to kinetic energy/charge ratio. Appropriate geometries combining the two separation devices (either the electric sector first, magnet second, so-called EB, or BE) achieve double-focussing, and hence improved resolution, and also permit the establishment of connectivities between precursor and product ions. Two approaches are possible. With either the EB or BE configurations, decompositions occurring in the first field-free region (FFR1, following the ion source) may be detected by linked scanning of B and E [1]. BE instruments may be used to study the decompositions of precursor ions, selected by B, in the second field-free region (FFR2, between B and E); product ions are scanned by using E to give the mass-analyzed ion kinetic energy spectrum (MIKES) [2, 3]. The limitations for the study of precursor/product ion relationships arise from the achievable resolution of precursor or product ions. Scanning of FFR1 decompositions is done with poor effective resolution of precursor ions whereas MIKES analyses have poor effective mass resolution of product ions. The simultaneous achievement of unit resolution or better of both precursor and product ions requires a tandem MS instrument incorporating additional separation devices.

In principle, any of the well known types of mass spectrometers can be connected in series to form tandem hybrid mass spectrometers. These include magnetic and electric sectors, quadrupole mass filters (Q), quadrupole ion traps, ion cyclotron res-



*M. L. Gross (ed.), Mass Spectrometry in the Biological Sciences: A Tutorial, 29–58.*

onance (ICR) cells, and time-of-flight (TOF) mass spectrometers. Various combinations of these analyzers have been implemented.

## 1.1 HYBRID SECTOR/QUADRUPOLE INSTRUMENTS

This chapter concerns the predominant general type, which is sector/quadrupole combinations. The reader is referred to the primary literature for discussion of other types of hybrids, such as the Q/FT-ICR hybrid developed by Hunt, McIver and co-workers [4] and the Q/TOF hybrid developed by Glish and co-workers [5, 6].

Hybrid sector/quadrupole instruments commonly incorporate an RF-only quadrupole (q) as a region for both metastable (spontaneous) decompositions and for collisionally activated dissociations (CAD). In the latter, the ion beam is allowed to collide with an inert gas such as argon or xenon to increase the internal energy of the ionic species and thus increase the degree of fragmentation. For metastable ion studies, the RF-only quadrupole provides a relatively long time frame (tens to hundreds of microseconds) during which spontaneous decompositions may occur. For CAD experiments, the RF-only quadrupole serves as a containment device for both the precursor ions and the product ions that result from energetic collisions, facilitating the transmission of fragment ions resulting from multiple collisions. These considerations were part of the original rationale for the use of an RF-only quadrupole for CAD experiments on a triple quadrupole (QqQ) tandem mass spectrometer [7, 8, 9]. For an excellent discussion of the mass dependency, ion containment and transmission characteristics of RF-only quadrupoles, the reader is referred to a recent treatment by Boyd [10].

The earliest hybrid sector/quadrupole instruments were intended for specialized physical chemical applications [11-18]. The first such instrument designed explicitly for general analytical tandem mass spectrometric applications was a magnetic sector/RF-only quadrupole/mass analyzer quadrupole (BqQ) combination designed and built by Glish and Cooks [19, 20]. The routine MS/MS operation of the BqQ instrument involved selection of source-formed precursor ions with B (which therefore served as the first mass spectrometer, MS1); a resolution of up to 5000 could be achieved. The precursor ions were transmitted into the RF-only quadrupole for collisional activation. The resulting product ions were then transmitted to the mass analyzer quadrupole (MS2), which was scanned to yield a product ion spectrum with unit mass resolution.

The principles established by those studies facilitated subsequent work by Cooks' group, including the extension of the BqQ instrument to the BEqQ design [21, 22, 23]. The studies involving both the BqQ and the BEqQ instruments addressed many of the ion optical problems involved in the coupling of a sector instrument, which operates with ion energies in the keV range, with a quadrupole instrument, where ion energies are usually confined to the tens to hundreds of eV range. A schematic of a BEqQ hybrid is presented in Figure 1A, and the corresponding potential energy diagram, describing the potential energy of transmitted ions at various stages of the instrument, is presented in Figure 1B. Referring to these Figures, one sees that ions are initially accelerated to keV ion energies from the ion source, so that the ions are moving with velocities appropriate for analysis by the sector portions of the instrument. These ions must then be decelerated by using a lens system, to give energies in the range of 10-200 eV, prior to their introduction into the RF-only quadrupole. Because of the large discrepancy in ion energy requirements, the quadrupole assemblies must be floated to a potential very close to, but somewhat less than, that of

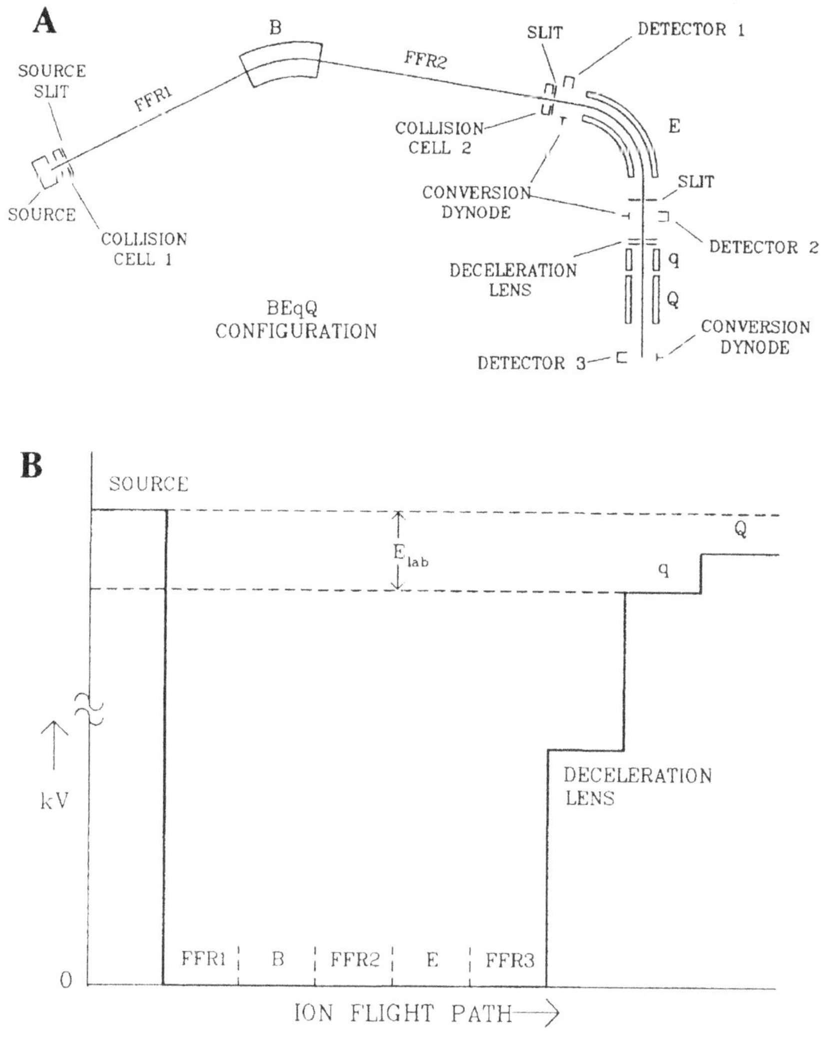


Figure 1. A) Schematic diagram of a BEqQ hybrid mass spectrometer. B) Potential energy diagram describing the electric fields experienced by positive ions traversing a BEqQ hybrid instrument. Note that the vertical axis is broken to permit visualization of the relatively small differences in the potentials of the ion source and the quadrupole assemblies.

the ion source. The exact float potential of q is determined by the collision energy desired in the RF-only quadrupole. The float potential of Q must be scanned with mass to ensure maximum transmission of the product ions [10]. The deceleration lens system is designed so that, during the deceleration process, the transmitted ion beam is reshaped from a ribbon cross-section, which is appropriate for the sector portion of the instrument, to a roughly circular cross-section, which is appropriate for transmission through the quadrupole assemblies.

Achievement of high ion transmissions through the hybrid instrument is clearly essential for analytical applications. The BEqQ hybrid built in Cooks' laboratory was satisfactory in this respect and was established to be a remarkably versatile instrument, with a great variety of possible scanning modes, yielding a variety of information. This design retained without compromise all of the features of a reverse geometry high resolution double focussing instrument, and provided access to both high (keV; in the first or second field-free regions) and low (1-500 eV) energy (laboratory frame of reference) collisional activation regimes for MS/MS and MS/MS/MS experiments. Versatility of operation is a feature shared by most hybrid mass spectrometric designs, and is one of the reasons for their continued development.

Further pioneering work in hybrid sector/quadrupole development was done in Beynon's laboratory with an EQ hybrid [24]. This instrument was built as a logical extension of a theoretical treatment of multi-analyzer combinations for ion dissociation studies [25]. Consisting of two relatively simple ion analyzers, this instrument was capable of producing conventional mass spectra, product ion spectra, precursor ion spectra, constant neutral loss spectra, ion kinetic energy spectra, and mass deconvoluted ion kinetic energy spectra, thus further illustrating the versatility possible with rational hybrid designs.

Another hybrid design recently developed by Glish and co-workers is of QEB geometry [26]. This instrument consists of a quadrupole mass analyzer connected in series with a conventional double focussing instrument of EB geometry, with an acceleration lens system separating the two portions of the instrument. Both low and high energy collisions are accessible between the quadrupole and the lens system, and high energy collisions are accessible between the lens and the electric sector. The original report pointed out some potential advantages of this configuration, which was also capable of a variety of scanning modes. Particular emphasis was given to its utility for the study of ion-molecule reactions.

### 1.2 EBqQ AND BEqQ DESIGNS

Although many different combinations of sectors and quadrupoles may be envisaged [27], two configurations have gained prominence. These are the EBqQ and BEqQ hybrid configurations, both of which are nominally a serial combination of a high resolution double focussing instrument and a quadrupole mass analyzer. These two configurations differ in the double focussing portion, incorporating either 'forward' (EB) or 'reverse' (BE) geometries. As stand-alone double focussing instruments, these configurations differ primarily in that the BE instrument is capable of performing MIKES analyses, an experiment that is not possible on the EB instrument [2]. However, when incorporated as either an EBqQ or BEqQ hybrid, the difference between the two configurations becomes magnified, so that the BEqQ instrument is considerably more versatile than the EBqQ. All of the scanning modes that can be implemented on the EBqQ can also be performed on the BEqQ, but the latter is capable of a number of additional informative scanning modes. The

TABLE I. SCAN MODES AVAILABLE ON HYBRID INSTRUMENTS OF BEqQ GEOMETRY[1]

| Scan mode | Acquisition: static | Acquisition: scanned | Decomposition region | Collision[2] energy | Effective mass resolution | Exemplary Applications |
|---|---|---|---|---|---|---|
| product ion scan (mode 1) | B and E | Q | q | 10-500 eV | >1000 (precursor)<br>unit (products) | structural analyses |
| *product ion scan (mode 2) | B | E and Q | FFR2 | keV | >1000 (precursor)<br>unit (products) | structural analyses |
| *mass deconvoluted MIKES | B and Q | E | FFR2 | keV | >1000 (precursor)<br>unit (products) | kinetic energy release studies |
| constant neutral loss scan | E | B and Q | q | 10-500 eV | >1000 (precursor)<br>unit (products) | compound class screening |
| precursor ion scan | Q and E | B | q | 10-500 eV | >1000 (precursor)<br>unit (products) | compound class screening |
| 2nd generation product ion scan (mode 1) | B and E | Q | FFR1 (1)<br>q (2) | keV (1)<br>10-500 eV (2) | ca 300 (precursor)<br>ca 800 (intermediate)<br>unit (products) | highly specific structural information |
| *2nd generation product ion scan (mode 2) | B and E | Q | FFR2 (1)<br>q (2) | keV (1)<br>10-500 eV (2) | >1000 (precursor)<br>ca 300 (intermediate)<br>unit (products) | highly specific structural information |
| Reaction intermediate scan (mode 1) | Q | B and E | FFR1 (1)<br>q (2) | keV (1)<br>10-500 eV (2) | ca 300 (precursor)<br>ca 800 (intermediates)<br>unit (products) | oligomer sequencing |
| *Reaction intermediate scan (mode 2) | B and Q | E | FFR2 (1)<br>q (2) | keV (1)<br>10-500 eV (2) | >1000 (precursor)<br>ca 300 (intermediates)<br>unit (products) | oligomer sequencing, mechanistic studies |
| Product-constant neutral loss scan (mode 1) | | B, E, and Q | FFR1 (1)<br>q (2) | keV (1)<br>10-500 eV (2) | ca 300 (precursor)<br>ca 800 (intermediates)<br>unit (products) | oligomer sequencing, mechanistic studies |
| *Product-constant neutral loss scan (mode 2) | B | E and Q | FFR2 (1)<br>q (2) | keV (1)<br>10-500 eV (2) | >1000 (precursor)<br>ca 300 (intermediates)<br>unit (products) | oligomer sequencing, mechanistic studies |
| Constant neutral loss-precursor ion scan | Q | B and E | FFR1 (1)<br>q (2) | keV (1)<br>10-500 eV (2) | ca 300 (precursor)<br>ca 800 (intermediates)<br>unit (products) | compound class screening |
| Consecutive neutral loss scan | | B, E, and Q | FFR1 (1)<br>q (2) | keV (1)<br>10-500 eV (2) | ca 300 (precursor)<br>ca 800 (intermediates)<br>unit (products) | compound class screening, mechanistic studies |

[1] This table does not include the various scans employing only the double focussing portion of the instrument.
[2] Metastable studies are possible for all scan modes.
* Scan modes which are not possible on a hybrid instrument of EBqQ geometry.

popularity of EBqQ instruments stems from the additional tandem MS capability afforded by the addition of the qQ combination to well-established forward geometry sector designs.

To illustrate these points, Table I presents a number of the scanning modes possible with a BEqQ hybrid; the scans not possible on an EBqQ instrument are indicated with an asterisk. Experiments involving high energy (keV) collisions on the BEqQ instrument exploit either FFR1 or FFR2. The low energy (10-500 eV) CAD regime is accessible with the RF-only quadrupole. The effective mass resolution of the ionic species at each stage of analysis for the various scanning modes is also listed in Table I. These parameters are important both for good experimental design and for appropriate selection of the scanning mode in those instances where more than one possibility exists. For instance, second generation product ion scans may be accomplished in two different fashions with the BEqQ hybrid. For some experiments, good resolution of the precursor may be more important than the resolution of the intermediate, in which case mode 2 (Table I), employing FFR2, would be chosen. Similarly, for those experiments where good resolution of the intermediate is more crucial than the resolution of the precursor, mode 1 would be selected. Examples of the analytical utility of several of the scan modes listed in Table I will be presented below.

With their original description of the use of a BEqQ hybrid, Cooks and co-workers introduced a symbolism to denote the scanning mode used [21]. Cooks' group has recently produced a comprehensive framework to describe the various possible scan modes in multidimensional mass spectrometry [28]. Table II presents the Cooks symbols for several of the scan types listed in Table I. With this symbolism, a targeted, or preselected, mass is indicated by a filled circle, and a scanned mass is designated by an unfilled circle. A predefined mass transition is indicated by a heavy arrow, whereas a variable mass transition is denoted by a thin arrow. A variable mass is emphasized by being presented in triplicate in the symbol. These symbols are employed in the figures presenting tandem mass spectrometric data below.

The Cooks symbolism fully describes those tandem MS experiments where all of the analyzers effectively act as mass filters. This symbolism, however, requires extension to accommodate experiments involving a significant variable other than mass. For instance, certain experiments on hybrid instruments can reveal the kinetic energy release during ion dissociation; these include the conventional MIKES experiment, the mass deconvoluted MIKES experiment and the second mode of reaction intermediate scanning (*vide infra*). For instance, in the mass deconvoluted MIKES experiment, both the mass and the kinetic energy of the precursor are predefined. The mass of the product ion is also predefined, but its kinetic energy is variable, and the electric sector voltage is what is actually scanned in this experiment to focus products of different kinetic energies. In the symbol denoting this scan shown in Table III, we have used a large unfilled circle surrounding a concentric filled circle to indicate an ion that is fixed in mass but variable in energy. Similarly, two concentric unfilled circles may be used to indicate ions that vary both in mass and in energy, such as the detected product ions in a conventional MIKES spectrum, and the intermediate ions in the second mode of reaction intermediate scanning (Table I). Symbols for these scans are shown in Table III.

TABLE II. COOKS' SYMBOLISM FOR VARIOUS TANDEM MASS SPECTROMETRIC EXPERIMENTS

| Scan mode | Symbolic* designation |
|---|---|
| product ion scan | 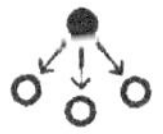 |
| constant neutral loss scan | 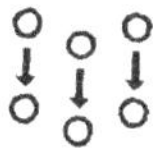 |
| precursor ion scan | 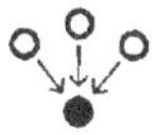 |
| 2nd generation product ion scan | 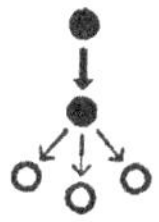 |
| Reaction intermediate scan (mode 1) | 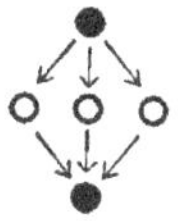 |
| Product-constant neutral loss scan | 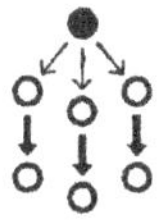 |
| Constant neutral loss-precursor ion scan | 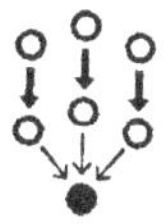 |
| Consecutive neutral loss scan | 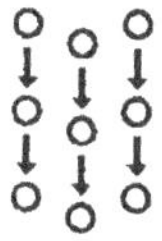 |

* ● Preselected mass ↓ Preselected mass transition

○ Variable mass ↓ Variable mass transition

TABLE III. EXTENSIONS TO COOKS' SYMBOLISM FOR TANDEM MS EXPERIMENTS DETERMINING KINETIC ENERGY RELEASE

| Scan mode | Symbolic* designation |
| --- | --- |
| conventional MIKES scan | 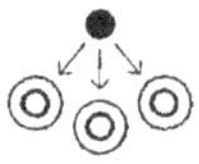 |
| mass deconvoluted MIKES scan (MAD MIKES) | 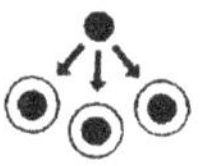 |
| Reaction intermediate scan (mode 2) | 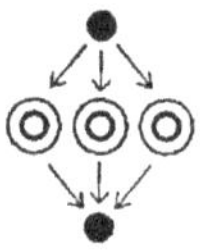 |

*  Preselected mass with variable kinetic energy

 Variable mass and kinetic energy

Other symbols as in Table II

### 1.3 COMPARISONS WITH OTHER TANDEM INSTRUMENTS

A detailed comparison of the design and performance of hybrid instruments and other prominent tandem MS designs is beyond the scope of this chapter but a few comments are appropriate.

The triple quadrupole (QqQ) is the simplest (and least expensive) of the major tandem mass spectrometer types. Triple quadrupoles capable of analyzing ions with a mass-to-charge ($m/z$) ratio of 4000 are now commercially available. Collision energies employed with this design are generally in the 10-500 eV range in the laboratory frame of reference. In principle, both mass analyzer quadrupoles can be operated at unit resolution, but either or both are frequently operated at lower resolution in order to increase the transmission of the ions of interest. Obviously, this practice can result in ambiguities in data interpretation, regardless of the type of instrument used. One of the advantages of quadrupole mass analyzers is their incorporation of a linear scan function, facilitating the implementation of the constant neutral loss scan. Quadrupole mass analyzers generally exhibit relatively poor transmission for high mass ions, although this is becoming less of a problem with the advent of high mass range quadrupoles. Quadrupoles are inherently low resolution devices as compared to sector mass spectrometers, so that, for example, precursor ions cannot be selected by MS1 at medium to high resolution in a triple quadrupole. The limitation of collision energies to less than 500 eV can be regarded as a disadvantage of the triple quadrupole because of the additional structural information which is frequently available at higher collision energies (i.e., 4-8 keV). Furthermore, the triple quadrupole has no facilities for MS/MS/MS (although in favorable cases sequential product ion information may be deduced by MS/MS analysis of presumed fragment ions formed in the ion source. MS/MS/MS analyses using only quadrupoles requires the pentaquadrupole design (QqQqQ).

Four-sector mass spectrometers consist of two double-focussing mass spectrometers connected in series (generally EBEB, BEEB or BEBE). Instruments with a mass-to-charge ratio range of more than 15,000 at full sensitivity are commercially available. With this instrument design, a collision cell is located between the two halves of the instrument, and the collision energies employed are generally in the 4-8 keV range, with helium typically employed as the collision gas. Both halves of the instrument are high resolution mass spectrometers, so that unit resolution of both precursor and product ions is readily achieved with little or no sacrifice in transmission. Newer four sector designs incorporate array detection, so that the signals from a wide mass range of ions may be detected simultaneously with unit resolution. This affords a large gain in sensitivity, and represents a significant advantage over instruments employing a quadrupole as the final stage of mass analysis, because these devices are necessarily limited to the detection of one $m/z$ value at a time. The decreasing efficiency of the collisional activation process with increasing mass of the precursor ion has limited the useful mass range of four-sector tandem MS experiments to ions of approximately 3000 u in most instances. Nevertheless, an expanding body of structurally informative data, particularly in the area of peptide analyses (reviewed elsewhere in this volume), attests to the value of the four-sector instrument and the application of high energy CAD analyses. The complexity of the relevant scan laws, however, has discouraged the application of precursor ion and constant neutral loss scanning, analytical approaches of established utility that are readily achieved on triple quadrupole and hybrid instruments.

## 2. Operational Modes of Hybrid Sector/Quadrupole Instruments

In this section, specific analytical applications of several of the scanning modes possible with hybrid sector/quadrupole instruments will be discussed to illustrate the different kinds of information that can be obtained. Some of the scan modes are applicable on either an EBqQ or a BEqQ hybrid, whereas others may be performed only with a BEqQ instrument. The MS/MS examples will be categorized according to the decomposition region involved, which determines whether low or high energy CAD can be employed. The MS/MS/MS modes, which employ two decomposition regions, will be discussed separately. For convenience, the specific examples presented in the following sections derive from the authors' work. This is not necessarily to imply, however, the authors' precedence in these areas; additional examples of the types of experiments described here exist in the literature.

### 2.1 MS/MS ANALYSES INVOLVING DECOMPOSITIONS IN q

The majority of the tandem mass spectrometric analyses performed using hybrids involve decompositions that occur in the RF-only quadrupole, either under

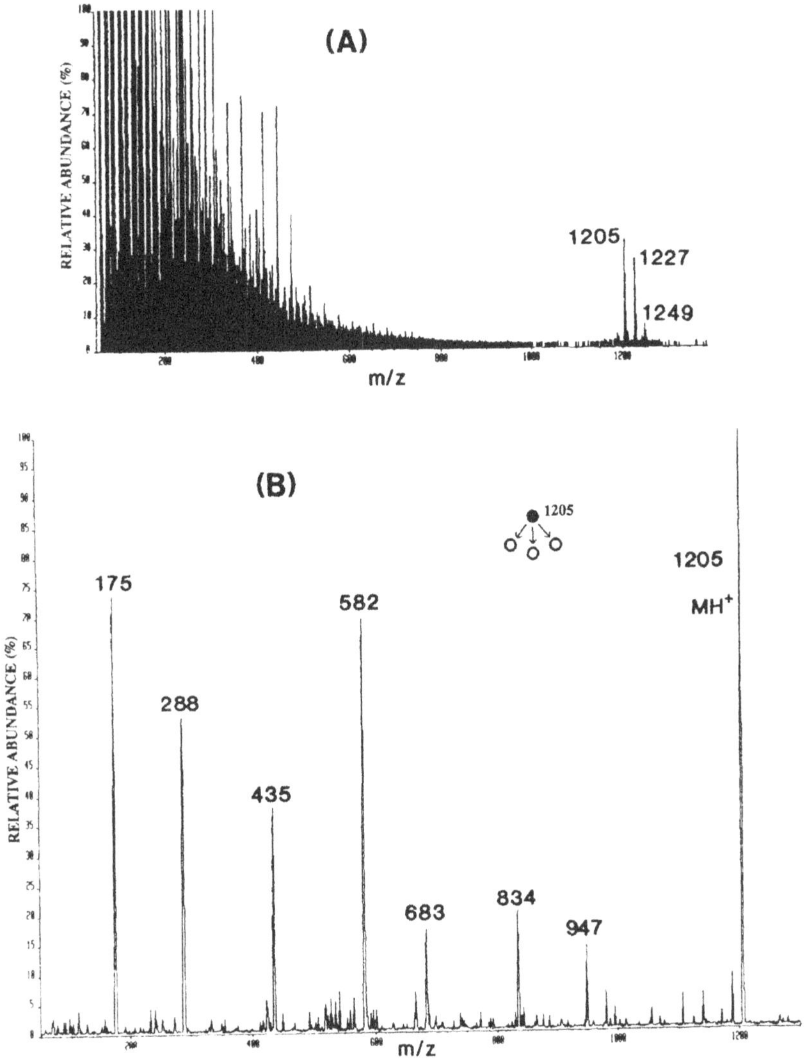


Figure 2. A) Conventional FAB spectrum of an oligopeptide isolated following tryptic hydrolysis of human apoprotein B-100. B) Product ion spectrum of the $m/z$ 1205 ion from the same sample, obtained on a BEqQ hybrid by scanning Q following low energy CAD ($E_{lab}$ = 30 eV, argon collision gas) in q. All major fragment ions are members of the $Y_n$" series [29].

metastable or low energy CAD conditions. The pertinent MS/MS modes of operation include the product ion scan, the precursor ion scan, the constant neutral loss scan, and selected reaction monitoring (see Tables I and II). The mode of operation used depends upon the type of information being sought. Both the EBqQ and the BEqQ hybrids can utilize these scanning modes with equal facility.

2.1.1 *Product Ion Scanning.* The product ion scan is the most commonly performed tandem mass spectrometric experiment, regardless of the type of instrument used. This scan is used for acquiring structural information for selected source-formed ions. The precursor ion is selected with the double focussing portion of the instrument (MS1) set to achieve a resolution of 1000 or greater, so that ions of a single nominal *m/z* value are transmitted to the RF-only quadrupole. Decomposition occurs in q, either spontaneously or following collisional activation, and the product ions are mass analyzed with Q (MS2). In this manner, a product ion spectrum can be obtained for each source-formed precursor ion generated from a complex mixture. Because MS1 excludes all ions except the selected *m/z* value, the product ion spectra obtained show a marked reduction in contributions from other ionic species generated in the ion source. This is a particularly significant consideration when fast atom bombardment and other forms of liquid secondary ion mass spectrometry are employed. Using these techniques, a significant signal is often generated at every *m/z* value from the liquid matrix alone so that minor analyte-derived fragment ions are frequently obscured. Furthermore, the assignment of fragment ions may be ambiguous; there is little direct evidence from conventional mass spectra to establish connectivity between precursor and product ions. Establishment of such connectivity represents the key feature establishing the enhanced selectivity of analysis with tandem MS.

These points are exemplified in Figure 2. The conventional FAB spectrum of an oligopeptide isolated following tryptic hydrolysis of human apoprotein B-100 is pre-

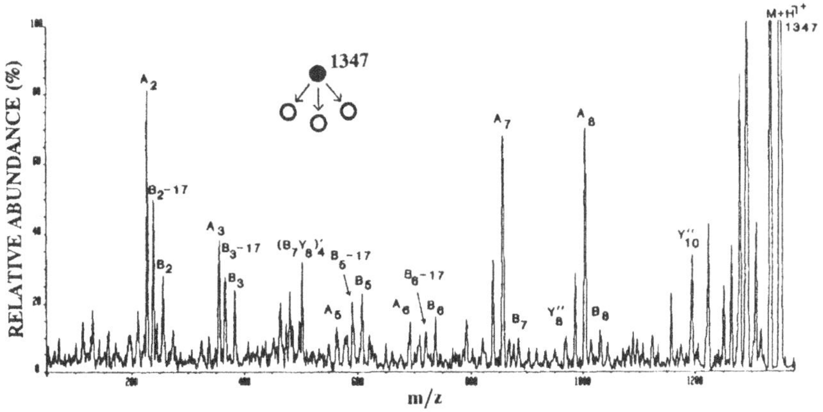


Figure 3. Low energy CAD product ion spectrum of the $[M + H]^+$ ion (*m/z* 1347) of the peptide substance P, obtained on a BEqQ hybrid by scanning Q following low energy CAD ($E_{lab}$ = 30 eV, argon collision gas) in q.

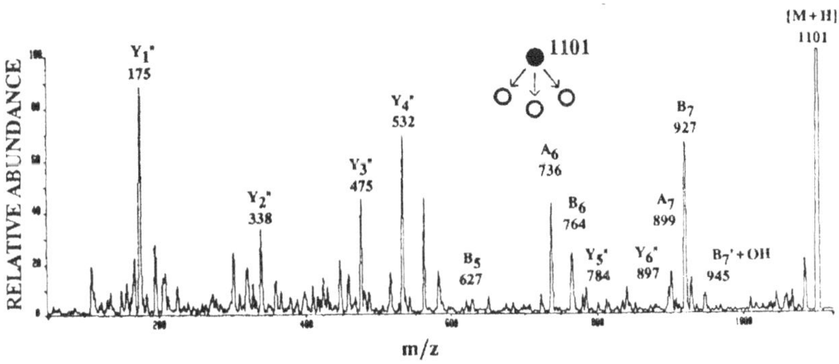


Figure 4. Product ion spectrum of the synthetic conjugate of acetaminophen and the octapeptide GFLCGHYR obtained by scanning Q on a BEqQ hybrid under low energy CAD ($E_{lab}$ = 25 eV) conditions.

sented in Figure 2A. Whereas the signals due to the $[M + H]^+$ (*m/z* 1205), $[M + Na]^+$ (*m/z* 1227) and $[M - H + 2Na]^+$ (*m/z* 1249) ionic species are clearly observed, any fragmentation information is obscured by the chemical background. This same sample was analyzed by product ion scanning; the $[M + H]^+$ precursor ion was selected by MS1 and subjected to collisional activation followed by scanning of the first generation products with MS2. A structurally informative product ion spectrum (Figure 2B) was obtained; it is unequivocal that the product ions were formed through decomposition of *m/z* 1225 ions [29]. A caveat is, however, required: not all the ions observed in a product ion spectrum necessarily derive through decomposition of a single precursor structure but may be the products of isobaric precursors [30]. Peaks attributable to decomposition of matrix ions are frequently observed in product ion spectra.

The utility of the product ion scanning experiment for peptide sequence analysis is well established from data generated with triple quadrupoles [31, 32], hybrids [33-36] and four-sector instruments [37-39]. Peptides tend to cleave at predictable sites along the backbone, although the relative yield of the predicted ionic species varies dramatically from peptide to peptide. The low energy CAD product ion spectrum of the undecapeptide substance P (amino acid sequence RPKPQQFFGLM-$NH_2$) is presented in Figure 3. The observed sequence ions are indicated using the Roepstorff and Fohlman nomenclature [40]. Extensive sequence information is contained in the spectrum. This spectrum illustrates the tendency for low energy CAD to generate internal cleavage products, exemplified by the $(B_7Y_8)_4$' ion, which (if recognized) can provide overlapping sequence information to facilitate structural assignment. Absent from this spectrum are the d and w type side chain fragments characteristically formed by high energy CAD, which can permit the differentiation of isobaric amino acid residues such as leucine and isoleucine [41]. This information is accessible (albeit with reduced sensitivity) on the BEqQ hybrid through the E/Q linked scan, as discussed below.

Product ion scanning is also useful for determining the structure of side chain modified peptides. Such an analysis of the synthetic conjugate formed between acetaminophen and the octapeptide GFLCGHYR is presented in Figure 4. Although several of the side chains of this peptide were potential sites of modification, it was anticipated that the thiol group of the cysteine residue would be the modification site. The overlapping sequence information obtained by product ion scanning confirmed that cysteine was the site of conjugation.

2.1.2 *Constant Neutral Loss and Precursor Ion Scanning.* Two other MS/MS scan modes, the constant neutral loss and the precursor ion scan, are particularly useful for compound class-selective detection. For the constant neutral loss scan, MS1 and MS2 are scanned simultaneously with a constant mass offset corresponding to the mass of the neutral fragment that is lost. For precursor ion scanning MS2 is set to transmit only ions equal in $m/z$ value to the specified product ion, and MS1 is scanned. The availability of these two scanning modes on triple quadrupole and hybrid instruments enables the use of powerful screening and characterization strategies directed toward detecting particular compound types, such as metabolites or derivatives of a particular drug, or compounds that fall into a particular structural class. Lee and Yost [42] noted that, because drug metabolites usually retain some of the structural features characteristic of the original drug, the metabolites are likely to lose some of the same neutral fragments and give rise to some of the same product ions as does the original drug. The neutral loss and precursor ion scanning capabilities of a triple quadrupole were used to screen complex physiological samples to detect possible metabolites of a particular xenobiotic; the candidate species identified in these scans were then subjected to product ion scanning for structural characterization.

An equivalent approach can be employed to detect various conjugates. In the negative ion mode, aliphatic steroid sulfates characteristically give rise to a common product ion of $m/z$ 97, corresponding to $HSO_4^-$. Therefore, a scan to detect all precursors of $m/z$ 97 in the negative ion mode selectively detects steroid sulfates, with reduced interference from other components in the sample [43]. Similarly, glucuronide conjugates tend to lose a neutral fragment 176 u, and thus a scan to detect ions which fragment through the loss of 176 u selectively detects glucuronide conjugates [44].

These techniques may be applied to the direct analysis of extremely complex mixtures, which are intractable to conventional MS analysis. Figure 5A depicts the conventional FAB spectrum of a crude bile extract from a rat which was administered a 1:1 mixture of unlabelled and [$^{13}C_1$]-labelled 2-furamide [45]. The objective was to identify and characterize glutathione conjugate metabolites of this compound, using the twin ion (stable isotope signature) technique to facilitate the recognition of metabolites derived from this compound. The conventional FAB spectrum of this biological sample is obviously complex, and no candidate metabolites can be readily discerned from this spectrum. Glutathione conjugates, however, characteristically lose a neutral fragment of 129 u, so that a constant neutral loss scan can be used to detect selectively these metabolites [46]. Such a scan of the bile extract is presented in Figure 5B. The twin ions at $m/z$ 417 and 418 were easily identified from this spectrum as due to candidate glutathione conjugates, particularly when this spectrum was compared to the corresponding one for control bile. (The ions of $m/z$ 308 and 613 represent endogenous reduced and oxidized glutathione.) These candidate

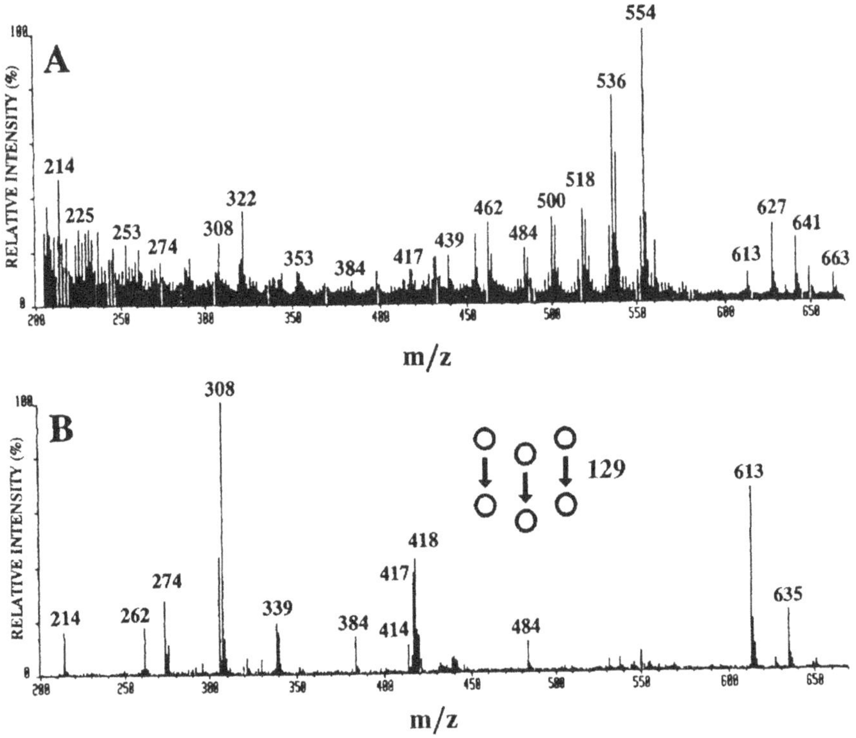


Figure 5. A) Conventional FAB mass spectrum of a crude bile extract from a rat dosed with a 1:1 mixture of unlabelled and [$^{13}C_1$]-2-furamide. B) Constant neutral loss spectrum (monitoring the loss of 129 u) of the same bile sample as in A. Signals in the constant neutral loss spectrum are labelled as the *m/z* ratios of the precursor ions. Spectrum B was obtained on a BEqQ hybrid by scanning B and Q simultaneously with a constant mass offset of 129 u, with low energy CAD conditions ($E_{lab}$ = 8 eV) employed in q. Reproduced with permission from reference 45.

ions were then subjected to first and second generation product ion scanning for structural characterization [45]. Analogous strategies were devised to detect and characterize derivatized glutathione conjugates [47].

The constant neutral loss and precursor ion scans are also useful in the analysis of isotopically labelled compounds [48]. The isotopic purity and the location of the label in stable isotope labelled compounds can be readily determined by tandem mass spectrometry. A constant neutral loss scan reflects the isotopic composition of the product ion that is ultimately detected; similarly, a precursor ion scan reflects the isotopic composition of the neutral fragment that is lost [49]. Either technique can be used to determine the isotopic purity of a labelled compound, and to give a direct indication of the portion of the molecule that contains the label. Exemplary data

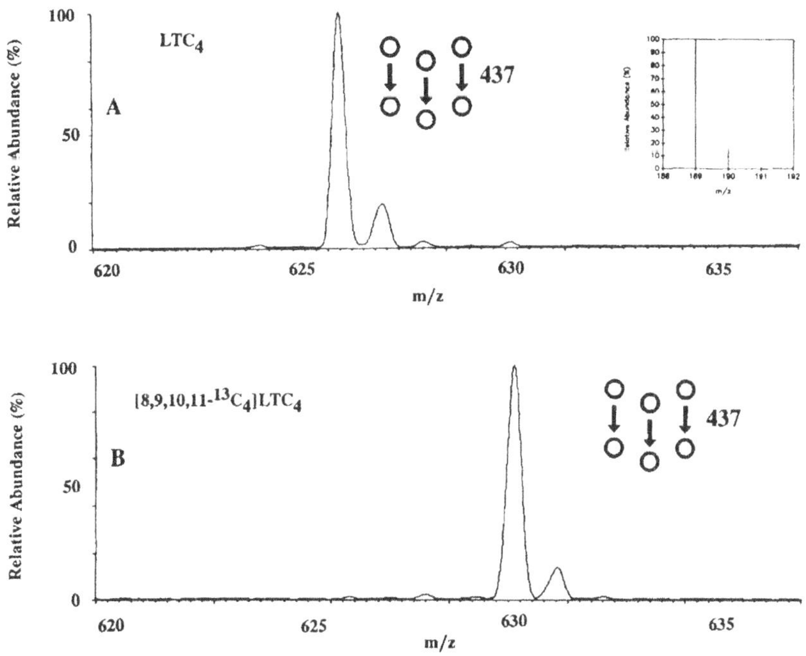


Figure 6. Constant neutral loss spectra of (A) unlabelled leukotriene $C_4$ and (B) [$^{13}C_4$]-leukotriene $C_4$. Both spectra were obtained on a BEqQ hybrid by scanning B and Q simultaneously with a constant mass offset of 437 u. Low energy CAD conditions ($E_{lab}$ = 15 eV) were employed in q. The fragmentation [50] represents cleavage of the lipid moiety with charge retention on the fragment retaining the [$^{13}$C]-label.

from constant neutral loss scanning analyses of leukotriene $C_4$ and its [$^{13}C_4$]-analogue are presented in Figures 6A and 6B [49]. In these scans, the product ions detected in the tandem mass spectrometer were of *m/z* 189 for the unlabelled material and *m/z* 193 for the labelled material. The horizontal scale indicates the mass of the precursor ions. As is obvious from these spectra, the isotopic purity is very high, with excellent agreement with the expected values for the analyte of natural isotopic abundance (Figure 6, inset). In this scan mode, a mass difference is observed between the labelled and unlabelled material if the isotopic label is located in the portion of the molecule represented by the detected product ion. Thus, the isotopic purity and an indication of the location of the label may be determined in a single analysis. Analogous considerations apply for precursor ion scans. Other techniques, such as product ion scanning and MS/MS/MS, can be used to pinpoint further the location of the label.

The constant neutral loss and precursor ion scans can also form the basis for quantitative trace analyses when used in conjunction with stable isotope dilution

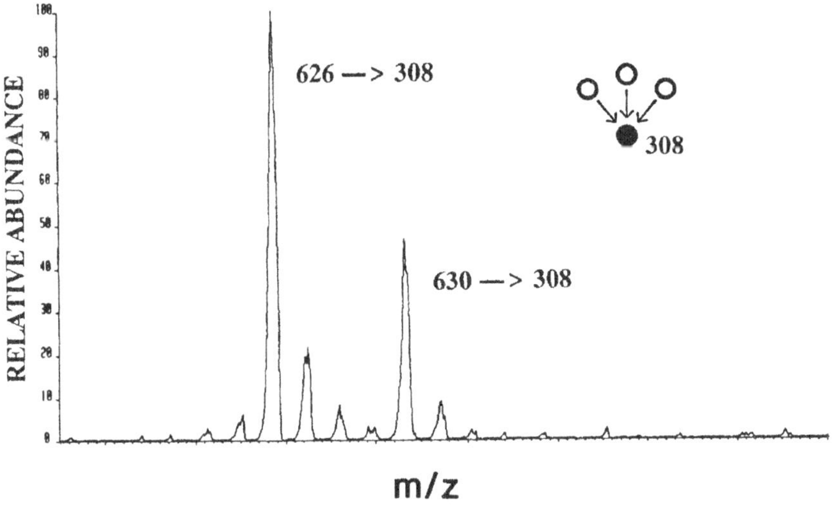


Figure 7. Limited mass range precursor ion scan (detecting precursors of $m/z$ 308) of an isolated fraction of rat bile after the animal had been administered endotoxin. The spectrum was obtained on a BEqQ hybrid by scanning B with Q set to transmit ions of $m/z$ 308 (low energy CAD conditions employed in q). Leukotriene $C_4$ and its $[^{13}C_4]$-analogue (the internal standard) were detected in a ratio of approximately 2:1.

techniques. For these analyses, the mass range of the scan is usually limited to a narrow range to increase the sensitivity of detection. Leukotriene $C_4$ and its $[^{13}C_4]$-analogue both give rise to a prominent product ion of $m/z$ 308 [50]. A scan for precursors of $m/z$ 308 will thus detect both leukotriene $C_4$ and the stable isotope labelled internal standard, and the relative abundances of the ions may be used for quantitation [51]. Such an analysis of a biological sample is presented in Figure 7, where leukotriene $C_4$ and its $[^{13}C_4]$-analogue were detected in a ratio of approximately 2:1. Similar techniques employing a hybrid instrument have been successfully applied to the quantitation of platelet activating factor (PAF) [52].

2.1.3 *Selected Reaction Monitoring.* The utility of selected reaction monitoring (SRM) as a quantitative method has been well demonstrated using either triple quadrupoles [53] or hybrids [54], but the technique remains under-utilized. A particular reaction is monitored by setting MS1 to transmit only specified precursor ions while MS2 transmits only specified product ions, and either metastable or CAD conditions are employed in the decomposition region. Multiple such transitions can be monitored concurrently under computer control, for example, during the course of a gas chromatographic separation. SRM provides an added dimension of molecular specificity over simple selected ion monitoring, where only the presence or absence of specified source-formed ions is monitored. With hybrid instruments, the facility for operating MS1 at medium to high resolution can further increase the se-

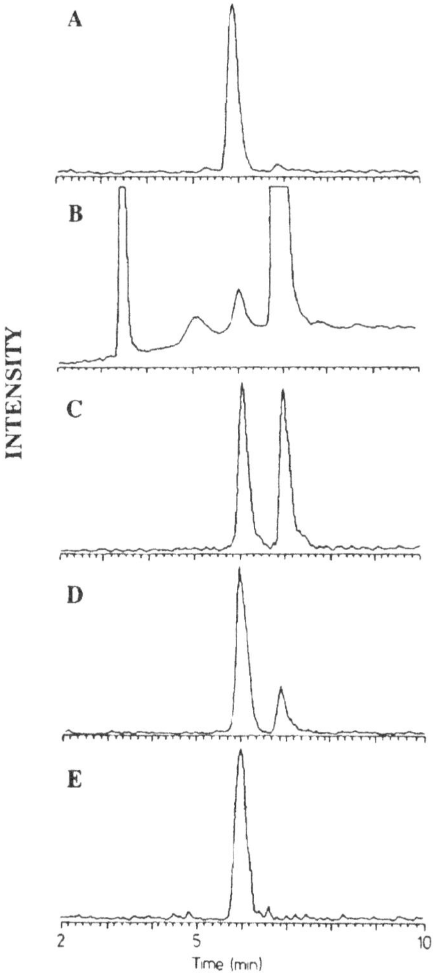


Figure 8. Gas chromatographic/mass spectral analyses for the bis-*tert*-butyldimethylsilyl ether of estradiol. (A) Selected reaction monitoring of *m/z* 500-->443 for a 25 pg standard. (B-E) Analyses of an extract of blood plasma. (B) Low resolution (resolving power 1000) selected ion monitoring of *m/z* 500, the molecular ion. (C) Selected reaction monitoring of *m/z* 500-->443 with a precursor ion resolution of 1000. (D) As in (C), but with a precursor ion resolution of 2500. (E) As in (C), but with a precursor ion resolution of 5000. The selected reaction monitoring analyses were performed on an EBqQ hybrid, with the EB portion set to transmit *m/z* 500 ions, and with Q set to transmit *m/z* 443 ions. Low energy CAD conditions were employed in q. Reproduced from Reference 54, with permission.

lectivity of the analyses. Figure 8 presents a series of gas chromatographic/mass spectral (GC/MS) analyses of estradiol as the bis-*tert*-butyldimethylsilyl ether [54]. Figure 8A shows an SRM analysis of 25 pg of authentic material, monitoring the *m/z* 500-->443 transition (M+--> $[M - C_4H_9]^+$). A low resolution (resolving power 1000) selected ion monitoring analysis (monitoring *m/z* 500) of a biological extract is shown in Figure 8B, where the peak attributable to estradiol is minor compared to other signals. Analysis of the same sample by SRM, with a precursor ion resolution of 1000, greatly simplified the result (Figure 8C) with detection of only two components, including the estradiol. The relative intensity of the other chromatographic peak was reduced by selecting precursor ions of *m/z* 500.35 at a resolving power of 2500 (Figure 8D), and the interference was completely eliminated when the precursor ion resolving power was 5000 (Figure 8E).

### 2.2 MS/MS ANALYSES INVOLVING DECOMPOSITIONS IN THE FIRST OR SECOND FIELD-FREE REGION

High energy collisional activation may be achieved on double focussing instruments in either FFR1 or FFR2. Both EB and BE type instruments can be used to generate product ion spectra resulting from FFR1 decompositions by means of the familiar B/E linked scan [1]. The double focussing portion of tandem sectors/quadrupole hybrid instruments is uncompromised, and the B/E scan is readily implemented with both EBqQ and BEqQ hybrids, as is the $B^2/E$ linked scan for producing precursor ion spectra for FFR1 decompositions [1]. Numerous examples of the use of these scans exist in the literature [27].

2.2.1 *MIKES Analyses*. Instruments incorporating the BE configuration (including the BEqQ hybrid) also have the capability of performing MIKES analyses. In these experiments, the precursor ion is selected with B, and E is scanned to detect product ions formed through metastable or CAD processes in FFR2. The widths of the product ion peaks observed in a MIKES spectrum reflect the kinetic energy released in the decomposition processes [55]. Kinetic energy release information for first field-free region decompositions is accessible on EB type instruments through the $B^2/E$ linked scan, but this experiment is subject to poor effective precursor ion resolution, which is less of a problem with the MIKES experiment [27]. The value of the availability of both low and high energy CAD regimes on a single instrument is illustrated in Figure 9 [56]. Figure 9A depicts a low energy CAD product ion spectrum of cholestanol sulfate. The only significant fragments are *m/z* 97 ions corresponding to $HSO_4^-$, but these were formed in very high abundance. Such a highly favored decomposition occurring in the RF-only quadrupole is highly useful for quantitative analyses employing selected reaction monitoring or precursor ion scanning techniques as described above. This spectrum, however, contains very little structural information. This is in contrast to the corresponding MIKES spectrum of cholestanol sulfate obtained on the same instrument (Figure 9B). This spectrum, obtained under 8 keV CAD conditions, compares closely with the previously reported MIKES spectrum of this compound obtained on an EBE instrument [57]. This spectrum is rich in structural information from the so-called charge remote fragmentation, but the overall yield of product ions is very low. Consequently, the high energy CAD regime would be unsuitable for the sensitive quantitative analysis of this compound.

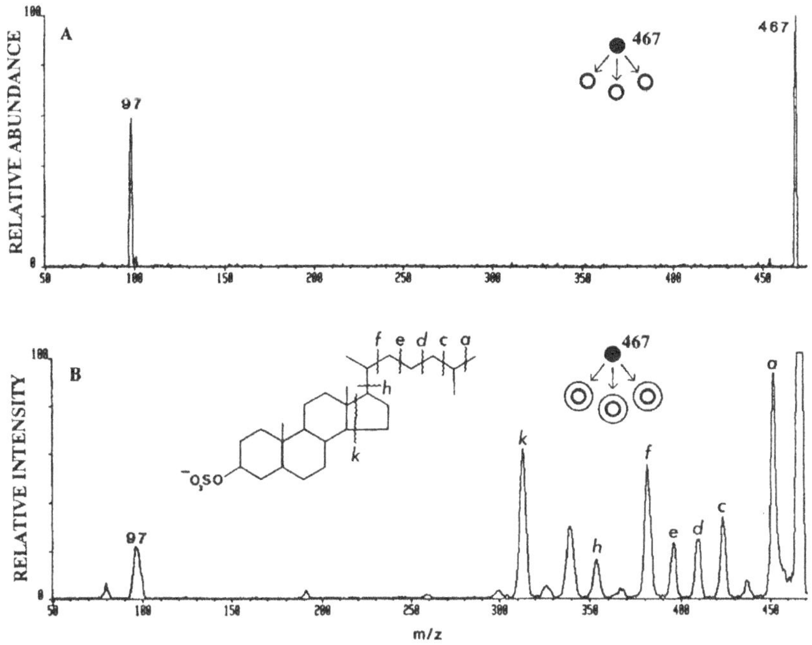


Figure 9. Low and high energy CAD product ion spectra obtained on a BEqQ hybrid following collisional activation of cholestanol sulfate anions (of *m/z* 467). (A) Low energy CAD in q, with the spectrum obtained by scanning Q. (B) High energy CAD in the collision cell between B and E; spectrum obtained by scanning E (MIKES analysis). Reproduced from Reference 56, with permission.

2.2.2 *E/Q-Linked Scanning.* As mentioned above, the B/E linked scan is one means of acquiring product ion spectra from FFR1 decompositions on a double focussing instrument. This type of scan gives good resolution of the product ions but relatively poor effective resolution of the precursors. Conversely, the MIKES experiment gives good resolution of the precursors but poor effective mass resolution of the product ions. Good resolution of both precursors and products from the high energy CAD regime may be achieved on the BEqQ hybrid through the E/Q linked scan. In this mode of operation, precursor ions are selected with B, and E and Q are scanned simultaneously with E/Q constant to detect product ions formed in the second field-free region. The RF-only quadrupole acts purely as a transmission device for this experiment. The mass analyzer quadrupole is operated at unit resolution to separate overlapping masses in the MIKES spectrum. As Boyd [10] has pointed out, this apparently simple scan mode in fact requires the simultaneous linear tracking of at least five potential fields with very high accuracy, so that small errors in tracking can lead to reduced transmission and consequently reduced sensitivity. Several solutions to this problem have been proposed. The simplest is to restrict the E/Q linked scan to a narrow range to minimize the effects of imperfect tracking of the various fields. Alternatively, to achieve a full mass range scan, E and Q may be scanned repetitively at different scan rates and in an asynchronous fashion. All scans are accumulated into a single data file so that the composite spectrum is equivalent to a single E/Q linked scan with accurate tracking [58]. An ingenious alternative approach, also developed by Boyd and coworkers [59], is to superimpose a modulation of E on an E/Q linked scan. Despite these improvements, the E/Q linked scan and equivalent approaches require further development and assessment. It should be noted that under no circumstance will the transmission of product ions of high energy collision in the hybrid instrument be equivalent to that achieved on a four-sector instrument in which some degree of energy focussing of product ions is achieved by the double focussing configuration of MS2.

Despite the technical difficulties associated with the E/Q linked scan, it can be used to good advantage to provide high energy CAD data with good resolution of both precursor and product ions. This is exemplified in Figure 10 [56]. The conventional MIKES spectrum, obtained under 8 keV CAD conditions, of isovalerylcarnitine is depicted in Figure 10A. As is typical of MIKES spectra, the peaks are broad and the presence of unresolved components is suggested by the shapes of several of the peaks. The corresponding spectrum obtained by linked scanning of E and Q is shown in Figure 10B. The *m/z* 100 and 102 components are well separated, and the cluster of peaks at *m/z* 57-60 are resolved into their individual components. Data

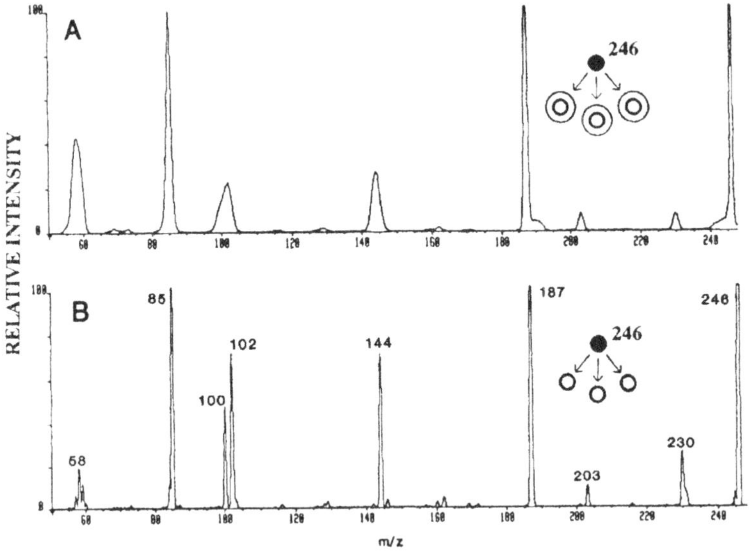


Figure 10. High energy (8 keV) product ion spectra obtained on a BEqQ hybrid following collisional activation of isovalerylcarnitine $[M + H]^+$ ions. (A) Conventional MIKES analysis obtained by scanning E, with CAD in the collision cell between B and E. (B) E/Q linked scan under the same CAD conditions, obtained by scanning E and Q simultaneously. Reproduced from Reference 56, with permission.

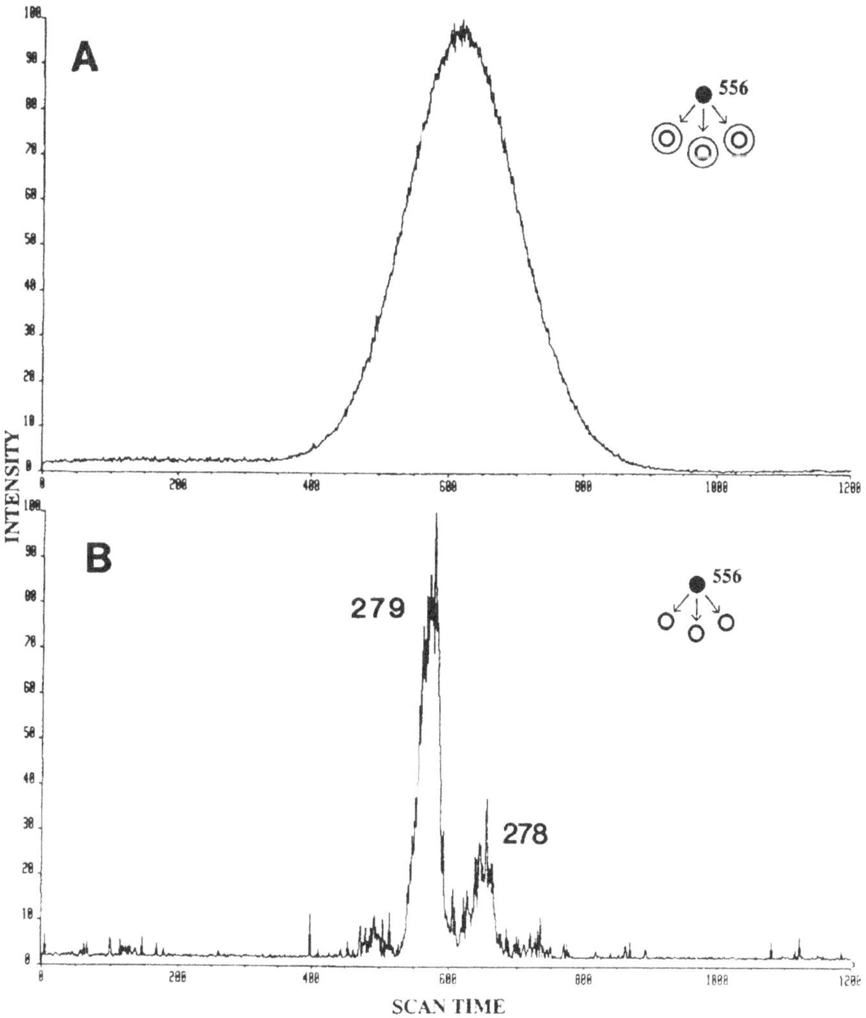


Figure 11. Metastable decompositions of leucine-enkephalin occurring in the second field-free region of a BEqQ hybrid. (A) Limited range MIKES analysis, obtained by scanning E. (B) Limited range E/Q linked scan obtained by scanning E and Q simultaneously, with E scanned over the same range as in (A).

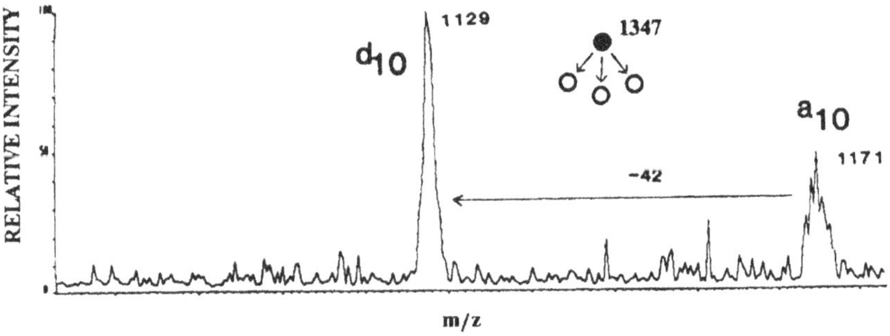


Figure 12. Limited mass range E/Q linked scan (BEqQ hybrid) illustrating the formation of d-type ions following high energy CAD of $[M + H]^+$ ions of substance P.

such as these facilitate correct assignment of high energy CAD product ions and assessment of the relative abundances of the individual product ions which overlap in the MIKES spectrum.

An additional example is provided by leucine-enkephalin. A portion of the metastable MIKES spectrum of this compound is presented in Figure 11A. Because the effective mass resolution of the MIKES analysis is poor, the peak in this spectrum could represent either a $B_3$ ion of *m/z* 278 or a $Y_2$" ion of *m/z* 279, and there is no obvious inhomogeneity of peak shape to suggest the presence of both species. The corresponding E/Q linked scan, presented in Figure 11B, clearly indicates the presence not only of both of these two ions but also *m/z* 280, attributable to complete loss of the leucine side-chain from the $Y_3$" species (*m/z* 336).

A final example of the utility of the E/Q linked scan involves the peptide substance P, the low energy CAD product ion spectrum of which failed to provide information for distinguishing the isobaric amino acid residues leucine and isoleucine (Figure 3). A narrow range E/Q linked scan analysis of this compound is presented in Figure 12, illustrating the formation of the d-type fragment ion under high energy CAD conditions. These data confirm the presence of a leucine, rather than isoleucine, at residue 10 [60].

2.2.3 *Mass Deconvoluted MIKES Analyses.* Although the E/Q linked scan provides good mass resolution for product ion scanning purposes, the kinetic energy release information contained in the MIKES spectrum is lost with the E/Q linked scan. Kinetic energy release information with good resolution of both precursor and product ions can be achieved on the BEqQ hybrid through the mass deconvoluted MIKES analysis (Table III), to which we lightheartedly refer as the MAD MIKES experiment. Cooks' group [23] reported the first such scan obtained on a BEqQ hybrid. A similar experiment, nominally a mass deconvoluted ion kinetic energy scan, was previously described and implemented by Beynon and co-workers [24] using an EQ hybrid. The MAD MIKES experiment is identical to the MIKES analysis except that the product ions passed by E are transmitted through q to the mass analyzer quadrupole, which is set to transmit ions of specified *m/z* value at unit mass resolution. Thus, although the E/Q linked scan involves scanning E and Q simultaneously, the MAD MIKES experiment involves scanning E with Q "parked" at a

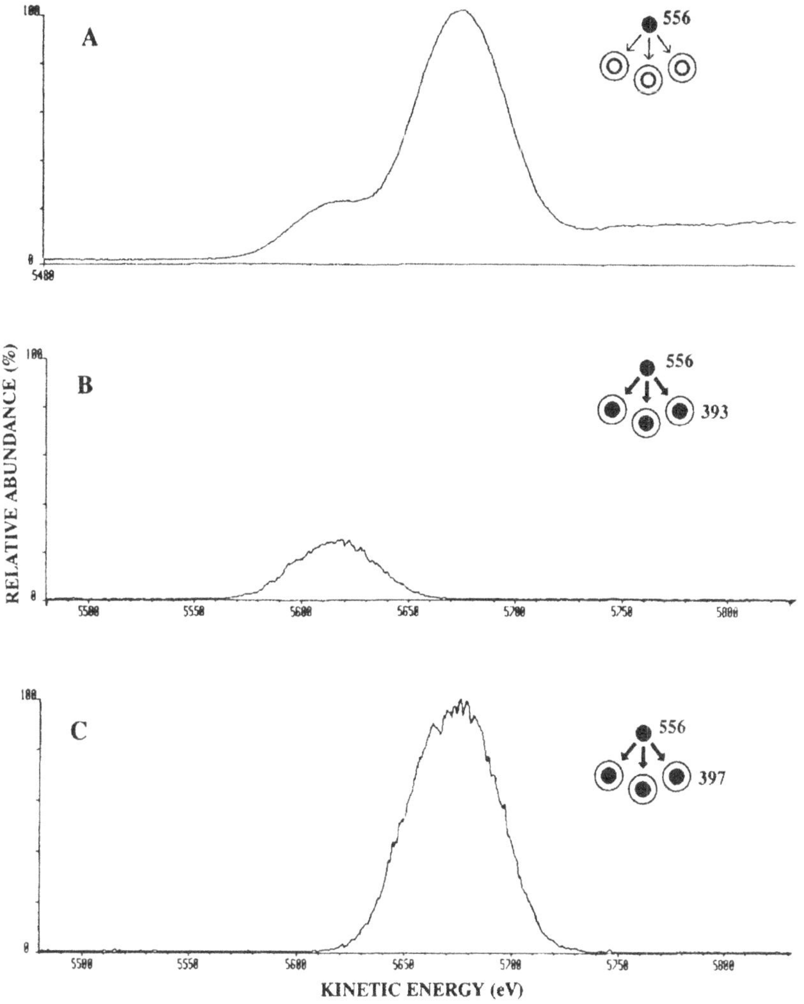


Figure 13. A) Portion of the metastable MIKES spectrum of leucine-enkephalin, obtained by scanning E on a BEqQ hybrid. (B) Mass deconvoluted MIKES (MAD MIKES) analysis over the same range as in (A), obtained by scanning E with Q set to transmit ions of *m/z* 393. (C) As in (B) but with Q set to transmit ions of *m/z* 397.

specified value. Leucine-enkephalin again provides an example of the utility of this experiment. Figure 13A presents a portion of the conventional metastable MIKES spectrum of this peptide. Two overlapping peaks are present, representing ions of $m/z$ 393 (the $Y_4''$ ion) and 397 (the $A_4$ ion). Because these peaks overlap, the kinetic energy released in the decomposition process, reflected in the widths of the peaks, cannot be accurately estimated. Figures 13B and 13C present the corresponding MAD MIKES spectra; the mass analyzer quadrupole was set to transmit ions of $m/z$ 393 and 397 respectively. The overlapping MIKES peaks are resolved into their individual components with retention of the MIKES peak shape, so that the peak widths, and therefore the kinetic energy release, may be accurately estimated.

## 2.3 SEQUENTIAL MASS SPECTROMETRY (MS/MS/MS)

Tandem mass spectrometers of the hybrid sectors/quadrupole design are capable of performing a variety of MS/MS/MS experiments. As indicated in Table II, instruments of the BEqQ configuration are capable of several of these operational modes that are not possible with EBqQ instruments. All of the sequential MS experiments listed in Table II have been implemented on BEqQ instruments. These scan modes involve an initial stage of decomposition occurring in one of the field-free regions with a second decomposition occurring in the RF-only quadrupole. These experiments are useful for a variety of mechanistic studies as well as for qualitative and quantitative analytical applications.

2.3.1 *Reaction Intermediate Scanning.* The utility of the reaction intermediate scan in peptide sequencing has been demonstrated by Cooks' group [61] using both a BEqQ hybrid and a pentaquadrupole instrument. With this scan, the precursor and second generation product ions are preselected, and the first generation product ions (intermediates) are scanned. This can be implemented on the BEqQ hybrid in two fashions. For the first of these, Q is set to transmit the second generation product ions, and a B/E linked scan is performed to scan for first generation product ions formed from the preselected precursor. This scan thus detects intermediate ions formed in the first field-free region from a selected precursor, which subsequently decompose in the RF-only quadrupole to form a selected second generation product. Examples of this scan exist in the literature [61]. Like all scans involving a B/E link, this mode of operation suffers from poor effective resolution of the precursor ions. With the second mode of reaction intermediate scanning possible with a BEqQ hybrid, good resolutions of the precursor and second generation product ions are achieved, but the resolution of the intermediates is degraded. For this scan, precursor ions are selected by B, the second generation product ions are selected by Q, and E is scanned to transmit intermediates formed in the second field-free region. Operationally, this is identical to an extended MAD MIKES experiment, except that collision gas is frequently employed in q for the reaction intermediate scan. Thus, the MAD MIKES experiment can be regarded as a special case of the second mode of operation of the reaction intermediate scan. Just as in the MAD MIKES spectra, the peak widths in the reaction intermediate spectra obtained in this fashion reflect the kinetic energy released in the first stage of decomposition. An exemplary reaction intermediate scan for leucine-enkephalin employing this mode of operation is presented in Figure 14. In obtaining this spectrum the magnetic field strength was set to transmit precursor $[M + H]^+$ ions of $m/z$ 556, the quadrupole was set to transmit ions of $m/z$ 397 (the $A_4$ ion), and the electric sector was scanned. All de-

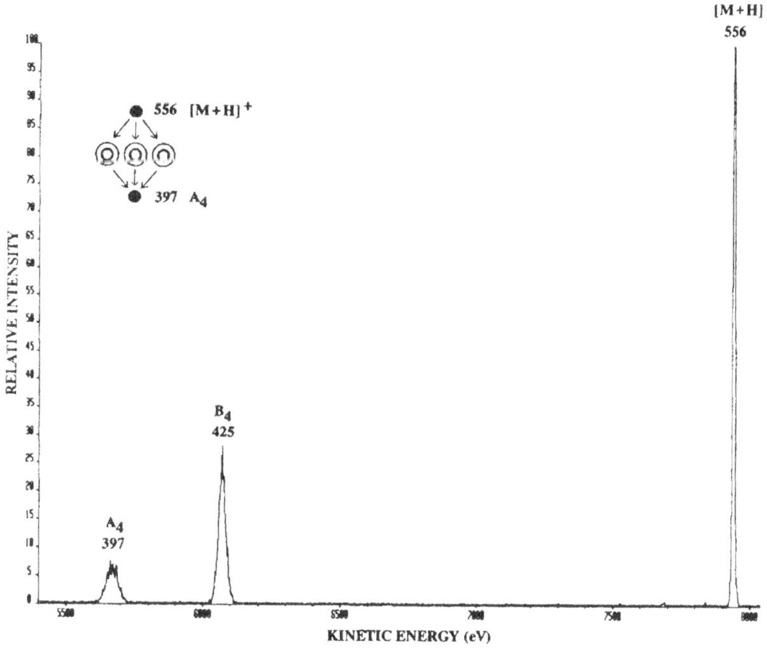


Figure 14. Reaction intermediate scan for leucine-enkephalin obtained on a BEqQ hybrid under metastable decomposition conditions. The spectrum was obtained with the magnet set to transmit precursor $[M + H]^+$ ions of *m/z* 556, and the quadrupole set to transmit ions of *m/z* 397 (the $A_4$ ion); the electric sector was scanned.

compositions were metastable. This spectrum, although apparently quite simple, is nevertheless rich in analytical information. The peak corresponding to the precursor ion of *m/z* 556, represents the 556-->397 transition occurring in the RF-only quadrupole, and the width of this peak reflects the kinetic energy spread in the main beam. The peak corresponding to the *m/z* 397 ion represents the 556-->397 transition occurring in the second field-free region, and the width of this peak reflects the kinetic energy released during the 556-->397 transition. This peak is identical to that in the MAD MIKES spectrum for this transition. The peak of *m/z* 425, corresponding to the $B_4$ ion, indicates that the $B_4$ ion can subsequently spontaneously decompose to form the $A_4$ ion. This peak, therefore, represents the *m/z* 556-->425-->397 transition, and the width of this peak reflects the kinetic energy released during the *m/z* 556-->425 transition. It is evident from this spectrum that the kinetic energy released during the metastable decomposition of *m/z* 556 ions to form $A_4$ ions is different from that released during the metastable formation of $B_4$ ions. This is the subject of continuing study. Similar spectra for leucine-enkephalin involving other mass transitions under high energy CAD conditions were reported by Alexander and Boyd [36].

2.3.2 *Second Generation Product Ion Scanning.* The second generation product ion scanning capability of hybrid instruments is useful for mechanistic studies and for obtaining highly specific structural information. Again, two modes of operation are possible for this scan with BEqQ hybrids, both of which have been employed successfully for a variety of applications [21, 45, 62-66]. The first of these, which may also be performed with EBqQ instruments, employs a static B/E link to transmit into the RF-only quadrupole selected first generation product ions formed in the first field-free region from selected precursor ions. The second stage of decomposition occurs in q, and mass analysis of the second generation product ions is accomplished by scanning Q. As discussed previously, this mode of operation is chosen

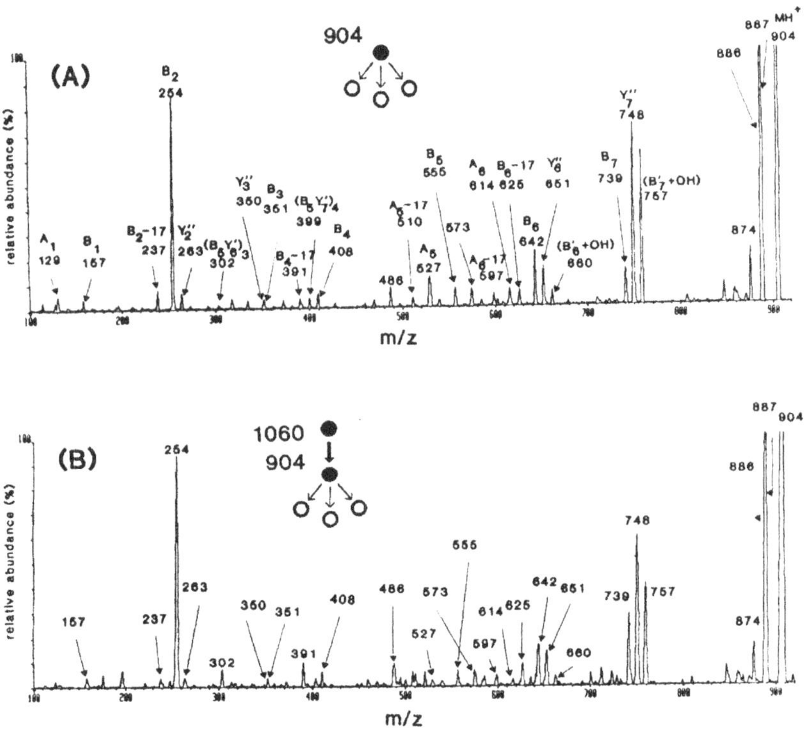


Figure 15. (A) First generation product ion spectrum (MS/MS) of Des-Arg$^9$-bradykinin $[M + H]^+$ ions, obtained by scanning Q. (B) Second generation product ion spectrum (MS/MS/MS) of a rearrangement ion (*m/z* 904) derived through decomposition of bradykinin $[M + H]^+$ ions. Spectrum obtained on a BEqQ hybrid with the magnet set to transmit ions of *m/z* 1060, E set to transmit *m/z* 904 ions, and Q scanned. Reproduced from Reference 65, with permission.

when the resolution of the first generation product ions is critical. For the alternative approach to second generation product ion scanning, source-formed precursor ions are selected with the magnet and allowed or induced to decompose in the second field-free region to yield the first generation product ions. These ions are selected with the electric sector and introduced into the rf-only quadrupole where the second stage of decomposition occurs. The second generation product ions thus formed are scanned by the mass analyzer quadrupole.

An example of the utility of this mode of operation is presented in Figure 15B, which depicts the second generation products derived from decomposition of a rearrangement ion (*m/z* 904), which was itself formed through decomposition of bradykinin $[M + H]^+$ ions. The amino acid sequence of bradykinin is RPPGFSPFR, and it was hypothesized that this rearrangement species was identical in structure to protonated Des-Arg$^9$-bradykinin, or RPPGFSPF. The first generation product ion spectrum of the $[M + H]^+$ ions of the latter compound is depicted in Figure 15A. The close similarity of these two spectra provided strong evidence of structural identity between the rearrangement ion and protonated Des-Arg$^9$-bradykinin [65]. Both of these approaches to second generation product ion scanning have been employed successfully with a minimum of sample purification for structural analyses involving samples of biological origin [45].

## 3. Conclusions

Tandem hybrid mass spectrometric techniques have proven to be useful for addressing a wide variety of analytical problems. The principal virtue of the sector/quadrupole design (particularly the BEqQ configuration) is the variety of operational modes that may be accessed and the consequent diversity of analytical information that may be obtained. Thus, important applications have been reported in the areas of compound class screening, structural and mechanistic studies, and trace analyses. In comparison with other common tandem instrument designs, both advantages and disadvantages are apparent. As a general rule, rapid switching between various MS/MS modes (product ion, precursor ion and constant neutral loss scanning) is more simply achieved on triple quadrupole instruments. Four sector mass spectrometers remain the instruments of choice if the principal objective is product ion scanning following high energy collisional activation, when the capabilities of good resolving power and array detection may be exploited. Maximum versatility of operation, however, including high and low energy CAD and techniques of MS/MS/MS, is achieved with hybrid instruments. A high proportion of the currently installed hybrid instruments are grossly underused as tandem mass spectrometers. Improved appreciation and understanding of their capabilities should promote their more effective use.

## 4. Acknowledgements

The authors' work is funded in part by the U.S. National Institutes of Health and by generous gifts from Glaxo Inc. and the Burroughs Wellcome Company.

## 5. References


1. Boyd, R. K. and Beynon, J. H. (1977) *Org. Mass Spectrom.* **12**, 163.
2. Beynon, J. H., Cooks, R. G., Amy, J. W., Baitinger, W. E., and Ridley, T. Y. (1973) *Anal. Chem.* **45**, 1023A.
3. White, F. A., Rourke, F. M., and Sheffield, J. C. (1958) *Appl. Spectrosc.* **12**, 46.
4. Hunt, D. F., Shabanowitz, J., Yates III, J. R, McIver Jr., R. T., Hunter, R. L., Syka, J. E. P., and Amy, J. (1985) *Anal. Chem.* **57**, 2728.
5. Glish, G. L. and Goeringer, D. E. (1984) *Anal. Chem.* **56**, 2291.
6. Glish, G. L., McLuckey, S. A., and McKown, H. S. (1987) *Anal. Instrum.* **16**, 191.
7. Yost, R. A. and Enke, C. G. (1978) *J. Am. Chem. Soc.* **100**, 2274.
8. Yost, R. A. and Enke, C. G. (1979) *Anal. Chem.* **51**, 1251A.
9. Yost, R. A., Enke, C. G., McGilvery, D. C., Smith, D., and Morrison, J. D. (1979) *Int. J. Mass Spectrom. Ion Proc.* **30**, 127.
10. Yost, R. A. and Boyd, R. K. (1990) in J. A. McClosky (ed.), Methods in Enzymology, Academic Press, Inc., New York, p 154.
11. von Zahn, U. and Tartarczyk, H. (1964) *Phys. Lett.* **12** 190.
12. Leventhal, J. J. and Friedmann, L. (1968) *J. Chem. Phys.* **49**, 1974.
13. Cosby, P. C. and Moran, T. F. (1970) *J. Chem. Phys.* **52**, 6157.
14. L'Hote, J. P., Abbe, J. Ch., Paulus, J. M., and Ingersheim, R. (1971) *Int. J. Mass Spectrom. Ion Phys.* **7**, 309.
15. Giardini-Guidoni, A., Platania, R., and Zocchi, F. (1974) *Int. J. Mass Spectrom. Ion Phys.* **13**, 453.
16. Thomas, T. F., Dale, F., and Paulson, J. F. (1977) *J. Chem. Phys.* **67**, 793.
17. Medley, S. S. (1978) *Rev. Sci. Instrum.* **49**, 698.
18. Armentrout, P. B. and Beauchamp, J. L. (1980) *J. Am. Chem. Soc.* **102**, 1736.
19. Glish, G. L. (1980) Ph.D. Thesis, Purdue University.
20. Glish, G. L., McLuckey, S. A., Ridley, T. Y., and Cooks, R. G. (1982) *Int. J. Mass Spectrom. Ion Phys.* **41**, 157.
21. Louris, J. N., Wright, L. G., Cooks, R. G., and Schoen, A. E. (1985) *Anal. Chem.* **57**, 2918.
22. Schoen, A. E., Amy, J. W., Ciupek, J. D., Cooks, R. G., Dobberstein, P., and Jung, G. (1985) *Int. J. Mass Spectrom. Ion Proc.* **65**, 125.
23. Ciupek, J. D., Amy, J. W., Cooks, R. G., and Schoen, A. E. (1985) *Int. J. Mass Spectrom. Ion Proc.* **65**, 141.
24. Harris, F. M., Keenan, G. A., Bolton, P. D., Davies, S. B., Singh, S., and Beynon, J. H. (1984) *Int. J. Mass Spectrom. Ion Proc.* **50**, 273.
25. Beynon, J. H., Harris, F. M., Green, B. N., and Bateman, R. H. (1982) *Org. Mass Spectrom.* **17**, 55.
26. Glish, G. L., McLuckey, S. A., McBay, E. H., and Bertram, L. K. (1986) *Int. J. Mass Spectrom. Ion Proc.* **70**, 321.
27. Busch, K. L., Glish, G. L., and McLuckey, S. A. (1988) Mass Spectrometry/Mass Spectrometry: Techniques and Applications of Tandem Mass Spectrometry, VCH Publishers: New York.
28. Schwartz, J. C., Wade, A. P., Enke, C. G., and Cooks, R. G. (1991) *Anal. Chem.*, in press.
29. Yang, C.-Y., Kim, T. W., Sparrow, J. T., and Gaskell, S. J. (1990) *J.Protein Chem.* **9**, 323.
30. Mathews, W. R., Runge, T. A., Haroldsen, P. E., and Gaskell, S. J. (1989) *Rapid Commun. Mass Spectrom.* **3**, 314.

31. Hunt, D. F., Bone, W. M., Shabanowitz, J., Rhodes, J., and Ballard, J. M. (1981) *Anal. Chem.* **53**, 1704.
32. Hunt, D. F., Yates III, J. R., Shabanowitz, J., and Winston, S., Hauer, C. R. (1986) *Proc. Nat. Acad. Sci. USA* **83**, 6233.
33. Gaskell, S. J., Reilly, M. H., and Porter, C. J. (1988) *Rapid Commun. Mass Spectrom.* **2**, 142.
34. Gaskell, S. J. and Reilly, M. H. (1988) *Rapid Commun. Mass Spectrom.* **2**, 188.
35. Leary, J. A., Williams, T. D., and Bott, G. (1989) *Rapid Commun. Mass Spectrom.* **3**, 192.
36. Alexander, A. J. and Boyd, R. K. (1989) *Int. J. Mass Spectrom. Ion Proc.* **90**, 211.
37. Biemann, K. and Martin, S. A. (1987) *Mass Spectrom. Rev.* **6**, 1.
38. Biemann, K. and Scoble, H. A. (1987) *Science* **237**, 992.
39. Carr, S. A., Anderegg, R. J., and Hemling, M. E. (1988) In C. J. McNeal (ed.), The Analysis of Peptides and Proteins by Mass Spectrometry, Wiley: New York, p. 95.
40. Roepstorff, P. and Fohlman, J. (1984) *Biomed. Mass Spectrom.* **11**, 601.
41. Johnson, R. S., Martin, S. A., and Biemann, K. (1989) *Int. J. Mass Spectrom. Ion Proc.* **86**, 137.
42. Lee, M. S. and Yost, R. A. (1988) *Biomed. Mass Spectrom.* **15**, 193.
43. Gaskell, S. J. (1988) *Biomed. Mass Spectrom.* **15**, 99.
44. Gaskell, S. J. and Haroldsen, P. E. (1989) In T.A. Bailie and J. R. Jones (eds.), Synthesis and Applications of Isotopically Labelled Compounds, Elsevier: Amsterdam, p. 135.
45. Ballard, K. D., Raftery, M. J., Jaeschke, H., and Gaskell, S. J. (1990) *J. Am. Soc. Mass Spectrom.* **2**, 55.
46. Haroldsen, P. E., Reilly, M. H., Hughes, H., Gaskell, S. J., and Porter, C. J. (1988) *Biomed Mass Spectrom.* **15**, 615.
47. Pearson, P. G., Howald, W. N., and Nelson, S. D. (1990) *Anal. Chem.* **62**, 1827.
48. Singleton, K. E., Cooks, R. G., and Wood, K. V. (1983) *Anal. Chem.* **55**, 762.
49. Gaskell, S. J., Ballard, K. D., Thorne, G. C., and Orkiszewski, R. S. (1990) Proceedings of the 38th ASMS Conference on Mass Spectrometry and Allied Topics, Tucson, Arizona, p. 1472.
50. Raftery, M. J., Thorne, G. C., Orkiszewski, R. S., and Gaskell, S. J. (1990) *Biomed. Environ. Mass Spectrom.* **19**, 465.
51. Raftery, M. J., Jaeschke, H, and Gaskell, S. J. (1990) Proceedings of the 38th ASMS Conference on Mass Spectrometry and Allied Topics, Tucson, Arizona, p. 1395.
52. Haroldsen, P. E. and Gaskell, S. J. (1989) *Biomed. Environ. Mass Spectrom.* **18**, 439.
53. Perchalski, R. J., Yost, R. A., and Wilder, B. J. (1982) *Anal Chem.* **54**, 1466.
54. Gaskell, S. J., Porter, C. J., and Green, B. N. (1985) *Biomed. Mass Spectrom.* **12**, 139.
55. Cooks, R. G., Beynon, J. H., Caprioli, R. M., and Lester, G. R. (1973) Metastable Ions. Elsevier: Amsterdam.
56. Gaskell, S. J. and Reilly, M. H. (1988) *Rapid Commun. Mass Spectrom.* **2**, 139.
57. Tomer, K. B., Jensen, N. J., Gross, M. L., and Whitney, J. (1986) *Biomed. Environ. Mass Spectrom.* **13**, 265.
58. Guevremont, R. and Boyd, R. K. (1988) *Int. J. Mass Spectrom. Ion Proc.* **84**, 47.
59. Boyd, R. K., Dyer, E. W., and Guevremont, R. (1989) *Int. J. Mass Spectrom. Ion Proc.* **88**, 147.

60. Orkiszewski, R. S., Thorne, G. C., Dourdeville, T. A. and Gaskell, S. J. (1989) Proceedings of the 37th ASMS Conference on Mass Spectrometry and Allied Topics, Miami Beach, Florida, p. 897.
61. Schey, K. L., Schwartz, J. C., and Cooks, R. G. (1989) *Rapid Commun. Mass Spectrom.* **3**, 305.
62. Guevremont, R. and Wright, (1987) *Rapid Commun. Mass Spectrom.* **1**, 12.
63. Guevremont, R. and Wright, (1988) *Rapid Commun. Mass Spectrom.* **2**, 50.
64. Thorne, G. C. and Gaskell, S. J. (1989) *Rapid Commun. Mass Spectrom.* **3**, 217.
65. Thorne, G. C., Ballard, K. D., and Gaskell, S. J. (1990) *J. Am. Soc. Mass Spectrom.* **1**, 249.
66. Ballard, K. D., Thorne, G. C., and Gaskell, S. J. (1990) Proceedings of the 38th ASMS Conference on Mass Spectrometry and Allied Topics, Tucson, Arizona, p. 946.

# TRIPLE QUADRUPOLE MASS SPECTROMETRY

V.H. Wysocki
*Department of Chemistry*
*Box 2006*
*Virginia Commonwealth University*
*Richmond, VA 23225-2006*

ABSTRACT. The use of triple quadrupole mass spectrometers for bioanalytical applications is introduced. The influence of collision energy and pressure on low-energy collisional activation spectra is described and the interpretation of a product ion spectrum for the structure determination of a peptide is illustrated.

## 1. Introduction

The first tandem quadrupole instruments built for general analytical use were introduced little more than a decade ago [1, 2, 3]. Today, there are approximately 400 mass spectrometers that utilize a quadrupole collision cell followed by a quadrupole mass analyzer [4]. The majority of these instruments are triple quadrupole mass spectrometers, although a few are hybrid instruments that combine conventional sectors with quadrupoles (see Chapter 2). The same factors that led to the widespread use of single quadrupole instruments also led to rapid growth in the number of tandem quadrupole systems (e.g., low cost, compact size, rapid scanning speed, excellent dynamic range, flexible sample introduction) [5]. In the most common MS/MS experiment performed in a triple quadrupole instrument, a mass-selected ion is passed into a pressurized rf-only quadrupole and allowed to collide at a given kinetic energy with a target gas. The products of this collision-activated dissociation step are then mass analyzed by the final quadrupole. Triple quadrupole instruments have had a significant impact on ion chemistry and analytical investigations. There are many excellent papers and books that describe both practical and theoretical considerations in the operation of quadrupole and tandem quadrupole mass spectrometers [4-13]. Numerous analytical applications have also been described [e.g., 1, 2, 12-21]. This chapter is not intended to be a substitute for those publications, but to serve as a brief guide to important experimental considerations and bioanalytical applications of triple quadrupole mass spectrometers. Necessary background information, such as the strong influence of collision energy and collision pressure on low-energy collisional activation spectra, is described. The complementary nature of triple quadrupole data and sector MS/MS data is discussed briefly. The interpretation of a product ion spectrum for the structure determination of a peptide is illustrated.



*M. L. Gross (ed.), Mass Spectrometry in the Biological Sciences: A Tutorial, 59–77.*

## 2. Ionization Methods

Quadrupole mass analyzers are compatible with many different sample introduction and ionization methods because of their utility up to relatively high pressures ($10^{-4}$ Torr) and their rapid-scan capabilities. Most commercial single and triple-stage quadrupole mass spectrometers are available with GC and LC interfaces. The instruments are normally designed to allow electron ionization (EI), chemical ionization (CI), and liquid secondary ion mass spectrometry (LSIMS) or fast-atom bombardment (FAB), with little mechanical manipulation required to switch between ionization methods. Thermospray [22] and electrospray [23] interfaces are also commercially available. The combination of electrospray with quadrupole mass spectrometry is particularly useful because the formation of highly charged ions, with their low mass-to-charge ratios, allows the analysis of molecules with molecular weights that are outside the mass range of commercial quadrupoles (4000 u) [23, 24, 25]. Early studies show that multiply charged ions can be fragmented in a triple quadrupole instrument to provide structural information, although the charge states of the fragment ions are not always readily apparent [25, 26].

## 3. Scan Modes

Miller and Denton [9] published an excellent article that provides a framework for visualization of the operation of quadrupoles. Other articles and books provide both qualitative and mathematical descriptions of quadrupole operation [5-13]. Briefly, a quadrupole mass analyzer consists of a set of four electrodes accurately positioned in a radial array. The electrodes may be of circular or hyperbolic cross section. The quadrupole acts as a tunable variable bandpass mass filter upon application of a both a dc voltage and an rf voltage, with adjacent rods having opposite polarities. The mass and charge of an ion determine whether it will follow a stable trajectory through the quadrupole at given dc and rf voltages. The ratio of the dc and rf voltages determine the width of the bandpass. In a triple quadrupole mass spectrometer, the first and third quadrupoles are operated in this mass-resolving manner. The second quadrupole is used as a reaction cell and is operated in an rf-only mode. This means that the quadrupole is operating as an ion containment device that will transmit a large range of ions of different mass-to-charge ratios. All three quadrupoles have an additional dc voltage applied to them, the offset voltage or the pole bias. The potential difference between the ion formation region and each offset voltage determines the kinetic energy of the ions entering each quadrupole. For the rf-only collision cell, this potential difference is the laboratory collision energy. Quadrupole performance is optimal when the collision energy is in the range of a few eV to a few hundred eV. This is in contrast to sector instruments, which are generally operated at energies of a few kiloelectron volts.

Figure 1 illustrates schematically a QQQ instrument and shows four different scan modes that can be performed on this type of instrument. These scan modes are (i) single-reaction monitoring, (ii) daughter ion scans, (iii) parent ion scans, and (iv) neutral loss (or neutral gain) scans [27, 28]. A parent ion is any ion that can be produced in the ion source and mass selected by the first mass analyzer (Q-1). A daughter ion, or product ion, is any ion that can be produced upon fragmentation of a parent ion. In the selected-reaction monitoring experiment, both mass analyzing quadrupoles are set to pass an ion of a single *m/z* ratio to monitor a particular reaction. For example, the determination of tetrachlorodibenzodioxin is made by monitoring the transition *m/z* 320 to *m/z* 257 [19]. The daughter ion scan (also recently called the product ion scan) is performed by mass-selecting an ion of interest with the first quadrupole, allowing the ion to collide with collision gas is the second quadrupole, and scanning the third quadrupole for all products that result from activation of the ion. Figure 2 provides an example of the application of the daughter ion scan for a small peptide [29].

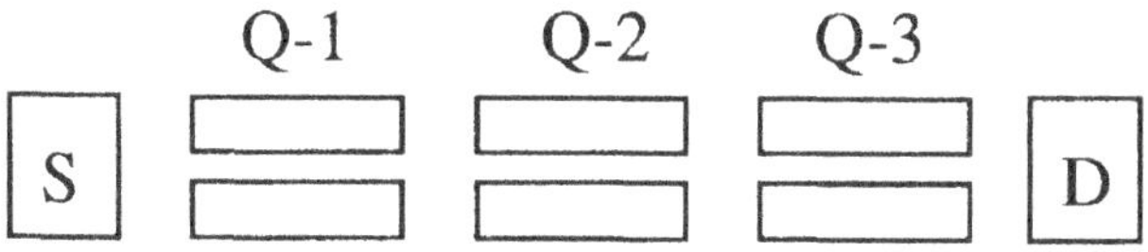


| Scan Type | Q-1 | Q-2 | Q-3 | Representation |
|---|---|---|---|---|
| Single-Reaction Monitoring | Fix | Reaction Region | Fix | ● ↓ ○ |
| Daughter Scan | Fix | Reaction Region | Scan | ● ↓ ○ |
| Parent Scan | Scan | Reaction Region | Fix | ○ ↓ ● |
| Neutral Loss (Gain) Scan | Scan | Reaction Region | Scan | ○ ⬇ ○ |

**Figure 1.** Schematic representation of a triple quadrupole mass spectrometer and summary of four different scan modes that can be performed (S and D represent the ion source and the detector, respectively). For example, in the daughter ion scan, the first quadrupole is set to pass an ion beam of a single *m/z*, quadrupole 2 is an rf-only collision cell, and quadrupole 3 analyzes product ions formed upon collision of the mass-selected ion with the target gas in quadrupole 2. This can also be represented conveniently with symbols (closed circle indicates a fixed mass, open circle indicates a scanned or variable mass, thin arrow represents a variable mass transition, thick arrow represents a fixed mass transition).

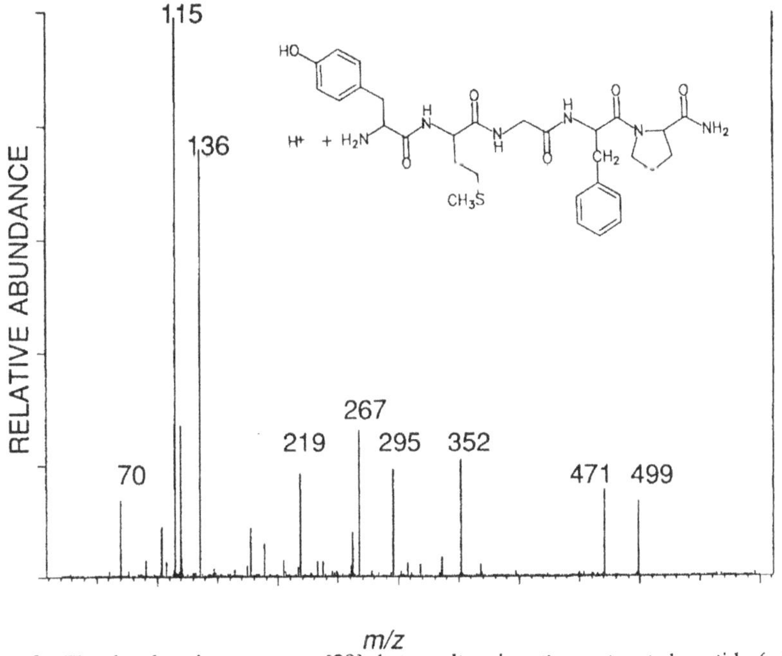


**Figure 2.** The daughter ion spectrum [29] that results when the protonated peptide (*m/z* 613), Tyr-D-Met-Gly-Phe-Pro-$NH_2$, is allowed to collide with Xe (0.5 mtorr) at 30 eV (Finnigan TSQ70).

The parent ion scan is similar to the daughter ion scan, except that the third quadrupole is fixed on an ion of a single *m/z* ratio while the first quadrupole scans. This scan provides information on all parent ions that can fragment to produce a certain product. For example, in a screening for phthalates, all protonated parent ions that fragment to produce protonated phthalic anydride (of *m/z* 149) can be detected by setting Q-3 to pass only ions of *m/z* 149 and scanning Q-1. Because no other compounds on the EPA priority pollutant list produce a fragment at *m/z* 149, a scan for all parent ions of *m/z* 149 can be used to detect phthalates in a mixture containing other pollutants [15].

The neutral loss scan involves scanning both mass-analysis quadrupoles (Q-1 and Q-3) at a constant mass difference. For example, all parent ions that can lose a methyl group are detected by a scan in which Q-3 is always offset in mass from Q-1 by 15 u. This scan, and the related constant neutral gain scan for ion/molecule reaction products, are part of a subset of scans that are more generally called functional relationship scans [28].

Another scan that can be performed on a triple quadrupole mass spectrometer steps one of the mass analyzers one mass at a time while the second analyzer scans the entire mass range quickly at each setting of the first. This scan relates each parent ion present in an analyte to its possible products and requires two mass axes and one intensity axis for representation [28].

## 4. Ion Activation

### 4.1. ENERGY DEPOSITION

For high- or low-energy collisional activation, the maximum energy that can be converted into internal energy is calculated by converting to the center-of-mass frame of reference, $E_{cm} = E_{lab}m_t/(m_t + m_p)$, where $E_{lab}$ is the laboratory collision energy and $m_t$ and $m_p$ are the masses of the target gas and the projectile ion. Note that this energy is the maximum available, and that the ion population will gain a range of internal energies with the most probable energy deposited being less than the maximum available energy. The distributions of internal energy deposited by collisional activation have been estimated [30-31]. Collisional activation at keV (lab) energies deposits a broad distribution of internal energies [30-31]. Although the majority of the ions activated by this method contain low internal energies, a small percentage of the ions gain very high internal energies as illustrated by the low-probability, high-energy tail of Figure 3a. Collisional activation at eV energies (under single collision conditions) deposits low internal energies at the lowest laboratory energies available [30]. This distribution then broadens and a percentage of the ions gain higher internal energies as the laboratory collision energy is increased. The difference between the energy distributions for high- and low- energy collisional activation is illustrated in Figures 3a and 3b. The distributions are compared with those associated with surface-induced dissociation (Figure 3c), which will be discussed in section 4.6.

The activation mechanisms for high- and low-energy collisional activation are summarized well in reference 12 and will not be described in detail here. The activation mechanisms for high- and low-energy collisional activation are different, although both involve excitation followed by energy randomization and unimolecular dissociation of the activated ion. High-energy collisional activation can involve both electronic and vibrational excitation, with vibrational excitation becoming more likely as the size of the projectile ion increases. The activation mechanism for low-energy collisional activation of polyatomic ions generally involves direct vibrational excitation [33,34] and at collision energies of a few eV can involve formation of a long-lived collision complex. Important features of low-energy collisional activation are that it is efficient, the cross-sections are high, and a large fraction of the available center-of mass energy is converted into internal energy of the ion [12, 13, 27, 34-36]. As discussed below, the internal energy deposition is easily controlled for low-energy collisional activation by varying the kinetic energy of the ions [20, 27, 30, 37]. Internal energy deposition can also be varied by changing the target gas pressure (section 4.3), although pressure-resolved plots may not correlate with energy-resolved plots obtained under single-collision conditions [30, 37, 38, 39].

### 4.2. COLLISION ENERGY EFFECTS

Low-energy collisional activation spectra can be aquired in multiquadrupole instruments, quadrupole ion traps [40], hybrid instruments [41-44], and Fourier transform mass spectrometers [45]. Recently, four-sector mass spectometers have also been operated in the eV range [46]. Definitions for the range of energies that correspond to low-energy collisional activation vary; for this paper, the phrase will refer to energies of less than 1 kiloelectron volt. In practice, triple

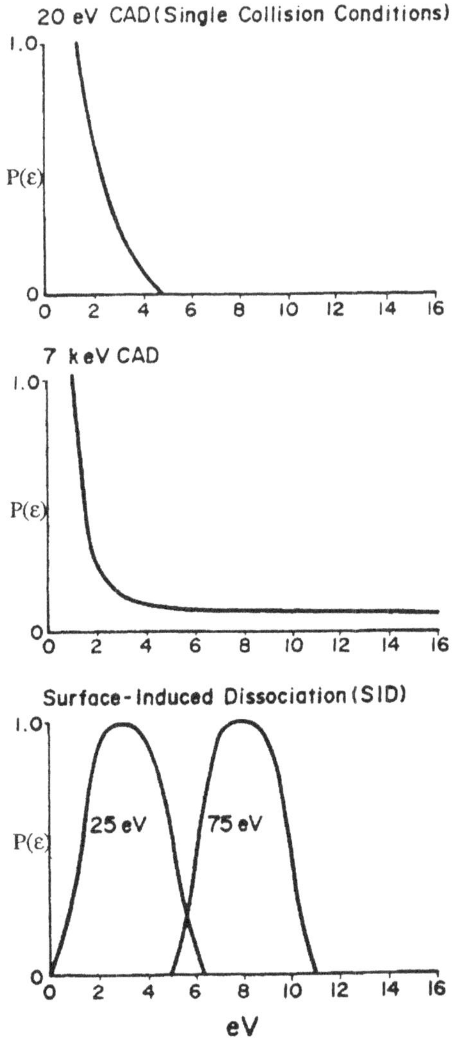


**Figure 3.** The internal energy distributions, P(ε), associated with activation of ions by (a) a 20 eV collision with Ar, (b) 7 keV collisions with Ar, (c) 25 eV and 75 eV collisions with a stainless steel surface. (Figure modified from reference 32.)

quadrupole spectra are rarely obtained at energies greater than 200 eV. In contrast to high-energy collisional activation spectra (commonly obtained at 3-10 keV), the low-energy collisional activation spectra are quite sensitive to small relative changes in ion kinetic energy [8, 27, 30, 36, 37]. This is one of the great advantages of low-energy collisional activation: changes in laboratory collision energy result in changes in the extent of fragmentation that occurs for a given compound. Breakdown curves (plots of relative ion abundance vs. internal energy) established by the quasiequilibrium theory of mass spectrometry and photoelectron photoion coincidence experiments (PEPICO) were compared with energy-resolved mass spectra obtained by varying the collision energy in low-energy collisional activation experiments [47-49]. These types of comparisons have established that changes in laboratory collision energy result in variation of the internal energy of the ion population. These types of experiments were used to differentiate between isomeric ions [38, 50, 51], to establish bond dissociation energies [52-58] or isomerization barriers [51], and in many cases, simply to control the extent of fragmentation. An example of energy-resolved mass spectra for two isomeric ions is shown in Figure 4 [51].

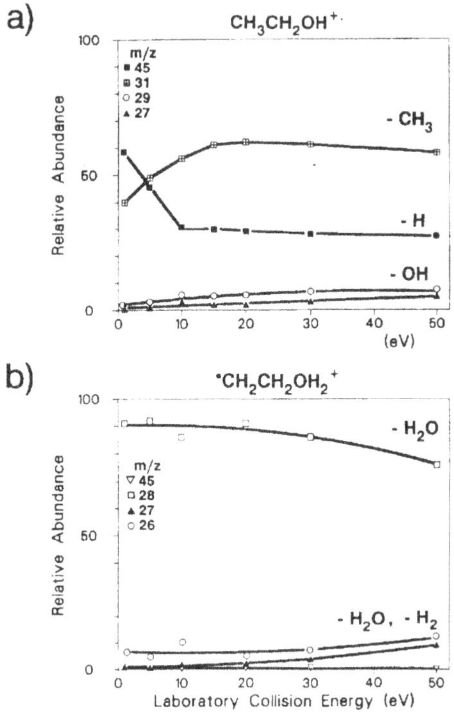


**Figure 4.** Collision-activated dissociation spectra plotted as a function of laboratory collision energy for (a) ionized ethanol and (b) its β-distonic isomer. These energy-resolved mass spectra were obtained by using a triple quadrupole mass spectrometer with an Ar pressure of 1 mtorr (multiple collision conditions). Reprinted with permission from reference 51.

### 4.3. COLLISION TARGET EFFECTS

Low mass targets are commonly used for high-energy collisional activation in sector instruments to minimize scattering of the product ions outside the acceptance angle of the instrument. Because rf-only quadrupoles are strong focusing devices, products scattered by high mass targets are collected more efficiently in this type of collision cell as compared to conventional collision cells. Heavier targets are advantageous in that they can be used to increase the maximum internal energy available upon collision. Another way to increase energy deposition is to increase the collision gas pressure [30, 38, 39]; significantly higher average and maximum energy depositions are available than if single-collision conditions are used. When target pressure is high enough, a parent ion may undergo several activating collisions. Moreover, fragment ions may themselves be activated and dissociate. Under these conditions, isomerization of the excited ions between collisions may occur. In spite of this possible complication, high target pressures have proven to be analytically useful in those cases where few structurally characteristic fragments are observed under single collision conditions. For the fragmentation of peptide ions, Hunt and coworkers [24] found that a mixture of Ar and Xe gives excellent results.

### 4.4. EXTENT OF FRAGMENTATION

Over the years, there have been numerous comparisons of low- and high-energy collisional activation spectra [18, 30, 41, 48, 59]. Some comparisons indicate more extensive fragmentation by high-energy collisional activation [30,48], others indicate more extensive fragmentation by low-energy collisional activation [30,35,36,39], and still others indicate similar extents of fragmentation by the two methods [18,30,38]. The energy requirements for fragmentation of a given projectile combined with the internal energy distributions imparted to the projectile ion under the chosen collisional activation conditions could lead to each of these situations [30]. Rather than question which of the two methods is "better", perhaps the complementary nature of the data obtained should be exploited as it is in hybrid sector-quadrupole mass spectrometers [41-44] (see Chapter 2). The high-energy collisional activation data represent extensive fragmentation by pathways that require a range of different activation energies, whereas the low-energy collisional activation spectra can be used in a manner that produces extensive fragmentation or in a manner that allows determination of relative energetics of dissociation. Both high- and low-energy collisional activation have been quite useful in the determination of biomolecules. The use of triple quadrupole mass spectrometers to sequence proteins is described in more detail below (section 5.1). Case studies describing the use of four sector mass spectrometers for protein sequencing are described later in this book.

### 4.5. REPRODUCIBILITY

Because low-energy collisional activation spectra are quite sensitive to the experimental conditions used and somewhat sensitive to the instrument design, misconceptions have arisen regarding the reproducibility of triple quadrupole MS/MS spectra. The spectra are highly reproducible as long as experimental and instrumental factors are reproducibly controlled. Early interlaboratory comparisons of data were performed before the strong influence of a small change in collision energy or pressure were well-understood [60]. Recently, Martinez [4, 61, 62] showed that it is possible to operate triple quadrupole mass spectometers in a kinetically well-controlled fashion. Another question that sometimes arises is whether the high (kV) rf voltages applied to collision quadrupoles make a significant contribution to the collision energy in triple quadrupole instruments. Both qualitative and quantitative data indicate that the rf field of multiquadrupole collision cells such as quadrupoles, hexapoles, and octapoles contibutes little to the laboratory collision energy (≤ 1 eV) [44, 52-54,63]. If this contribution were significant, the strong agreement of triple quadrupole energy-resolved mass spectra with theoretical or experimental (PEPICO) breakdown curves would not be observed, and threshold energies for reactions [52, 53] could not be determined.

There are several other experimental factors that must be controlled to varying extents in triple quadrupole instruments depending on the type of information desired. For example, most manufacturers provide some means of ramping the offset voltage of Q-3 as a function of daughter ion mass to account for the partitioning of kinetic energy between the ionic and neutral products formed in a given reaction [64]. Some manufacturers allow the operation of their instuments to provide linkage of the Q-2 rf voltage to either Q-1 (constant $q$ parent transmission mode) or to Q-3 (constant $q$ parent transmission mode) [8]. The Q-2 rf-level can be controlled by the operator in some instances. This can be important because the relative abundances of ions collected from an rf-only quadrupole are dependent on ion mass and on the amplitude of the rf voltage applied to the quadrupole as well as on kinetic energy and the entrance and exit conditions of the quadrupole. Some manufacturers now use hexapole or octapole collision cells because these cells have wider stability regions than quadrupoles, and ion transmission in these cells is not as dependent on various operating parameters such as rf-level and collision energy.

## 4.6. ION/SURFACE COLLISIONS

There are examples where neither low- or high-energy gas-phase collisional activation leads to the extent of fragmentation desired. A new technique, in which a mass-selected projectile ion is allowed to collide with a surface, has been developed [65]. This method has been implemented in several types of mass spectrometers including tandem quadrupole instruments [66, 67, 68]. For tandem quadrupole instruments, it was shown that the experiment can be performed with the mass analysis quadrupoles placed at 90°(see Figure 5) or in-line [66-68]. The in-line arrangement is attractive for users of triple quadrupole instruments because an in-line surface-induced dissociation device can be placed into an existing triple quadrupole instrument [67, 68]. Surface-induced dissociation causes more extensive fragmentation than does collision-activated dissociation [65-68]. The distribution of energies deposited is narrower than those associated with collisions with a target gas (see Figure 3c). These points are illustrated qualitatively by Figure 6, which compares the energy-resolved surface-induced dissociation data for $W(CO)_6$ with that obtained by energy-resolved low-energy collisional activation and by high-

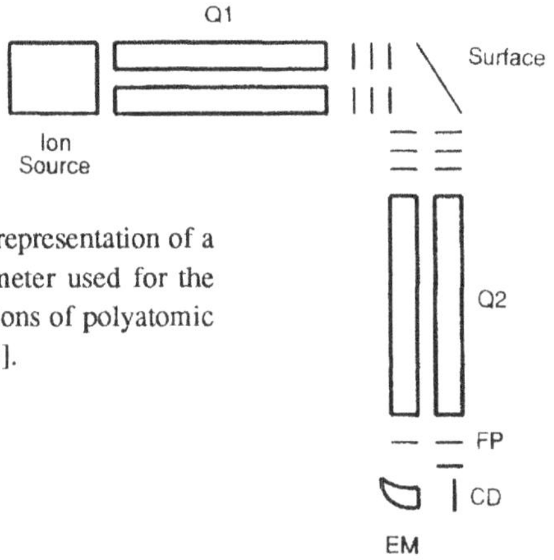


**Figure 5.** Schematic representation of a tandem mass spectrometer used for the investigation of collisions of polyatomic ions with surfaces [68].

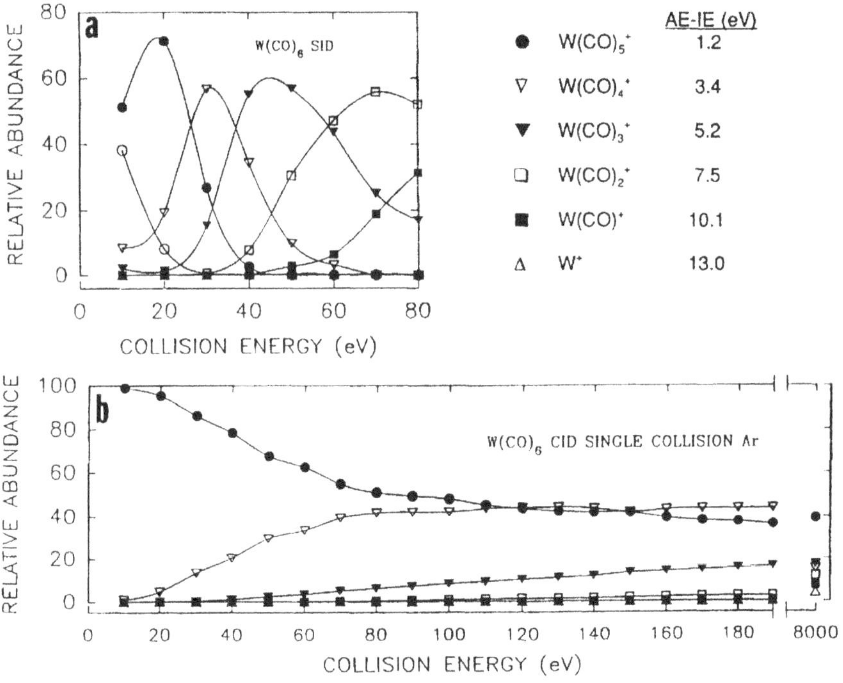


**Figure 6.** Energy-resolved MS/MS spectra of $W(CO)_6$ obtained upon (a) ion/surface collision, (b) collision-induced dissociation with Ar at 10-200 eV (single collision conditions) and collision-induced dissociation with He at 8000 eV. Modified from reference 68.
○ $W(CO)_6^{+\cdot}$; ● $W(CO)_5^{+\cdot}$; ∇ $W(CO)_4^{+\cdot}$; ▼ $W(CO)_3^{+\cdot}$; ❐ $W(CO)_2^{+\cdot}$; ■ $W(CO)^{+\cdot}$

energy collisional activation. Although the majority of early surface-induced dissociation studies involved relatively low mass ions, some higher mass ions have been investigated. The spectra for peptides show that sequence information can be obtained from both singly- [65, 69] and multiply-charged peptide ions [69]. The types and relative abundances of the sequence ions change with the collision energy of the surface-induced dissociation experiment. *Reaction* products at *m/z*-ratios greater than the *m/z*-ratio of the singly-charged peptide are detected in addition to surface-induced dissociation fragment ions. As these investigations are extended to higher mass ions, it will be determined whether the high average internal energies available by surface-induced dissociation lead to more extensive fragmentation than is available by low- or high-energy collisional activation and whether greater control of the fragmentation is possible and advantageous.

## 5. Applications

From its early days, triple quadrupole mass spectrometry has been applied to a wide array of analytical and fundamental ion chemistry problems. For example, complex mixtures have been analyzed [12, 13, 15, 17, 18], isomers have been differentiated [50], relative proton affinities have been determined [70], and ion/molecule association reactions have been utilized to distinguish molecules that are difficult to fragment [71]. This discussion will focus on protein sequencing by triple quadrupole mass spectrometry, because the determination of biomolecules is the major topic of this book.

### 5.1. PROTEIN SEQUENCING

In 1981, Hunt and coworkers [14] were the first to report collisional activation spectra of oligopeptide ions formed by fast-atom bombardment. Since then, they have continued to develop instrumentation and chemical methods for the characterization of the primary structure of proteins. The series of steps involved in this characterization consists of both wet chemistry and both single stage and tandem mass spectrometry, and the subject was reviewed recently [16]. Hunt's work in this area is unique because he uses triple quadrupole mass spectrometry and quadrupole-FTMS in the protein characterization, while the majority of other workers in the field utilize sector or hybrid mass spectrometers [72, 73] (see Chapters 1 and 2). The value of the approach that utilizes the relatively simple and inexpensive quadrupole instruments is attested to by the large number of successful characterizations that have been performed [16, 24, 74-84].

The molecular weight of the protein or large peptide is first determined by electrospray mass spectrometry. The series of steps required to sequence the peptide or protein then varies with the complexity of the problem. Scheme 1 shows the general order of events. Initially, the protein is digested with site specific reagents such as cyanogen bromide and proteolytic enzymes. Different reagents are added to individual samples of the intact protein to produce a number of overlapping peptide sequences. The digests are then separated by HPLC into a number of fractions, each containing only a few peptides. Each fraction is submitted to electrospray mass spectrometry or particle bombardment mass spectrometry (high-energy beam of cesium ions) to determine the *m/z* ratios of each of the peptides in each fraction. Each peptide in a fraction is then mass-selected by the first quadrupole of the triple quadrupole mass spectrometer and allowed to collide with target gas (2/1 Ar/Xe mixture at 3 mtorr). Fragment ions that are produced are then analyzed by the third quadrupole and detected by a conversion dynode (15 kV) electron multiplier. Oligopeptides of greater than 20 residues can be routinely analyzed at the 5-100 picomole level by this mass-selection/activation process. The sequence information obtained for each oligopeptide of each fraction of each digest is then pieced together to provide as much of the overall sequence of the protein as possible. When necessary, additional simple chemical steps are performed, such as methylation, acetylation, and prolylation.

**Digestion with Site-Specific Reagents**

↓

**HPLC**

↓

**Determination by Mass Spectrometry of the *m/z* Ratios of Each Peptide in Fraction**

↓

**Selection of Individual Peptide by Q-1**

↓

**Collisional Activation of Selected Peptide**

↓

**Mass Analysis of Fragment Ions by Q-3**

↓

**Interpretation of Fragmentation Spectrum of Selected Peptide**

↓

**Determination of Protein Sequence by Piecing Together Overlapping Sequences**

Scheme I.

In these experiments, mass selection of a protonated peptide is performed by fixing the rf and dc voltages of Q-1 to pass only the ion of interest. Often, it is desirable to maximize ion transmission so a *m/z* window of 3-9 u around the ion of interest is actually selected by Q-1; Q-3 may also be operated at less than unit resolution. Even though this generally does not hinder interpretation of the collisional activation spectra, those who are accustomed to seeing unit resolution MS/MS spectra (with no isotope peaks present - see Figure 2) find it disconcerting. One drawback of the triple quadrupole approach is that it has not yet been possible to differentiate between leucine and isoleucine in peptides. The side chain cleavages that occur upon high-energy collisional activation and allow this distinction to be made are not detected by the triple quadrupole method described above. (Preliminary experiments show that side chain cleavages are promoted by ion/surface collisions in the eV range [69].) Although the oligopeptide spectra obtained by low-energy (e.g., triple quadrupole) collisional activation and those obtained by high-energy collisional activation (sector instruments) are different, both types of spectra can be used effectively to sequence peptides and proteins.

Successful identification of the novel synthetic test peptides provided by the protein soceity illustrate the power of the triple quadrupole approach. For example, approximately three nanomoles of Symposium Test Peptide-3 (SPT-3) were sent to 150 groups. Of the *three* groups who successfully sequenced the unknown, Hunt's group was the only successful group that used tandem mass spectrometry [81]. Hunt's group also successfully identified Symposium Test Peptide-4 (STP-4) by using a combination of chemical steps, Fourier transform mass spectrometry, triple quadrupole mass spectrometry, and two cycles of Edman degradation [84]. All fragment ions produced upon collision of the singly-protonated peptide (STP-4) with target gas are shown in Figure 7. The steps involved in identification of the unknown are summarized in Table 1. (Ions of type *b* contain the amino terminal residue and have the general structure $H\text{-}[NHCH(R)CO]_n^+$. Ions designated as type *a* correspond to type *b* ions minus a carbon monoxide. Ions of type *y* contain the C-terminal residue and have the general structure $H\text{-}[NH_2CH(R)CO]_nOH^+$.)

## 6. Related Instrumentation

There are related types of mass spectrometers that have features in common with triple quadrupole mass spectrometers. The ease with which quadrupoles can be linked together to provide useful analytical information is illustrated by the recent success of Cooks and coworkers [85] in the rapid construction and application of a pentaquadrupole mass spectrometer. This instrument has been used to perform several types of MS/MS/MS scans in the analysis of polymers and peptides. Hybrid sector-quadrupole mass spectrometers such as BEQQ instruments combine the advantages of sector instruments with those of triple quadrupole instruments and also allow MS/MS/MS experiments (see Chapter 2). Reports that state that compound X can't be fragmented by low-energy collisional activation in a hybrid mass spectrometer should be read with some caution until the instrument manufacturers and operators have carefully addressed instrumental and experimental factors that contribute to early poor performance of these complicated instruments. Quadrupole ion traps [40], whose operation may be described by the same stability diagrams that represent quadrupole operation, allow several stages of mass spectrometry to be performed tandem-in-time as opposed to tandem-in-space. These small, inexpensive mass spectrometers are seen by many as defining the direction of mass spectrometry for the future [40], and are covered in Chapter 4. Striking advances in the mass range and resolution of these sensitive instruments have been made recently [40] and the instruments have been successfully interfaced to several external ion sources, including electrospray.

| | | | | | | | | | | | | | | | | |
|---|---|---|---|---|---|---|---|---|---|---|---|---|---|---|---|---|
| 72.0 | 185.1 | 332.2 | 469.3 | 526.3 | 682.4 | 781.4 | 868.5 | 1054.6 | 1125.6 | 1256.6 | 1403.7 | 1500.8 | 1614.8 | 1671.8 | 1817.9 | $b_n$ |
| Ala | Lxx | Phe | His | Gly | Arg | Val | Ser | Trp | Ala | Met | Phe | Pro | Asn | Gly | Lys | |
| | 1746.9 | 1633.8 | 1486.7 | 1349.7 | 1292.6 | 1136.5 | 1037.5 | 950.4 | 764.4 | 693.3 | 562.3 | 415.2 | 318.2 | 204.1 | 147.1 | $y_n$ |

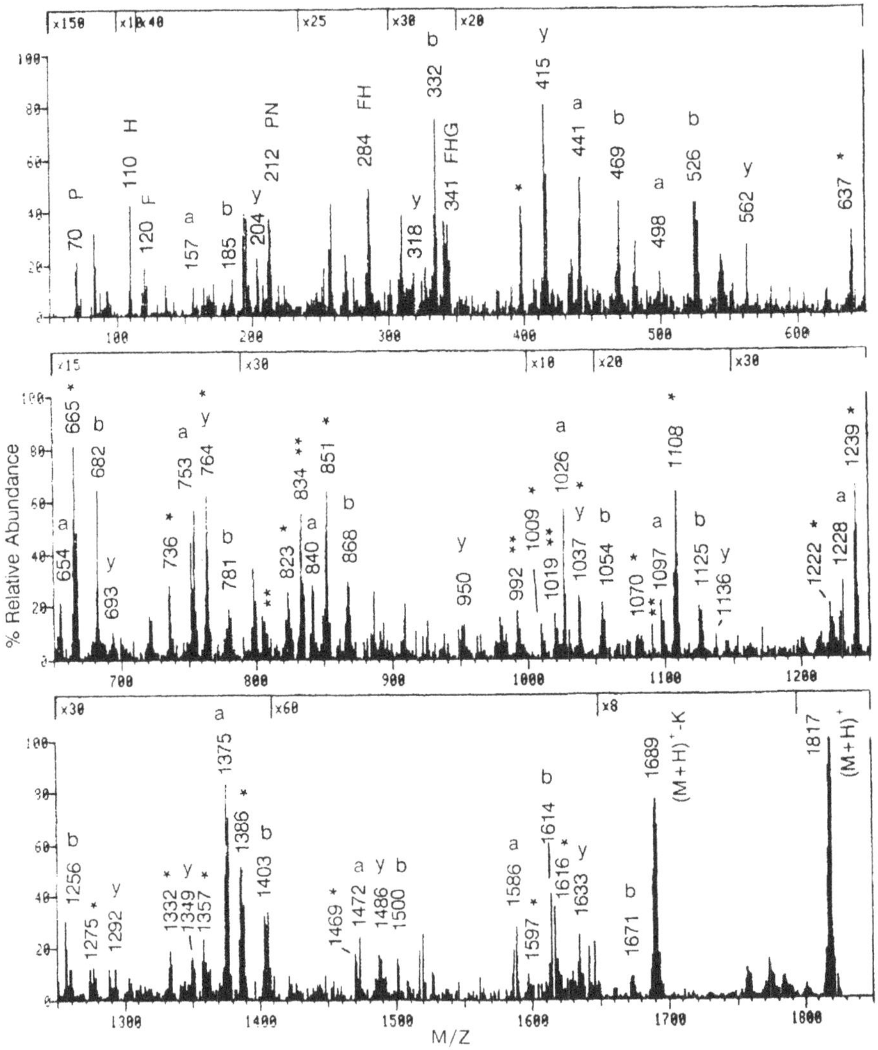


**Figure 7.** Spectrum that resulted when protonated STP-4 ions were mass-selected and subjected to collision activated dissociation on a triple quadrupole mass spectrometer (20 eV; 3-5 mtorr Ar). See Table 1. Reprinted with permission from reference 84.

**Table 1.** Summary of the determination of Sympsium Test Peptide 4 (STP-4).

| EXPERIMENT/OBSERVATION | RESULT/CONCLUSIONS |
|---|---|
| Record mass spectra of STP-4 by<br>LSIMS on FTMS<br>Electrospray on QQQ | <br>$(M + H)^+$ = 1817.6, so MW = 1816.6<br>$(M + 3H)^{+++}$ ~ 606 - 607, confirm MW |
| STP-4 acetylated at free amino groups; product MW determined by electrospray | MW shifted by 94 u (2 acetyl groups), so STP-4 contains 2 free amino groups - amino terminus plus Lys or 2 Lys plus blocked amino terminus |
| STP-4 converted to methyl ester; product MW determined by electrospray | MW shifted by 14 u (1 group methylated), so STP-4 contains free carboxy terminus <u>or</u> a single acidic residue |
| STP-4 ionized and $(M + H)^+$ mass-selected and fragmented by collisional activation with Ar. Spectrum interpreted as follows: | |
| Identify those pairs of ions separated by 28 Da (b - 28 = a). | b ions indicated at *m/z* 469, 526, 682, 781, 868, 1054, 1125, 1256, 1403, 1500, 1614; Δ*m/z* between adjacent pairs of b ions gives partial sequence Gly Arg Val Ser Trp Ala Met Phe Pro Asn |
| Identify y ions that correspond to proposed b ions $[(M + H)^+ - b + 1] = y$ | y ions indicated at *m/z* 1349, 1292, 1136, 1037, 950, 764, 693, 562, 415, 318, 204 |
| Find additional y ions. Look for unassigned pairs of ions separated by 17 Da (y and y - $NH_3$). Confirm by identifying corresponding b ions. | y - b pairs found at the following *m/z* values y=1486, b=332; y=1633, b=185; extends sequence to *Phe His* Gly Arg Val SerTrp Ala Met Phe Pro Asn |
| The Δ*m/z* between the y ion at 1633 and $(M + H)^+$ is 184. | N-terminus corresponds to Ala-Leu/Ile or Ser-Pro. |
| Two cycles of Edman degradation and mass spectra of shortened peptides provide MW's of 1746.3 and 1633.6. | N-terminus identified as Ala-Lxx. Leu and Ile not differentiated. |
| Identify $1689^+$ (an unassigned ion of high relative abundance) as $[(M + H)^+ - 128]$ | ion of *m/z* $1689^+$ is $[(M + H)^+ - Lys]$ or $[(M + H)^+ - Gln]$; Lys consistent with acetylation results. |
| Assigned residues account for all of $(M + H)^+$ except for 57 Da. | Unidentified residue is Gly. (Consistent with assignment of the ion of *m/z* 1671 as the next b ion from the ion of *m/z* 1614.) Completes sequence determination |

Ala Lxx Phe His Gly Arg Val Ser Trp Ala Met Phe Pro Asn *Gly* Lys

## 7. References


1. Yost, R. A. and Enke, C. G. (1978) *J. Am. Chem. Soc.*, **100**, 2274.
2. Hunt, D. F., Shabanowitz, J., and Giordani, A. B. (1980) *Anal. Chem.*, **52**, 386.
3. Zakett, D., Cooks, R. G., and Fies, W. J. (1980) *Anal. Chem.*, **119**, 129.
4. Martinez, R. I. and Ganguli, B. (1989) *Rapid Commun. Mass Spectrom.*, **3**, 377.
5. Brunnée, E. (1987) *Int. J. Mass Spectrom. Ion Processes*, **76**, 121.
6. Dawson, P. H. (ed.) (1976) Quadrupole Mass Spectometry and Its Applications, Elsevier, Amsterdam.
7. Dawson, P. H. (1986) *Mass Spectrom. Rev.*, **5**, 1.
8. Dawson, P. H., French, J. B., Buckley, J. A., Douglas, D. J., and Simmons, D. (1982) *Org. Mass Spectrom.*, **17**, 205.
9. Miller, P. E. and Denton, M. B. (1986) *J. Chem. Ed.*, **63**, 617.
10. Miller, P. E. and Denton, M. B. (1986) *Int. J. Mass Spectrom. Ion Processes*, **72**, 223.
11. Campana, J. E. (1980) *Int. J. Mass Spectrom. Ion Processes*, **33**, 101.
12. Busch, K. L., Glish, G. L., and McLuckey, S. A. (1988) Mass Spectrometry/Mass Spectrometry: Techniques and Applications fo Tandem Mass Spectrometry, VCH Publishers, Inc., New York.
13. McLafferty, F. W., ed. (1983) Tandem Mass Spectrometry, Wiley and Sons, Inc., New York.
14. Hunt, D. F., Bone, W. M., Shabanowitz, J., Rhodes, J., and Ballard, J. M. (1981) *Anal. Chem.*, **54**, 1704.
15. Hunt, D. F., Shabanowitz, J., Harvey, M. T., and Coates, M. (1985) *Anal. Chem.*, **57**, 525.
16. Hunt, D. F., Shabanowitz, J., Yates, R., Griffin, P. R., and Zhu, N. Z. (1990) in McEwen, C.N and Larsen, B.S (Eds.) Mass Spectrometry of Biological Materials, Marcel Dekker, Inc., New York, p. 169.
17. Hayward, M. J., T. Kotiaho, Lister, A. I., Cooks, R. G., Austin, G. D., Narayan, R., and Tsao, G. T. (1990) *Anal. Chem.*, **62**, 1798.
18. Ciupek, J. D., Zakett, D., Cooks, R. G., and Wood, K. V. (1982) *Anal. Chem.*, **54**, 2215.
19. Voyksner, R. D., Hass, J. R., Sovocool, G. W., and Bursey, M. M. (1983) *Anal. Chem.*, **55**, 744.
20. Fetterolf, D. D. and Yost, R. A. (1982) *Int. J. Mass Spectrom. Ion Processes*, **44**, 37.
21. Johnson, J. V. and Yost, R. A. (1985) *Anal. Chem.*, **57**, 758A.
22. (a) Blakely, C. R., McAdams, M. J., and Vestal, M. L. (1978) *J. Chromatogr.*, **158**, 264.
    (b) Vestal, M. L. (1985) in Burlingame, A. L. and Castagnoli, N., Jr. (Eds.), Mass Spectrometry in the Health and Life Sciences, Elsevier, Amsterdam, p. 99.
23. (a) Fenn, J. B., Mann, M., Meng, C. K., Wong, S. F., and Whitehouse, C. M. (1989) *Science*, **246**, 64. (b) Covey, T. R., Bonner, R. F., Shushan, B. I., and Henion, J.

D. (1988) *Rapid Commun. Mass Spectrom.*, **2**, 249. (c) Loo, J. A., Udweth, H. R., and Smith, R. D. (1988) *Rapid Commun. Mass Spectrom.*, **2**, 207. (d) Chowdhury, S. K., Katta, V., and Chait, B. T. (1990) *Rapid Commun. Mass Spectrom.*, **4**, 81.
24. Griffin, P. R., Martino, P. A., McCormack, A. L., Shabanowitz, J., and Hunt, D. F. (1990) in J. J. Villafranca (ed.), Current Research in Protein Chemistry: Techniques, Structure, and Function, Academic Press, Inc., p. 117.
25. Smith, R. D., Loo, J. A., Baringa, C. J., Edmonds, C. G., and Udseth, H. R. (1990) *J. Am. Soc. Mass Spectrom.*, **1**, 53.
26. Hunt, D. F., Zhu, N. -Z., and Shabanowitz, J. (1989) *Rapid Commun. Mass Spectrom.*, **3**, 122.
27. Yost, R. A. and Enke, C. G. (1979) *Anal. Chem.*, **51**, 1251A.
28. Schwartz, J. C., Wade, A. P., Enke, C. G., and Cooks, R. G. (1990) *Anal. Chem.*, **62**, 1809.
29. Wysocki, V. H., Ross, M. M., and Callahan, J. H. unpublished result.
30. Wysocki, V.H., Kenttämaa, H. I., and Cooks, R. G. (1987) *Int. J. Mass Spectrom. Ion Processes*, **75**, 181.
31. (a) McLafferty, F. W., Bente III, P. F., Kornfeld, R. , Tsai, S.-C., and Howe, I. (1973) *J. Am. Chem. Soc.*, 95, 2120; (b) Kim, M. S. (1983) *Int. J. Mass Spectrom. Ion Processes*, 189.
32. Cooks, R. G., Mabud, Md. A., Horning, S. R., Jiang, X. Y., Paradisi, C., and Traldi, P. (1989) *J. Am. Chem. Soc.*, **111**, 859.
33. Douglas, D. J. (1982) *J. Phys. Chem.*, **86**, 185.
34. Shukla, A. K., Qian, K., Anderson, S., and Futrell, J. H. (1990) *J. Am. Soc. Mass Spectrom.* **1**, 6. These authors have shown that low-energy collisional activation of small molecules can involve electronic excitation.
35. Nacson, S. and Harrison, A. G. (1985) *Int. J. Mass Spectrom. Ion Processes*, **63**, 325.
36. Dawson, P. H. and Sun, W.-F. (1982) *Int. J. Mass Spectrom. Ion Processes*, **44**, 51.
37. Dawson, P. H., French, J. B., Buckley, J. A., Douglas, D. J., and Simmons, D. (1982) *Org. Mass Spectrom.*, **17**, 212.
38. Kenttämaa, H. I. and Cooks, R. G. (1985) *J. Am. Chem. Soc.*, **107**, 1881.
39. Kenttämaa, H. I. (1985) *Org. Mass Spectrom.*, **20**, 703.
40. (a) Cooks, R. G. and Kaiser, R. E., Jr. (1990) *Acc. Chem. Res.*, **23**, 213. (b) Cooks, R. G., Glish, G. L., McLuckey, S. A., and Keiser, R. E. (1991) *Chem. Eng. News*, 26.
41. Gaskell, S. J. and Reilly, M. H. (1988) *Rapid Commun. Mass Spectrom.*, **2**, 139.
42. Gaskell, S. J., Haroldsen, P. E., and Reilly, M. H. (1988) *Biomed. Environ. Mass Spectrom.*, **16**, 31.
43. (a) Schoen, A. E., Amy, J. W., Ciupek, J. D., Cooks, R. G., Dobberstein, P., Jung, G. (1985) *Int. J. Mass Spectrom. Ion Processes* **65**, 125. (b) Ciupek, J. D., Amy, J. W., Cooks, R. G., Schoen, A. E. (1985) *Int. J. Mass Spectrom. Ion Processes* **65**, 141.
44. Alexander, A. J. and Boyd, R. K (1989) *Int. J. Mass Spectrom. Ion Processes*, **90**, 211-240.

45. Buchanan, M. V. (1987) Fourier Transform Mass Spectrometry, ACS Symposium Series 359, Washington, DC.
46. Orlando, R., Murphy, C., Fenselau, C, Hansen, G., and Cotter, R. J. (1990) *Anal. Chem.*, **62**, 125.
47. McLuckey, S. A., Sallans, L., Cody, R. B., Burnier, R. C., Verma, S., Freiser, B. S., and Cooks, R. G. (1982) *J. Am. Soc. Mass Spectrom.*, **44**, 215.
48. McLuckey, S. A., Ouwerkerk, C. E. D., Boerboom, A. J. H. , and Kistemaker, P. G. (1984) *Int. J. Mass Spectrom. Ion Processes*, **59**, 85.
49. Zwinselman, J. J., Nacson, S., and Harrison, A. G. (1985) *Int. J. Mass Spectrom. Ion Processes*, **67**, 93.
50. Verma, S., Ciupek, J. D., Cooks, R. G., (1984) *Int. J. Mass Spectrom. Ion Processes* **62**, 219.
51. Wysocki, V. H. and Kenttämaa, H. I. (1990) *J. Am. Chem. Soc.*, **112**, 5110.
52. Graul, S. T. and Squires, R. R. (1990) *J. Am. Chem. Soc.*, **112**, 2517.
53. Graul, S. T. and Squires, R. R. (1990) *J. Am. Chem. Soc.*, **112**, 2506.
54. Magnera, T. F., David, D. E., Stulik, D., Orth, R. G., Jonkman, H. T., and Michl, J. (1989) *J. Am. Chem. Soc.*, **111**, 5036.
55. Hanley, L. and Anderson, S. L. (1987) *J. Phys. Chem.*, **91**, 5161.
56. Aristov, N. and Armentrout, P. B. (1986) *J. Phys. Chem.*, **90**, 5135.
57. Loh, S. K., Hales, D. A., Lian, L, and Armentrout, P. B. (1989) *J. Phys. Chem.*, **90**, 5466.
58. Dawson, P. H. (1982) *Int. J. Mass Spectrom. Ion Processes*, **43**, 195.
59. Wysocki, V. H. and Ross, M. M. (1991) *Int. J. Mass Spectrom. Ion Processes*, **104**, 179.
60. Dawson, P. H. and Sun, W. -F. (1983/84) *Int. J. Mass Spectrom. Ion Processes*, **55**, 155.
61. Martinez, R. I. (1988) *Rapid Commun. Mass Spectrom.* **2**, 8.
62. Martinez, R. I., Ganguli, B. (1988) *Rapid Commun. Mass Spectrom.* **2**, 41.
63. Yost, R. A., Enke, C. G., McGilvery, D. C., Smith, D., Morrison, J. D. (1979) *Int. J. Mass Spectrom. Ion Phys.* **30**, 127.
64. Shushan, B., Douglas, D. J., Davidson, W. R., and Nacson, S. (1983) *Int. J. Mass Spectrom. Ion Processes*, **46**, 71.
65. Cooks, R. G., Ast, T., and Mabud, Md. A. (1990) *Int. J. Mass Spectrom. Ion Processes*, **100**, 209.
66. Bier, M. E., Amy, J. W., Cooks, R. G., Syka, J. E.P., Ceja, P., and Stafford, G. (1987) *Int. J. Mass Spectrom. Ion Processes*, **77**, 31.
67. Bier, M. E., Schwartz, J. C., Schey, K. L., and Cooks, R. G. (1990) *Int. J. Mass Spectrom. Ion Processes*, **103**, 1.
68. Wysocki, V. H., Ding, J.-M., Jones, J.L., Callahan, J. H., and King, F. L. (1991) *J. Am. Soc. Mass Spectrom.*, in press.
69. Wysocki, V. H. and McCormack, A. L., (1991) Proceedings of the 39th ASMS Conference on Mass Spectrometry and Allied Topics, Nashville, Tennessee.

70. Wright, L. G., McLuckey, S. A., Wood, K. V., and Cooks, R. G. (1982) *Int. J. Mass Spectrom. Ion Phys.*, **42**, 115.
71. (a) Kenttämaa, H. I. and Cooks, R. G. (1989) *J. Am. Chem. Soc.*, **111**, 4122. (b) Pachuta, R.R., Kenttämaa, H. I., Cooks, R. G., Zennie, T. M., Ping, C., Chang, C., and Cassady, J. M. (1988) *Org. Mass Spectrom.*, **23**, 10.
72. Biemann, K. (1988) *Biomed. Environ. Mass Spectrom.*, **16**, 99.
73. Johnson, R. S., Martin, S. A., and Biemann, K. (1988) *Int. J. Mass Spectrom. Ion Processes*, **86**, 137.
74. Griffin, P. R., Kumar, S., Shabanowitz, J., Charbonneau, H., Namkung, P. C., Walsh, K. A., Hunt, D. F., and Petra, P. H. (1989) *J. Biolog. Chem.*, **264**, 19066.
75. Moore, C. R., Yates, J. R., III, Griffin, P. R., Shabanowitz, J., Martino, P. A., Hunt, D. F., and Cafiso, D. S. (1989) *Biochemistry*, **28**, 9184.
76. Erickson, A. K., Payne, D. M., Martino, P. A., Rossomando, A. J., Shabanowitz, J., Weber, M. J., Hunt, D. F., and Sturgill, T. W. (1990) *J. Biol. Chem.* 265, 19728.
77. Groome, J. R., Tillinghast, E. K., Townley, M. A., Vetrovs, A., Watson, W. H., III, Hunt, D. F., Griffin, P. R., Alexander, J. E., and Shabanowitz, J. (1990) Peptides, 3, 205.
78. Krishnamurthy, T., Szafraniec, L., Hunt, D. F., Shabanowitz, J., Yates, J. R., III, Hauer, C. R., Carmichael, W. W., Skulberg, O., Codd, G. A., and Missler, S. (1989) *Proc. Natl. Acad. Sci., USA*, **86**, 770.
79. Hunt, D. F., Yates, J. R., III, Shabanowitz, J., Bruns, M. E., and Bruns, D. E. (1989) *J. Biolog. Chem.*, **264**, 6580.
80. Hunt, D. F., Yates, J. R., III, Shabanowitz, J., Winston, S., and Hauer, C. R. (1986) *Proc. Natl. Acad. Sci., USA*, **83**, 6233.
81. Hunt, D. F., Krishnamurthy, T., Shabanowitz, J., Griffin, P. R., Yates, J. R., III, Martino, P. A., McCormack, A. L., and Hauer, C. R. (1990) in D. Desiderio, (ed.), Mass Spectrometry of Peptides, CRC Press, p. 139.
82. Bieber, A. L, Becker, R. R., McParland, R., Hunt, D. F., Shabanowitz, J., Yates, J. R., III, Martino, P. A., and Johnson, G. R. (1990) *Biochim. Biophys. Acta*, 413.
83. Hunt, D. F., Shabanowitz, J., Moseley, M. A., McCormack, A. L., Michel, H., Martino, P. A., Tomer, K. B., and Jorgenson, J. W. (1991) in H. Jornall, J.O.Hogg, and A.-M. Gustavsson, (eds.) Methods in Protein Sequence Analysis, Birkhauser Verlag, Boston, USA, pp. 257-266.
84. Hunt, D. F., Alexander, J. E., McCormack, A. L., Martino, P. A., Michel, H., and Shabanowitz, J. (1991) in J.J. Villafranca, (ed.), Protein Chemistry Techniques II, Academic Press, New York, pp. 455-465.
85. (a) Schwartz, J. C., Schey, K. L., and Cooks, R. G. (1990) *Int. J. Mass Spectrom. Ion Processes*, **101**, 1. (b) Schey, K. L., Schwartz, J. C., and Cooks, R. G. (1989) *Rapid Commun. Mass Spectrom.*, **3**, 305.

# QUADRUPOLE ION TRAP MASS SPECTROMETERS FOR STUDIES OF BIOMOLECULES

R. CAMERON DOREY
*Department of Chemistry*
*Wichita State University*
*Wichita, KS 67208*

ABSTRACT. The technique of ion trap mass spectrometry is reviewed with special emphasis on applications to molecules of biological importance. A brief mathematical background of ion trapping in quadrupole felds is presented to understand methods used in mass range extension and MS/MS. Applications of ion trap analysis using both gas chromatographic and other sample inlets are given, which show the potential of this instrument in studies of substances ranging from volatile organic molecules to proteins and nucleotides.

## 1. Introduction

The quadrupole ion trap mass spectrometer was described in Paul's original patent covering quadrupole devices in 1956 [1]. A quadrupole ion trap uses the same principles of ion optics as the more familiar quadrupole mass filters, but the two mass spectrometers have quite different characteristics in practice. A quadrupole mass filter confines ions in two dimensions, so that ions must be accelerated through the length of the device by external electric fields. This necessitates an external ion source and fixes the length of time that ions reside within the quadrupole field. Quadrupole ion traps, however, are able to confine ions in a three dimensional volume for extended lengths of time, so that they can act as ion source, ion-molecule reaction chamber, and mass analyzer. Although most applications to date generate ions within the trap, ions can also be produced externally and injected into the ion trap volume for reaction or analysis. These characteristics give the quadrupole ion trap great versatility in the types of samples that potentially can be analyzed and the experiments that can be performed on the ions produced.

It was not until the development of the Finnigan Ion Trap Detector ("ITD") in 1983 and the subsequent introduction of the Ion Trap Mass Spectrometer ("ITMS") with its expanded capabilities that the quadrupole ion trap received significant attention as a general use mass spectrometer. Because of this late start, there have been few papers portraying quadrupole ion traps as analytical instruments relative to those using quadrupole filters. Most papers that describe an ion trap other than as a gas chromatography detector have emphasized the potential of the instrument for complex mass spectrometric experiments, rather than its routine use in such applications. As familiarity with the experimental characteristics of the ion trap grows, however, the advantages of this design in high sensitivity mass spectrometry will certainly lead to its expanded use.



*M. L. Gross (ed.), Mass Spectrometry in the Biological Sciences: A Tutorial,* 79–92.

An ion trap mass spectrometer can be operated in the same manner as a quadrupole filter, storing a single ion at a time, which is then ejected to an external detector. This mode, called "mass-selective stability," does not present significant advantages over the operation of a conventional quadrupole filter in sensitivity or speed of operation. A second mode of operation of ion traps, termed "mass-selective instability," was developed in 1983 [2] and used in the ITD and ITMS as their "analytical scan" mode. Here, all masses above a cutoff point are stored simultaneously, and ions are ejected sequentially in the order of increasing mass. In this technique, the ion creation and storage time can be lengthened relative to the ion ejection (analysis) time, so that sensitivity can be greatly enhanced. Operation of the ion trap as a mass-selective stability device is used to isolate single mass ions for MS/MS or for ion-molecule reactions, as will be presented later. Other operating conditions, not directly applicable to quadrupole filters, have been used to extend the capabilities of ion traps to the analysis of high mass ions.

Since their introduction, much progress has been made in interfacing the ITD and ITMS to various ion sources and extending the mass range of the instruments, both necessary for the determination of biomolecules. This paper shall describe the basic operation of quadrupole ion traps, concentrating on the ITD and ITMS, and the recent research extending the capabilities of the ion trap, with special regard to biochemical applications.

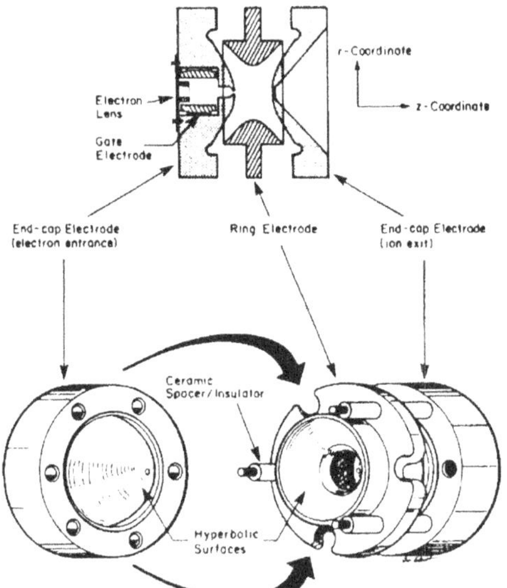


Figure 1. Ion-trap electrodes. An RF field is established between the ring electrode and the two end-cap electrodes. The assembled structure is shown in cross-section. Reprinted from Reference 6 with permission.

## 2. Fundamentals of ion storage in quadrupole ion traps

Quadrupole ion traps utilize high frequency oscillating electric fields to contain ions between electrodes. The most common geometry of quadrupole ion trap is one in which there are hyperbolic endcap electrodes above and below a ring electrode (also with a hyperbolic inner surface), as shown in Figure 1. In the ITD and ITMS, a sinusoidal radio frequency field ("RF", approx. 1 MHz) is applied to the ring electrode, while the endcap electrodes are grounded (in the ITD) or may be biased with either a constant ("DC", may also be zero) or oscillating potential ("AC", both potentials available in the ITMS).
Ion motion within an ion trap is governed by the Mathieu equation, and has been described in detail in several publications (see, for example, 3, 4). The basic equations will be introduced here only to help explain methods that researchers have used to make extended mass ranges, MS/MS, or other techniques possible. Usually, diagrams of ion stability within a trap are plotted in Mathieu "*a,q* space." If the radius of the ring electrode is twice the distance between the endcaps, as is common, a stability diagram for ions of a particular m/e ratio within the trap can be constructed as in Figure 2, where *a* and *q* are defined by the following relations:

$$a_z = -2a_r = -16 \frac{e}{m} \frac{U}{r_o^2 W_{RF}^2} \qquad (1)$$

$$q_z = -2q_r = -8 \frac{e}{m} \frac{V}{r_o^2 W_{RF}^2} \qquad (2)$$

The subscripts *z* and *r* represent motion between and perpendicular to the endcaps, respectively, U is the DC bias on the endcap electrodes (if any), V is the RF potential applied to the ring electrode, $r_o$ is the radius of the ring electrode, and $W_{RF}$ is the RF angular frequency. Ions whose *a* and *q* values lie inside the crosshatched envelope have stable trajectories inside the trap.

The parameter ß in the figure denotes the fundamental frequency of oscillation of an ion at a particular *a* and *q* as a fraction of $W_{RF}$,; there are three important regimes for values of ß. At ß approaching zero, an ion is essentially unaffected by the quadrupole field, and can drift out of the trap. At ß equal to one, an ion oscillates in resonance with the field, it absorbs kinetic energy, its oscillation magnitude increases, and the ion escapes the trap or collides with an electrode.
When ß is between zero and one, an ion can be stored within the trap, where it oscillates in a periodic stable fashion, and its average kinetic energy is roughly dependent on the value of ß.

The ITD and ITMS are operated with a low residual pressure of helium inside the trap, about 0.1 Pa [2]. When low concentrations of ions are confined at a low ß in this helium atmosphere, they collapse into the center of the trap through collisional cooling. They do not react upon collisions with the helium gas, but they may participate in exothermic reactions with other neutral molecules inside the trap. The reactions are slow because the concentrations of

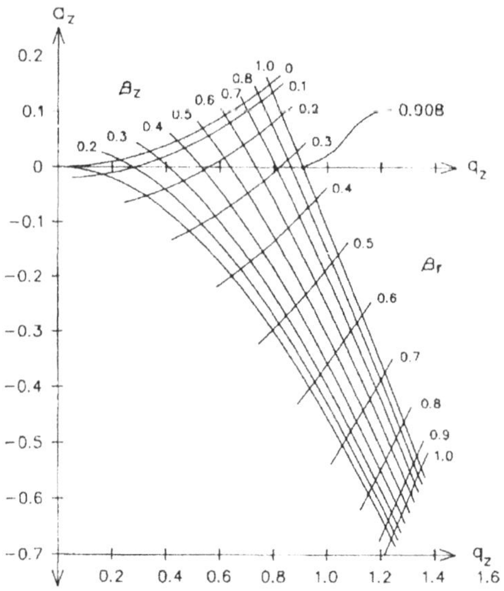


Figure 2. Stability diagram for the three-dimensional quadrupole ion trap with iso-ß lines. Reprinted from Reference 3 with permission.

both reactants are very low. When ions are raised to a high ß in the helium background, they may have sufficient kinetic energy to fragment upon collision with a helium atom. By using a high ion concentration or raising the ß value of specific ions, one can use trapped ions to perform chemical ionization, low-energy collision-activated dissociation, or other ion-molecule experiments [5, 6, 7, 8].

Ion traps are commonly operated in three basic ways, according to the potentials applied between the endcap and ring electrodes. First, when the trap is operated at a fixed RF potential without a DC bias potential between the endcap and ring electrodes (the mode of operation of the ITD and the normal analytical scan of the ITMS), all ions above a cutoff m/e (such that $q_z$ is greater than 0.908) have stable trajectories inside the trap. In this mode, there is no clear upper m/e cutoff, so that even high mass ions are trapped. Electron ionization commonly takes place in this mode, where all ions are formed within the desired m/e range of the experiment are trapped. When the RF potential is raised, the cutoff m/e for stability is raised, and ions are ejected. Thus, by increasing the RF potential at a controlled rate, ions can be sequentially ejected from the trap in the order of low to high mass. This is the "mass-selective instability" mode of operation. In this mode of analysis, the mass range is limited by the maximum RF potential that can be applied between the electrodes (6000 V zero-to-peak). Ions of higher m/e remain inside the trap until the RF potential is dropped to zero, at which time all ions drift out with the bath gas, and the next scan is initiated. Selected-ion-monitoring is performed in this mode by a short (one u) scan of the ion of interest, followed by rapid emptying of the trap and repeat.

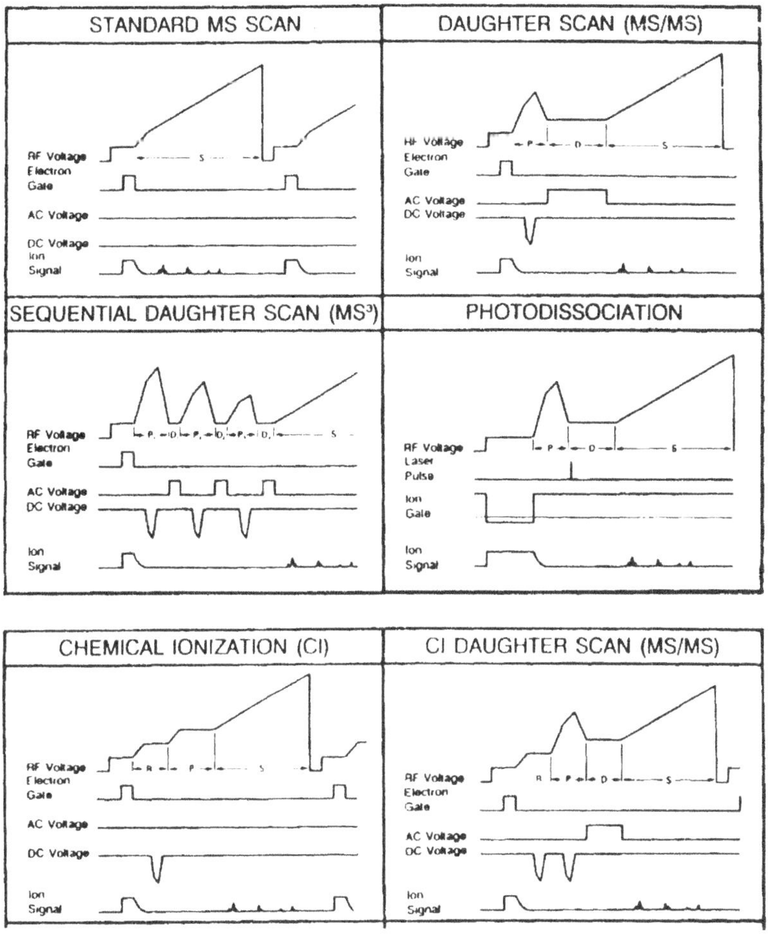


Figure 3. Ion-trap scan programs used to obtain standard mass spectra, daughter spectra, and $MS^n$ spectra. The selection of the parent ion occurs in the interval P, daughters are generated during period D, and the mass analysis scan is indicated as S. Reprinted from Reference 4 with permission.

In the second mode, the trap is operated by using a nonzero DC potential applied to the endcap electrodes, as is possible in the ITMS. This situation will generally result in both a low and high cutoff m/e for ions. This situation can be described using a line from the origin whose slope is the RF/DC ratio. At any particular set of voltages, the line with nonzero slope will intersect the boundaries of the stability diagram at two points (if it intersects at all). The line segment between the two points determines the range of ion m/e ratios that will be trapped. It is then possible to isolate ions of only one nominal m/e for high-sensitivity selected-ion-monitoring or further chemical reactions (such as MS/MS or reagent ion-selected CIMS).

The third manner of using an ion trap involves adding kinetic energy selectively to an ion by applying an auxiliary oscillatory field between the endcap electrodes, and this mode is used for low energy MS/MS experiments. If the frequency of this auxiliary field matches an ion's fundamental frequency $\beta_z$ (between the endcaps), the ion will absorb kinetic energy from the field, and its oscillation amplitude will increase. The ITMS is generally operated with a low pressure of helium (about 0.1 Pa) inside the trap as described above, and the ions that have absorbed kinetic energy in this way may fragment owing to collisions with neutral helium atoms (low-energy CAD). Because of the very long times during which ions can be trapped in the ITMS, very low potentials can be used for this auxiliary field, normally less than one volt peak-to-peak. With this small amplitude auxiliary field, ions will gain kinetic energy slowly, and will generally undergo a fragmenting collision before they escape the trap, yielding nearly 100% MS/MS efficiency in some cases.

This ion-selective kinetic energy addition can also be used for selective ejection of undesired ions from an ion trap, such as solvent-derived ions from liquid or supercritical fluid injection, or mtrix-related ions in FAB or LSIMS. A constant frequency auxillary field at a high voltage between the endcaps during the ionization period will selectively eject a single ion. Space charge saturation from the undesired ions can then be minimized, and the trap can be scanned in the normal manner for analysis.

## 3. GC/MS and GC/MS/MS applications

The normal configuration for the ITD had been a gas chromatographic detector, utilizing the GC carrier gas to stabilize ions inside the ITD and maintain unit mass resolution. Electron ionization (EI) was the first ionization method used, with electrons entering the trap from an external filament through a "gate electrode" and an orifice in an endcap electrode. The gate electrode regulates the length of the ionization period, and is biased positive relative to the filament only during this time. Ions are then expelled sequentially from the trap after the ionization period by increasing the potential on the ring electrode ("mass-selective instability"). The voltages applied to the gate, ring, and endcap electrodes as a function of time (called the "scan function") are shown in Figure 3. Ion trap EI mass spectra are reproducible and similar enough to quadrupole EI mass spectra that library search routines can be used for identification of compounds by using the NBS/EPA mass spectral library, as was demonstrated for the analysis of fatty acid methyl esters [9]. Adams [10] has compiled an extensive volume of EI mass spectra of terpenes taken on an ITD. One significant difference, however, in ion trap EI mass spectra is an exaggerated $(M + 1)^+$ ion signal intensity observed at high concentrations of analytes, for instance 50 ng pyrene used in an environmental study [11]. This is usually interpreted as the result of having too high a concentration of ions in the trap ("space charge" saturation), that decreases the resolution of the resulting mass spectrum. Space-charge saturation also sets a maximum on the number of ions that are able to be stored in the trap, so that high ion concentrations will be underestimated. A software technique ("Automatic Gain Control") has been developed to minimize the quantitative problem, but increased $(M + 1)^+$ ion signals are sometimes still seen, particularly with molecules that produce an abundant $M^{+\cdot}$ ion. Increased $(M + 1)^+$ ion concentrations can also arise chemically from "self-CI" processes from proton transfer from analyte ions to neutral molecules entering the trap. This situation is particularly prevalent in the analysis of esters, such as fatty acid methyl esters, which otherwise give normal EI mass spectra [9, 12].

For molecules suitable for gas chromatography, ion traps commonly have comparable or better quantitative detection limits than quadrupoles; for example, 50 pg (injected) of amphetamine can be detected in a full-scan mode [13]. Volatile compounds in human breath have also been quantitated at very low levels; for example, five picograms of ethanol can be detected in a selected ion monitoring mode of operation [14]. In the normal selected ion monitoring mode, all ions of mass above the ion of interest are stored, but are not ejected during the analytical scan. For a relatively clean sample, as the analysis of ethanol in breath, this is not a significant problem. In the analysis of very complex samples with extremely low analyte levels, however, where compounds that ionize to more massive ions than the analyte would coelute, "space charge" saturation of the trap volume reduces the sensitivity through ion expulsion. Using the ITMS in a selected ion storage and selected ion monitoring scheme (isolating only one ion during the ionization step, Figure 3) prevents storage of the higher mass ions, restoring the sensitivity lost from this space-charge expulsion of analyte ions. This has been used in the GC/MS and GC/MS/MS analyses of crude biological extracts of tryptamine (as the bis-pentafluoropropionamide derivative) from rat brain [15] and eicosanoids (as the trimethylsiloxyl derivative) from an unidentified source [16].

The already high reliability in the identification of compounds by GC/MS can be further increased by using an infrared detector in tandem with the mass spectrometer. This has been demonstrated with a Nicolet light-pipe FTIR placed between a capillary GC and the open-split interface on an ITD [17]. Both IR and mass spectra library searches can be used for the identification of components from mixtures.

Chemical ionization methods are performed on an ion trap by the direct introduction of a low pressure (about 0.1 mPa) of a CI reagent gas and the trapping of ions formed from the reagent gas to react *in situ* with analyte molecules. Low mass ions of the CI reagent are preferentially stored at low trapping potentials, the trapping potential is raised to accelerate these ions and trap the products of the CI reaction, and then the product ions are analyzed by using the "mass-selective instability" mode (Figure 3). One parameter that is varied in ion trap CI and not in conventional CI is the length of time for the reagent ions to react with analyte molecules because the neutrals are in very low abundance compared to those in a conventional CI source (about 100 Pa) [5]. A similar situation pertains to CI in FTMS instruments, as is discussed in the chapter by Amster. Ion trap CI has proven to be useful for toxicological studies, especially because selective CI reagents can be used in an ion trap. These reagents more difficult to use on a conventional mass spectrometer owing to their relatively low volatility (i.e., acetonitrile CI of the stimulant amphetamine [13] and a cholinergic drug, arecoline [18]).

## 4. Alternative sample inlets for ion traps

Non-gas chromatographic sample inlets are necessities for the analysis of many biomolecules because biomolecule ions have high mass or a large number of permanent charges or both.

Because of this, much work has gone into the coupling of ion traps with alternative sample inlets. This has been accomplished in two ways: first, introduction of a sample, which is then ionized within the trap, and second, injection of preformed sample *ions* from an external source (discussed in the next section). Sample introduction techniques, such as laser desorption/ionization of solids [19, 20] and supercritical fluid chromatography (using $CO_2$ as mobile phase) [21] have been used without extensive modification of the basic instrument, and external ion sources such as FAB or liquid SIMS [22], thermospray [23], atmospheric pressure ionization [24], and electrospray [25] have all been demonstrated with ion traps.

One of the earliest non-GC sample molecule inlets used on the ITD and/or ITMS for nonvolatile samples was laser desorption (LD). This was accomplished by drilling holes in the ring electrode opposite each other, inserting a sample probe into one and directing a $CO_2$ laser through the other (Figure 4). In one case, the acquisition of LD mass spectra of sugars, a small peptide, and a quaternary ammonium salt were demonstrated with an ITD [19], and in another, an ITMS was used to produce both LD mass spectra and LD tandem mass spectra on another quaternary ammonium salt [20]. It was shown that the desorption of ions continued for a finite (milliseconds) period after termination of the laser pulse. The sensitivity of the technique for neutral compounds could be improved by simultaneously firing the laser and the electron beam used in a normal EI experiment. The signal intensity from cationized molecules was apparently not increased by the post-desorption ionization, implying that non-laser-desorbed molecules were not efficiently volatilized by the electron beam [19].

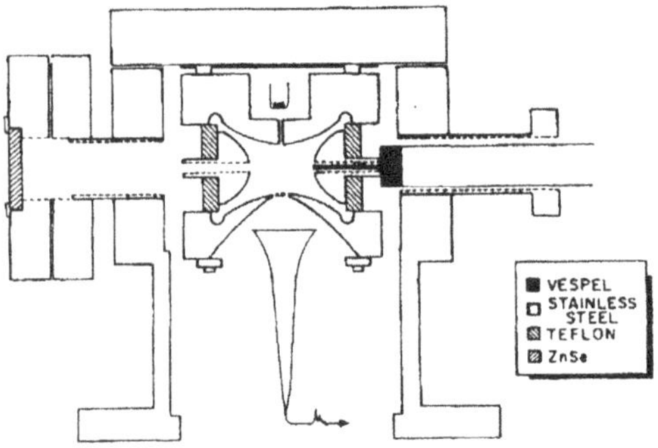


Figure 4. Cross-section of the ITD-probe combination, viewed from the side. Reprinted from Reference 19 with permission.

One report of the coupling of an ITD to a capillary supercritical fluid chromatograph was published [21]. In this study, the total effluent of the chromatograph was introduced into the ITD, and the mobile phase ($CO_2$) served also as the CI reagent. A second stream of helium had to be introduced to increase sensitivity and resolution, as the relatively massive $CO_2$ molecules scattered ions out of the trap. The resulting spectra resemble EI mass spectra for the series of aromatic compounds studied; this is because the ionization was produced by $CO_2$ charge exchange.

**5. External ion sources for ion traps**

Several methods of sample ion injection into ion traps have been investigated. This capability is necessary if common "soft" ion sources used for large biomolecules, such as FAB or LSIMS are to be interfaced to these mass spectrometers. Louris et al [22] showed the feasibility of ion injection from an external EI ion source, the ion entry to the trap being controlled via a potential at a gate electrode between the ion source and the trap volume (Figure 5). This is similar to the gate used to admit electrons for EI in the ion trap, although in EI the electrons are not trapped. In most of the external ion sources demonstrated to date, preformed ions enter the trap through the endcap electrode EI source orifice. Once externally-generated ions can be trapped and analyzed, most, if not all, common ion sources used for biomolecules become feasible for use with an ion trap.

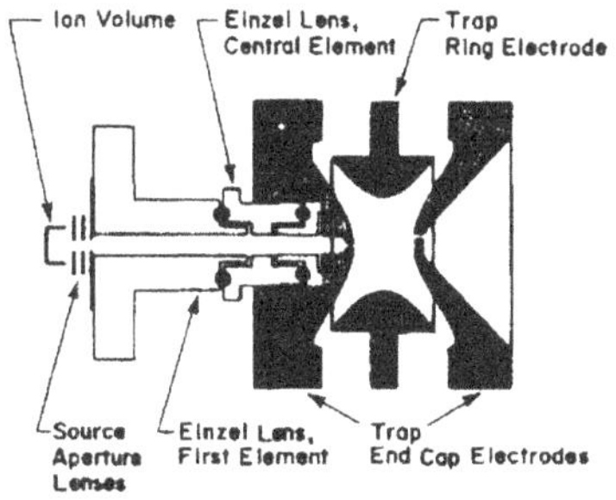


Figure 5. External ion source, interface, and the ion trap shown in cross-section. Reprinted from Reference 22 with permission.

A thermospray ion source has also been interfaced to an ITMS, through a modification of the external EI source to include a repeller electrode and a skimmer [23]. One traditional weakness of thermospray is a lowered sensitivity and quantitative precision for some molecules, which can be compensated for by accumulating ions inside the trap for extended periods before accumulating the mass spectrum. In this system, a methanol/water solvent system was used, and no chromatography was performed. Since a thermospray source continually forms ions, ion injection into the trap was again controlled by a gating lens system [22], preventing ions from entering during the analytical scan. Also, before the start of the analytical scan, space-charge effects were reduced by eliminating trapped solvent ions by resonant ejection, applying a supplementary oscillating field to the endcap electrodes.

An atmospheric-pressure ionization source was successfully used for the monitoring of ambient air for nitroaromatics and other electronegative compounds as their radical anions [24]. This ion source was recently modified to interface between an ion spray (pneumatically-assisted electrospray) ion source and an ITMS for molecular weight determination and MS/MS of peptides and proteins [25]. Albumin and the peptide bradykinin were easily detected when approximately 100 femtomoles were injected into the ITMS. A CAD spectrum of the (M +

$4H)^{4+}$ ion of the 26-residue peptide mellitin was obtained from a four picomole (injected) sample.

The above ion source was subsequently used as an LC interface to the ITMS [26]. Using this arrangement, the authors made mass measurements for cytochrome *c* ($MW_{avg}$ = 12220, measured 12200 $\pm$ 22, Figure 6) and myoglobin ($MW_{avg}$ = 16951, measured 16960 $\pm$ 30) when the ITMS mass range was extended to 3900 by resonance ejection techniques (see below). With this system, the elution of human serum albumin ($MW_{avg}$ = 66000) was detected, but the *m/z* ratios of the multiply-charged protein were not reproducible enough to estimate the molecular weight. This was presumably because of the small *m/z* ratio differences between ions differing by one positive charge and space-charging effects at even low concentrations. The multiple tandem MS capability of the ITMS for multiply-charged peptides ionized by means of ion spray was demonstrated with MS/MS/MS of a tryptic digest peptide ion of *m/z* 477 $(M + 2H)^{2+}$, through an *m/z* 717 (singly-charged) daughter ion, to an *m/z* 251 (singly-charged) granddaughter ion.

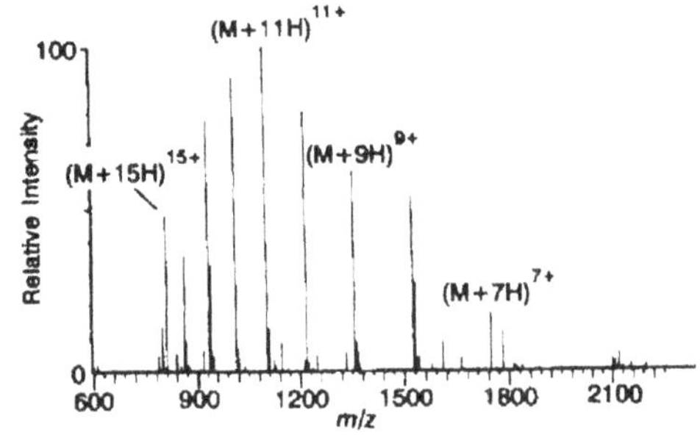


Figure 6. Electrospray mass spectrum of cytochrome *c*. Reprinted from Reference 26 with permission.

An ion trap has also been used in a hybrid (BE/Trap) MS/MS instrument [27]. In this case, the sector instrument was used as a mass filter to inject ions of a selected mass into the trap and a decelerating lens system adjusted the ions' kinetic energy entering the trap. Dissociation of the injected ions increased with increasing injection kinetic energy, as the ions collided with helium bath gas inside the trap and the walls of the trap. A reduced-size ion trap was used to increase the mass range to 2500 Da (see below), so that ions above mass 650 could be analyzed by MS/MS. The daughter spectrum of the $(M + H)^+$ ion (*m/z* 556) of leu-enkephalin at 50 eV was shown as a demonstration of the system for peptide analysis (Figure 7).

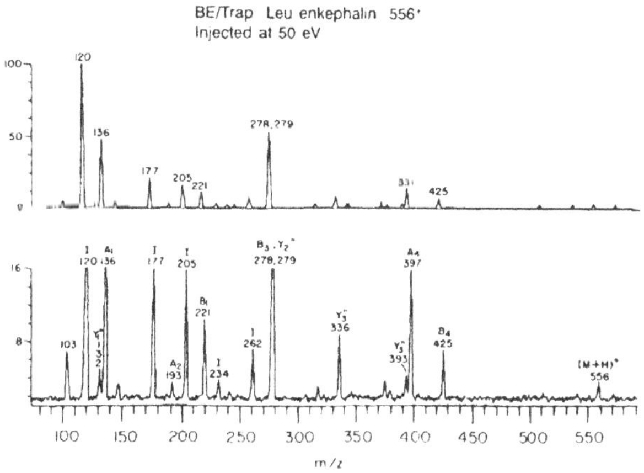


Figure 7. Daughter spectrum of protonated *leu*-enkephalin, m/z 556, 50 eV injection energy. Reprinted from Reference 27 with permission.

## 6. Analysis of high-mass range molecules

The most serious limiting feature of the ion trap for the analysis of biomolecules had been the limited mass range (650 u), which is set by the maximum peak-to-peak output of the RF generator. Ions must be ejected from a quadrupole ion trap to an electron multiplier to be detected, which means that the voltages applied to the ring and/or endcaps must be sufficient to increase the kinetic energy of an ion to force it to escape in the normal analytical scan mode. From the equations for mass analysis in the ITD (equation 2 and Figure 2), ions are ejected in the normal, analytical scan mode (no applied endcap voltages) when $q_z$ is equal to 0.908. This problem has been addressed with success on two main fronts.

From equation 2, the mass of the ion rejected at a particular RF voltage can be increased by decreasing the size of the trap ($r_0$). Because $q$ is inversely proportional to $r_0^2$, a twofold decrease in the size of the trap increases its mass range by a factor of four. This has been demonstrated by Cooks and colleagues [28], to yield a mass range of over 2500 Da without significant losses in sensitivity or mass resolution.

A second strategy to increase mass range involves ejection of ions at a point on the $q_z$ line below 0.908. This can be done through application of a strong auxiliary oscillatory field between the endcaps at a frequency lower than $W_{RF}$. As the main RF field potential is raised, ions will come into resonance with this *auxiliary* RF frequency and be rejected at the value of $\beta_z$ equal to $W_{aux}/W_{RF}$ ("resonance ejection"). In this case, the auxiliary RF field is typically 10 V peak to peak, or much larger than the voltage used in MS/MS studies, so that a resonating ion can be ejected within a very few RF cycles before it undergoes a fragmenting collision with the helium bath gas. This was used by Kaiser *et al* [29] to extend the mass range of an ITMS to over 20,000 Da, and the range of a reduced-size trap to over 45,000 [30], although such high mass range extensions were accomplished at reduced resolution and sensitivity. The technique appears to be applicable to the analysis of ions below 6500 Da, where ions differing by one mass unit can be resolved [27, 31]. This has been particularly important in studies using electrospray ionization, because the mutiply-charged ions formed in electrospray often exhibit *m/z* ratios between 800 and 2000 [24, 32]. Peptides typical of trypsin fragments also fall within this mass range, and quadrupole ion traps present a viable alternative to sector instruments for their determination at unit mass resolution.

## 7. Conclusions

Quadrupole ion trap mass spectrometers have a great potential for use, especially by scientists interested in ion-molecule interactions because of their unique configuration. Extremely high sensitivity analysis is possible with these instruments because they have the capability for storing ions for long periods. Because they are only now finding widespread use among mass spectrometrists, much of the current work is focussed on expanding their general capabilities. One expects that the ion trap will be developed along similar lines as FTMS instruments (next chapter) because both are traps. Unlike the FT mass spectrometer, however, the quadrupole ion trap does not require a magnet and low operating pressure. Through innovative mass scanning techniques, observation of singly-charged molecular ions above 45,000 Da has been demonstrated, and efficient MS/MS analyses of ions, heretofore only useable with tandem sector instruments, are readily achievable. Multiple stages of tandem mass spectrometry are also possible without physically adding to the mass spectrometer. Sample inlets and ion sources used on quadrupole and sector mass spectrometers are being adapted to ion traps to address the demanding requirements of biomolecular research involving labile, nonvolatile, and highly polar samples. Perhaps their greatest advantage, however, is their simplicity of construction and low cost, which suggests that the capabilities of mass spectrometry can be put in the hands of many scientists, as has occurred for GC/MS instruments.

To date, there has not been a great amount of work performed on quadrupole ion trap mass spectrometers dealing specifically with biomolecules. As researchers become more familiar with the advanced techniques being developed on these instruments, they are sure to become indispensable tools in biomolecular mass spectrometry.

## 8. References

1. Paul, W. and Steinwedel, H. (1956) German Patent 944,900.
2. Stafford, G. C., Kelley, P. E.,Syka, J. E. P., Reynolds, W. E., and Todd, J. F. J. (1984) 'Recent advances in and analytical applications of advanced ion trap technology', *Int. J. Mass Spectrom. Ion Proc.*, **60** 85-98.
3. March, R. E. and Hughes, R. J. (1989) *Quadrupole Storage Mass Spectrometry*, John Wiley & Sons.
4. Nourse, B. D. and Cooks, R. G. (1990) 'Aspects of recent developments in ion-trap mass spectrometry', *Anal. Chim. Acta*, **228** 1-21.
5. Brodbelt, J. S., Louris, J. N., and Cooks, R. G. (1987) 'Chemical ionization in an ion trap mass spectrometer', *Anal. Chem.*, **59** 1278-1285.

6. Louris, J. N., Cooks, R. G., Syka, J. E. P., Kelley, P. E., Stafford, G. C., and Todd, J. F. J. (1987) 'Instrumentation, applications, and energy deposition in quadrupole ion trap tandem mass spectrometry', *Anal. Chem.*, **59** 1677-1685.
7. Brodbelt, J. S., and Cooks, R. G. (1988) 'Ion-molecule reactions investigated by using tandem mass spectrometry in an ion trap: halomethylation', *Anal. Chim. Acta*, **206** 239-251.
8. Berberich, D. W., Hail, M. E., Johnson, J. V., and Yost, R. A. (1989) 'Mass-selection of reactant ions for chemical ionization in quadrupole ion trap and triple quadrupole mass spectrometers', *Int. J. Mass Spectrom. Ion Proc.*, **94** 115-147.
9. Horman, I. and Traitler, H. (1989) 'Routine gas chromatographic/mass spectrometric analysis of fatty acid methyl esters using the ion trap detector', *Biomed. Environ. Mass Spectrom.*, **18** 1016-1022.
10. Adams, R. P. (1989) *Identification of Essential Oils by Ion Trap Mass Spectrometry*, Academic Press.
11. Eichelberger, J. W. and Budde, W. L. (1987) 'Studies in mass spectrometry with the ion trap detector', *Biomed. Environ. Mass Spectrom.*, **14** 357-362.
12. McLuckey, S. A., Glish, G. L., Asano, K. G., and VanBerkel, G. J. (1988) 'Self chemical ionization in an ion trap mass spectrometer', *Anal. Chem.*, **60** 2312-2314.
13. Lim, H. K., Sakashita, C. O., and Foltz, R. L. (1987) 'Toxicological applications of the ion trap detector', *Proc. 35th ASMS Conf. Mass Spectrom. Allied Topics*, 696-697.
14. Ghoos, Y., Hiele, M., Rutgers, P., and Vantrappen, G. (1989) 'Porous-layer open tubular gas chromatography in combination with an ion trap detector to assess volatile metabolites in human breath', *Biomed. Environ. Mass Spectrom.*, **18** 613-616.
15. Strife, R. J. and Simms, J. R. (1989) 'Ion trap mass spectrometry of tryptamine. Tissue extracts and isotope dilution using combined radio frequency and direct current voltages', *Anal. Chem.*, **61** 2316-2319.
16. Simms, J. R. and Strife, R. J. (1989) 'Ion trap tandem mass spectrometry (MS/MS) for the determination of eicosanoids', *Proc. 37th ASMS Conf. Mass Spectrom. Allied Topics*, 1264-1265.
17. Olson, E. S. and Diehl, J. W. (1987) 'Serially interfaced gas chromatography/Fourier transform infrared spectrometer/ion trap mass spectrometer system', *Anal. Chem.*, **59** 443-446.
18. Shetty, H. U., Daly, E. M., Soncrant, T. T., Greig, N. H., and Rapoport, S. I. (1990) 'Chemical ionization in an ion trap mass spectrometer: a suitable detection method for routine plasma monitoring of neuroactive molecules', *Proc. 38th ASMS Conf. Mass Spectrom. Allied Topics*, 762-763.
19. Heller, D. N., Lys, I., Cotter, R. J., and Uy, O. M. (1989) 'Laser desorption from a probe in the cavity of a quadrupole ion storage mass spectrometer', *Anal. Chem.*, **61** 1083-1086.
20. Glish, G. L., Goeringer, D. E., Asano, K. G., and McLuckey, S. A. (1989) 'Laser desorption mass spectrometry and MS/MS with a three-dimensional quadrupole ion trap', *Int. J. Mass Spectrom. Ion Proc.*, **94** 15-24.
21. Todd, J. F. J., Mylchreest, I. C., Berry, A. J., Games, D. E., and Smith, R. D. (1988) 'Supercritical fluid chromatography/mass spectrometry with an ion trap detector', *Rapid Commun. Mass Spectrom.*, **2** 55-58.

22. Louris, J. N., Amy, L. W., Ridley, T. Y., and Cooks, R. G. (1989) 'Injection of ions into a quadrupole ion trap mass spectrometer', *Int. J. Mass Spectrom. Ion Proc.*, **88** 97-111.
23. Kaiser, R. E., Williams, J. D., Lammert, S. A., and Cooks, R. G. (1991) 'Thermospray liquid chromatography-mass spectrometry with a quadrupole ion trap mass spectrometer', *J. Chromatog. Biomed. Applic.*, **562** 3-11.
24. McLuckey, S. A., Glish, G. L., and Asano, K. G. (1989) 'Coupling of an atmospheric-sampling ion source with an ion-trap mass spectrometer', *Anal. Chim. Acta*, **225** 25-35.
25. Van Berkel, G. J., Glish, G. L., and McLuckey, S. A. (1990) 'Electrospray ionization combined with ion trap mass spectrometry', *Anal. Chem.*, **62** 1284-1295.
26. McLuckey, S. A., van Berkel, G. J., and Glish, G. L. (1991) 'Ion spray liquid chromatography/ion trap mass spectrometry determination of biomolecules', *Anal. Chem.*, **63** 375-383.
27. Schwartz, J. C., Kaiser,R. E., Cooks, R. G., and Savickas, P. J. (1990) 'A sector/ion trap hybrid mass spectrometer of BE/trap configuration', *Int. J. Mass Spectrom. Ion Proc.*, **98** 209-224.
28. Kaiser, R. E., Cooks, R. G., Moss, J., and Hemberger, P. H. (1989) 'Mass range extension in a quadrupole ion trap mass spectrometer', *Rapid Commun. Mass Spectrom.*, **3** 50-53.
29. Kaiser, R. E., Louris, J. N., Amy, J. W., and Cooks, R. G. (1989) 'Extending the range of the quadrupole ion trap using axial modulation', *Rapid Commun. Mass Spectrom.*, **3** 225-229.
30. Hemberger, P. H., Moss, J. D., Kaiser, R. E., Louris, J. N., Amy, J. W., Cooks, R. G., Syka, J. E. P., and Stafford, G. C. (1989) 'Mass range extension in quadrupole ion traps', *Proc. 37th ASMS Conf. Mass Spectrom. Allied Topics*, 60-61.
31. Cooks, R. G., and Kaiser, R. E. (1990) 'Quadrupole ion trap mass spectrometry', *Acc. Chem. Res.*, **23** 213-219.
32. McLuckey, S. A., van Berkel,G. J., and Glish, G. L. (1990) 'Reactions of dimethylamine with multiply charged ions of cytochrome *c*', *J. Am. Chem. Soc.*, **112** 5668-5670.

FOURIER TRANSFORM MASS SPECTROMETRY: FEATURES, PRINCIPLES, CAPABILITIES, AND LIMITATIONS.

CLIFFTON B. JACOBY, CHRISTOPHER L. HOLLIMAN, and MICHAEL L. GROSS
*Midwest Center for Mass Spectrometry*
*Department of Chemistry*
*University of Nebraska-Lincoln*
*Lincoln, Nebraska 68588-0362*

ABSTRACT. Although a comparably young area of mass spectrometry, Fourier transform mass spectrometry is developing into an important analytical tool capable of ion storage, manipulation and remarkable resolution and sensitivity. Unfortunately, these attributes have been particularly challenging to demonstrate in the determination of large biomolecules. This chapter is presented to highlight the features, capabilities, methods of sample introduction and ionization, as well as applications in the area of biochemistry. Included is a fundamental discussion of operational theory and the areas of instrument development that must be pursued to establish a role for Fourier transform mass spectrometry in the analysis of biomolecules.

## 1. Introduction

Fourier transform mass spectrometry (FTMS), since its introduction by Comisarow and Marshall [1-5], has been heralded as the quintessential mass spectrometer capable of ultra-high mass resolution, high upper mass limit, and high sensitivity due the multiplex or Fellgett advantage. Although these characteristics have been achieved in several applications [6-8], the realization of these goals has not being accomplished in routine work.

As is true for any relatively new technique, many problems exist that were not anticipated. In FTMS these problems include pressure requirements, space charge effects, electric field inhomogeneities, and excitation and detection requirements, all of which will be discussed later. Although all of these problems may not be understood in exacting detail, the fundamentals are understood sufficiently well to allow for the commercial production of instrumentation capable of solving problems in analytical sciences.

This review is written as an introduction to FTMS. Although the principles and applications of FTMS can be diverse and highly specialized, it is the intent of the authors to present a general review of the field of FTMS to provide both a basic working knowledge and some direction for those interested in a particular application of FTMS. There are other reviews of FTMS that may cover certain aspects of this technique in greater detail or present different applications [9-12]. A book has also been recently published which deals with the theory of Fourier transform and its applications to mass spectrometry [13].

This review is divided into five sections. The first discussion will cover the features of Fourier transform mass spectrometry. That will be followed by principles of operation, capabilities and limitations, applications to biomolecules, and future prospects. This format is the basis of the often cited 1984 review by Gross and Rempel [14], and the present chapter serves to update that earlier review.



*M. L. Gross (ed.), Mass Spectrometry in the Biological Sciences: A Tutorial,* 93–116.

## 2. Features

The very diverse fields of chemistry are producing more difficult samples for analytical analysis. As the awareness of environmental, medical, toxicological, energy, and waste issues increases, the samples for analysis becomes more difficult and demanding. These samples may contain mixtures of simple synthetic organic materials, thermally labile, highly ionic compounds, polymers, or high mass peptides. As the complexity of the samples increase, the demands on the chemical analysis also increase.

It is the features of FTMS that make this technique a strong candidate in the field of chemical analysis for now and in the future. Some of these features are (*i*) mechanical simplicity of design, (*ii*) ease of operation, (*iii*) high mass resolution and accurate mass measurement, (*iv*) high upper mass limits, (*v*) multichannel, or Fellgett advantage, (*vi*) high sensitivity, (*vii*) tandem MS capabilities, and (*viii*) ion storage.

### 2.1. MECHANICAL SIMPLICITY

The above mentioned features of the FT technique belie its rather simple mechanical design. The heart of the instrument is an ion trap that acts as the ion source, mass analyzer, and detector, its functions separated temporally rather than spatially. Early trap designs were almost exclusively composed of six electrically isolated stainless steel electrodes to form a cubic cell of ca. 2.54 cm. on edge (see Figure 1), but today several geometries and variations exist. The requirements of the technique dictate that this cell be placed in a high vacuum region (routinely $<10^{-8}$ torr) and a uniform magnetic field. Initially, the magnetic field was provided by an electromagnet, but, as is becoming more common, superconducting magnets are used to produce a more stable field at a higher magnetic induction.

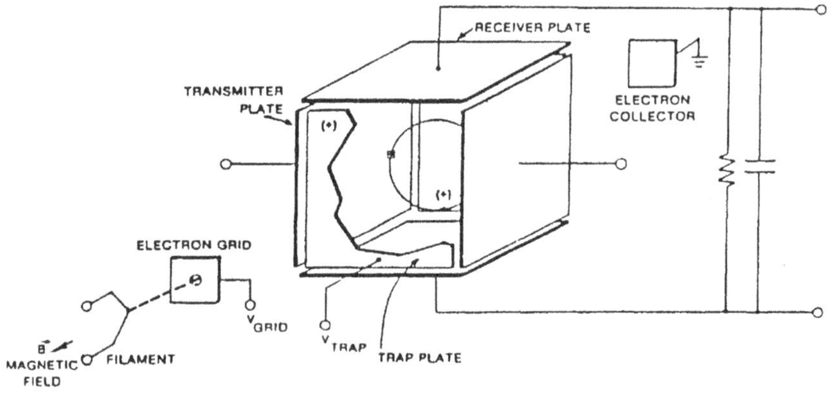


Figure 1. The basic FTMS single cell with excitation and detection scheme. Reprinted from reference [15] with permission.

There are examples of FT instruments that consist of more than a single cell. Examples are dual cell instruments [16-18], designs with external sources located outside the magnetic field, and instruments with ports allowing for the use of lasers for ionization. By appropriate use of these design modifications, FTMS has been combined with fast atom bombardment (FAB) [19, 20], multi-photon ionization (MPI) [21, 22], plasma desorption (PD) [23], electrospray ionization (ESI) [24], and secondary ion mass spectrometry (SIMS) [8]. Yet, even these more complex FT instruments remain more simple in design then sector instruments with comparable capabilities.

### 2.2. EASE OF OPERATION

As the design of a basic FT instrument is simple, so is the operation of the instrument. As mentioned above, the single cell constitutes the essential elements of any mass spectrometer: ion source, mass analyzer, and detector, separated in time instead of space. The sequence of the analysis event is typically performed via the application of differing voltages and rf signals to appropriate plates of the cell at precise times. This precision as well as the need to do the Fourier transform rapidly require that the instrument be computer intensive.

The use of the computer for instrument control also introduces other advantages to the FTMS technique. Owing to the exacting control of time afforded by the computer, other events can be introduced: examples are multiple resonance or ion ejection experiments, pulsed chemical ionization (CI) [25], multi-photon ionization (MPI), photodissociation (PD) [26], gas chromatography FTMS (GC/FTMS) [27-29], and gas-phase reaction kinetics [30]. A few of these experiments will be discussed in greater detail later.

The use of high speed computers also provides for another very important feature of FTMS; that is "fast scan" capabilities. For many type of experiments performed with an FTMS, an experiment repetition rate can approach 100 Hz for a full scan over a range of several hundred mass units.

### 2.3. HIGH MASS RESOLUTION AND ACCURATE MASS MEASUREMENT

The analysis of compounds under high mass - resolution conditions has been one of the most emphasized advantages of FTMS. The basic FTMS equation (equation 1) is

$$\omega_c = \frac{qB}{m} \tag{1}$$

where $\omega_c$ is an experimentally measurable cyclotron frequency, $q$ is the charge of an ion in coulombs, $B$ is the magnetic induction in tesla, and $m$ is the mass of the ion in kilograms. We see that the relation between frequency and mass is inversely proportional. This proportionality indicates that the accuracy and precision with which we can measure frequency should determine the accuracy and precision of the mass measurement. By taking advantage of the high speed capabilities of modern electronics, it is possible to measure frequencies to nine figures. This corresponds to a potential mass measurement with accuracy of 1 part per billion (ppb). This has not yet been achieved because the relationship between mass and frequency, at the ppm accuracy level, is more complex than given by equation 1.

Nevertheless, examples of very high resolution have been reported. Bamberg, Allemann, and Wanczek [7] reported a mass resolution of greater than 2 X $10^8$ for the argon ion of *m/z* 40 at a magnetic field induction of 7.02 T. This resolution corresponds to a peak width (FWHH) of <14 mHz for a resonance frequency of 2.72 MHz (see Figure 2).

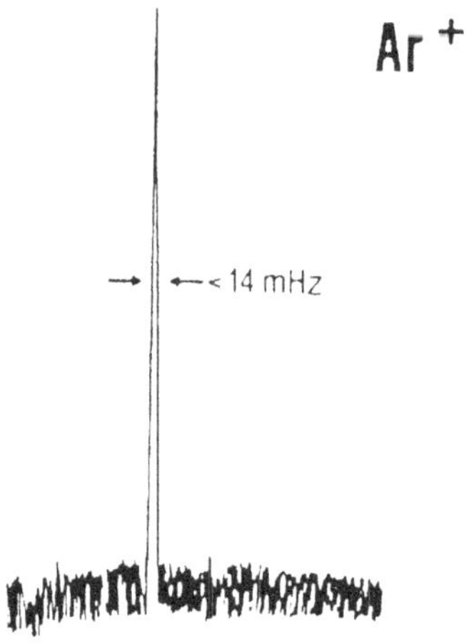


Figure 2. The high resolution (2 X $10^8$) spectrum of $Ar^+$ at 7.02 T magnetic field strength and a pressure of $10^{-11}$ mbar. Reprinted from reference [7] with permission.

Another dramatic example of the remarkable resolution attainable with FTMS can be seen in Figure 3. In this work, Cody et al. [18] show the isotopic clusters of $SnBr_3^+$ at a nominal mass of *m/z* 356. The spectrum shows shows a resolving power (FWHH) of 833,000.

In sector mass spectrometry, the ability to achieve high resolution is taken as a measure of the ability to achieve high accuracy mass measurement. For FTMS, the ability to obtain resolution leads, by orders of magnitude, the capability for accurate mass measurements. This disparity will be discussed later.

### 2.4. HIGH UPPER MASS LIMITS

The analysis of high mass ions (>10,000) is one of the current foci of mass spectrometry. Some recent reports in high mass analysis can be exemplified in work done by Hillenkamp and Karas [31] in which a fragment of the peptide jack bean urease (275,000 u) was analyzed by matrix-assisted laser desorption on a reflectron time of flight mass spectrometer. Another example of high mass analysis is demonstrated in the work of Loo et al. [32] who show the electrospray ionization mass spectrometry of the dimer of bovine albumin at a mass of 133,000 (the interested reader is directed to the chapters on laser desorption and electrospray).

It is necessary to point out that these analyses, while very informative, were performed on instruments that are not optimum for high resolution. The ability to analyze high mass compounds under high resolution conditions would be beneficial for purposes of obtaining

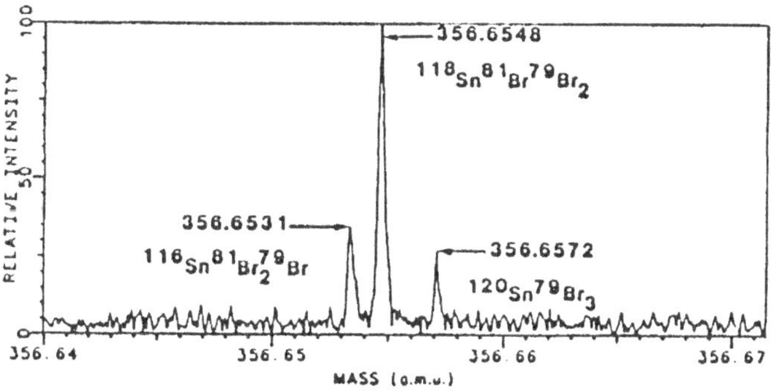


Figure 3. The high resolution spectrum of $SnBr_3^+$ isotopic triplet at a resolution of 8.33 X $10^5$. Reprinted from reference [16] with permission.

accurate mass assignments. Some progress in the area of high mass analysis of singly charged ions via FTMS can be seen in the work of Hunt, Russell, McLafferty, and others [33-36]. The highest mass singly charged ion detected via FTMS to date is the $Cs(CsI)_{122}^+$ ion at *m/z* 31,830 reported by McIver and coworkers [37]. Although this accomplishment does not yet equal those with TOF instruments, it is a noticeable improvement over high resolution sector instruments for which the detection of a singly charged ion of *m/z* 30,000 is very difficult.

From equation 1, an obvious way to increase the high mass capabilities of FTMS is to increase *B*, the magnetic field induction. Although this judgement is correct, we are bound by modern technology and cost. Currently, superconducting magnets with field strengths of up to 7 tesla are available for about $100,000. Perhaps in the near future, with an increase in technology in superconducting materials, magnets may be built at reasonable cost with field strengths greater than 10 tesla.

From equation 1, we should also see that another more practical alternative to raising the magnetic field strength would be to increase the number of charges on the molecule or ion. This is the strategy of electrospray ionization (ESI). The combination of ESI and FTMS was first demonstrated by McLafferty, Shabanowitz, Hunt, and coworkers [38] who detected multiply-charged ions of compounds with molecular weights up to 29,000 u. More recently, Henry and McLafferty [8] showed the electrospray ionization FT mass spectra of bovine albumin with a molecular weight of 66,267 u.

## 2.5. MULTICHANNEL ADVANTAGE

As for any Fourier technique, FTMS has the ability to detect a range of frequencies, or a time domain signal, over a large bandwidth. This ability is known as the multichannel or Fellgett advantage. The obvious difference compared to conventional ion detection is that an entire mass range is detected simultaneously. In a scanning instrument, one *m/z*-ion strikes the detector for a brief period of time while ions of all other *m/z* values are lost. The multichannel ability of FTMS allows one to examine a broad mass range of ions at low resolution or a narrow mass range at high resolution. The highest resolution is achieved by employing a frequency mixer mode, or heterodyne mode of detection, which is necessary because of computer limitations.

The multichannel ability of mass detection is unique to only FTMS and time of flight instruments (TOF). The difference between the two techniques is that FTMS incorporates a non-destructive ion detection scheme whereas TOF detection requires that the ions strike the detector and be destroyed. An advantage of non-destructive detection is that the ions are available for remeasurement, as will covered more thoroughly later. The ability to detect simultaneously ions over a wide mass range is necessary for high sensitivity work. Thus, a considerable effort has been spent on the incorporation of simultaneous detection (array detectors) on high performance magnetic sector instruments. Although this concept is sound, it has shown success over rather small mass ranges [39-42].

## 2.6. HIGH SENSITIVITY

To discuss the sensitivity of a mass spectrometer, one must specify operating conditions: low or high resolution. The detection of ions under high resolution conditions in FTMS actually results in an increase in sensitivity over detection at low resolution [43]. On the other hand, the simultaneous achievement of high resolution and high sensitivity with magnetic sector instruments is impossible owing to the necessity of having narrow slit openings for high resolution ion beam definition. For FTMS, an increase in resolution requires low pressures and an increase in detection time, but the ions do not need to be constrained to a narrow beam. Comisarow [44] claimed that about 10 ions are required to produce a detectable signal in FTMS. Therefore, one might consider that FTMS is a less sensitive technique when compared to sector instruments that are capable of single ion detection. One must remember, however, that many of the ions produced in the source of a sector mass spectrometer are not detected because of the constraints of the narrow slit openings. Furthermore, to recognize an ion mass, one needs sufficient ions to define a peak envelope.

To envision the ultimate in sensitivity, recall that FTMS detection is a nondestructive event. The ions are detected by the production of image currents on the receiver plates of the cell and do not strike a detector. McLafferty and coworkers [45] have taken advantage of this unusual characteristic to propose the multiple detection of the same group of ions. An extension of their work may allow for the repeated detection of a single ion under high resolution conditions.

## 2.7. TANDEM MS CAPABILITIES

The ability to perform tandem mass spectrometry, or MS/MS, is of paramount importance for structural elucidation. The use of FTMS for MS/MS analysis highlights a few of the characteristics of this Fourier technique. Two notable advantages of FTMS for MS/MS experiments are: 1) high resolution product ion analysis and 2) capability for multiple stages of MS/MS or $MS^n$.

An ion of interest can be excited to a larger radius in the FTMS cell by application of the appropriate rf voltage, causing the selected ion to gain energy and dissociate by collisions with neutral molecules. Alternatively, the ion may be activated by photon absorption or by collision with a surface. The product ions are now excited by a wide band rf signal and detected. As was discussed earlier, the ability to detect the product ions under high resolution conditions is still retained.

The sequence can be extended to additional stages of activation in which product ions are forced to dissociate to form third generation product ions. Multi-stage MS or $MS^n$ has been demonstrated as high as $MS^5$ by Kleingeld et al. [46]. This accomplishment is remarkable when compared to those of sector instruments that require an additional sector (or at least collision cell) for each stage of analysis.

### 2.8. ION STORAGE

FTMS theory suggests that ions in a uniform magnetic and electric field are capable of being stored for long periods of time. Storage times of up to $5 \times 10^4$ seconds have been achieved by Allemann, Kellerhals, and Wanczek [47]. It is this storage ability that allows FTMS to detect ions for extended periods and results in high resolution detection. Moreover, ion storage also permits studies of ion reactivity, ion-molecule reactions, ion-ion interactions, photodissociation, and ion kinetics. These types of experiments are difficult on non trapping instruments because ion storage times are normally limited to the microsecond time scale.

## 3. Principles of Operation

The FTMS experiment involves a temporal sequence of operations performed in an ion trap. These operations are described in terms of five broad categories: (i) ion formation, (ii) ion storage, (iii) ion manipulation, (iv) ion excitation and detection, and (v) quench (see Figure 4). Owing to the temporal nature, many different experiments can be performed by simply modifying the order of operations.

### 3.1. ION FORMATION

There are two desirable characteristics for the ionization source in FTMS. The first is that the ionization step is routinely pulsed so that the temporal relationship of the operation is maintained. The second requirement is that the ion formation must not significantly degrade the high vacuum of the ion trap because collisional dampening of the ions will reduce the amount of time they may be observed. In the most basic FTMS experiment, neutral molecules are leaked into the cell at low pressure and ionized by a pulsed electron beam admitted along the magnetic field lines. Chemical ionization, which is an inherently high pressure ionization technique, can be achieved through the use of pulsed valves to admit the reagent gas, [48] and long reaction events. Castro and Russell [49] demonstrated that static $Cs^+$ SIMS is possible by the ionization of vitamin $B_{12}$ from a gold mesh located at the entrance of the ion trap. Ionization techniques such as pulsed laser desorption and $^{252}Cf$ plasma desorption are compatible with FTMS and were successfully coupled [50-52]. Both are optimized by multichannel detection. These ionization techniques along with static SIMS do not produce a significant amount of background; therefore, the high vacuum of the trap region is maintained.

One strategy to resolve the pressure differential between the ion source and trap is to

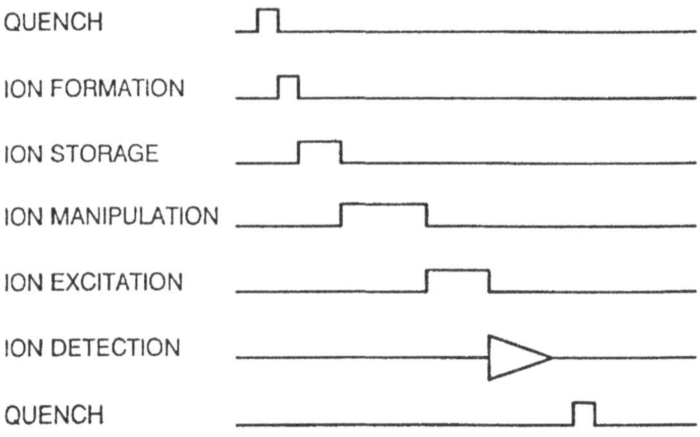


Figure 4. The sequence of events for a basic FTMS experiment. The time of each event and order of events are under computer control.

spatially separate them. Ghaderi and Littlejohn [53] and Cody et al. [18] in two publications report joining two cubic cells to a common wall to form a dual cell configuration. The two cells are differentially pumped so that a pressure ratio between the source and analyzer sides of 1000 or greater is maintained. The common wall has a hole (called a conductance limit) of approximately 2 mm in diameter to allow ions to be transferred between cells.

The idea of separating the ion source and analyzer cell can be extended to the extreme of locating the ion source outside the magnetic field region. McIver et al. [20] describe the coupling of external ion sources by several stages of differential pumping and two sets of quadrupoles. In addition to guiding the ions into the FTMS, the quadrupoles permit a larger separation of source and analyzer so that more differential pumping can be utilized and the analyzer can be coupled with fast atom bombardment and other high background sources. Wanczek, Kellerhals, and coworkers [19] described an external source with one stage of differential pumping which maintains the pressure in the analyzer a factor of 1000 lower than the source pressure . Unlike the McIver instrument however, there are no quadrupoles to separate the ions before they reach the analyzer cell. Cytochrome c (12,230) and myoglobin (MW 16,950) were analyzed with resolving powers greater than 60,000. The strategy of separating source and analyzer was recently employed to couple electrospray and FTMS [54].

### 3.2. ION STORAGE

Once an ion is formed, it is confined to the trap by the combined forces exerted by electric and magnetic fields. If the magnetic field lines are defined to be parallel with the z axis of the trap, an ion will cyclotron in the x-y plane about the z-axis. The basic equation which relates mass to the ion's motion is found by equating the inward and outward forces, and solving for frequency to give equation 2 (Same as equation 1).

$$\omega_c = \frac{qB}{m} \tag{2}$$

Although equation 2 is usually referred to as the "cyclotron" equation, it does not account for all the forces on an ion in the ion trap or describe the cyclotron frequency explicitly. Because there is no force confining the ion along the z axis, two parallel trapping plates with a small applied voltage (usually 1 V) are placed orthogonal to the magnetic field lines. In accordance with Gauss' law, the inward electric fields of the trapping plates also create an additional outward force upon the ion. If one now derives the cyclotron frequency in the same manner as previously described, equating the inward and outward forces, two solutions for frequency result.

$$\omega_+ = \frac{qB + \sqrt{q^2B^2 - 4mqE_0}}{2m} \tag{3}$$

$$\omega_- = \frac{qB - \sqrt{q^2B^2 - 4mqE_0}}{2m} \tag{4}$$

where

$$E_0 = \frac{2\alpha V_T}{a^2} \tag{5}$$

and $\alpha = 1.386$ for the cubic trap, $V_T$ is the potential applied to the trap plates, and $a$ is the length of the cell. The larger of the two terms, $\omega_+$, more accurately describes the frequency of the cyclotron motion and gives the best measure of mass. The smaller value, $\omega_-$, is the frequency of an another motion called the magnatron motion. This comes about by a precessing of the cyclotron motion about the z-axis. The reader should note that both frequencies are a function of the trap voltage and that the new cyclotron equation is equal to equation 2, when the trapping plates are removed, or the applied voltage is zero. Finally, there is a third resonant motion known as the z-oscillation, where the ions oscillate between the trapping plates at the trapping frequency given by equation 6 (see Figure 5).

$$\omega_T = \frac{2}{d}\sqrt{\frac{qV_T}{m}} \tag{6}$$

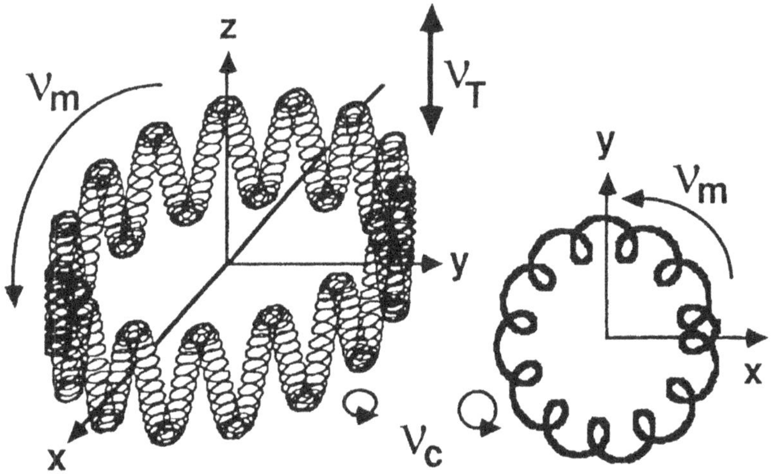


Figure 5. Diagram showing the three fundamental oscillations of an ion in an FTMS cell. The magnitude of the magnatron oscillation, $\nu_M$, is exaggerated with respect to the cyclotron oscillation, $\nu_C$, and the z-mode (trapping) oscillation $\nu_T$. Reprinted from reference [14] with permission.

The trapping plates are not the only source of outward electric fields. There also exists outward forces on the ion, called space charge effects, that are due to coulombic interactions between ions in the cloud. The magnitude of these forces increase with increasing cloud density, and because they operate in opposition to the magnetic force, cause $\omega_c$ to shift to a lower frequency [55]. The shift due to space charge effects is not as severe as that due to the trapping potential, but it is a concern for high accuracy mass analysis because the extent of space charge varies from experiment to experiment. Several trap geometries such as elongated [56], cylindrical [57], and hyperbolic [58], as well as modifications such as segmenting the trap plates [59], placing grounded screens just inside the trapping plates [60], or incorporating them as extensions of the excitation and receiver plates [61], are attempts to moderate those electrical effects that are results of trapping and space charge fields.

### 3.3. ION MANIPULATION

Two of the most useful means of ion manipulation are obtained by exciting the cyclotron mode. The first, called ion ejection or double resonance [62], involves excitation to the point where the excited ions collide with a cell electrode and are neutralized. Ions are ejected to establish relationships between reactants and products in ion-molecule reaction sequences (called double resonance) or to rid the cell of unwanted ions so that an ion of interest can be "isolated" and the dynamic range expanded. Unwanted ions may be neutralized in the dual cell by exciting them in the source cell to a radius that is greater than the conductance limit and then transferring the desired ions to the analyzer cell. The second method is excitation of the ions to promote interactions with a collision gas, but *not* to eject. If the translational energy is high enough, collisional activation followed by fragmentation will occur. The spectrum of the resulting fragment ions can be obtained as usual and used as a structural probe.

Some adducts of ion-molecule reactions are formed with considerable internal energy, and decompose before they can be isolated. Due to the low operating pressures of FTMS, they are not readily observed and must be formed in a high pressure regime, such as a chemical ionization source. Rempel and Gross [63] have demonstrated collision stabilization at approximately $10^{-3}$ torr by inserting an RF-only-mode event in the FTMS sequence. During the event, an RF frequency of sufficient amplitude is applied to the trapping plates such that the ion trap behaves as a QUISTOR, or quadrupole ion trap [64]. After stabilization, neutrals are pumped away, and the trap is returned to the conventional FTMS mode of operation for mass analysis.

### 3.4. ION EXCITATION AND DETECTION

Detection of the ions is achieved by using two of the six sides of the cell as "receiver plates" that are joined by a parallel RC circuit. This arrangement converts the image current induced on the plates by the passing ions, into a voltage, which can be amplified to a level necessary for analog-to-digital conversion. The magnitude of the signal induced upon the plates is a function of the orbiting ion's distance from the plates (their cyclotron orbit radius) and independent of mass [65]. Excitation of the cyclotron mode is used to observe the quantity and mass-to-charge ratio of various ion species by excitation to an orbit that passes within close proximity to the receiver plates. Following excitation, the image currents (called the time domain signal) will be superimposed on the receiver plates, converted to image voltages, amplified, digitized, and stored in a computer for discrete Fourier transformation. This simultaneous observation of all the ion species gives the multichannel advantage. Fourier transformation gives the frequency domain, and from the spectral intensities and frequency centroids of the peaks in the resulting frequency spectrum, one can deduce the quantity and mass of each ion species.

Before excitation, the ions of a given *m/z* ratio will exist along or near the z-axis with the same cyclotron frequency but with different phases, or at different positions about the same circular orbit. This results in no net image current on the receivers. The application of an excitation waveform to the cell's transmitter plates (the remaining two plates) excites the ions and creates a coherent "packet" as the cyclotron radius is increased. The radius at which the ions are excited to is a function of the magnitude of the waveform and its duration but is independent of the ion's mass [66]. To generate a signal that accurately represents all *m/z*-values of all the ions and their relative abundances, a waveform that has a constant amplitude over the desired frequency range should be chosen. This can be referred to as a flat frequency domain magnitude mode spectrum.

The standard FTMS excitation waveform is the frequency sweep or "chirp" [2]. This involves a rapid scan over a frequency range of constant amplitude. Chirp excitation offers the advantage that a broad range of *m/z* values can be excited with a reasonably low rf amplitude.

Although chirp excitation has been widely adopted, several shortcomings of the waveform have been recognized, and alternative waveforms have been developed to address the problems. Gross and coworkers [67] showed that in addition to expanding the ion's radius in the x-y plane, chirp excitation will also increase motion along the z-axis. z-Excitation is a mass dependent mechanism that is in part due to the period of excitation (ms) being longer than the period of the ion's z-oscillation. The authors proposed modifications of the chirp waveform with respect to sweep rate, excitation time, and amplitude to minimize z-excitation. McIver, Hunter, and Baykut [68] addressed z-excitation problems with a short (5 us) high amplitude (120 V) impulse excitation waveform. Here all ions are abruptly moved towards one of the transmitter plates. The authors describe impulse excitation as, "conceptually the same as when a piano is hit hard with a hammer, causing all the strings to vibrate simultaneously." Impulse excitation is a nonselective excitation, and therefore would not be useful for double resonance experiments. To our knowledge, the first use of impulse excitation was used in the development of the hyperbolic cell [58], despite claims to the contrary.

Perhaps the most versatile waveform is the stored waveform inverse Fourier transform (SWIFT) method developed by Marshall and coworkers [69]. The desired excitation waveform is defined in the frequency domain. By using an inverse Fourier transform, it is converted to the corresponding time domain form, which is then applied to the transmitter plates. This results in a nearly flat magnitude spectrum with the ability to define precisely which ion, or ensemble of ions, is excited. SWIFT excitation is especially valuable for double resonance and ion chemistry studies.

In addition to application of the excitation waveform to the two transmitter plates, application to the four plates orthogonal to the trapping plates (quadrature excitation) [70], or the trapping cones of the hyperbolic cell (parametric excitation) [58] were demonstrated.

Just as quadrature excitation is possible, quadrature detection can be implemented by using the four plates orthogonal to the trapping plates [71], or by electric orthogonization of the two transmitter plates [72]. Quadrature detection offers advantages such as doubling the bandwidth for the same transform size and increasing the signal to noise ratio by a factor of $\sqrt{2}$.

A repeated detection scheme takes advantage of the nondestructive nature of FTMS. McLafferty and coworkers [45] reported that it is possible to excite and detect repeatedly the same ion population of large molecules such as gramicidin A, thus increasing sensitivity. After the original excitation and detection sequence, a small collision gas (i.e., helium) is used to remove energy from the cyclotroning ions and focus them back to the center of the trap where the sequence may be repeated. Although it is difficult to understand how the ions relax to the center of the cell, it is clear that the sensitivity may be increased, even to that of an electron multiplier.

### 3.5. QUENCH AND REPEAT

At the end of the sequence, the cell is usually purged of all ions by applying a large potential to one of the trap electrodes to eliminate the trapping potential well or by inverting the polarity of the trap potential. The sequence of ion formation, manipulation, excitation, and detection is repeated for signal averaging, if necessary.

Due to the temporal nature of FTMS, various operations can be combined to perform different experiments such as mass spectrometry, ion-molecule reactions, double resonance, and mass spectrometry-mass spectrometry (MS/MS). For example, the "mass spectrometry sequence" occurs as ion formation, ion excitation, detection, and quench. The "ion-molecule reaction sequence" is identical to the MS sequence except that a variable time delay is added between ion formation and detection. Addition of one or more ejection events during the time delay between ion formation and detection gives the "multiple resonance sequence". If all mass ions but one are ejected, the remaining ions are excited for collisional decomposition, and the resulting product ions excited and detected, the MS/MS experiment is realized.

## 4. Capabilities and Limitations

### 4.1. MASS RESOLUTION

As for other Fourier transform techniques, there is an exponential decay of the time domain signal intensity in FTMS, and the rate of decay limits the amount of time that the transient may be observed. In FTMS the major mechanism for signal decay is collisions of the excited ions with neutrals. Collisions alter the ion trajectory, resulting in a loss of phase coherency. With low background pressures, the decay is slow, and with long detection times, FTMS becomes a high-resolution technique [4]. Resolution of at least 3 X $10^8$ for the $O^+$ ion was obtained by observing a transient response for 100 s at the nonroutine base pressure of $10^{-13}$ torr and a magnetic field induction of 7.02 T [7]. In general, resolution, $R$, is limited by the time of observation $t$ (equation 7) and the collisional damping frequency, $\xi/n$ (equation 8),

$$R = m/\Delta m \leq kqBt/m \tag{7}$$

$$R = m/\Delta m \leq k'qB/(mP)(\xi/n) \tag{8}$$

where $q$ is in multiples of electronic charge, $B$ is the magnetic field strength (T), $P$ is the pressure (torr), $k$ is 1.3 X $10^{-7}$ u $T^{-1}$ $s^{-1}$, $k'$ is 8.6 X $10^{-10}$ u $cm^{-3}$ torr $T^{-1}$ $s^{-1}$ per molecule, and $\xi/n$ is typically 1.0 X $10^{-9}$ $cm^3$ $s^{-1}$ per molecule. One should note, that as the mass-to-charge increases, or the pressure increases, resolution decreases.

Several new source designs and source housing designs have been developed to maintain the high-resolution potential. For example, electrospray ionization places several charges on the same molecule, and that decreases its mass-to-charge ratio. High speed pumping systems, that utilize turbo and oil diffusion pumps are also employed to establish low pressure. Experiments undertaken in the low $10^{-8}$ torr range give transient responses that last for approximately 1 s, and result in resolving powers of about 100,000 for $m/q$ 1000 at 5 T [73].

Equations 7 and 8 predict that resolution in FTMS should decrease linearly as mass increases. This in fact is often not the case. As mass is increased, the relative strength of the magnetic field decreases, and electric field effects become more important. Therefore, resolution tends to decrease precipitously for bioions having mass-to-charge ratios greater than 2000. Electric fields in the cell are nonideal, and their effect on mass resolution is not completely understood. Hunt et al. [34] have reported that large ion populations of the cesium iodide cluster of $m/z$ 5848.7 can be detected at resolutions in excess of 30,000. When the same system is used to detect horse cytochrome $c$, which is more difficult to ionize and produces smaller ion populations, the observed resolution is less than 100. Some insight into this dichotomy of results has been revealed with high mass modeling, or scaling experiments [74]. Scaling models the behavior of high mass ions at high magnetic fields with the behavior of low mass ions at appropriately scaled lower magnetic fields. This approach has demonstrated that an increase in ion number moderates nonideal electric effects in the trap and results in an increase in resolution. Trap designs that were developed to moderate the electric field effects and increase resolution include splitting the trapping electrodes into multiple segments so that a differential of voltages may be applied (the compensated cell) [75] or placing grounded screens in front of the trapping electrodes to remove as much as possible the electric field inside the cell (the screened cell) [60]. The screened cell, however, does not adress the problems of space charge effect whereas the compensated cell was designed precisely for that purpose.

### 4.2. HIGH MASS LIMIT

The advent of high mass ionization techniques has focused attention on the high mass capabilities of all mass spectrometers. Although observation of the $Cs^+$ cluster at *m/z* 31,831 is impressive, this does not define the upper mass limit in FTMS. The highest mass that can be trapped is defined as the critical mass, $m_c$. For a cubic cell, $m_c$ is given by equation (6)

$$m_c = qB^2a^2/8V_{eff}\alpha \tag{9}$$

where $a$ is the length of the cell, $V_{eff}$ is the "effective" trapping voltage (i.e., the applied value corrected for space charge effects), and the geometry constant $\alpha$ is 1.3869 [55]. It can be seen from equation 9 that the critical mass can be raised by increasing the magnetic field strength or lowering the trapping voltage. As the trapping voltage is increased, so is the strength of the outward radial electric field. Increasing the outward force has the same effect as decreasing the magnetic field and, therefore lowers the critical mass. Elongated and screen traps attempt to decrease this radial electric field in the region where ions are detected. A calculation of $m_c$ for an instrument equipped with a 2.5-cm cubic cell operated with a trap of 1 V in a 13-T magnetic field yields a value of 950,000 u for a singly charged ion. It is our opinion that at least 10 percent of the critical mass limit should be useful for FTMS. If this is true, FTMS will not be the instrument of choice for mass analyzing very large (>100,000) bioions.

### 4.3. DYNAMIC RANGE

The upper end of the ion abundance dynamic range is dictated by space charge effects and the number of ions the cell can hold. For studies of ion chemistry or other situations that require only low resolution spectra, the ion trap is operated with ca. $10^6$ ions. Exact mass measurements, however, are best done with ca. 2 X $10^4$ ions for a 2.5-cm cubic cell in a magnetic field of 1.2 T. The lower end of the dynamic range is limited by amplifier noise: usually 50 to 100 ions are necessary to produce a good signal, depending on the instrument. Thus, a crude estimate of the dynamic range for refined experiments involving exact mass determination with the conventional FTMS sequence is ca. 100. One way to reduce space charge effects, and hence increase dynamic range is to selectively eject ions that are not being detected. Marshall and coworkers [76] demonstrated this approach with SWIFT multiple ion ejection to enhance the sensitivity for minor peaks in a perfluorotri-*n*-butylamine spectrum.

Laude and Beu [77] reduce space charge effects by introducing a new event, which they call suspended trapping. This event grounds the trapping plates for 1-2 ms after the ionization and initial trapping event. The ion cloud undergoes a self regulating diffusion process along the z-axis to a density where space-charge effects are negligible. This event has been successfully incorporated for GC/FTMS experiments and has shown an increase in the dynamic range of the cell by four orders of magnitude for injections of samples of peppermint oil and gasoline [78]. These results approach the dynamic range of conventional mass spectrometers equipped with electron multipliers, where the dynamic range is greater than $10^5$ under favorable conditions.

### 4.4. ION STORAGE AND DYNAMICS

The literature is replete with examples in which FT mass spectrometers are used to study ion chemistry. Here, one takes advantage of ion storage and ion manipulation to control the ion population. Neutral populations can be manipulated by using pulsed valves and dual cells. High resolving power is available if needed. This is an area where FTMS excels, and most of the users today are ion chemists. The development of the rf-only-mode event to study ions that require collisional stabilization and were not previously observed in FTMS will augment the utility of FTMS in ion chemistry studies.

### 4.5. ACCURATE MASS ASSIGNMENT

The potential of highly accurate mass assignment is derived from the ability to measure the observed ion frequencies to nine significant figures. Variations in space charge effects due to variations in ion number, however, have made an accurate relationship between observed frequency and mass difficult to define. A detailed description of space charge theory in the cubic trap was presented by Jeffries, Barlow, and Dunn [55]. On the basis of their theory, the following mass calibration law was derived by Ledford et al. [79]

$$m = a / f_{obs} + b / f_{obs}^2 \tag{10}$$

where $m$ is mass, $f_{obs}$ is the observed ion frequency, and $a$ and $b$ are calibration constants. Utilization of the law requires only two reference masses to define $a$ and $b$, a situation that is understandably simpler than that for sector magnet mass spectrometers. Presented with this calibration was the suggestion that mass calibration errors will decrease with the square of the magnetic field strength.

Wang and Marshall [80] extended this calibration law into a four term equation by considering interactions of negative frequency peaks in the presence of ion-molecular collision. This extended law offers approximately a factor of two improvement over the previous law, a small return for additional complexity.

In addition to mathematical manipulation, instrument modification is an effective route to mass accuracy improvement. Mass errors in the nearly perfect quadrupolar potential of the hyperbolic trap were shown to be less sensitive to space charge effects than those for the cubic cell [58]. The elongated trap of McIver and coworkers [56] attempts to decrease space charge effects by physically increasing the volume of the trap. The trapping potential of this trap drops very quickly as one moves away from the trapping plates. The result is smaller radial fields at the trap center and a smaller susceptibility of frequency shift to trapping voltage.

Possibly the ultimate extension of this route is the screened trap, which is believed to create a nearly particle in a box trapping potential [60]. A comparison of frequency shift with respect to trapping voltage by using benzene revealed values of 64.6 Hz/V for a 5-cm. cubic trap versus 0.67 Hz/V for the same 5-cm. screened trap. Once again it is noted that as mass is increased, the effects of the electric fields become more important. Scaling experiments suggest that as the trapping well deviates from the parabolic trapping potential of the hyperbolic trap, the ability to trap and store high mass ions is severely degraded [81]. Therefore, the screen trap may reduce the electric field at the expense of ion storage capability, especially for high mass ions.

Accurate mass assignments are routinely performed on sector instruments; however, the most accurate analysis requires a standard about every 14 u. This is usually accomplished by leaking perfluorokerosene or a similar mass standard into the ion source along with the sample. FTMS is capable of reasonably high mass accuracy (approx. 10 ppm) without external calibration [82]. This feature should make FTMS a valuable tool for accurate measurements with ion sources for which extensive use of calibrants is difficult, such as those for chemical and desorption ionization.

## 5. Sample Introduction and Ionization

Prior to discussion of applications of FTMS, one must first consider a method of sample introduction into the instrument and subsequent ionization of that sample. This initial process can be broken down into three realms. These are the introduction and ionization of volatile species, non-volatile species, and in true analytical fashion, chromatographic introduction.

### 5.1. VOLATILE SPECIES

When considering sample introduction and ionization, one must always remember that to achieve the best possible results with FTMS, it is imperative to maintain a low pressure (< $10^{-8}$) in the cell of the instrument. This requirement suggests the need for an inlet system capable of fine control of vapor admission. For samples initially in the gas phase, or for volatile liquids or solids, this is readily accomplished by use of a molecular leak valve or a high speed pulsed valve. For less volatile solid samples, a heated solid probe inlet has been successfully used.

For such samples, the ionization process is very straightforward. In electron ionization (EI), a gated electron beam is used on both commercial and custom-built instruments. The control of the electron beam parameters such as the numbers produced, energy, and the time for admission to the cell are easily accomplished by a few simple computer commands.

Chemical ionization in an FT cell is also a very feasible technique. In its simplest form, this requires the production of a protonated species from a sufficiently low proton affinity gas, this so called reagent ion transfers a proton, or other charged species, to the analyte. This process of chemical ionization in FTMS is covered in greater detail in the chapter by Amster. A recent example of chemical ionization FTMS was shown in a study by Geribaldi and coworkers [83] in which they used negative ion chemical ionization combined with collisionally induced dissociation to differentiate diastereomeric terpenoid alcohols.

The last method of ionization of volatile species to be covered in this chapter is multiphoton ionization (MPI). The process of MPI is discussed in a comprehensive book edited by Lubman [84]. Multiphoton ionization is considered to be a very versatile technique for mass spectrometry. That is, MPI can be altered to be a hard or soft ionization event. Soft ionization occurs when the abundance of the molecular ion is large with respect to fragment ions. By contrast, a hard ionization process is one which results in extensive fragmentation.

Typically, photoionization occurs when a region of gaseous molecules is irradiated with a beam of photons. Usually this is accomplished with a laser of variable or fixed wavelength. Upon illumination, gas-phase molecules will absorb either multiple photons (MPI) or, if the energy of the photon is equal to the ionization energy, a single photon will cause ionization. When the laser light frequency corresponds to an allowed transition, this process is referred to as resonance photon ionization (RPI).

The first MPI mass spectrometry experiments were reported by Chin [85] for $I_2$, $D_2O$, and $CCL_4$, and Berezhetskaya et al. [86] for $H_2$. It was not until 1982 that McIver et al. [87] coupled an excimer laser operating at 222, 249, and 308 nm with an FTMS to show its feasibility with FTMS analysis. In a recent report by Williams and McLafferty [22], the use of laser photoionization and photodissociation was discussed as a means to differentiate the structural isomers of $C_7H_8$, $C_7H_9N$, $C_7H_7F$, and $C_8H_{10}$.

## 5.2. NON-VOLATILE SPECIES

The analysis of a non-volatile sample usually relies on a method of ionization that incorporates some feature of desorption. The desorption ionization techniques include fast atom bombardment (FAB), secondary ion mass spectrometry (SIMS), electrospray ionization (ESI), laser desorption (LD), and matrix assisted laser desorption (MALD), field desorption (FD), and plasma desorption (PD).

Perhaps the most common method employed for desorption and ionization of non-volatile samples involves the use of highly accelerated particles to bombard the surface of the sample probe (FAB or SIMS). The application of a desorption/ionization process involving particle bombardment combined with FTMS as a detection scheme was demonstrated by Castro and Russell [49] in 1984. In this work, the authors compare spectra of non-volatile, thermally labile biomolecules obtained via SIMS/TOF and SIMS/FTMS. The authors concluded that SIMS/FTMS spectra may differ from those obtained with other instruments because the longer FTMS time scale allows slow decompositions to occur before detection. Shortly thereafter, Hunt, McIver, and coworkers [21] extended the use of SIMS/FTMS to the analysis of peptides. More recently, Ijames and Wilkins [25] showed SIMS/FTMS to be compatible with high mass analysis (over 8000 u) and also to be capable of high resolution.

One method of sample introduction and ionization with high promise is electrospray ionization (ESI). Electrospray ionization provides a way to introduce non-volatile and thermally labile compounds to the gas phase while keeping the *m/z* value low by multiple charging of the analyte molecule.

An alternative for large molecule work is laser desorption. Laser desorption of an analyte is accomplished by focussing the output of a pulsed laser onto the surface of a probe on which the sample has been deposited. The desorbed or laser ablated ions are then detected by mass analysis. Although LD has been successfully performed on several mass spectrometers, it is best-suited to either TOFMS or FTMS owing to the pulsed nature of the ion production and detection.

To achieve maximum information from each laser pulse, FTMS is the detection method of choice, owing to the high mass and high resolution capabilities of the instrument. McCrery, Ledford, and Gross [88] reported the first application of LD/FTMS in the analysis of citric acid, glycyltyrosine, and other organic compounds. Extensions of this initial work can be seen in the analysis of copolymers, polycyclic aromatic hydrocarbons, polypeptides, and polymers, as well as for the coupling of MPI to LD/FTMS for the ionization of desorbed neutral molecules [23, 89-91].

Matrix-assisted laser desorption was first reported by Karas and Hillenkamp [92] in 1988. Although the details of matrix-assisted laser desorption is not completely understood, it appears to be an extension of true LD. The difference of the two techniques is that the energy of the photons striking the sample surface is absorbed by a matrix molecule. It is believed that this energy causes a rapid desorption of matrix molecules that carry along high mass analyte ions. The work of Hettich and Buchanan [93] is the first to show the application of MALD and FTMS. The application was not to a particularly high mass species, substance P. In the normal LD spectrum of substance P we see very little evidence of a molecular ion. Under MALD conditions, however, the molecular ion is readily observed (see Figure 6).

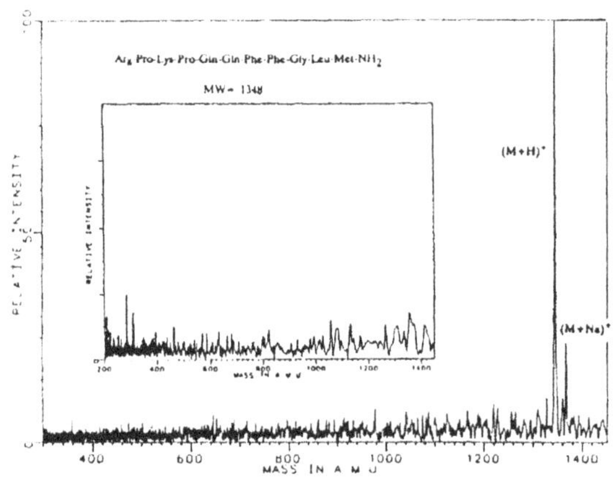


Figure 6. Matrix-assisted laser desorption (MALD) of substance P (266 nm, nicotinic acid matrix). Inset shows spectrum without nicotinic acid matrix. Reprinted from reference [91] with permission.

The scarcity of reports on high mass systems point to the serious problems that confront FTMS and its use for high performance mass spectral analysis of biomolecules. Nevertheless, FTMS should play a role in the MALD analysis of medium-sized biomolecules ($M_r$ > 30,000) provided the problems of mass resolution can be solved.

### 5.3. CHROMATOGRAPHIC INTRODUCTION

When one thinks of analytical chemistry, chromatography is among the most commonly used techniques that come to mind. With the recognition that FTMS possesses superior resolution and "fast scan" capabilities over other mass spectrometers, it was not long after the report of the first instrument that Gross, Wilkins, and coworkers [27] reported the limited success of coupling a capillary gas chromatograph (GC) to FTMS. This coupling was not a trivial matter because to achieve quality FTMS spectra, low pressure ($<10^{-7}$ Torr) requirements must be met. Initially, to maintain a low pressure in the FT cell, a jet separator was used to limit the load of the carrier gas. Though mildly successful, background pressures were limited to 5 X

$10^{-7}$ Torr, which in turn limited the resolving power to 23,000 (FWHH) for the benzene radical cation.

In 1983, Sack and Gross [29] reported the use of a high speed pulsed valve as a method to maintain low background pressure in the FT cell during GC/FTMS experiments. The incorporation of the pulsed valve enabled them to achieve a resolution of 103,000, normalized to *m/z* 78 and 1.2 T, for the molecular ion of naphthalene at *m/z* 128.

Most recently, dual cell FTMS instruments have been used to improve the capabilities of GC/FTMS [16]. With a dual cell instrument, the GC eluent is admitted into the source side of the dual cell at pressures of about $10^{-5}$ Torr, where it is ionized. Periodically, the potential of the conductance limit is dropped to allow a portion of the ions to pass into the analyzer side of the dual cell, where the pressure is ca. $10^{-8}$ Torr, for high resolution mass analysis.

Although GC/FTMS is now capable of producing high resolution spectra at low pressures, little effort was exerted until recently to increase the sensitivity and detection limits of the technique. In 1988, Hogan and Laude [78] proposed a method of suspended trapping to help maintain the number of ions below the limit of space charge effect, which was described in Section 4.3. A potential improvement of 12 to 24 times for limits of detection was estimated for incorporation of this new event.

Another method of improving the limits of detection for GC/FTMS was developed by Jacoby, Rempel, and Gross [94, 95] in 1990. The problem with pulsed GC methods is that the duty cycle is low; most of the sample is sent to waste. A new approach was designed to overcome duty cycle restrictions by concentrating the eluent. An eluting GC peak was concentrated into a narrow width of approximately 25 ms. The concentration of the analyte was accomplished as the eluent passed through a portion of stainless steel tubing that had been cooled via thermal contact with liquid $N_2$. Under computer control, the steel tubing was then resistively heated to flash vaporize the trapped eluent and pass the sample into the FT cell for analysis. This effectively increases the number of molecules in the cell and available for ionization by a factor of about 100 times when compared to conventional pulsed GC/FTMS. Results thus far indicate improvements of 12 to 14 times in the detection limits of the instrument.

Supercritical fluid chromatography SFC/FTMS was also successfully demonstrated in instruments of dual cell design, as first reported by Henion, Cody and coworkers [96]. In a more thorough investigation performed by Laude, Wilkins, and coworkers [97], mixtures of up to seven different compounds at concentrations of 56 ng per compound were successfully analyzed (see Figure 7).

Although liquid chromatography (LC) mass spectrometry is an effective technique for the analysis of polar, non-volatile compounds normally not amendable to GC, no reports on the coupling of HPLC to FTMS were made until recently. The first report was presented by Moini [98] in 1991. In this study, the coupling between the HPLC and the FTMS was accomplished via a moving belt interface. Although no results were reported, there may be important applications.

## 6. Applications to Biomolecules

The determination of biomolecules is one of the most active areas of research in mass spectrometry. Biomolecules are comprised of both low mass and high mass species. LD/FTMS was already shown to be an attractive method of analysis for low mass molecules [99]. Accordingly, most recent work has been focussed on the analysis of large biomolecules. Hunt and coworkers [34] reported the detection of horse cytochrome c at *m/z* 12,384; however, they were unable to obtain high resolution results. Because good resolution could be achieved for inorganic cluster ions, some thought that the problem was with the size and shape of

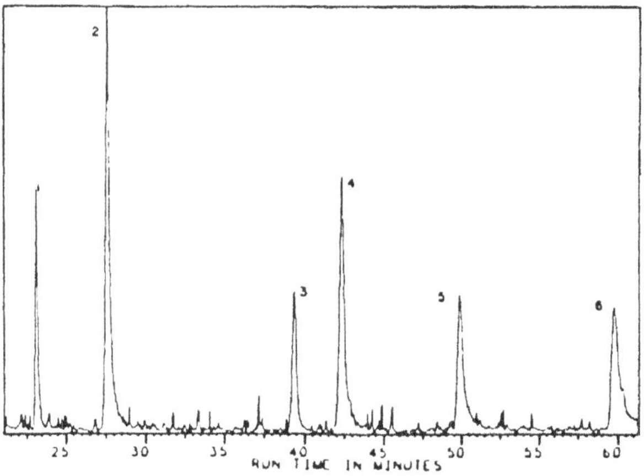


Figure 7. Segment of reconstructed SFC/FTMS chromatogram (150-230 u integration window) from a six-component substituted aromatic hydrocarbon mixture of: 1) *o* - dichlorobenzene, 2) naphthalene, 3) 1 - bromonaphthalene, 4) fluorene, 5) anthracene, 6) pyrene. Reprinted from reference [97] with permission.

organic ions and nothing inherent in the FTMS method.

In 1988, Ijames and Wilkins [33] reported the first successful demonstration of high resolution LD/FTMS of high mass organic ions. In this work, the authors used a laser to desorb molecular ions of several polyethylene glycol polymers of mass over 9000 u; resolution of 60,000 for *m/z* 5922 were achieved. Thus, the problem does not lie with the organic nature of the sample ions.

Gross and coworkers [74, 81] have used a method of scaling to study the high mass problems in FTMS, which they suggest are due to cell geometries, trapping potentials, and electric field inhomogeneities. Although their work does indeed show that the problems for high mass detection are related to electric field effects in the trap, the use of compensated traps have showed some favorable results in solving the problems. Improvements in cell design and control of the ion cloud shape appear to offer hope for improving FTMS for measuring the molecular weights of large biomolecules.

The other need for biomolecular mass spectrometry is instrumentation for determining structure. Therefore, studies involving tandem mass spectrometry of large biomolecules are also of importance. Some progress has also been made here as is evident in the number of published studies. These studies show how the use of several activation methods can be used to induce structurally informative fragmentation. A few of these methods are collisional activation [24, 100, 101], chemical ionization [102], photodissociation [103], and surface induced dissociation [104]. A promising method developmentis found in the determination of carbohydrates by Comisarow et al. [105] and the direct LD of peptides from agarose gels following electrophoretic separation [106]. Hunt and coworkers [104] have also made some excellent progress in their work on protein sequencing. One strategy that has had high effectiveness in their hands has been the use of FTMS to measure the molecular weights of small proteins and smaller polypeptides. Peptide sequencing is then accomplished by using triple stage quadrupoles (see chapter by Wysocki). In an attempt to use FTMS for MS/MS, photodissociation was successfully implemented. The problem is the fragmentation induced by photoactivation is too extensive.

These accomplishments point to good potential for FTMS in molecular weight and structural studies. Although the number of applications are small thus far, the capabilities of ion trapping and multichannel detection should not be overlooked. More instrument and method development is still needed.

## 7. Future

The future challenge for mass spectrometry is the determination of biomolecules. This challenge subdivides into molecular weight measurement at high mass and structural determination at lower masses (up to 3000 u). FTMS, more than any other method, should be able to provide molecular weights at unit resolution or better for ions of *m/z* 30,000 or lower. Fulfillment of this prediction requires that the problems of poor mass resolving power at *m/z* 2000 or greater be solved. Improvements in cell design and ion cloud control offer promise.

Structural studies require that ions be submitted to either high energy collisional activation (CA) or multi-collision low energy CA or photodissociation. Different cell designs are needed if we are to count on the first possibility. The capability to trap ions in the rf-only mode (i.e., The ability to operate the FT trap as a quadrupole ion trap) opens the door to multi-collisional activation as well as ion cloud shape manipulation. Photodissociation will also require better ion cloud control. It is noted that ion clouds are poorly shaped especially when ions are formed in the cell by LD or injected from external sources. High resolution, high sensitivity product ion analysis is the capability that motivates further developments in the structural area.

The ability to provide a routine, user-friendly instrument is a factor that will clearly affect the race to develop better mass spectrometers. Although FT mass spectrometers are relatively simple to build and operate, their expense is largely determined by costs of computers, electronics, and superconducting magnets. The future is auspicious in these areas because, if the The past is a reliable guide, capabilities in these three areas will continue to increase as costs decrease.

## 8. Acknowledgement

This review was prepared with support of the U.S. National Science Foundation (Grant CHE - 9017250). We acknowledge helpful discussion with D. L. Rempel.

## 9. References

1. Comisarow, M. B., Marshall, A. G. (1974) *Chem. Phys. Lett.* **25**, 282.
2. Comisarow, M. B., Marshall, A. G. (1974) *Chem. Phys. Lett.* **26**, 489.
3. Comisarow, M. B., Marshall, A. G. (1975) *J. Chem. Phys.* **62**, 293.
4. Comisarow, M. B., Marshall, A. G. (1976) *J. Chem. Phys.* **64**, 110.
5. Marshall, A. G., Comisarow, M. B. (1979) *J. Chem. Phys.* **71**, 4434.
6. Alleman, M., Kellerhals, H., Wanczek, K. P. (1983) *Int. J. Mass Spectrom. Ion Proc.* **46**, 139.

7. Bamberg, M., Allemann, M., Wanczek, K. P. (1987) Proceedings of the 35th ASMS Conference on Mass Spectrometry and Allied Topics, Denver, Colorado, p. 1116.
8. Henry, K. D., McLafferty, F. W. (1990) *Org. Mass Spectrom.* **25**, 490.
9. Marshall, A. G., Grosshans, P. B. (1991) *Anal. Chem.* **63**, 215A.
10. Wanczek, K. P. (1989) *Int. J. Mass Spectrom. Ion Proc.* **95**,1.
11. Laude, D. A.,Johlman, C. L., Brown, R. S., Weil, D. A., Wilkins, C. L. (1986) *Mass Spectrom. Reviews* **5**, 107.
12. Wilkins, C. L., Chowdhury, A. K., Nuwaysir, L. M., Coates, M. L. (1989) *Mass Spectrom. Reviews* **8**, 67.
13. Marshall, A. G., Verdun F. R. (1990) Fourier Transforms in NMR, Optical, and Mass Spectrometry: A User's Handbook, Elsevier, Amsterdam.
14. Gross, M. L., Rempel, D. L. (1984) *Science* **226**, 260.
15. Nicolet Operation Manual, (1985) Madison, Wisconsin.
16. Nicolet Analytical Instruments. "Nicolet Analytical Instrument Guide"; Madison, WI, 1985.
17. Cody, R. B., Kinsinger, J. A. (1987) in "Fourier Transform Mass Spectrometry", Buchanan, M.V., Ed.; American Chemical Society Symposium Series, No. 359: Washington, D.C. p. 59.
18. Cody, R. B., Kinsinger, J. A., Ghaderi, S., Amster, I. J., McLafferty, F. W., Brown, C. E. (1985) *Anal. Chim. Acta* **178**, 43.
19. Kofel, P., Allemann, M., Kellerhals, Hp., Wanczek, K. P. (1985) *Int. J. Mass Spec. Ion Proc.* **65**, 97.
20. McIver, R.T., Hunter, R.L., Story, M.S., Syka, J., Labunsky, M. (1983) Proceedings of the 31st ASMS Conference on Mass Spectrometry and Allied Topics, Boston, Massachusetts, p. 44.
21. Hunt, D. F., Shabanowitz, J., McIver, R. T., Hunter, R. L., Syka, J. E. P. (1985) *Anal. Chem.* **57**, 768.
22. Williams, E. R., McLafferty, F. W. (1990) *J. Am. Soc. Mass Spectrom.* **1**, 361.
23. Zimmerman, J. A., Watson, C. H., Eyler, J. R. (1991) *Anal. Chem.* **63**, 361.
24. Williams, E. R., McLafferty, F. W. (1990) *J. Am. Soc. Mass Spectrom.* **1**, 427.
25. Ijames, C. F., Wilkins, C. L. (1990) *J. Am. Soc. Mass Spectrom.* **1**, 208.
26. See Chapter In This book by Wysocki, V. H.
27. Ledford, E. B., White, R. L., Ghaderi, S., Wilkins, C. L., Gross, M. L. (1980) *Anal. Chem.* **52**, 2450.
28. White, R. L., Wilkins, C. L. (1982) *Anal. Chem.* **54**, 2443.
29. Sack, T. M., Gross, M. L. (1983) *Anal. Chem.* **55**, 2419.
30. Nibbering, N. M. M. (1990) *Acc. Chem. Res.* **23**, 279.
31. Hillenkamp, F., Karas, M. (1989) Proceedings of the 37th ASMS Conference on Mass Spectrometry and Allied Topics, Miami Beach, Florida, p. 1168.
32. Loo, J. A., Udseth, H. R., Smith, R. D. (1989) *Anal. Biochem.* **179**, 404.
33. Ijames, C. F., Wilkins, C. L. (1988) *J. Am. Chem. Soc.* **110**, 2687.
34. Hunt, D. F., Shabanowitz, J., Yates, J. R., Zhu, N., Russell, D. H., Castro, M. E. (1987) *Proc. Natl. Acad. Sci. U.S.A.* **84**, 620.
35. McIver, R. T., Hunter, R. L., Baykut, G., (1989) *Anal. Chem.* **61**, 489.
36. Amster, I. J., McLafferty, F. W., Castro, M. E., Russell, D. H., Cody, R. B., Ghaderi, S. (1986) *Anal. Chem.* **58**, 485.
37. Lebrilla, C. B. Wang, D. T.-S., Hunter, R. L., McIver, R. T. (1990) *Anal Chem.* **62**, 878.
38. Henry, K. D., Williams, E. R., Wang, B. H., McLafferty, F. W., Shabanowitz, J., Hunt, D. F. (1989) *Proc. Natl. Acad. Sci. U.S.A.* **86**, 9075.

39. Hill, J. A., Martin, S. A., Biller, J. E., Biemann, K. (1988) *Biomed. Environ. Mass Spectrom.* **17**, 147.
40. Rollins, K., Scrivens, J. H., Jennings, R. C. K., Morden, W. E., Welby, J. K., Bateman, R. H. (1990) *Rapid Comm. Mass Spectrom.* **4**, 454.
41. Kaur, S., Hollander, D., Hass, R., Burlingame, A. L.(1990) ACS Sym. Ser. 420 (Liq. Chromatogr./Mass Spectrom.), p. 270.
42. Burlingame, A. L. (1988) Proceedings of the 36th ASMS Conference on Mass Spectrometry and Allied Topics, San Francisco, California, p. 727.
43. White, R. L., Ledford E. B., Ghaderi, S., Wilkins, C. L. Gross, M. L. (1980) *Anal. Chem.* **52**, 1525.
44. Comisarow, M. B. (1982) In K. P. Wanczek (ed.), Ion Cyclotron Resonance Spectrometry, Springer-Verlag Press, Berlin.
45. Williams, E. R., Henry, K. D., McLafferty, F. W. (1990) *J. Am. Chem. Soc.* **112**, 6157.
46. Kleingeld, J. C. (1984) *Thesis*, Univ. Of Amsterdam.
47. Allemann, M., Kellerhals, Hp., Wanczek, K. P. (1980) *Chem. Phys. Lett.* **75**, 328.
48. Carlin, T. J., Freiser, B. S. (1983) *Anal. Chem.,* **55**, 571.
49. Castro, M. E, Russell, D. H. (1984) *Anal. Chem.,* **56**, 578.
50. McCrery, D. A., Gross, M. L. (1985) *Anal. Chem. Acta,* ***178***, 91.
51. Tabet, J., Rapin, J., Poretti, M., Gaumann, T. (1986) *Chemia,* **40**, 169.
52. Weller, R. R., Viswanadham, S. K., Sheetz, M. A., Giam, C. S., Hercules, D. M. (1986) Proceedings of the 34th ASMS Conference on Mass Spectrometry and Allied Topics, Cincinnati, Ohio, p. 160.
53. Ghaderi, S., Littlejohn, D. P., (1985) Proceedings of the 33rd ASMS Conference on Mass Spectrometry and Allied Topics, San Diego, California, p. 727.
54. Henry, K. D., Quinn, J. P., McLafferty, F. W., (1991) *J. Am. Chem. Soc.* **113**, 5447.
55. Jeffries, J. B., Barlow, S. E., Dunn, G.H. (!983) *Int. J. Mass Spectro. Ion Process* **54**, 169.
56. Hunter, R. L., Sherman, M. G., McIver, R.T. (1983) *Int. J. Mass Spectrom. Ion Phys.* **50**, 259.
57. Lee, S.-H., Wanczek, K. P., Hartmann, H. (1980) *Adv. Mass Spectrom.* **88**, 1645.
58. Rempel, D. L., Ledford, E. B., Huang, S. K., Gross, M. L. (1987) *Anal. Chem.* **59**, 2527.
59. Rempel, D.L. (1987) Proceedings of 35th ASMS Conference on Mass Spectrometry and Allied Topics, Denver, Colorado, p. 1124.
60. Wang, M., Marshall, A. G. (1989) *Anal. Chem.* **61**, 1288.
61. Beu, S. C., Laude, D. A. (1991) Proceedings of the 39th ASMS Conference on Mass Spectrometry and Allied Topics, Nashville, Tennessee.
62. Anders, R. L., Beauchamp, J. L., Dunbar, R. C., Baldeschwieler, J. D., (1966) *J. Chem. Phys.,* **45**, 1062.
63. Rempel, D. L., Gross, M. L., (1991) Proceedings of the 39th ASMS Conference on Mass Spectrometry and Allied Topics, Nashville, Tennessee.
64. March. R. E., Hughes, R. J. (1989) Quadrupole Storage Mass Spectrometry, John Wiley & Sons, New York.
65. Comisarow, M. B. (1978) *J. Chem. Phys.* **69**, 4097.
66. Marshal, A. G., Roe, D. C. (1980) *J. Chem. Phys.* **73**, 1581.
67. Huang, S. K., Rempel, D. L., Gross, M. L. (1986) *Int. J. Mass Spectrom. Ion Processes,* **72**, 15.
68. McIver Jr. R. T., Hunter, R. L., Baykut, G. (1989) *Anal. Chem.* **61**,489.
69. Marshall, A. G., Wang, T.-C. L., Ricca, T. L. (1985) *J. Am. Chem. Soc.* **107**, 7893.
70. Verdun, F. R., Ricca, T. L., Marshall, A. G., Mullen, S. L. (1989) in P. Longevialle (ed.), Advances in Mass Spectrometry, Heydon and Son Press, London, **11A**, p. 670.

71. Verndun, F. R., Mullen, S. L., Ricca, T. L., Marshall, A. G. (1988) Advances In Mass Spectrometry, Vol. 11A, P. 670.
72. Wachter, E. A., Farrar, T. C., Kontney, M. J. (1991) *Int. J. Mass Spectrom. Ion Proc.* *103*, 169.
73. Johlman, C. L., White, R. L., Wilkins, C. L. (1983) *Mass Spec. Rev.* **2**, 389.
74. Rempel, D. L., Grese, R. P., Gross, M. L. (1990) *Int. J. Mass Spectrom. Ion Processes.* **100**, 381.
75. Huang, S. K., Rempel, D. L. (1987) Proceedings of the 35th ASMS Conference on Mass Spectrometry and Allied Topics, Denver, Colorado, p. 167.
76. Wang, T-C. L., Ricca, T. L., Marshall, A. G. (1986) *Anal. Chem.* **58**, 2935.
77. Laude Jr. D. A., Beu, S. C. (1989) *Anal. Chem.* **61**, 2422
78. Hogan, J. D., Laude Jr. D. A. (1990) *J. Am. Soc. Mass Spectrom* **1**, 431.
79. Ledford, E. B., Rempel, D. L., Gross, M. L. (1984) *Anal. Chem.* **56**, 2744
80. Wang, M., Marshal, A. G. (1988) *Int. J. Mass Spectrom. Ion Processes* **86**, 31.
81. Holliman, C., Rempel, D.L., Gross, M.L. *J. Am. Soc. Mass Spectrom.* submitted.
82. White, R.L., Onyuruyka, E. C., Wilkins, C. L. (1983) *Anal. Chem.* **55**, 339.
83. Geribaldi, S., Decouzon, M., Gal, J.-F., Rouillard, M., Sturla, J.-M. (1990) *Org. Mass Spectrom.* **25**, 312.
84. Lubman, D. M. (1988) *Mass Spectrom. Rev.* **7**, 535-54,559-92.
85. Chin, S. L. (1971) *Phys. Rev. A* **4**, 992.
86. Berezhetskaya, N. K., Voronov, G. S., Delone, G. A., Delone, N. B., Piskova, G. K. (1970) *Soviet Phys. JETP* **31**, 403.
87. Irion, M. P., Bowers, W. D. Hunter, R. L., Rowland, F. S., McIver, R. T. (1982) *Chem. Phys. Lett.* **93**, 375.
88. McCrery, D. A., Ledford, E. B. Gross, M. L. (1982) *Anal. Chem.* **54**, 1437.
89. Nuwaysir, L. M., Wilkins, C. L., Simonsick, W. J. (1990) *J. Am. Soc. Mass Spectrom.* **1**, 66.
90. Greenwood, P. F., Strachan, M. G., Willett, G. D., Wilson, M. A. (1990) *Org. Mass Spectrom.* **25**, 353.
91. Wilkins, C. L., Weil, D. A. Yang, C. L. C., Ijames, C. F. (1985) *Anal. Chem.* **57**, 520.
92. Karas, M., Hillenkamp, F. (1988) *Anal. Chem.* **60**, 2299.
93. Hettich, R. L., Buchanan, M. V. (1991) *J. Am. Soc. Mass Spectrom.* **2**, 22.
94. Jacoby, C. B., Rempel, D. L., Gross, M. L. (1990) Proceedings of the 38th ASMS Conference on Mass Spectrometry and Allied Topics, Tucson, Arizona. p. 840.
95. Jacoby, C. B., Rempel. D. L., Gross, M. L. (1991) Proceedings of the 39th ASMS Conference on Mass Spectrometry and Allied Topics, Nashville, Tennessee.
96. Lee, E. D., Henion, J. D., Cody, R. B., Kinsinger, J. A. (1987) *Anal. Chem.* **59**, 1309.
97. Laude, E. D., Pentoney, S. L. Griffiths, P. R., Wilkins, C. L. (1987) *Anal. Chem.* **59**, 2283.
98. Moini, M. (1991) Proceedings of the 39th ASMS Conference on Mass SPectrometry and Allied Topics, Nashville, Tennessee.
99. Fung, E. T., Wilkins, C. L. (1988) *Biomed. And Environ. Mass Spectrom.* **15**, 609.
100. Yang, L.-C., Wilkins, C. L. (1989) *Org. Mass Spectrom.* **24**, 409.
101. Sheil, M. M., Guilhaus, M., Derrick, P. J. (1990) *Org. Mass Spectrom.* **25**, 671.
102. Speir, J. P., Gorman, G. S., Cornett, D. S. Amster, I. J. (1991) *Anal. Chem.* **63**, 65.
103. Williams, E. R., Furlong, J. J. P., McLaffery, F. W. (1990) *J. Am. Soc. Mass Spectrom.* **1**, 288.
104. Williams, E. R., Henry, K. D., McLafferty, F. W., Shabanowitz, J., Hunt, D. F. (1990) *J. Am. Soc. Mass Spectrom.* **1**, 413.
105. Lam, Z., Comisarow, M. B.,Dutton, G. S., Parolis, H., Parolis, L. A. S., Bjarnason, A., Weil, D. A. (1990) *Anal. Chim. Acta* **241**, 187.
106. Dunphy, J. C., Busch, K. L., Hettich, R. L., Buchanan, M. V. (1991) Proceedings of the 39th ASMS Conference on Mass Spectrometry and Allied Topics, Nashville, Tennessee.

# TIME–OF–FLIGHT MASS SPECTROMETERS

H. WOLLNIK, U. GRÜNER AND G. LI

*II. Physikalisches Institut der Universität Gießen*

*6300 Gießen (Germany)*

ABSTRACT. Time–of–flight mass spectrometers are discussed in view of classical scanning mass spectrometers. Emphasis here is placed on systems that use grid–free ion reflectors. Ion detectors and ion sources connected to time–of–flight mass spectrometers are also addressed as well as some applications and recently achieved results.

## 1. Introduction

Today many mass analyzing instruments rely on the magnetic deflection of widely monoenergetic ion beams. Because ions are always produced with a certain spread of energies, a very important step was the development of mass analyzers that combined momentum–dispersive magnetic fields with energy–dispersive electrostatic fields [1, 2] yielding purely mass–dispersive systems. By using properly designed magnetic and electrostatic sector fields – often combined with additional quadrupoles or rotationally symmetric lenses – such instruments can focus initially divergent ion beams to small spots even if the energies of the ions deviate a little from each other [3].

At first all mass analyzers were used as mass spectrographs in which all initially produced ions were recorded on a photographic plate with ion positions characterizing ion masses. Because of the complexity in handling photographic plates and evaluating the different positions and intensities, the use of single point, electrical recording instruments became favored. In those cases, the magnetic field is scanned, and the variation in time of the intensity behind a narrow exit slit is recorded electronically [4]. This method considerably simplifies the procedure of obtaining a mass spectrum. This was achieved, however, at the expense of sensitivity because at each moment only a small fraction of the produced ions is recorded. In a mass analyzer of mass resolving power $R = m/\Delta m$, this fraction is approximately $a/R$, where $a$ is the ratio between the width of the exit slit and the width of the mass line under consideration. This ratio is typically $a \approx 1/3$ or $a \approx 1/5$, so that commonly less than 0.01 % or 0.1 % of all initially formed ions can be finally recorded. This loss of sensitivity is a distinct disadvantage of mass spectrometers as compared to mass spectrographs in which, at least in principle, all initially formed ions are recorded finally. Time–of–flight mass spectrometers have been used for approximately 40 years [5]. In such instruments, one starts all the ions simultaneously and records their arrival times at some ion detector after a flight path which in the simplest case is a field–free region.



*M. L. Gross (ed.), Mass Spectrometry in the Biological Sciences: A Tutorial*, 117–131.

The mass resolving power $R$ of such instruments is proportional to the length $l$ of this field-free region and to the square root of the ion mass $m$ (see equation (2)), provided all ions have the same energy and the length of the initial ion packet is mass independent. In most cases, time-of-flight mass spectrometers were developed for special applications, and generally the achieved mass resolving powers were limited. In most cases, however, all initially produced ions were finally recorded. At least in principle, the sensitivity of longitudinally dispersive time-of-flight mass spectrometers is comparable to that of laterally dispersive mass spectrographs. Because for time-of-flight mass spectrometers, the form of

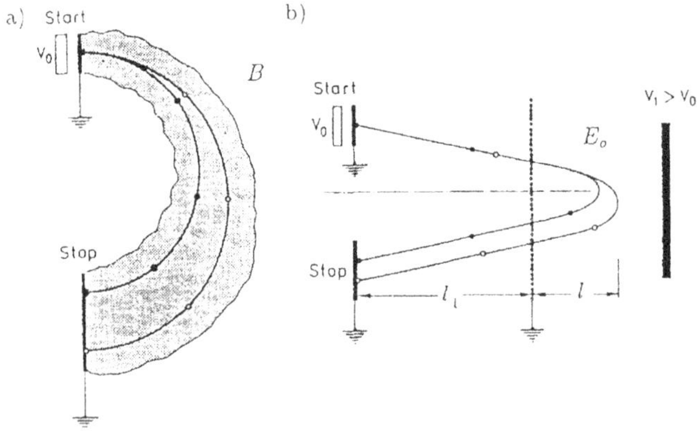


Figure 1: There are at least two well known ways to achieve equal arrival times for ions of equal mass-to-charge ratios but different velocities [14]:
1. by use of a homogeneous magnetic field,
2. by use of a deflecting homogeneous electrostatic field combined with a field-free flight region of specified length $l_1$ [6].

an initially produced ion pulse directly influences the achievable mass resolving power, it is of utmost importance for such systems that the ion source be carefully matched to the time-of-flight system. Here it is important to ensure that

1. ions of equal masses but different energies – and thus different velocities – reach the ion collector at the same time (see Figure 1) [6, 7, 8],

2. either all ions are produced continuously and stored for some time until they are released simultaneously [9], or

3. all ions are formed more or less simultaneously. Here within a few nanoseconds, as many ions must be produced as would be formed in a DC-ion source over the time between two such ion pulses [10, 11, 12, 13].

## 2. Time-of-Flight Mass Spectrometers with Energy-, Angle- and Position-independent Mass Resolving Powers

Usually, one describes an arbitrary ion of mass $m$, kinetic energy $K$ and charge $q$ relative to a reference ion of mass $m_o$, kinetic energy $K_o$ and charge $q_o$ as:

$$\frac{K}{q} = \frac{K_o}{q_o}(1+\delta_K), \qquad \frac{m}{q} = \frac{m_o}{q_o}(1+\delta_m). \tag{1}$$

Neglecting relativistic effects these two ions have velocities of $v_o = \sqrt{2K_o/m_o}$ and $v = \sqrt{2K/m}$. To move through a field-free region of length $l_1$, the reference ion needs the time $T_o = l_1\sqrt{m_o/2K_o}$ and the arbitrary ion the time $T = T_o(1+\delta_t) = l_1\sqrt{m/2K}$. Consequently, ions of equal masses but different kinetic energies have different flight times. A classical way to avoid the energy dependence in ion-flight times is to send the ions passing along half circles in a magnetic field (see Figure 1a) because here the more energetic ions move along correspondingly larger circles. A similar independency of flight times for ions of somewhat different kinetic energies can be achieved in more complex sector field systems in which other advantages can be obtained at the same time [7, 8]. Another effective method to make the ion-flight times independent of energy differences is to use a homogeneous electrostatic retarding field [6] of strength $E_o$ behind a field-free region of length $l_1$ (see Figure 1b). Into this field ions of energies $K_o$ and $K$ will penetrate to a depth $l_o = K_o/E_o$ and $l = K/E_o$, respectively, before they turn around. The two ions under consideration in equation (1) then have moved over distances $2(l_1 + l_0)$ or $2(l_1 + l)$ with flight times of $T_{o1} = 2(l_1 + 2l_o)/v_o$ or $T_1 = 2(l_1 + 2l)/v$ where the factors 2 of $l_o$ and $l$ result from the linearly decreasing ion energies in the electrostatic field. For an arbitrary ion of mass $m$ and energy $K = K_o(1+\delta_K)$, one calculates the flight time T, as

$$\begin{aligned} T &= \frac{2\sqrt{m/m_o}}{v_o\sqrt{(1+\delta_K)}}\left[l_1 + 2\frac{K_o}{E_o}(1+\delta_K)\right] \\ &\approx \frac{2}{v_o}\sqrt{m/m_o}\left[(l_1 + 2\frac{K_o}{E_o}) - (l_1 - 2\frac{K_o}{E_o})\frac{\delta_K}{2} + (3l_1 - \frac{2K_o}{E_o})\frac{\delta_K^2}{8}\cdots\right]. \end{aligned} \tag{2}$$

By choosing $l_1 = 2K_o/E_o = 2l_o$ in equation (2), the linear term in $\delta_K$ vanishes so that to first order the ion flight times are independent [6] of energy deviations. Alternatively choosing $l_1 = 2K_o/(3E_o) = 2l_o/3$ in equation (2) causes the quadratic term in $\delta_K$ to vanish. Thus, by using such a retarding field under no circumstances can the ion flight times become independent of energy deviations to first and second order simultaneously. This can be achieved, however, if two retarding fields of different field strengths are placed in series [15, 16] or if specially designed grid-free retarding fields are used [9, 14, 17, 18, 19, 20]. Because of the equipotential lines partially reaching through the meshes of the grid, the limits of fields separated by a grid are not flat but rather "bumpy" as shown in Figure 2. This causes the same effect as an array of small lenses placed side by side, which forbids a clear image formation for any particle beam at the end. Thus, it seems reasonable to form the retarding field by a number of grid-free diaphragms (see Figure 3) at different electrostatic potentials. In this case, the ion trajectories of a beam can all be described by the same potential dis-

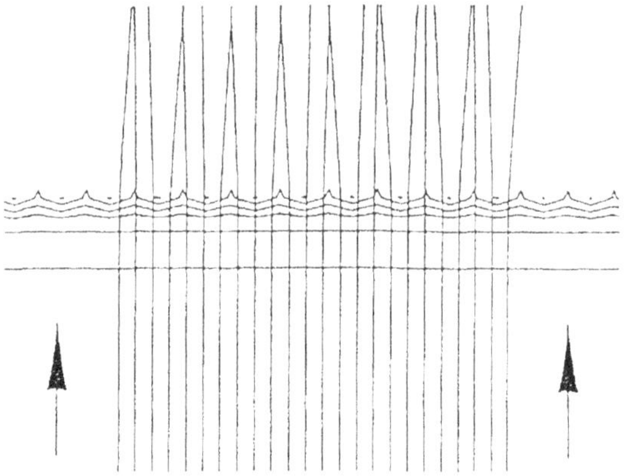

Figure 2: Calculated "diffraction" of an initially parallel beam passing through a grid that separates regions of different field strengths.

tribution expanded in a power series [19, 20] similarly as in sector fields or quadrupole lenses [3, 7, 8].

In all these cases one can describe the trajectory of an arbitrary ion of mass $m$, charge $q$ and energy $K$ or momentum $p$ relative to the trajectory of a reference ion of the same mass and charge and the energy $K_o$ or momentum $p_o$, respectively. In detail it suffices to describe at any position $z$ the distances $x(z)$, $y(z)$, the momentum components $p_x(z)$, $p_y(z)$, as well as the flight time deviations $T - T_o = \Delta T(z) = T_o\delta_t$. Here it is assumed that the z-axis is the generally curved trajectory of a specific reference ion and that the reference ion has used the flight time $T_o$ to reach the position $z$. Very generally then, the quantities $x_1 = x(z_1)$, $a_1 = p_x(z_1)/p_o$, $y_1 = y(z_1)$, $b_1 = p_y(z_1)/p_o$ at the position $z = z_1$ of the ion collector are found [3] to first order from the corresponding quantities: $x_o = x(z_o)$, $a_o = p_x(z_o)/p_o$, $y_o = y(z_o)$, $b_o = p_y(z_o)/p_o$ at the ion start position $z = z_o$:

$$\begin{pmatrix} x_1 \\ a_1 \\ \delta_{t1} \\ y_1 \\ b_1 \\ \delta_{K1} \end{pmatrix} = \begin{pmatrix} (x|x) & (x|a) & 0 & 0 & 0 & (x|\delta_K) \\ (a|x) & (a|a) & 0 & 0 & 0 & (a|\delta_K) \\ (\delta_t|x) & (\delta_t|a) & 1 & 0 & 0 & (\delta_t|\delta_K) \\ 0 & 0 & 0 & (y|y) & (y|b) & 0 \\ 0 & 0 & 0 & (b|y) & (b|b) & 0 \\ 0 & 0 & 0 & 0 & 0 & 1 \end{pmatrix} \begin{pmatrix} x_0 \\ a_0 \\ 0 \\ y_0 \\ b_0 \\ \delta_K \end{pmatrix} \tag{3}$$

It should be mentioned also that the ion collector is assumed here to be at the same potential as the ion start position. Because all ion trajectories must be solutions of the canonical equations of motion, one finds to first order additionally (see for example [3]):

$$(x|x)(a|a) - (x|a)(a|x) = 1 \tag{4a}$$

$$(y|y)(b|b) - (y|b)(b|y) = 1 \tag{4b}$$

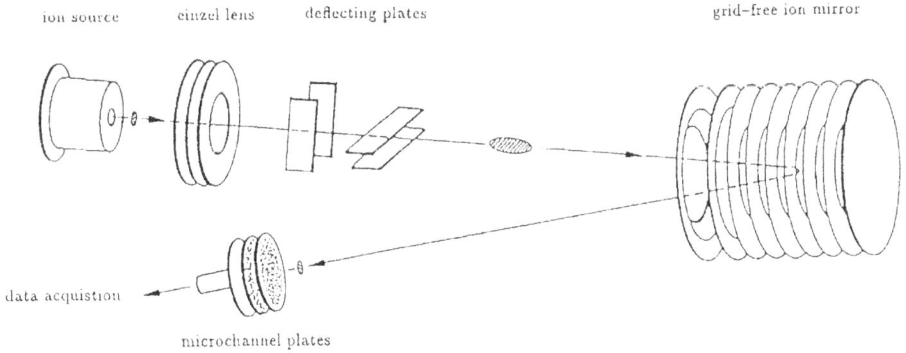


Figure 3: Schematic of the ion–optical system of a time–of–flight mass spectrometer. Ideally neither the ion deflectors nor the einzel lens as shown are necessary. These elements are mainly used for adjustment purposes. Indicated is also a packet of ions of one mass at four positions in the system:

1. at the takeover point downstream from the ion source, where a longitudinally and laterally small ion packet exists;
2. before the ion–reflector, where the ion packet diverges laterally and longitudinally;
3. downstream from the ion–reflector, where the ion packet is laterally and longitudinally convergent after a lateral intermediate focus in the reflector (see also Figure 4);
4. shortly before the ion detector, where the ion packet is longitudinally as well as laterally quite small.

as well as in particular [3, 8]:

$$(\delta_t|x) = (p_o/K_o)\left[(a|x)(x|\delta_K) - (x|x)(a|\delta_K)\right] \tag{4c}$$

$$(\delta_t|a) = (p_o/K_o)\left[(a|a)(x|\delta_K) - (x|a)(a|\delta_K)\right]. \tag{4d}$$

The equations (4c, 4d) show that an optimization of the longitudinal coefficients $(\delta_t|\cdots)$ is only possible by a proper modification of the lateral coefficients $(x|\cdots), (a|\cdots)$. Furthermore it follows from equations(4a,4b) that the first–order coefficients of equation (3) are not independent (i.e., additional cross relations must be fulfilled). Noted here that in a fully symmetric system that has an intermediate $x$–image (see Figure 4) one always finds: $(x|a) = (x|\delta_K) = (\delta_t|x) = 0$ and that if one makes $(a|\delta_K) = 0$ additionally one finds also $(\delta_t|a) = 0$ [see equations (4c, 4d)]: Note also that in an only almost symmetric system (as shown in Figure 3) these relations are fulfilled only approximately. By properly forming the field distribution, one always can achieve additionally $(\delta_t|\delta_K) = 0$. Consequently to first order, $\delta_t$ is independent of the width of the energy distribution $\delta_K$ and at least approximately of the start position $(x, y)$ as well as of the start angle $(a, b)$. Furthermore, at least in x-dir–

ection, all ions are laterally focused at the end of the system. If one can achieve additionally that also $(y|b)$ vanishes, the ions are focused in x–and y–directions simultaneously to the ion collector. Thus such a system can be used as an ion microprobe.

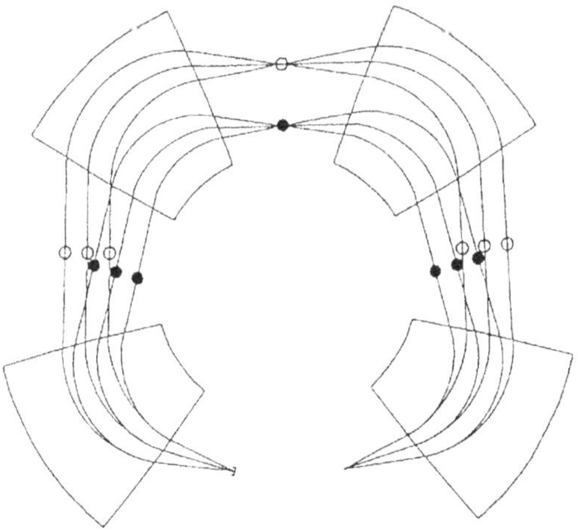

Figure 4: Schematic of trajectories of ions of equal masses and charges but different kinetic energies in (TOFI) a "Time–of–Flight Isochronous Spectrometer"[21]

Similarly, because there are relations between first–order elements (see equations(4)) owing to the fact that the particle trajectories must be solutions of canonical equations, there are relations between second– and first–order elements. Note that these relations determine about half of the second–order coefficients. Analogously there also are relations between third–, second– and first–order elements. In detail, one finds that except for the coefficients of $\delta_K$, $\delta_K^2, \ldots$, all other longitudinal coefficients $(t|\ldots)$ are determined from the lateral ones of equal and lower order [3, 22]. Thus, also due to the higher–order coefficients, an optimization of the longitudinal focusing properties is only possible by a suitable, however, usually complex optimization of the lateral coefficients.

In the case of a time–of–flight mass spectrometer as shown in Figure 3, one for example can optimize the system by properly choosing the distribution of the retarding field in the ion–reflector [18]. To do this, one can trace all ions to that $z$–position $z_P$ at which the ion of lowest energy turns around (see Figure 5). The ion of second–lowest energy reaches this point a little earlier than the ion of lowest energy with the time difference being $\Delta T_1$. At this point $z_P$, one then can modify the field strength such that the ion of second–lowest energy reaches its turn–around point just within the time $\Delta T_1$. At this turn–around point, one then can modify the field strength such that the ion of third–lowest energy, arriving there a time $\Delta T_2$ earlier than the ion of second–lowest energy, reaches its turn–around point just within the time $\Delta T_2$. Going through this procedure for ions of all energies, one has chosen the retarding field strength along the whole z–axis. This potential distribution has the property that for the complete time–of–flight system one finds: $(\delta_t|\delta_K) = (\delta_t|\delta_K^2) = \cdots = 0$.

Note, however, that this potential distribution depends very much on the initially chosen potential distribution up to the point $z_P$. Varying this portion of the potential distribution one can iteratively try to achieve good longitudinal and lateral focusing simultaneously.

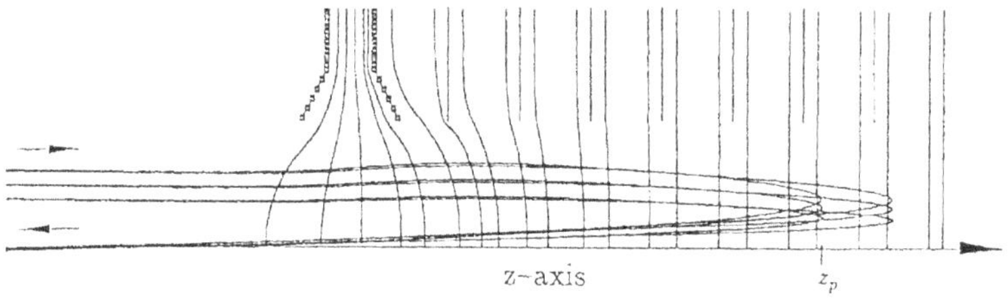


Figure 5: Upper half of a grid-free ion reflector, showing equipotential lines as well as ion trajectories. These trajectories are assumed to enter the system in parallely, starting at three different positions with two different energies.

With growing ion energy, the focal length of an electrostatic round lens increases. On the other hand the focal length of a grid-free ion reflector in most cases decreases for ions of growing energy because here the more energetic ions pass through additional regions of the ion reflector, which have additional refractive powers. By properly distributing the refractive power in the ion reflector, one can achieve a balance between the reduced refractive power of the electrostatic round lens and the increased refractive power of the ion reflector for ions of higher energies so that the overall refractive power of a system as shown in Figure 3 can be made to be widely independent of ion energy deviations. Because in general one also can achieve a reasonable correction of geometrical aberrations, it is possible in most cases to achieve longitudinal and lateral focusing simultaneously for ion bundles of large angular divergence and large energy spread [19, 20].

In Figure 5 an experimental ion reflector is shown together with some ion trajectories. This ion reflector features a slow but distinctively non-linear potential increase in its later part and a quite rapid potential increase at its entrance. For an optimization of the lateral focusing properties the first two ring electrodes additionally were formed to be conical [19]. The properties of this ion reflector are:

1. the total ion flight time is widely independent of the energy deviation ($\delta_K$), the ion start position (x,y) and the ion start angle (a,b),

2. a sufficiently achromatic lateral focusing is achieved by using an additional electrostatic einzel lens,

3. the maximum diameter of the ion bundle stays within limits, thus guaranteeing high transmission.

## 3. Generation of Ion Pulses

### 3.1. PULSED ION PRODUCTION

To generate short ion pulses one commonly fires pulsed laser beams [13], single fission fragments [10] or short pulses of $Cs^+$ or $Ar^+$ ions [11, 21] to a target surface. During this short bombardment time, one then produces secondary ions of surface adsorbants. For this type of ion formation, the mass resolving power in a time–of–flight mass analyzer increases with $\sqrt{m}$ (see equation (2)). For intense laser beams or intense primary ion pulses, however, the mass resolving power increases significantly less with increasing mass and often it even decreases. This can be explained by assuming that during the ionization process, a dense plasma cloud exists into which the acceleration field can only penetrate after the plasma cloud has expanded and thus the plasma density has been diluted sufficiently. Up to this time, however, all ions – independent of their masses – have moved with the same collective plasma velocity. For ions of small masses, this collective velocity is only a small fraction of the finally achieved velocities after acceleration. In contrast, for ions of larger masses, it is a remarkable large fraction of the final ion velocities.

### 3.2. PULSED ION EXTRACTION

Alternatively to a pulsed ion production one can form ions continuously and accelerate them only during short time intervals. In such cases it is most effective to store the produced ions between extractions [9]. To achieve this storage effect, different potentials were applied to three parallel grids or diaphragms $G_{2-1}, G_3, G_4$ (see Figure 6), thus forming an electrostatic potential well. In this case, the ions were formed by continuous electron ionization.

By reducing the potential of the middle grid $G_3$, a potential well is formed and the produced ions are trapped until this potential is raised to expel the ion cloud. This expelling field causes the ions formed in the upstream portion of the storage volume to obtain a little more energy then the ions formed in the downstream portion. Together with a DC-post acceleration, this causes all ions to reach a certain takeover point simultaneously [9]. In detail the ions are extracted during about 1 $\mu sec$ after a 100 – 1000 $\mu sec$ storage time and are bunched to less than 10 $nsec$ at the takeover point as illustrated in Figure 7. This causes a significant increase of the signal–to–noise ratio and allows the takeover point to be the actual start point for the time–of–flight mass analyzer. Characteristic for this ion source is the relatively wide energy distribution of the ions ($\approx 50\ eV$, i.e., $\delta_K \approx \pm 0.05$ if the post acceleration voltage of 500 $V$). This requires the associated time–of–flight mass analyzer to be highly energy achromatic (i.e. the flight times must be quite independent of the ion–energy spread). In other words, all energy dependent aberration coefficients should be eliminated: $(\delta_t|\delta_K)\ \delta_K = (\delta_t|\delta_K^2)\ \delta_K^2 = ... = 0$, which is widely achieved by a good grid–free ion reflector [17, 18].

All ions of equal masses that are at rest at the moment of extraction reach the takeover point at almost exactly the same time. If, however, they have a certain initial thermal energy spread and thus start with different velocities, they arrive at the takeover point with a time spread $\Delta T$. This time spread approximately equals the time an ion needs to reverse its velocity if it had started to move against the suddenly applied extraction field. If such

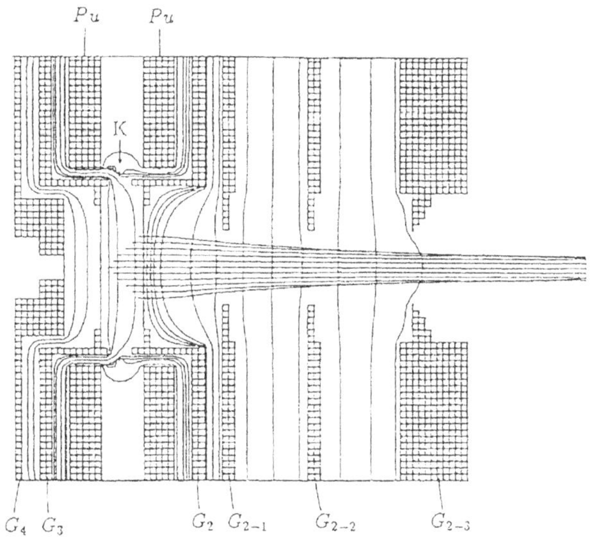


Figure 6: Storage ion source showing the potential distribution during the extraction phase and corresponding ion trajectories. $G_4, G_3, G_2$ = potential determining diaphragms/grids, $K$ = cathode, $Pu$ =electronpusher–electrodes.

an ion of mass $m$ and charge $q$ had an initial thermal energy $K_{th}$, this time $\Delta T$ in an extraction field $E_{ex}$ is:

$$\Delta T = 2\frac{\sqrt{2mK_{th}}}{qE_{ex}}. \tag{5}$$

As one sees from equation (5), this time $\Delta T$ is proportional to $\sqrt{m}$. Because in the following time–of–flight mass analyzer, the flight time T in equation (2) is also proportional to $\sqrt{m}$ (see equation (2)), one finds that the finally achieved mass resolving power is independent of the ion mass. From equation (5) it also can be established that $\Delta T$ decreases with an increasing strength of the extraction field $E_{ex}$. On the other hand, one should note that an increased $E_{ex}$ also increases the left over time–of–flight differences $(\delta_t|\delta_K^n)\ \delta_K^n$ with $n = 1, 2, 3, \cdots$ due to the increased ion–energy spread, which poses larger and larger restrictions on the design of the ion reflector.

## 4. Ion storage

In a storage ion source the ion losses should be small between extractions. It is rather complicated, however, to determine the storage effectiveness (i.e. the ratio of extracted to overall produced ions) because the potential distribution in the source, and thus the ionization efficiency, are very different for the continuous and for the pulsed mode. The ionization efficiency also depends on the residual gas pressure in the source. Experimentally

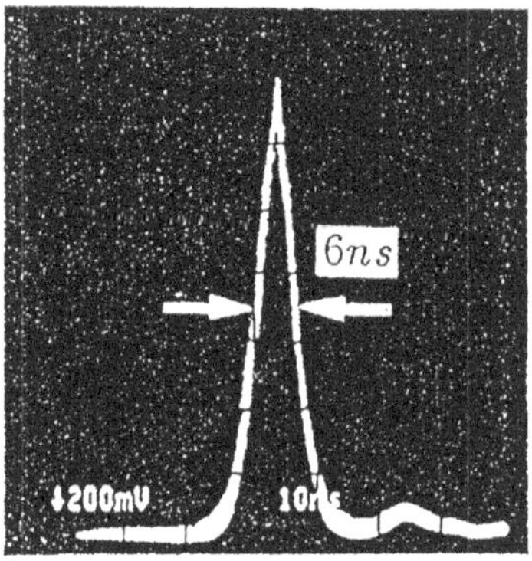


Figure 7: Oscilloscope signal of the $N_2^+$-Peaks [23]; $\Delta t(FWHM) = 6\ ns$.

obtained efficiencies are shown in Table I. As one can see the sensitivity $S$ increases with decreasing residual gas pressure $P$ in the source, where $S\sqrt{P}$ is approximately constant. Consequently such a source is very effective for determining mass spectra of residual gases at low pressures (see also Figure 8).

Because ions of lower masses generally are much more abundant than ions of higher masses, the storage efficiency depends mainly on the number of already stored ions of low masses. To increase the storage efficiency for high mass ions, it is advantageous to enhance the storage losses for low mass ions. Such losses tend to occur on the surface of the storage volume by recombinations of ions with electrons or by simply impinging such ions on electrode surfaces. Because ions of equal energies need times $t \sim 1/v \sim \sqrt{m}$ to cross the storage volume, ions of lower masses reach the outer region of the storage volume more frequently and thus have a higher loss rate as compared to ions of higher masses.

Figure 9 shows the experimentally obtained dependence of the intensities of ions of three masses for different storage times. For the ions of $m = 2\ u$ and $m = 18\ u$ maximal intensities are obtained after storage times of $0.25\ ms$ and $0.5\ ms$. For ions of $m \approx 220\ u$ the optimal storage times are above $10\ ms$. The maxima in the intensity distributions result from the limited storage capacity of the source because there is always an equilibrium between newly formed and lost ions. Longer storage times always result in an enhancement of high mass ions and a reduction of low mass ions, which generally is advantageous for the sensitive detection of molecular ions. For an additional a reduction of low mass ions one also can apply an AC voltage at the middle electrode $G_3$ shown in Figure 6. This periodically reduces the size of the storage volume and thus increases the storage losses for low mass ions without noticeably changing the storage losses for heavy ions [25].

| pressure $[mbar]$ | ion current | sensitivity $[A/mbar]$ |
|---|---|---|
| $2 \times 10^{-5}$ | $\approx 100\ pA$ | $2\ \times 10^{-6}$ |
| $4 \times 10^{-7}$ | $\approx 10\ pA$ | $2\ \times 10^{-5}$ |
| $5 \times 10^{-9}$ | $<\ pA$ | $\approx 10^{-4}$ |

Table I: Dependency of the obtained ion current as function of the residual gas pressure [24]

## 5. Applications

In a time-of-flight mass spectrometer as described in Section 2, equipped with an electron ionization ion source as described in Section 4, mass resolving powers of $m/\Delta m \approx 20,000$ have been achieved (see Figure 10) [17, 18]. Such mass resolving powers are comparable to the ones of sector field instruments. Time-of-flight mass spectrometers, however, have advantages over sector field systems because:

1. their mass range is theoretically unlimited which helps to investigate large molecular ions or cluster ions [12, 24],

2. their sensitivity is high owing to the simultaneous collection of ions of all masses.

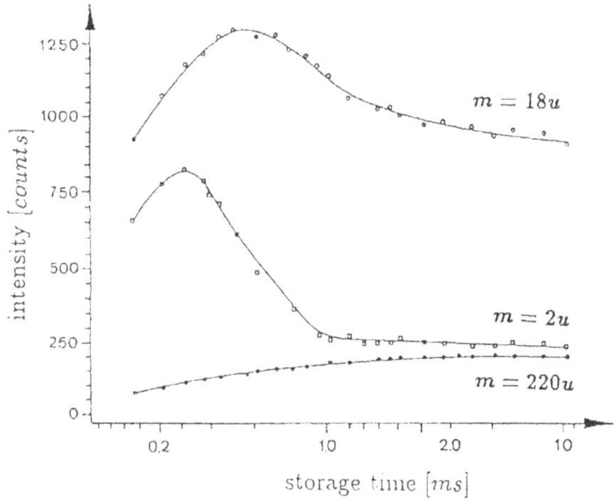


Figure 8: Storage of ions of different masses (2 $u$, 18 $u$, 220 $u$) as function of storage time.

For the detection of ion packets, one commonly uses microchannel plates, which have rise times of several 100 psec for single ions. As for all ion detection systems that rely on the production of secondary electrons, however, the detection efficiency unfortunately decreases with reduced ion velocities, which is detrimental for the detection of large molecular ions. To reduce this defect, we have developed an ion electron converter in which the secondary electrons are guided isochronically over 180° by a magnetic field to a microchannel plate (see Figure 11) [27]. One of the advantages of this method is that the converter surface can be covered by thin layers of CsJ, for example, for which the yield of secondary electrons is especially high [28, 29].

An additional advantage of time-of-flight mass spectrometry is the short time required to record a full mass spectrum. Especially in connection with electron ionization ion sources,

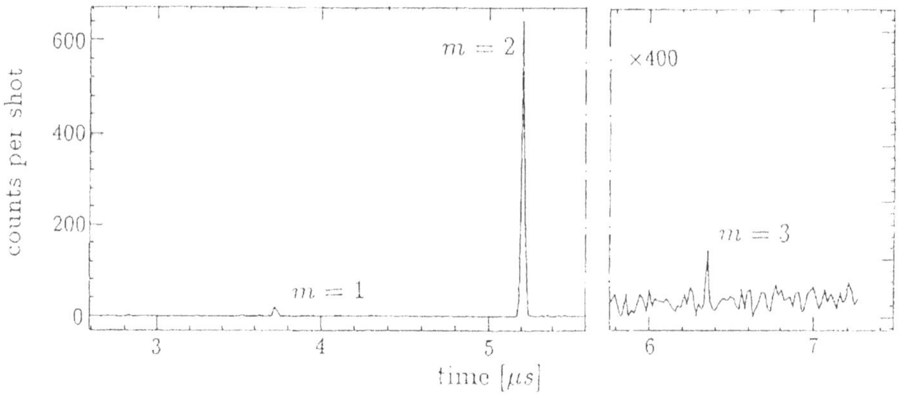


Figure 9: $H^+$–, $H_2^+$– and $HD^+$–abundances in the spectrum of a residual gas at pressures of $2 \times 10^{-9}$ *mbar* [24].

a time–of–flight mass spectrometer is ideally suited as a mass selective detector for a gas chromatograph. Within one second, typically several 1000 mass spectra can be recorded and if desired added to enhance the statistics.

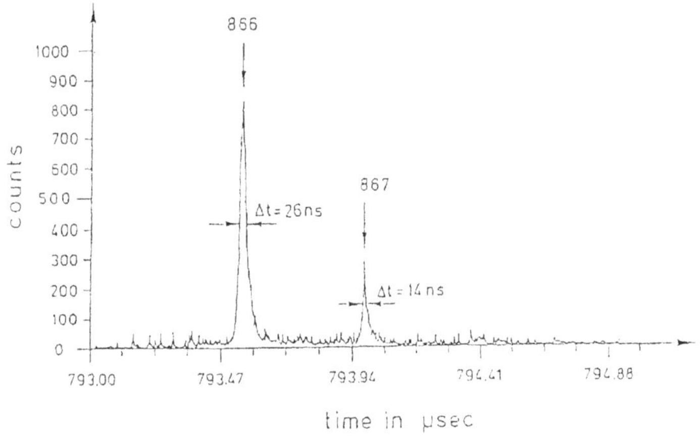


Figure 10: Ions of 866 $m/z$ and 867 $m/z$ of 2,4,6–Tris–(pentadecafluorheptyl)–1,3,5–triazine. The peak widths (FWHM) of 26 *nsec* at 866 $m/z$ and 14 *nsec* at 867 $m/z$ correspond to a mass resolving power of $\approx 20,000$.

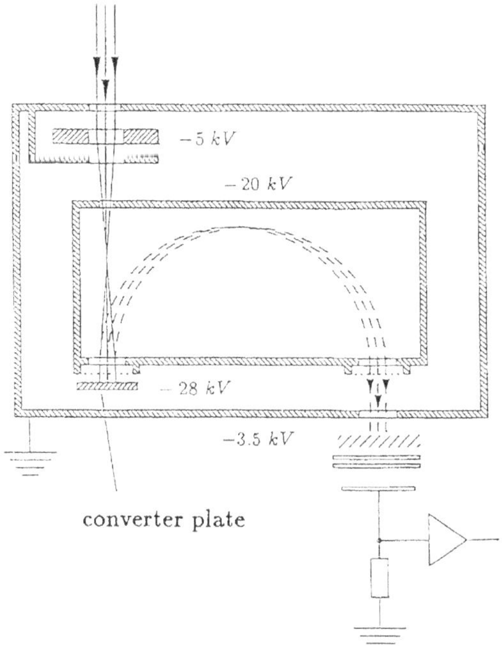


Figure 11: Ion–electron converter that allows both a postacceleration of the ions and a coating of the converter surface with appropriate materials.

Furthermore, it should be noted that the required 100 $\mu sec$ for the recording of one mass spectrum is very short as compared to the GC–peak width. Thus, all effects that are due to the finite scan velocity of quadrupole– or sector–field mass spectrometers [30, 31] are avoided (i.e., mass spectra recorded in the rise or the fall of a GC–peak are identical to those recorded at the maximum of the GC–peak). Thus, no corrections of the recorded abundances of different mass ions are necessary. Monitoring the intensities of ions of several masses as functions of time one finds the GC–spectrum as the sum of all ion signals. While single ion recordings characterize the GC–spectra of a selection of substances, combing such spectra one can mathematically separate unresolved GC peaks effectively. For the time being, GC–peaks in capillary chromatography are approximately one second wide.

In principle, such peak widths could be reduced to less than 10 $ms$ [32, 33] which would reduce the time necessary to record a gas chromatrogram considerably. Scanning mass spectrometers would then only be of limited use for mass analysis. By using a time–of–flight mass spectrometer on the other hand, 30 to 50 mass spectra could still be recorded for each elution component. This would allow new applications of GC–MS, for example, in fast process control systems or applications in which very many substances must be determined.

## 7. References

1. F.W. Aston (1919)*Phil.Mag.*, **38**, 709.

2. J. Mattauch and R. Herzog(1934) *Z. Phys.*, **89** 447, 786

3. H. Wollnik(1987) *Optics of Charged Particles*, Acad.Press, Orlando

4. A.J. Dempster (1935)*Am.Phil.Soc.*, **75** 755, 760

5. A.E. Cameron and D.F.Eggers (1948)*Rev.Sci.Instrum.*, **19**

6. V.I. Karataev, B.A. Mamyrin and D.V. Schmikk (1972) *Sov. Phys.-Tech. Phys.*, **16**, 1177

7. W. Poschenrieder (1972) *Int.J.Mass Spectro.Ion Phys.*, **9**, 357

8. H. Wollnik and T. Matsuo (1981) Int. J. Mass Spectro. Ion Phys., **37**, 209

9. R. Grix, U. Grüner, G. Li, H. Stroh and H. Wollnik (1989) *Int. J. Mass Spectro. Ion Proc.*, **93**, 323-330

10. R.D. MacFarlane and D.F. Torgerson (1976) *Science*, **191**

11. B.T. Chait and F.H. Field (1985) *Int. J. Mass Spectr. Ion Proc.*, **65**, 169

12. A. Bennighoven et al. (Eds.) (1986) *SIMS V, Springer Series in Chem. Phys.*, **44**

13. M. Karas, U. Bahr and F.Hillenkamp (1989) *Int. J. Mass Spectr. Ion Proc.*, **92**, 231

14. H. Wollnik (1987) *Anal. Instrum.*, **161**, 15

15. B.A. Mamyrin, V.J. Karatajev, D.V. Shmikk and V.A. Zagulin (1973) *Sov. Phys. JETP*, **37**, 45

16. W. Gohl, R. Kutscher, H.-J. Laue and H. Wollnik (1983) *Int. J. Mass Spectr. Ion Phys.*, **48**, 411

17. H.Wollnik, R.Grix, G.Li, R.Kutscher and P.Feigl (1987) Proc. 2nd Joint Japan–China Symp. on Mass Spectrom.; H.Matsuda and X.Liang (Eds.), Bando Press, Osaka, 181

18. R. Grix, R. Kutscher, G. Li, U. Grüner and H. Wollnik (1988) *Rapid Comm. In Mass Spectr.*, **25**, 83

19. R. Kutscher, R. Grix, G. Li and H. Wollnik (1991) *Int. J. Mass Spectr. Ion Proc.*, **103**, 117

20. J.Trötscher, H. Irnich and H. Wollnik (1990) *Nucl. Instr. and Meth.*, **A298**, 480

21. H. Wollnik and M. Berz (1985) *Nucl. Instr. and Meth.*, **321**, 127

22. J.M. Wouters, D.J. Vieira, H. Wollnik, G.W. Butler, R.H. Kraus,Jr. and K. Vaziri (1987) *Nucl. Instr. and Meth.*, **B26**, 286

23. R. Grix (1988) Dissertation, Justus–Liebig–Universität Gießen

24. H. Stroh (1989) Diplomarbeit, Justus–Liebig Universität Gießen

25. G.E. Yefchak, M.A. Puzycki, C.G. Enke, R. Grix, J.F. Holland, G. Li, Y. Wang, H. Wollnik; (1990) 38th ASMS Conference on Mass Spectrometry and applied Topics, Tucson, AZ,

26. T. Lange, T. Bergmann, H. Göhlich, H. Schaber und T. P. Martin (1989) *Physik in unserer Zeit*, **6**, 172

27. A.M. Zebelman, W.G. Meyer, K. Halbach, A.M. Poskanzer, R.G. Sextro, G. Gabor and D.L. Landis (1977) *Nucl. Instr. and Meth.*, **141**, 439

28. S. Dela–Negra, N. Dumaid and Y. Lebeyec (1985) *IPN-Orsay Report*, IPNO-DRE-85-33

29. H.L. Seifert, D.J. Vieira, H. Wollnik and J.M. Wouters (1990) *Nucl. Instr. and Meth.*, **A292**, 533

30. K. Biemann (1962) Mass Spectr. Organic Chemical Appl., McGraw Hill Publ. Co., New York, 217

31. J.E. Biller and K. Biemann (1974) *Analytical Letters*, **7** , 515

32. P.A. Leclerq and C.A. Cramers (1988) *J. High Resol. Chromat.*, **11**, 845

33. R. Grix, R.E. Tecklenburg, J.F. Holland, H. Götz, U. Grüner and H. Wollnik (1990) 2nd Int. Symp. on Appl. Mass Spectr. in Health Sci., Barcelona

# ION DETECTION IN MASS SPECTROMETRY

PAUL W. GENO
*Southwest Research Institute*
*6220 Culebra Rd., P.O. Drawer 28510*
*San Antonio, Texas 78228-0510 USA*

ABSTRACT. A review of the principles of ion detect is discussed. These principles include secondary electron emission, analog and pulse-counting detection, basic and advanced detector designs. In addition, factors that influence the detection of molecular ions of biomolecules, as well as some of the methods that have been employed to improve detection efficiency of these large ions, are described.

## 1. Introduction

The detector is as important as the mass analyzer and ionization source when optimizing the mass range, sensitivity and dynamic range of a mass spectrometer for the analysis of biological molecules. A wide variety of detector designs are currently in use. Most are based on the electron multiplier principle and depend upon the efficient production of secondary electrons upon ion impact onto a surface. Many of the improvements realized in the last few years have involved the enhancement of secondary electron production. Another improvement has been the development of array detectors, which spatially resolve and simultaneously detect ions over a wide mass range. The purpose of this chapter is to familiarize the beginning mass spectrometrist and the non-instrumentalist with the principles of ion detection and factors that influence ion detection efficiency. Although this is by no means an exhaustive review, many of the common ion detectors and their characteristics, as well as approaches that can be used to enhance sensitivity and detection efficiency, will be discussed.

## 2. Secondary Electron Emission

Secondary electron emission occurs when the amount of energy imparted to the surface by collision with an ion is greater than the energy required to eject an electron from the surface. This energy can be from the neutralization of the ion (i.e., potential emission) or from the kinetic energy that is transferred to the surface from the solid during penetration of the ion (kinetic emission). In most cases kinetic emission is the dominant process [1].

Investigations into the process of secondary electron emission by many researchers have determined that the average number of electrons emitted per incident ion–or the secondary electron coefficient ($\gamma$)–increases with the velocity of the incident ion. This relationship is an important one for the detection of large molecules because the velocity is proportional to the inverse of the square root of the mass of the ion when all ions are accelerated to a given kinetic energy. This



*M. L. Gross (ed.), Mass Spectrometry in the Biological Sciences: A Tutorial,* 133–141.

means that the detection efficiency will decrease as the ion mass increases. A typical plot of the variation of the secondary electron coefficient as a function of incident velocity can be divided into three regions [2]. The first region, from zero velocity to a "threshold" value, is where the incident ion velocity is insufficient for electron emission to occur. Beuhler [3] and Sundqvist [4] predicted a velocity threshold of $1.8 \times 10^4$ m/s, which would correspond to a mass limit of 11,910 daltons at 20 kV acceleration. The second or low-energy region is where $\gamma$ is a linear function of ion energy. The third, or high energy region, is where $\gamma$ is a linear function of ion velocity. A variety of models have been proposed to account for this observed behavior. Most include contributions from inner-shell and valence band excitation as well as nuclear stopping power [1].

## 3. Analog vs. Pulse-counting Detection

Data collection in a mass spectrometer can be accomplished in either the analog or pulse-counting mode. Each of these modes has its own desirable detector characteristics [5]. The most widely used mode of detection in mass spectrometry is the analog or discrete acquisition. In this mode, an output current or voltage is measured as a function of time and recorded by an oscilloscope, strip-chart recorder or multichannel analyzer as the mass spectrometer is scanned. Figure 1a shows a typical time spectrum acquired in analog mode prior to mass calibration. The analog

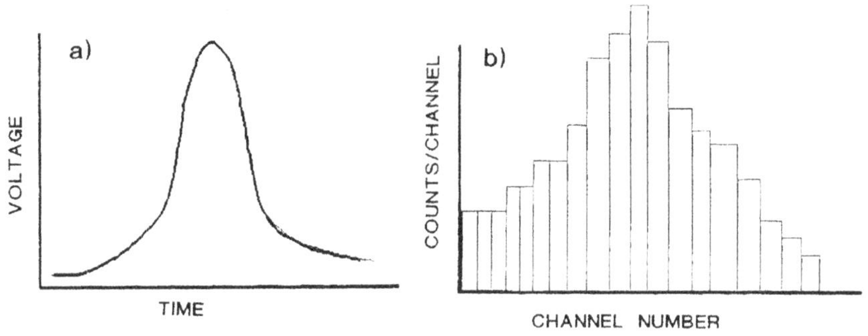


Figure 1. Uncalibrated mass spectral peaks acquired in (a) analog and (b) pulse-counting mode.

mode of acquisition is best used in scanning mass spectrometers such as sector instruments and quadrupole mass filters. An alternate approach to analog data acquisition is pulse or single-ion counting acquisition. The pulse counting mode is most useful when very low ion currents are produced and when events can be correlated to a "start" event such as $^{252}$Cf fission or a laser pulse. In this mode, an ion impinging on the detector produces a fast pulse signal that is sent to a time-to-digital converter (TDC). The TDC measures the time between the ion signal and the correlated event and sends the information to a data system that stores the information in an appropriate time channel. Other pulses are processed, stored, and added to acquired data until suitable statistics are obtained. A spectrum isthen produced from the stored data by plotting counts per channel versus channel number as shown in Figure 1b. A detailed description of the pulse counting process for $^{252}$Cf plasma desorption mass spectrometry was published by Macfarlane [6].

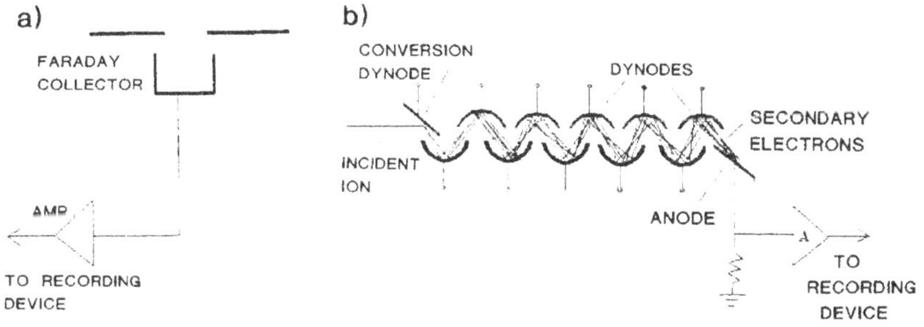


Figure 2. Schematic diagram of a (a) Faraday cup and (b) electron multiplier detector.

Owing to the nature of data acquisition, the analog and pulse counting modes each have their own desirable detector characteristics [7]. For analog signal processing, the detector output current should be proportional to the number of ions incident on the detector. For most scanning mass spectrometers, response times on the order of milliseconds are sufficient. Time-of-flight mass spectrometers operating in an analog mode, however, can require detector response times on the order of a few nanoseconds. For pulse-counting mass spectrometry, detectors are operated so that the pulse height is fairly uniform for each detected ion. This makes it possible to use discriminators to separate ion signals from random noise pulses and thus achieve higher signal-to-noise ratios. For pulse counting, detectors with response times on the order of a nanosecond are desirable.

## 4. Basic Detector Systems

The simplest of all detectors used in mass spectrometry is the Faraday collector (see Figure 2a). The Faraday collector consists of a metal plate or cup connected to an amplifier or current-to-voltage converter. Ions striking the collector induce a current in the circuit equal to the incident ion current. The Faraday collector has a minimum detectable current on the order of $10^{-8}$ to $10^{-14}$ A [8].

Another basic detector is the electron multiplier (see Figure 2b). Ions are accelerated into the conversion dynode, resulting in the emission of several secondary electrons per incident ion. The actual number of electrons ejected depends upon several factors including incident ion velocity, incident ion mass and dynode material. These electrons are accelerated into the next dynode, which results in the emission of several electrons per incident electron. This "electron multiplication" is continued through n dynodes until the electrons are collected at the anode as in the case of the Faraday collector. The measured current (I′) is given by

$$I' = IG = I\gamma\delta^n , \tag{1}$$

where I is the incident ion current, G is the detector gain, $\gamma$ is the secondary electron coefficient and $\delta$ is the average number of electrons ejected per incident electron. A typical gain for the electron multiplier ranges from of $10^4$ to $10^8$ and detectable ion currents can be as low as $10^{-20}$ A [5,8].

The channel electron multiplier (CEM) continuous dynode detector has become popular for a wide range of mass spectrometry configurations [7]. These detectors are compact, stable when exposed to air, have excellent signal-to-noise ratio characteristics and can be used in both analog and pulse-counting modes. A partial diagram of a straight CEM is shown in Figure 3a. The CEM is constructed from specially formulated lead-doped glass which has a high resistance–typically $10^8$ to $10^9$ ohms–and a high secondary electron coefficient. Nichrome or Iconel contacts are applied to each end of the channel. A high voltage potential (1-2 kV) is applied across to the contacts in order to replenish ejected electrons and to attract the electrons to the output end of the channel. Secondary ion emission and multiplication occur in an analogous manner to the traditional electron multiplier.

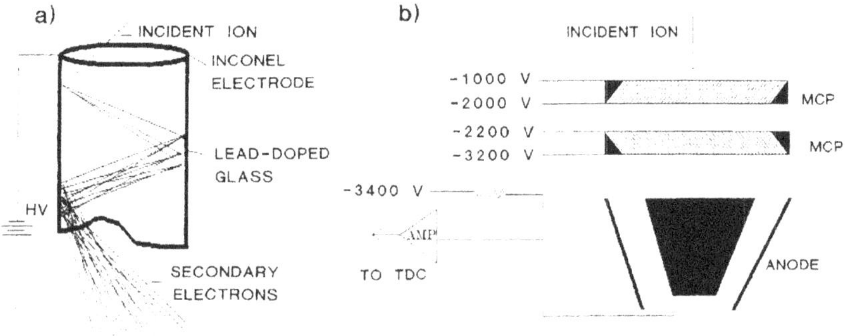


Figure 3. Schematic diagram of (a) a portion of a straight channel electron multiplier and (b) a chevron microchannel plate detector.

For straight CEMs, the gain is limited to ca. $10^4$ owing to ion feedback. The large electron currents in the channel tend to ionize residual gas molecules, which are accelerated toward the channel entrance. These ions then strike the channel wall and begin a new cascade of electrons. The feedback process can be significantly reduced by curving the channel and thereby limiting the distance the residual gas ions can travel. Gains on the order of $10^8$ and pulse widths of 20 ns are typical for curved CEMs. Kurz [7] published a detailed account of channel electron multipliers and their characteristics.

The gain of a CEM is not dependant upon the length or diameter of the channel but upon the ratio of the two parameters. This means that short channels can be constructed that have high gain provided their diameter is very small. This is the principle of the microchannel plate (MCP) or channel electron multiplier array (CEMA) detector system [9]. The MCP consists of an array of $10^4$ to $10^7$ channels having diameters from 10 to 100 µm. The channels are arranged at an angle 8° from perpendicular to the plate. Like the CEM, the MCP is made of lead-doped glass and each face of the plate is coated with Nichrome or Iconel for electrical contact. Each of the channels can act independently, which makes the MCP useful for spatially resolved array detection. Most conventional applications use two MCPs combined in a "chevron" configuration (see Figure 3b) to reduce ion feedback and provide more gain than a single MCP. MCPs with curved microchannels are also available. The chevron detector is often desired for its fast response time characteristics. Pulse rise-times and pulse widths of less than a nanosecond and gains of $10^8$ are typical.

## 5. Detection Efficiency of Peptide Molecular Ions

Macfarlane's group [10] used the principles of secondary electron emission to investigate the detection efficiency of peptide molecular ions on pulse-counting MCPs. They found that for peptide molecular ions, $\gamma$ is exponentially dependent on the incident ion velocity and linearly dependent on the incident ion mass such that

$$\gamma = 2.58 \times 10^{-7}\, m\, [\exp(2.31 \times 10^{-4}\, v)]\ , \tag{2}$$

where m is in amu and v is in m/s. For pulse counting, the electron multiplicity can be approximated by a Poisson distribution function such that

$$P = 1 - \exp(-\gamma)\ , \tag{3}$$

where P is the probability of detection. At values much less than one, P and $\gamma$ are approximately equal. When $\gamma = 1$ then the detection probability is 63% and increases to 95% when $\gamma = 3$.

By combining Equations 2 and 3, the detection probability for a peptide molecular ion of any mass and velocity can be predicted. The utility and validity of this approach is shown in Figure 4, which is a plot of theoretical and experimental pulse-counting detection probability as a function of ion velocity for the serum bovine insulin $(M + H)^+$ ion. The plot shows that at a combined acceleration and post-acceleration energy of 26 keV (i.e., the maximum potential that could be applied while maintaining stable operation) only 65% of the produced insulin ions were detected. This figure also shows that to detect these ions with a probability near unity, a total energy of 40 keV would be required.

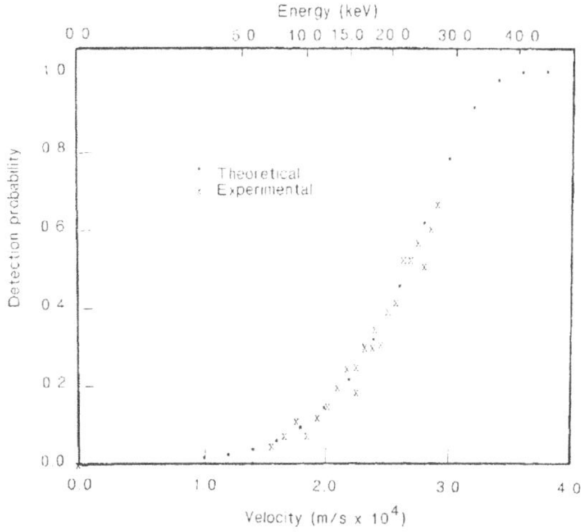


Figure 4. Plot of theoretical and experimental detection probability as a function of ion velocity for the molecular ion of bovine insulin. From reference [10] with permission.

In the above-mentioned secondary electron emission studies, molecular ions of the peptide bradykinin were detected at velocities as low as $1.56 \times 10^4$ m/s [10], clearly below the velocity threshold of 1.8 × $10^4$ m/s predicted by others. Similar studies have demonstrated the detection of cesium iodide cluster ions with velocities below 1.0 × $10^4$ m/s [11]. Recently, time-of-flight mass spectrometers using chevron detection systems recorded spectra of singly charged ions having m/z values as large as 200,000 and total ion energies of ca. 20 keV (corresponding to a velocity of 4.4 × $10^3$ m/s) produced by matrix assisted UV-laser desorption [12]. In this sub-threshold velocity region, there are clearly processes taking place that are not predicted by conventional secondary electron emission models. Calculation of the secondary electron coefficient by using Equation 2 for very heavy ions gives interesting results. Figure 5 is a theoretical plot of the secondary electron yield as a function of incident ion mass for ions accelerated to an energy of 20 keV. The secondary electron coefficient decreases as ion mass increases up to 50 kDa. For ions having masses above 50 kDa, the linear contribution from the mass overcomes the decrease in velocity and an overall increase in $\gamma$ results. This suggests that a 160 kDa peptide could possibly be detected with the same efficiency as a 20 kDa peptide and may account for the detection of ions produced by matrix-assisted laser desorption that have velocities well below the predicted threshold [12].

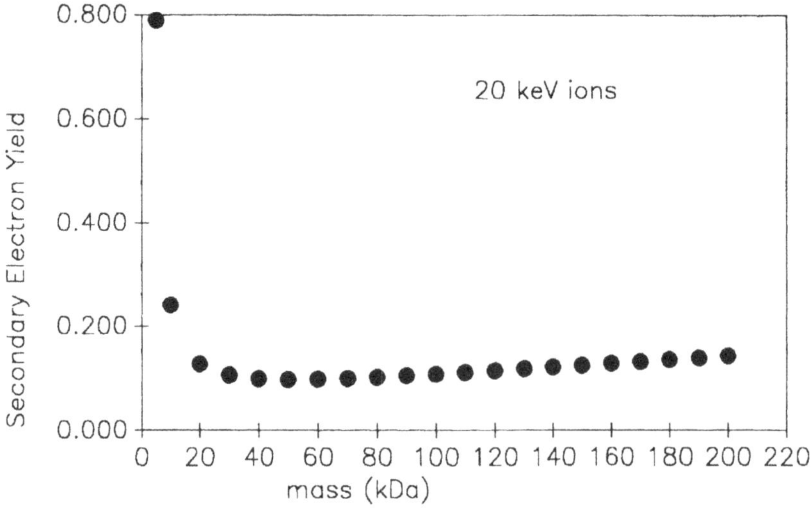


Figure 5. Plot of the secondary electron coefficient as a function of ion mass for ions accelerated to 20 keV.

One explanation which has been proposed to account for the detection of peptide ions in the sub velocity-threshold region is the production of fragments of ions by surface induced dissociation. These small fragment ions could be accelerated and induce electron emission in a more efficient manner than that of an intact molecular ion. Surface induced dissociation (SID) using a microchannel plate has been demonstrated by Aberth [13], and others have proposed secondary ion emission as a means of improving detection efficiency, but few data are available on the utility of SID as a method of enhancing detection efficiency of large biomolecules [14].

## 6. Advanced Detector Designs

As described above by Equation 2, secondary electron production increases linearly with incident ion mass but increases exponentially with velocity. This that means the simplest method of improving detection efficiency for heavy ions is to increase their velocity prior to impact by using a "post-acceleration" field. An example of a post-acceleration design for a chevron detector was described by Cotter et al. [15] for a CVC 2000 time-of-flight laser desorption mass spectrometer. In this design, the field-free region is maintained at -3 kV and the front face of the chevron is maintained at a potential of -12 kV, giving the ions an additional 8 keV of energy prior to detection (see Figure 3b). A lens has been incorporated to focus the ions onto the detector for improved sensitivity. This easily implemented modification can increase the detection probability of bovine insulin molecular ions (5735 m/z) from 1.5% at 3 keV to 20% at an energy of 12 keV.

One disadvantage of simple post-acceleration detectors is that it is often difficult to maintain the high negative potentials on delicate detector components. An alternate approach is to post-accelerate ions into a conversion dynode. The conversion dynode is typically a plate made of a material that has a high secondary electron coefficient. The dynode is typically maintained at high negative potential (e.g., -10 to -15 kV). Positive ions are post-accelerated into the dynode, resulting in secondary electron emission. Secondary electrons are then accelerated and transmitted into a detector that is maintained at an acceptable potential. One example of this arrangement was described by Jungclas et al. [16]. In the Jungclas design, secondary electrons are transported in a circular trajectory from the conversion dynode to a chevron detector by using a magnetic field. Macfarlane's group [17] constructed a conversion dynode detector for use in a reflectron time-of-flight mass spectrometer. Their design allows the reflectron to be operated in a linear mode without an annular detector.

Electron multiplier detectors have limited lifetimes and can become saturated when very high incident ion currents are used [5]. This has led to the development of hybrid detectors for mass spectrometry. Photomultiplier tubes can be utilized as ion detectors when coupled with a phosphor and a conversion dynode. Bateman et al. [18] constructed an off-line detector for a double focusing mass spectrometer based upon this principle. Their design, shown in Figure 6, eliminates the need for deflector plates and does not require a change in polarities for negative ion detection. In addition, the detector lifetime is longer than for an equivalent electron multiplier since the photomultiplier is sealed in a vacuum tube and not exposed to analyte or pump oil vapors, or air when the instrument is vented. Houk [19] constructed a conversion dynode-scintillator-photomultiplier detector system for an inductively coupled plasma/mass spectrometer which has many of the same advantages of the dynode/photomultiplier design.

## 7. Array Detectors

Array detectors were originally designed as positionally sensitive detectors for ion microscopy and other imaging applications [9]. Most designs use a microchannel array to produce electrons which are then converted to photons by a phosphor screen. Conversion of photons to an image is then accomplished either by a vidicon camera or a Redicon photodiode array [20]. A schematic diagram of a typical array detector is shown in Figure 7.

Mass spectrographs of the Matauch-Herzog geometry originally used a photographic plate located along the plane of spatial focusing to detect simultaneously ions ranging from 20 to over 400 m/z. These instruments lost favor because their resolution is poor ($m/\Delta m = 200$) and

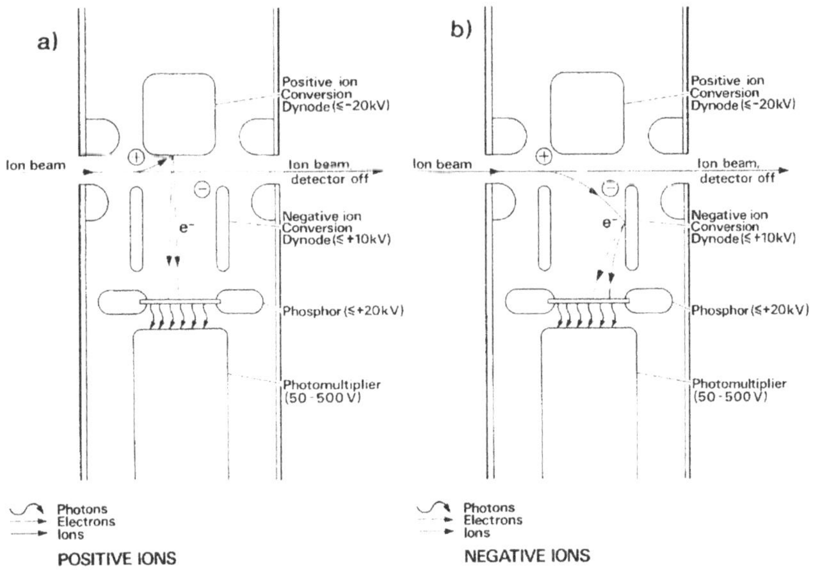


Figure 6. Hybrid photmultiplier detector for the detection of positive and negative ions. Used by permission from reference [18].

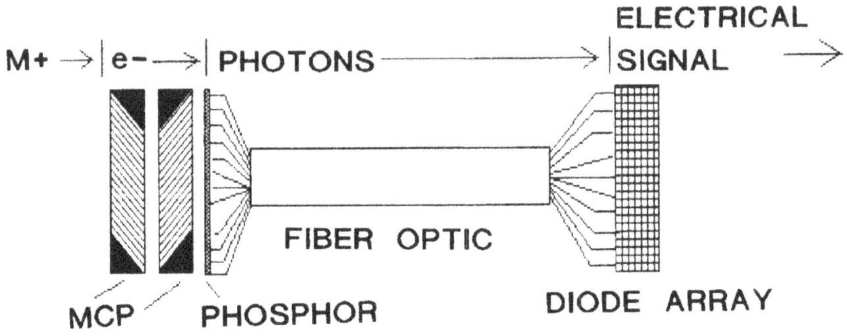


Figure 7. Schematic diagram of an array detector system.

breaking vacuum to retrieve the plate is inconvenient. Murphy and Mauerberger [21] have reported a array detector system for an instrument of this type which provides improved sensitivity for trace analyses of gases.

Over the past 25 years, high-resolution instruments of the Nier-Johnson geometry have become popular. Double focusing instruments have a very small point of double focus at the intersection of the spatial-focusing and energy focusing planes. Cottrell and Evans [20] constructed an array detector for a double focusing instrument, which can simultaneously detect a mass range of 4% about a central mass (e.g., 500-520 m/z). In this configuration, acquisition of a mass spectrum over a larger mass range requires that the 4% window be scanned in stepwise increments.

Biemann's group reported the incorporation of a pair of quadrupole lenses and an array detector into a JEOL HX110/HX110 tandem mass spectrometer [22]. The lens is used to change the dispersion and angle of the focal plane (see Figure 8). In the wide-range mode, with a focal plane angle of 40°, the detector can simultaneously acquire over a mass range of 40% of the central mass (1:1.4) at a resolution of 500. In the high-resolution narrow-range mode (1:1.055), corresponding to a focal plane angle of 17°, a resolution of up to 10,000 is possible. A 50-100 fold improvement in sensitivity has been reported with this design. Many commercial manufacturers now offer array detectors as optional equipment for their sector instruments.

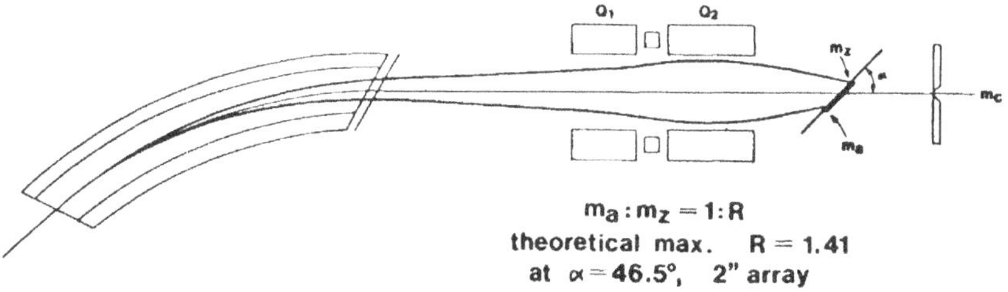


Figure 8. Array detector and double quadrupole lens system for the JEOL HX110/HX110 tandem mass spectrometer. Used by permission from reference [22].

## 8. Conclusion

There have been significant advances in the mass spectrometry of biological molecules in the last ten years. Even on sector instruments, the molecular weight determination of peptides with masses up to 10 kDa is now commonplace. Detector technology has had an important role in this development. There are still problems, however, with ion detection and these problems have not been solved. Array detection on sector instruments is still in the developmental stages. Simultaneous detection of high-resolution spectra over a wide mass range is still not possible. The process of ion detection in the sub velocity-threshold region is not yet understood, and the detection of ions by secondary ion production from surface-induced dissociation has not been thoroughly investigated. With recent exciting developments in mass spectrometry such as matrix-assisted laser desorption [12] and electrospray ionization [23], it is unclear what the ultimate mass range and limit of detection of mass spectrometry will be for the molecular weight determination of peptides and proteins. One thing that is certain, however, is detectors will be important in determining what those limits will be.

## 9. References


1. Carter, G. and Colligen, J.S. (1968) Ion Bombardment of Solids, Elsevier, New York.
2. Parilis, E.E. and Kishinevskii, L.M. (1961) *Sov. Phys.-Solid State*, **3**, 885.
3. Beuhler, R.J. and Friedman, L. (1980) *Nucl. Instrum. Meth.*, **170**, 309.
4. Hedin, A., Hakansson, P., and Sundqvist, B.U.R. (1987) *Int. J. Mass Spectrom. Ion Processes*, **75**, 275.

5. Ingle, J.D. and Crouch, S.R. (1988) Spectrochemical Analysis, Prentice Hall, Englewood Cliffs, NJ, Chap. 4.
6. Macfarlane, R.D. (1983) *Anal. Chem.*, **55**, 1247A.
7. Kurz, E.A. (1979) *Amer. Labor.*
8. Benninghoven, A., Ruedenauer, F.G. and Werner, H.W. (1987) Secondary Ion Mass Spectrometry: Basic Concepts, Instrumental Aspects, Applications and Trends, Wiley, New York, pp. 548-555.
9. Wiza, J.L. Nucl. (1979) *Instrum. Meth.*, **162**, 587.
10. Geno, P.W. and Macfarlane, R.D. (1989) *Int. J. Mass Spectrom. Ion Processes*, **92**, 195.
11. Geno, P.W. (1988) $^{252}$Cf Plasma Desorption Mass Spectrometry in a Low-voltage Regime, Doctoral Dissertation, Department of Chemistry, Texas A&M University, College Station, TX.
12. Karas, M. and Hillencamp, F. (1989) *Int. J. Mass Spectrom. Ion Processes*, **92**, 231. (and this volume).
13. Aberth, W. (1988) Presented at the 36th ASMS Conference on Mass Spectrometry and Allied Topics, San Francisco, CA.
14. Mahoney, J.F., Perel, J., Lee, T.D. and Legesse, K. (1988) in C.J. McNeal, (ed.), The Analysis of Peptides and Proteins by Mass Spectrometry, Wiley, New York.
15. Olthoff, J.K., Lys, I., Demirev, P. and Cotter, R.J. (1987) *Anal. Instrum.*, **16**, 93.
16. Danigel, H., Jungclas, H. and Schmidt, L. (1983) *Int. J. Mass Spectrom. Ion Processes*, **52**, 223.
17. Geno, P.W. and Macfarlane, R.D. (1987) *Int. J. Mass Spectrom. Ion Processes*, **77**, 75.
18. Bateman, R.H. and Bott, P.A. (1987) Presented at the 35th ASMS Conference on Mass Spectrometry and Allied Topics, Denver, CO.
19. Huang, L.Q., Jiang, S.J. and Houk, R.S. (1987) *Anal. Chem.*, **59**, 2316.
20. Cottrell, J.S. and Evans, S. (1987) *Anal. Chem.*, **59**, 1990.
21. Murphy, D. and Mauersberger, K. (1987) *Int. J. Mass Spectrom. Ion Processes*, **76**, 85.
22. Biemann, K. and Hill, J.A. (1989) Presented at the 37th ASMS Conference on Mass Spectrometry and Allied Topics, Miami Beach, FL.
23. Fenn, J.B., Mann, M., Meng, C.K., Wong, S.F. and Whitehouse, C.M. (1990) *Mass Spectrom. Rev.*, **9**, 37.

# Part II
# Methods in Mass Spectrometry

# ELECTROSPRAY MASS SPECTROMETRY

M. Mann
*Department of Molecular Biology*
*Odense University, Odense, Denmark*
S. Shen and J. B. Fenn
*Department of Chemical Engineering*
*Yale University, New Haven, CT USA*

ABSTRACT. The technique of Electrospray Ionization for mass spectrometry of non-volatile species is described and some typical results are examined.

## 1. Introduction:

In this chapter we will describe the relatively new technique of electrospray (ES) ionization for mass spectrometry (MS). ES has been under development since the early '80s but has attracted widespread interest only since 1988 when it was shown to be capable of producing intact ions from very large biomolecules [1]. Along with Laser Desorption (LD) [2,3], ES is now an ionization method of choice for MS analysis of very large molecules in the mass range above 10 kDa. It is also rapidly finding new uses in the lower mass range and seems destined to become an indispensable tool in the emerging discipline of biological mass spectrometry [4]. Here we will sketch the history and describe the basic features of ESMS. Several recent reviews [5-8] provide the interested reader with additional technical information, and more complete coverage of the literature as well as some comparisons of ES with other ionization techniques.

The basic challenge for mass spectrometry of large molecules is to transform them intact from their initial state in a sample to ions in vacuum ready to be mass analyzed. The lack of methods that would provide sufficiently gentle or 'soft' ionization has confined MS to the low molecular mass range for most of its history. Recent experimental progress in lifting this constraint has resulted from advances in techniques based on rapid energy deposition [5] or strong electrostatic fields [9]. Energy sudden techniques rely on an idea originally proposed by Buehler et al. [10]. If energy is applied sufficiently rapidly to a sample, analyte molecules may vaporize (and ionize) before decomposition has a chance to take place. Such fast application of energy can be achieved in various ways. Some techniques use the impact of a high velocity, particle which may be an atom, as in fast atom bombardment, (FAB) [11], an ion in so-called secondary ion mass spectrometry (SIMS), or a fission product of $^{254}$Cf as in plasma desorption (PD) [12]. These techniques have in the last ten years made possible a steady expansion of MS into the biosciences.



*M. L. Gross (ed.), Mass Spectrometry in the Biological Sciences: A Tutorial*, 145–163.

Then in 1988, by remarkable coincidence, at almost the same time as ES became noticed, matrix assisted laser desorption (LD) emerged as a very effective tool for the MS of very large biomolecules [2,3]. In this technique the analyte molecules are suspended in a matrix that efficiently absorbs the energy of a short laser pulse. Time-of-flight (TOF) analysis of the released and ionized molecules produces a mass spectrum (see the chapter by Hillenkamp and coworkers in this volume). Matrix assisted LD has spectacularly increased the mass range of energy sudden methods into the region above 200 kDa.

In contrast to these energy sudden approaches, high field methods involve no local high energy density but rely instead on producing free ions with the help of intense electric fields. In one class of these methods, the ions are produced in vacuum. In the other, which is quite different in both principle and practice, the ions are produced in a bath gas. Members of the first class include what is known as field desorption (FD) [13] as well as electrohydrodynamic ionization (EH) [14]. FD requires considerable experimental skill in emitter preparation and is currently not widely practiced. Experimental problems such as large energy spread and vacuum system fouling have likewise discouraged widespread application of EH. The other field techniques are sometimes known as the "spray methods" because they obtain the required electric field by producing small, highly charged droplets of solution from which ions desorb into a bath gas. In these spray methods the bath gas provides enthalpy for vaporizing solvent, thereby preventing a freeze drying of the droplet. The gas also moderates the kinetic and internal energies of the desorbed ions, limiting them to the thermal level of the gas itself.

The spray techniques can best be categorized in terms of the methods used to produce the charged droplets. In what could be called "Aerospray" (AS), a liquid is nebulized by the action of a stream of air (as in a perfume atomizer). Some of the resulting small droplets become charged by the statistical variations in the distribution of cations and anions in small volume elements of a liquid. Iribarne and Thomson [15] were the first to apply this as technique to mass spectrometry. In practice AS is almost never applied in the absence of an electric field so that both electrostatic and aerodynamic forces are involved in a hybrid technique sometimes called Ionspray, which will be discussed later. Thermospray (TS), developed by Vestal [16], is one of the most popular of the spray methods. It is based on passing sample liquid through a small heated tube. Rapid evaporation results in a jet of droplets in vapor emerging into a region of superheated vapor where the droplets continue to evaporate, having become charged by the same statistical fluctuations as in AS. There are some experimental difficulties associated with TS due to the requirement for careful control of relatively high temperatures. In practice they must often be adjusted by trial and error for different compounds. Thermospray is discussed more fully in the chapters by Arpino and by Freas and coworkers.

The idea for using electrostatic forces to produce the charged droplets actually precedes both AS and TS, having been introduced by Dole, et al., in the late '60s [17]. In practice, ES consists in passing a sample solution through a hypodermic needle held at high voltage. The resulting high field at the needle tip disperses the emerging liquid as a fine spray of charged droplets. Dole reasoned that if a solution of macromolecules were dilute, the droplets would on average contain only one macromolecule that would become a macro ion as the last of the droplet solvent evaporated. By analyzing these ions with a mass spectrometer he could thus obtain mass spectra of very large and non volatile species. Because no mass analyzer then available could "weigh" molecules with the anticipated m/z ratios, Dole used the simple "retarding potential" device shown in Figure 1. The ions produced by the spray pass through supersonic jet expansion into a vacuum [18]. The terminal gas velocity in free jets is readily calculated. Consequently, Dole thought he could determine the mass m of the ions from the relation $mv^2/2 =$ Vs where v is the terminal velocity of the free jet and Vs is the electric potential required to keep the ions from reaching the detector (see Figure 1).

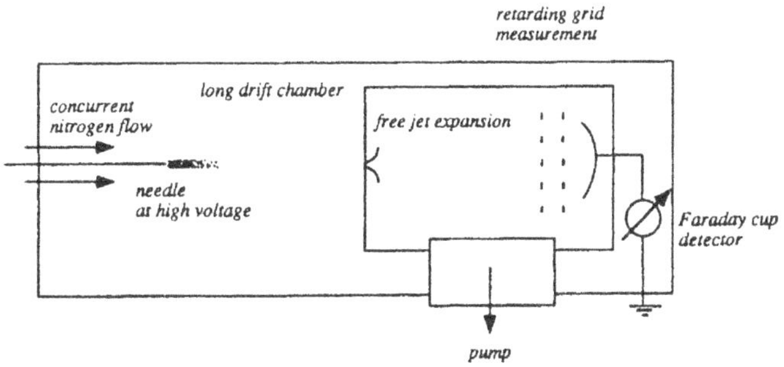


Figure 1. Schematic Diagram of Dole Apparatus

We now think that because of the actual experimental conditions and some faulty assumptions in the analysis, Dole's interpretation of his results is probably questionable [5]. Nevertheless, it is he who led the way to what has become a powerful technique.

## 2. Electrospray Methodology

In the early '80s our group at Yale started the first ES experiments with a conventional mass analyzer (i.e. a quadrupole mass filter) [19]. Figure 2 shows a sketch of the apparatus.

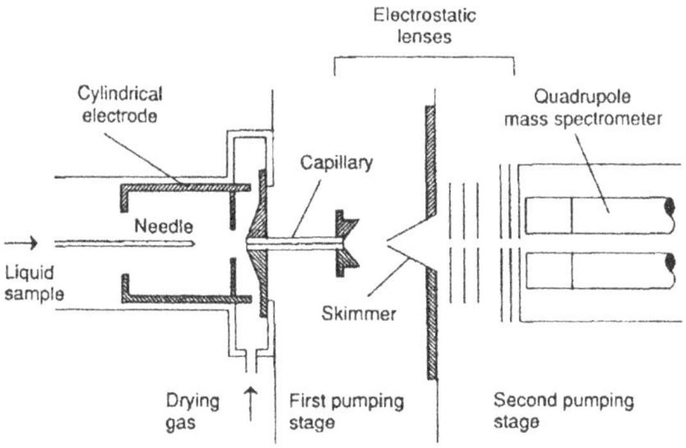


Figure 2. Schematic Diagram of ESMS System at Yale

Sample solution at flow rates typically of a few uL/min is introduced into the source through a hypodermic needle. Between the needle and the surrounding electrodes, an applied potential difference of a few kV produces a field at the needle tip that disperses the emerging liquid into a fine spray of charged droplets. The sign of the applied voltage determines the polarity of the droplets and the ions they produce, both of which are driven by the field toward the end plate of the chamber located a few centimeters away from the needle. Countercurrent to their motion is a flow of drying gas, usually warm dry nitrogen at atmospheric pressure, that helps to vaporize the droplets, thus producing ions by a mechanism to be described. Some of these ions are entrained in the gas flow that enters the capillary to emerge into the vacuum system in a free jet expansion. A core portion of the free jet passes through a skimmer and delivers ions into a second vacuum stage containing a mass analyzer, in this case a quadrupole mass filter. The glass capillary constitutes a dielectric channel that allows one to choose the potential of its inlet end independently of the potential at the exit end that is required to inject ions into the mass analyzer [20]. Thus, the hypodermic needle and the sample source can be kept at ground potential while the end plate is maintained at a sufficiently "below" ground to produce the field at the needle tip that is required to disperse this emerging liquid. The viscous forces due to the gas flowing through the capillary "drag" the ions from the "potential well" at the capillary inlet back up to any desired potential, even up to the 10 kilovolts or so that are needed for injection into magnetic sector instruments.

The ESMS method sketched above can be thought of as comprising three successive stages: (1) spraying the liquid; (2) forming ions from the highly charged droplets; (3) sampling and mass analyzing the ions in a mass analyzer in vacuum. These stages will be discussed in what follows.

## 2.1 SPRAYING THE LIQUID

The phenomenon of electrospray or "electrostatic nebulization of liquids" has been studied for a long time and has had a variety of applications [21]. In ESMS practice the visual spray pattern has been found to assume several distinct "modes" [5,22]. At low needle voltages, relatively large droplets form at the needle tip before being pulled off by the electric field. As the voltage is increased, the droplets become smaller until in the so-called "rain mode" they emerge into what looks like a steady stream of liquid, but is really a rapid succession of tiny droplets. When the voltage reaches a critical "onset" value, the "rain" abruptly changes to a fine mist of micron sized monodisperse droplets in what is called the "fog" mode. At the right combination of flow rate and voltage, the fog mode becomes the "rainbow" mode, a name that refers to the colorful higher order Tyndall spectra (HOTS), which can be seen when a very stable and monodisperse spray is illuminated with white light. The fog mode and especially the rainbow mode give rise to the most stable ion currents from the mass analyzer. As the voltage is further increased, the spray becomes unstable owing to formation of a corona discharge that distorts and disrupts the electric field at the needle tip. In practice, so called ‘mode hopping’ can occur in which two or more modes may seem to be coexisting. For example, a fine mist may be seen together with a center steam of liquid. In reality, the spray is then oscillating between the two modes that are readily distinguished in oscilloscope traces of the total spray current. To obtain a stable mass spectrometric signal, one needs to maintain the spray in the fog or rainbow mode, preferably the latter, for substantial periods of time, even while spray variables such as solution composition are changing. Recent investigations [23] into the mechanism of electrospray have focused on the role of the short liquid stem at the very tip of the needle from which the droplets emanate - - the so-called Taylor Cone [24]. The geometry of this cone is influenced by solution flow rate, conductivity and surface tension, among other variables. To maintain a stable Taylor cone geometry and hence a stable spray, these variables must remain within certain bounds that impose constraints on flow rate and composition of the liquid. For the simple configuration in Figure 2, solution conductivity must not exceed that equivalent to millimolal NaCl and the flow rate must be less than 20 uL/min. Various approaches have been tried to achieve stable sprays under a wider range of conditions. In the scheme reported by Bruins et al. [25], an annular flow of high velocity gas concurrent with and surrounding the emerging liquid provides a pneumatic assist to the electrostatic nebulization. This device increases usable flow rates to about 100 uL/min and provides stable sprays over a wider range of solution composition, probably by removing excess liquid from the needle tip so that large droplets cannot form. This technique was named "ion spray" by the authors, an unfortunate term because it implies a new ionization mechanism. "Pneumatically assisted electrospray" would be more accurate and meaningful. Vestal and coworkers [26] recently reported that good results can be obtained by raising the temperature of the needle and its surroundings, perhaps because the effective surface tension of the liquid is decreased. Another effective approach was introduced by Smith et al. [27] in their use of ESMS detection in electrophoresis (CZE). They provide an annular sheath of "well behaved" liquid around the flow of sample liquid. The original purpose was to provide a good electrical connection with the liquid emerging from the quartz capillary in which the electrophoresis takes place. It turns out that the electric field at the needle tip can effectively disperse the mixture of sheath and core flow of sample liquid even when a stable spray cannot be formed from the sample liquid alone.

With these and other experimental developments, it is now possible to spray liquids with a wide variety of properties over a wide range of conditions. Even pure water containing trifluoroacetic acid, a long time challenge to pure ES, can now be accommodated. However, high concentrations (i.e., over 10-2 M) of nonvolatile ionic buffers still present a problem as they do in all other ionization methods, with the possible exception of LD. They may result in such high solute concentrations during droplet evaporation that the solvent vaporization process is suppressed, thus inhibiting ion formation.

Because it allows, indeed requires, continuous infusion of liquid, ES is a 'natural' on-line technique for use with continuous sources of sample liquid. With the above mentioned improvements in liquid composition and flow rate compatibility, there are now good prospects for the practical combination of chromatographic methods with ESMS. In fact, successful coupling of ES and CZE has been achieved by Smith et al. [28] by means of the above "sheath flow" interface and later by Lee et al. [29] who used a simple supplementary flow. Microbore HPLC is a more attractive partner for ES than is standard bore because its flow rates are more compatible. It can be run with flow rates in the range of 100 uL/min that can be injected directly when ES is pneumatically assisted. However, sensitivity is improved and much of the sample stream can be available for other purposes if the flow is split so that only a fraction enters the ES leg. Even though coupling of low-flow LC to ES has been demonstrated [30], it is not yet a routine technique. Actually, the field is developing so rapidly that this picture may well be out of date by the time this chapter is in print.

## 2.2 ION PRODUCTION FROM CHARGED DROPLETS

After the small and highly charged droplets have been formed by the spray process, the analyte molecules must somehow be separated from the solvent as ions. This ion formation step, as in many of the soft ionization techniques, is probably the least understood of the component processes in ES. Extensive studies with poly(ethylene glycol) (PEG) as a model analyte have provided some insight on the ionization mechanism [31, 32, 22]. PEG is a well characterized polymer [33] whose repeating unit has a mass of 44 Da. When the chain is in its zig zag or stretched configuration the distance between oxygen atoms, the probable charge sites, is 3.5 Angstroms. PEG samples are readily available in different nominal molecular weights (i.e., mixtures of oligomers with molecular weights distributed around the most probable or "nominal" value). Therefore, the influence of mass on the desorption process can be separated from the effects of other variables. Figure 3 shows ESMS spectra of PEG samples with four different nominal molecular weights.

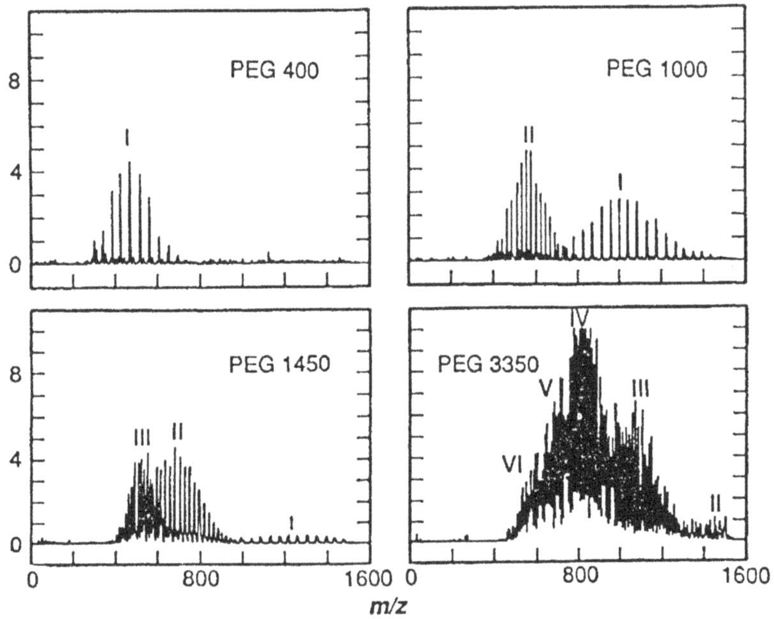


Figure 3. ES Spectra for PEG Samples of Various Molecular Weights in Methanol-Water Solutions

In the case of PEG 400, the spectrum comprises one band of peaks, each corresponding to singly charged protonated molecules formed by attachment of one sodium ion to a PEG oligomer of particular mass. The spectrum is typical of those obtained by ESMS of small molecules in that it shows very few of the background peaks that appear in great abundance in FAB spectra and are due to the matrix. The ES spectra show no fragmentation of the parent ions and little or no discrimination in mass (i.e., the distribution of peak heights agrees well with the manufacturer's data on the distribution of oligomer molecular weights). With PEG 1000 a strong band of peaks appears for doubly charged ions that have been formed by the attachment of two $Na^+$ to one PEG oligomer. As the molecular weight is increased further, the peaks for more highly charged molecules become more abundant at the expense of those for singly charged species. In the case of PEG 3350, it becomes difficult to distinguish individual peaks because of the overlap of two distributions - - the mass distribution around the most probable molecular weight and the charge distribution around the most probable charge state. As long as drying gas flow is adequate, there is no current due to charge carriers with m/z values above 1500, the upper limit for the quadrupole employed in these studies. This assertion is based on retarding potential measurements similar to those of Dole [34]. Further increases in PEG molecular weight continue the trend shown in Figure 2. For PEG 20M (average molecular weight 17,500 Da), the most probable number of charges on the most abundant oligomer is around 23. Such extreme multiple charging is an inherent feature of ES ionization and is responsible for the characteristic peak distributions in the ESMS spectra of large molecules.

It is interesting to note that the charge distribution of ES ions does not depend appreciably on flow rate or analyte concentration of the sample liquid. This fact has profound implications for the ion formation mechanism [8]. In the charged residue model (CRM) originally envisioned by Dole, the ions are formed simply by evaporation of the solvent from droplets containing only a few analyte molecules, each of which might retain some of the droplet charge as the last of the solvent leaves. If this "passive" model of charging were true, we would expect a dependence of the average charge state on analyte concentration, at least as long as there are more charging sites on a molecule than net charges available. For example, doubling the PEG concentration might have been expected to halve the average charge state of an ion. However, only a very weak dependence of charge state on analyte concentration is experimentally observed. Furthermore, on comparing the net charge flux into the spray (i.e., the current leaving the needle) to the flux of analyte ions into the source, one finds that there can be a large excess in the number of analyte molecules relative to the number of available charges, even when the average number of charges in the product ions is still very high (i.e., greater than 20 in the case of PEG 20M). These experimental findings do not support the Dole charged residue model, which cannot explain how a small fraction of the analyte molecules can sequester almost all of the available charges.

A more satisfactory explanation for the observed behavior can be found in the desorption of individual ions from the droplet under the influence of the strong electric field resulting from the surface charge. In this picture, which has its roots in the "ion evaporation model" proposed by Iribarne and Thomson [15], ions leave the surface when they have collected enough charges so that in the field at the surface, the lifting force overcomes the surface barrier due to the solvation forces. After a substantial fraction of the net charge on the droplet has been carried off in this way, the rest of the analyte molecules and other involatile solutes form a weakly charged residue as the last solvent evaporates. In this desorption model the number of charges required for ion ejection from the surface depends only on the properties of the solution gas interface and on the structure of the desorbing species. Therefore, within a fairly broad range, the average charge state of the desorbed ions should not depend appreciably on the flow rate or analyte concentration of the solution.

Iribarne and Thomson put forward a detailed analysis for "ion evaporation" of singly charged ions from charged droplets. However, the relevance of that model has been questioned on various counts. Among other things, it predicts that ion desorption cannot occur until droplet radii are much smaller than can be the case for the larger PEG oligomers. PEG 20M, by no means the largest molecule to be produced by ES, could be wrapped twice around the circumference of a droplet having the electric field required by the analysis of Iribarne and Thomson. As for the current state of ES "theory", we can conclude that the experimental evidence seems to favor a desorption mechanism but that the detailed processes involved are still shrouded in mystery.

One of the questions that can be asked concerns the maximum number of charges that an analyte molecule, can hold. With PEG as a model molecule an analysis was carried out that calculates this maximum number in terms of the coulombic repulsion forces generated between the charges and the bonding energy of each charge to the molecule. The essence of this model is that the electrostatic potential energy of the centermost charge [31] must be equal to or less than the known energy of binding to its site. For a PEG oligomer ion with $Na^+$ as the adduct charge, the site is an O atom and the binding energy is 2.05 eV. As shown in Figure 4 this model leads to a near linear relationship between molecular weight and the maximum number of charges.

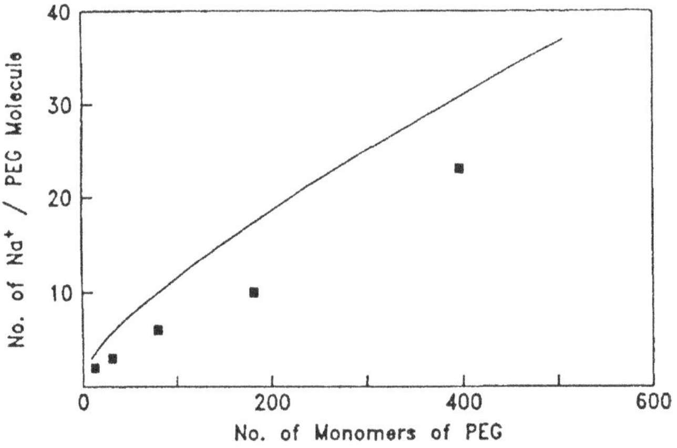


Figure 4. Dependence of Charge State for PEG Molecules on the Number of Monomers Per Oligomer

Figure 4 also shows that the observed number of charges is always less than the maximum predicted. The desorption model can account for this discrepancy by saying that the combination of surface field and ion charge reaches "lift-off" at surface charge densities below those required to achieve the maximum number of charges on an ion.

Apart from these general considerations on the desorption mechanism, a number of specific experimental observations regarding ES ionization can be made:

(1) Ion desorption occurs at moderate temperatures (around 340 K), and desolvation happens in bath gas at atmospheric pressure. When desorption from the droplet surface occurs, there is probably at least one solvation shell present around the ion so that the process is extremely soft. When fragmentation at covalent bonds has been observed it seems almost always to be attributable to energetic collisions with bath gas molecules during ion transport into the MS. However, some complexes with non-covalent bonds can probably be dissociated by electrostatic repulsion forces when an ion has extensive charge multiplicity. In the case of proteins such stretching may result in a loss of tertiary and quaternary structural features.

(2) There is very little dependence of the ionization mechanism on analyte molecular size. In particular, there seems to be no upper mass limit in sight for the ionization step. Recent work suggests that PEG oligomers with molecular weights of 5 million can be ionized [35]. However, there are indications that at these ultrahigh masses, the ionization may occur by the charged residue model of Dole, rather than by the desorption model of Iribarne and Thomson.

(3) Ionization efficiency in ES can be very high. In any consideration of such efficiency, it is useful to distinguish between when charge availabilty may be a limiting factor and when it is not. It is probably limiting when there are more adduct charge sites on the analyte molecules in the spray than there are charges in the spray current. The spray current is set by the spray characteristics and is under normal instrumental conditions on the order of about 100 nA. Therefore, charge limitation sets in at about 0.1 g/L if the flow rate is one uL/min and an average molecule can retain about one unit charge per 1000 Da. In this charge limited case, the charging efficiency of ES can be conveniently defined as the total number of charges on all the analyte ions divided by the number of net charges supplied. Clearly, this ratio can then vary from 0 in the worst case to unity in the best. In the non-charge limited circumstance, a more appropriate definition for ionization efficiency is the fraction of analyte molecules introduced that become ions in the source. Experience in different laboratories suggests that both of these efficiencies can be very high, approaching substantial fractions of unity. These considerations relate to the spray as a whole. The fraction of ions produced that actually reach the analyzer is always much smaller.

(4) Figure 5 shows the measured signal for doubly charged ions of the peptide gramicidin S and its dependence respectively on liquid flow rate and concentration.

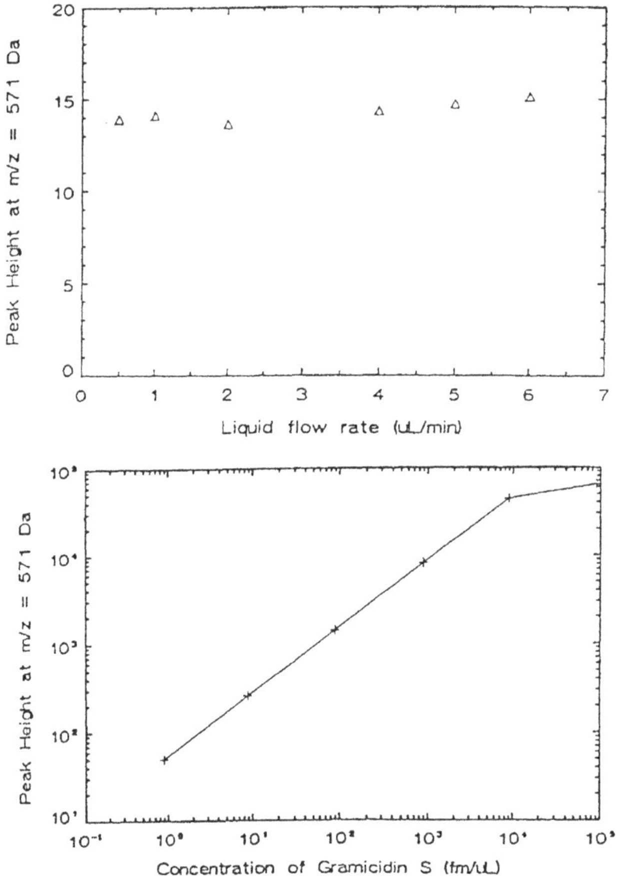


Figure 5. ESMS Peak Height for Gramicidin S ($2^+$): (a) Dependence on Sample Flow Rate; (b) Dependence on Analyte Concentration

The behavour depicted in the figure was noted early on [36] and demonstrates that the ES signal is linear in concentration but not in total amount of sample supplied. The relatively wide range over which the dynamic response is nearly linear in concentration together with its excellent absolute reproducibility make ES a good candidate for quantitative studies. When drawing quantitative conclusions from signals of multiply charged ions, one should sum the peak heights of all charge states to minimize errors due to any shift in charge distribution under different experimental conditions. The deconvoluted peak height (see below) can conveniently be used for this purpose when the number of charge states is greater than two or three.

(5) It is early for firm conclusions, but ESMS results thus far indicate little interference between components of mixtures, much less than in techniques such as FAB and PD.

(6) Although little systematic work has been done on determining what kinds of molecules are amenable to ES ionization, available data indicate that it works with most important classes of biomolecules [37]. These include peptides and proteins, carbohydrates, nucleotides and glycoproteins. In general, molecules which have a sufficient number of polar sites to result in a m/z values under 3000 can be ionized. However, there is no positive evidence that ions with even larger m/z values cannot be produced.

## 2.3 SAMPLING AND MASS ANALYZING THE IONS

To take full advantage of the high efficiency of ES ionization, it is crucial that as large a fraction of the ions formed as possible be transported into the mass spectrometer. Because the mass analyzer must operate in vacuum, the total gas flow that can enter its chamber is determined by the pumping speed. Therefore, adding additional pumping stages between the source and analyzer chamber can improve transport efficiency only if the ratio of ions to neutrals in the gas that enters the analyzer chamber can be increased by such staging. In the region where pressure is high enough so that ion trajectories are dominated by viscous forces, the difference in momentum between the carrier gas and the analyte ions can result in some separation of light carrier gas molecules from relatively heavy analyte ions. In low pressure regions, electrical fields can focus the ions on the center line whereas the trajectories of neutral species continue to diverge. It turns out that these separation phenomena can be quite effective. Thus, insertion of two rotary pumped stages between the ES chamber and the first "high vacuum" chamber has increased overall transmission of the ions from the ES chamber at atmospheric pressure into the mass analyzer chamber from a fraction of $10^{-4}$ or less in the apparatus of Figure 2 to perhaps $3 \times 10^{-2}$.

The region of the supersonic free jet expansion into vacuum is of particular importance. Applying an acceleration potential between the nozzle and the first skimmer can "heat" the ions by increasing the energy of their collisions with bath gas molecules. Thus, their internal energy can be varied, in principle, from below room temperature - because of cooling by the free jet expansion - to "temperatures" high enough to break covalent bonds. If the gas density is still moderately high as it enters the next skimmer, the kinetic energy seen by the mass spectrometer is independent of the electrostatic acceleration in the free jet region and is largely determined by the voltage at which the second skimmer is maintained. Some workers have used such acceleration between the nozzle and first skimmer to provide additional desolvation of the ions, an especially important option if counter current bath gas flow is not used in the ES source. It has also been used to fragment parent ions, thus achieving some MS/MS advantages in a single analyzer system.

Because ES produces highly charged ions that fall into a moderate m/z range regardless of molecular weight, quadrupole mass filters have been used for almost all of the early ESMS work. Figure 6 shows examples of typical spectra for multiply charged protein ions obtained at sample flow rates of 1 to 2 uL/min with a quadrupole having an upper limit for m/z of 1500.

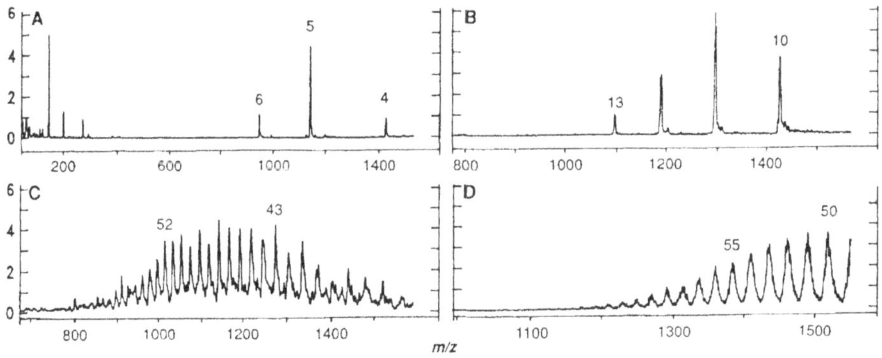


Figure 6. ESMS Spectra for: (A) Insulin, M=5730; (B) Lysozyme, M=14,300; (C) α -amylase, M=54,700; (D) Conalbumin, M=76,000

Quadrupoles are relatively cheap, easy to use and capable of excellent mass accuracy. Therefore, they will probably continue for quite some time to be the most popular mass spectrometer in ESMS analysis. However, they have limited resolution and the principal disadvantage of being dispersive analyzers. That is to say, at any one time only a very small fraction of the ions actually contributes to the read out signal. For some applications, therefore, other types of mass analyzers may be advantageous. In the past year, several groups have reported the combination of ES with double focussing sector instruments [39]. Interface design is somewhat more complicated in these combinations because the intermediate pressure region is prone to corona discharges at the high voltages required for injection of ions into the analyzer. It may also be difficult to incorporate the additional pumping in existing source housings. Nevertheless, results with a sensitivity approaching that of quadrupoles and with much higher resolution have already been achieved with ES ionization and sector analyzers.

Another interesting mass analyzer for ES sources is the quadrupole ion trap, sometimes called a "quistor" or "quadupole ion storer" [40]. In principle, a non-dispersive device it offers the prospect of very high sensitivity. Cooks and his colleagues along with workers at Finnigan MAT, its manufacturer [41, 39e], and others, have recently achieved dramatic improvements in mass range, sensitivity, MS/MS capability, the introduction of ions from external sources and especially in resolution. ES has been interfaced with ion traps to produce very promising results [42]. This combination seems in many ways ideal and will doubtless become a very important technique once instrumentation becomes commercially available.

ES has also been successfully interfaced with Fourier Transform Ion Cyclotron Resonance systems to accomplish so-called FTMS analysis [43]. However, it remains difficult to introduce enough ions from a source at atmospheric pressure into an FTMS cell that must be maintained at ultra high vacuum. The practical application of FTMS awaits solution of this not inconsiderable problem.

### 3. Interpretation of Spectra

As Figure 6 shows, the ES spectrum of even a single substance has an unfamiliar look in that it comprises a whole series of peaks for a single species. However, the use of some simple relations makes it quite easy to interpret this characteristic series of peaks for an unknown substance [44]. Suppose a molecule of a pure substance of molecular weight M has been transformed into a multiply charged ion by attachment of i adduct charges each with mass $m_a$. Then

$$K_i = (M + im_a)/i = M_2/i + m_a$$

or

$$(K_i - m_a) = K'_i = M/i \tag{1}$$

where $K_i$ is the measured location of a peak on the m/z scale of the analyzer and i is the number (unknown) of charges on the peak. For any two peaks separated from one another by j charges (for example j = + 1 for a peak whose ions have one more charge, that is, the peak immediately to the left of any i th peak in Figure 5.) we get from Eq. 1

$$i = jK'_{i+j}/(K'_i - K'_{i+j}) \tag{2}$$

That is to say, because we know that the ions of any one peak differ by one charge from those of either adjacent peak, the charge on ions of any peak can be determined from measured m/z values at any two measured peak locations. The adduct ion mass $m_a$ can in principle be determined just as easily from Eq. 1

$$m_a = (1/j)\,[(i + j)\,K_{i+j} - iK_i)] \tag{3}$$

but the accuracy required for $K_i$ and $K_{i+j}$ in this determination is very high. Therefore, one is often forced to assume the identity of the adduct charges, which for proteins seem almost always to be protons but may vary widely for other substances, depending upon the solution composition.

From equations 1 to 3, we easily get an estimate of the molecular weight M of the molecules

$$M = \left(\frac{1}{n_0}\right) \sum_i i K'_i \tag{4}$$

where the sum is over all the peaks included in the average, and $n_o$ is the number of those peaks.

Working with proteins of known molecular weight, one soon realizes that the mass estimate derived from one of the middle peaks in a distribution is often better than from one near the extremes. The reason is, of course, that the $K_i$ values for the peaks in the middle of the distribution are obtained from stronger signals. To get the best possible mass estimate, it is therefore advantageous to weight the contributions of the peaks in the series, for example by their relative height. A more subtle way of weighting the peaks makes use of the coherence of the series. From equation (1) we see that the ratio of $K_i$' values for the peaks in the middle of the distribution are obtained from stronger signals. To get the best possible mass estimate it is therefore advantageous to weight the contributions of the peaks in the series, for example by their relative height. A more subtle way of weighting the peaks makes use of the coherence of the series. From equation (1) we see that the ratio of Ki' values at any two peak locations must be a ratio of small integers. This feature can be used to weight their proximity to a line based on that criterion. Ref. 44 gives an algorithm for obtaining such a weighted average. (Note that Eq. (6) in that paper should read $W^{i}-1$ =... and that the second part of the formula is simply a statement that the sum of the weighting factors is unity.) In a slightly modified form, this algorithm can also be used to determine with confidence whether a given peak belongs to one or another series of peaks or whether it is due to some singly charged species.

For many applications, it would be a great advantage to be able to transform the series of peaks into a singly charged peak of mass M. It turns out that such a mathematical transformation is indeed possible. One starts with the following function F(M)

$$F(M^*) = \sum_i f\left(\frac{M^*}{i} + m_a\right) \tag{5}$$

where M* is the x variable of the parent mass spectrum, the sum runs from i = 1 to some maximal value such that M*/i is smaller than the lowest measured peak in the spectrum, and f represents the measured mass spectrum. One finds that equation 5 will give rise to a parent peak at the desired mass M. The properties of this so called "deconvolution algorithm" are described in detail in ref. 44. Figure 7 illustrates its application to the measured spectrum of the protein cytochrome C (horseheart).

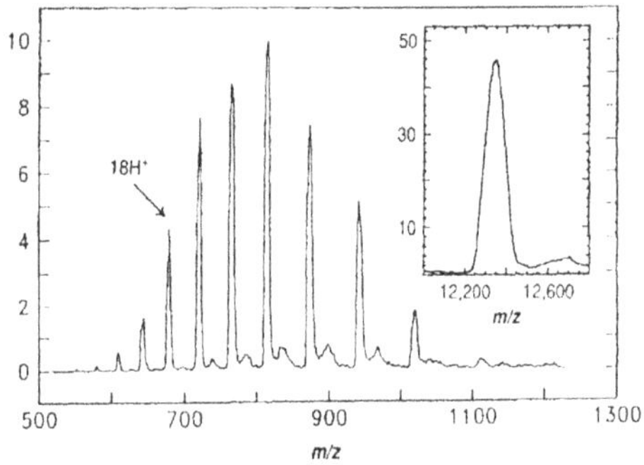


Figure 7. ESMS Spectrum for Cytochrome C. Insert Shows the Results of Deconvolution to the Same Ordinate Scale.

The main feature of the figure shows the measured ES mass spectrum of cytochrome C. The inset panel shows its deconvolution. Interestingly, no assumptions need be made about the spectrum except for a value of the adduct ion mass. In practice it can be useful as a routine measure to deconvolute experimental spectra to get a direct visual indication of the parent ion mass and the peak shape.

The deconvolution algorithm of Figure 5 is particularly useful for interpreting the spectra of mixtures because it can be very difficult to identify visually the peaks belonging to the sequence for a particular parent species, especially if they are numerous and unknown a priori. A single run of the algorithm, or an iterative procedure based on it, can greatly simplify the assignment of m/z values, even to components at a low concentration in a mixture. Finnigan has developed a commercial version of the algorithm with additional features. An even more flexible variation has been worked out by Labowsky and will be available from Analytica of Branford, CT.

These data processing techniques are very helpful in interpreting spectra of mixtures and can resolve and identify several unknown protein components but as mixture complexity increases so does the difficulty of interpretation. This difficulty along with the reality that biomolecules in the mass range above 50 kDa are rarely pure, put a premium on the resolving power of the mass analyzer. Consequently resolution obtainable with quadrupole mass filters may often not be sufficient. Thus, sector instruments become very attractive because they can achieve resolutions of 2000 to 3000 (10% valley definition) with relatively high sensitivity. Instruments with sufficiently high resolution allow one to determine the charge state of an ion directly from the separation of m/z values for parent ions with differing $^{13}C$ contents.

One of the attractive features that ESMS shares with LDMS is the high mass accuracy that it makes possible even for species of high molecular weight. Before they arrived on the scene, molecular weight determination for an unknown protein (short of finding its primary structure and adding up the contributions of component amino acids) was limited to accuracies not better than 5 percent. By ESMS protein masses up to 20 kDa and higher have been determined within less than one dalton corresponding to an accuracy in the vicinity of 0.01%. For example, using a sector instrument, Gallagher and Chapman [45] of Kratos Analytical have measured the mass of a trypsinogen (M > 23 kDa) to within a few tenths of a Da. Part of the reason for this astonishing accuracy is the fact that the mass spectrometer only needs to work at low m/z values where measurements can be more precise and calibrations are easier than in the high m/z range. Furthermore, each of the many peaks in an ESMS spectrum constitutes an independent measurement of the parent mass so that averaging over them can substantially reduce random error. Extremely precise calibration is required to reach very high mass accuracy because an error in absolute peak location is multiplied by the number of charges on its ions. For example, a systematic error of 0.05 units on the m/z scale can translate into a 5 unit mass error if the molecule has 100 charges. Although comparison of the mass determinations from the different peaks in a multiplet gives an estimate of the random error, the absolute error includes contributions from both random and calibration errors.

The high mass accuracy achievable in ESMS is extremely valuable in many biotechnological and biomedical problems. For example, it can provide in a single and nearly instantaneous observation, a reliable determination of the nature of a post-translational modification in a predicted protein structure.

## 4. ES MS/MS

During the last decade so-called tandem mass spectrometry (MS/MS) has begun to play an increasing role in determining the identity and structure of unknown species (see other chapters in this volume). Part of the reason for this development is the great need in biological sciences for structural elucidation of very small amounts of biologically active molecules. Mass spectrometry should in principle have advantages in speed, sensitivity and versatility over more traditional 'wet chemistry' techniques that now dominate biochemical analysis. Although much valuable work has been accomplished, there are a number of problems in MS/MS that ES ionization may help solve. First, ES easily adds the temporal dimension to MS/MS because it is a natural 'on-line' method. FAB can also be adapted to on-line use but requires a substantial amount of experimental skill and experience (see the chapter by Caprioli) and is often time consuming. There are, furthermore, several factors that severely inhibit the application of FAB MS/MS to large parent molecules. One is the declining efficiency of ion production with increasing mass. Another is that the efficiency of collisionally activated dissociation (CAD) also drops with increasing parent mass, due in part to the smaller fraction of energy in the collision center of mass and in part to the greater number of internal degrees of freedom among which the collision energy can be partitioned. The multiple charging of ES ions is such that the charge/mass ratio, and therefore the fraction of collision energy in the center of mass, remain about constant with increasing parent mass. The overall efficiency of ES ionization actually seems to increase somewhat with increasing parent mass.

It is to be noted, however, that for the ions that ES produces from large molecules, the total charge is distributed over an increasing number of charge states as the molecular weight increases. Therefore, the number of ions in a particular charge state available for selection and fragmentation does show a roughly linear decrease with increasing molecular weight. However, the overall ionization efficiencies with large species are so much higher with ES than with other soft ionization techniques that much larger selected ion currents are obtained in spite of "dilution" because of charge multiplicity. Moreover, this multiplicity means that each parent ion can produce a plurality of fragment ions whereas each singly charged parent ion produced by other techniques can provide only one fragment ion. Further to be noted, as mentioned above, is that multiply charged ions seem to be pre-stretched by coulombic repulsion so that less additional energy is needed to break bonds. All these factors combine to give a very high fragmentation efficiency for large ions produced by ES. On the other hand, as long as there is no independent measure of the charge states of the daughter fragment ions, interpretation of MS/MS spectra from multiply charged parents will be difficult, maybe impossible. This unknown may not be a problem in the peptide mass range where rules for identification of the fragments of doubly charged and triply charged ions are already being developed. For very large peptides and proteins, however, immediate applications of ES MS/MS may be confined to 'fingerprinting' [46] and looking for variations of known structures.

An interesting feature of ES MS/MS noted earlier is that one stage of fragmentation can be performed in the free jet expansion region by the above mentioned 'frictional heating'. This technique was first used by the Leningrad group [38] and allows very efficient fragmentation. Experimental requirements are then simplified as only one MS is required. Meng et al. [47], for example, have used this technique to determine almost the complete sequence of human renin substrate (Mr = 1706) with only one double focusing sector instrument. Note that this technique also avoids two disadvantages of conventional MS/MS: the low overall transmission between MS1 and MS2 and the reduced resolution of MS2. Of course, single species in particular charge states cannot be selected for fragmentation so that interpretation of daughter spectra remains complicated. Another need is for improved reproducibility of ES MS/MS spectra.

## 5. Summary and Outlook

The principles and practice of ESMS have been reviewed. The three main component processes include: the first or "spraying" step whose characteristics determine compatibility with the source of sample liquid including whatever species separation may be required (e.g., LC or CZE). As a result of new developments, a wide range of conditions in flow rate and solution composition can now be accommodated in this spray step. The second step, ion formation, is far from understood, but important features of its nature have emerged from experimental studies so far. Among the most significant of these are the extreme softness of the ionization process, the apparent absence of an upper mass limit, and the applicability to broad classes of molecules. In the third or mass analysis step, the ions are transported by way of a free jet expansion to the mass analyzer that perforce must be in vacuum. This free jet region provides an opportunity to control the internal energy of the ions and indeed to bring about fragmentation. Because ES is inherently continuous, all kinds of mass analyzers can and have been used with it.

Accuracy and sensitivity at reasonable instrumental complexity and cost, are alone enough to secure a solid place for ESMS in the biomedical field. Major advances in achieving its coupling with separation techniques such as CZE and HPLC have already been demonstrated in research settings and can be expected to become routine. Further downstream in conjunction with MS/MS techniques, it is to be expected that a major contribution of ES ionization will lie in structural elucidation of biomolecules by tandem mass spectrometry.

## 6. Acknowledgements:

Most of the research on which this account is based was sponsored by the Institute of General Medical Sciences of the National Institutes of Health under Grant 501 GM316605. The authors deeply appreciate the interest, cooperation and support of many colleagues in their respective institutions. Special thanks are also due to Craig M. Whitehouse and his associates of Analytica of Branford, CT.

## 7. References

1. Meng, C. K., Mann, M., and Fenn, J. B. (1988) *Zeit. Phys. D* **10**, 361.
2. Tanaka, K., Waki, H., Idao, Y., Akita, S., Yoshida, Y. and Yoshida, T. (1988), *Rapid Commun. Mas Spectrom*, **2**, 151.
3. (a) Karas, M. and Hillenkamp, F. (1989) Proc. 11th Int. Mass Spec. Conference, Bordeaux, France. *Adv. Mass Spectrom* (1989) **Vol. 11A**, p. 416, (b) Karas, M. and Hillenkamp, F. (1988) *Anal. Chem.* **60**, 2299.

4. Burlingame, A L., Millington, D. S., Norwood, D. L. and Russell, D. H. (1990) *Anal. Chem.* **268R**.
5. Fenn, J. B., Mann, M., Meng, C. K., Wong, S. F. and Whitehouse, C. M. (1990) *Mass. Spec. Rev.* **9**, 37.
6. Fenn, J. B., Mann, M., Meng, C. K., Wong, S. F. and Whitehouse, C. M. (1989) *Science* **246**, 64.
7. Smith, R. D., Loo, A , Edmonds, C. G., Barinaga, C. J. and Udseth, H. R. (1990) *Anal. Chem.* **62**, 882.
8. Mann, M. (1990) *J. Organic Mass Spectrom.* **25**, 575.
9. Field desorption is used here as a general term meaning the vaporization and ionization of molecules by action of a strong field, rather than the narrower term designating the specific ionization technique described by Inghram, M. and Gomer, R. J. (1954) *J. Chem. Phys.* **22**, 1279.
10. Buehler, R. J., Flanigan, E., Green, L. J. and Friedman, L. (1974) *J. Am. Chem. Soc.* **96**, 3990.
11. Barber, M., Bordoli, R. S., Elliot, G. J., Sedgwick, R. D. and Tyler, A. N. (1981) *J. Chem. Soc. Chem. Commun.* 325.
12 (a) Thorgerson, D. F., Skowronski, R. P. and Macfarlane, R. D. (1974) *Biochem. Biophys. Res. Commun.* **60**, 616. (b) Roepstorff, P., and Sundqvist, B. (1986) In Biochemical Applications of Mass Spectrometry, Gaskell, S. ed., J. Wiley and Sons Ltd: Chichester, p 269.
13. Beckey, H. D., (1969) *J. Mass. Spectrom. Ion. Phys.* **2**, 500; "Principles of Field Ionization and Field Desorption" (1977) *Mass Spectrom.* Pergamon, Oxford.
14. Simons, D. S., Colby, B. N. and Evans, Jr., C. A. (1974) **Int. J. Mass Spectrom.** *Ion Phys.* **15**, 291; Cook, K. D. (1986) *Mass Spec. Rev.* **5**, 467.
15. (a) Iribarne, J. V. and Thomson, B. A. (1976) *J. Chem. Phys.* **64**, 2287; (b) Thomson, B. A. and Iribarne, J. V. (1979) *J. Chem. Phys.* **71**, 4451; (c) Iribarne, J. V., Dziedzic, P. J. and Thomson, B. A. (1983) *Int. J. Mass Spetrom Ion Phys.* **50**, 331.
16. Blakley, C. R., McAdams, M. J. and Vestal, M. L. (1978) *J. Chromatogr.* **158**, 264; Blakley, C. R., Carmody, J. J. and Vestal, M. L. (1980) *Anal. Chem.* **52**, 1636; Blakely, C. R., Carmody, J. J. and Vestal, M. L. (1980) *Clin. Chem.* **26**, 1467; Blakely, C. R., Carmody, J. J. and Vestal, M. L. (1980) *J. Am. Chem. Soc.* **102**, 5931; Vestal, M. L. (Elsevier: Amsterdam, 1985) in Mass Spectrometry in the Health and Life Sciences, Burlingame, A. L. Castagnoli, Jr., N. eds.
17. (a) Dole, M, Mach, L. L., Hines, R. L., Mobley, R. C., Ferguson, L. P. and Alice, M.B. (1968) *Phys.* **49**, 2240; (b) Mach, L. L., Kralik, P., Rheude, A. and Dole., M. (1970) *J. Chem. Phys.* **52**, 4977; Clegg, G. A. and Dole, M. (1971) *Biopolymers* **10**, 821; (c) Teer, D. and Dole, M. (1975) *J. Polym. Sci.*, **13**, 985.
18. Anderson, J. B., Andres, R. P. and Fenn, J. B. (1966) "Supersonic Nozzle Beams" in *Advances in Chemical Physics* **10**, 275.
19. (a) Yamashita, M. and Fenn, J. B. (1984) *J. Phys. Chem.* **88**, 4451; (b) Yamashita, M. and Fenn, J. B. (1984) *J. Phys. Chem.* **88**, 4471.
20. Whitehouse, C. M., Yamashita, M., Meng, C. K. and Fenn, J. B. (1985) Proceedings of the 14th International Symposium on Rarefied Gas Dynamics, Oguchi, H. ed., (University of Tokyo Press, Tokyo), p 857.
21. Bailey, Adrian G. (1988) Electrostratic Spraying of Liquids, Research Studies Press LTD., Taunton, Sommerset, England and John Wiley & Sons Inc., New York.

22. Mann M. (1989) "Quantitative Aspects of Electrospray Mass Spectrometry, Ph.D. Thesis, Yale University, (University Microfilms).
23. (a) Hyati, I., Bailey, A. and Tadros, Th. T. (1986) *Nature* **319**, 41; (b) Hyati, I, Bailey, A. and Tadros, Th. T. (1987) *Coll. Interf. Sci.* **117**, 205; (c) Hyati, I, Bailey, A. and Tadros, Th. T. (1987) *Coll. Interf. Sci*. **117**, 222.
24. G. I. Taylor Proc. R. Soc. (1964) **A280**, 383.
25. Bruins, A. D., Weidolf, L. O. G. and Henion, J. D. (1987) *Anal. Chem.* **59**, 2647.
26. Allen, M. H., Field, F. H. and Vestal, M. L. (1990) Proceedings 38th Annual ASMS Conference on Mass Spectrometry and Allied Topics, p. 431 (1990).
27. Smith, R. D., Barinaga, C. J. and Udseth, H. R. (1988) *Anal. Chem.* 1948.
28. Loo, J. A., Udseth, H. R. and Smith, R. D. (1989) *Anal. Biochem.* **179**, 404.
29. Lee, E. D., Henion, J. D. and Covey, T. R. (1989) *Biomed. Environ. Mass Spectrom.* **18**, 253.
30. Huang, E. C., Wachs, T., Conboy, J. J. and Henion, J. D. (1990) *Anal. Chem.* **62**, 713 A.
31. Wong, S. F., Meng, C. K. and Fenn, J. B. (1988) *J. Phys. Chem.* **92**, 546.
32. Meng, C. K., Wong, S. F., Mann, M. and Fenn, J. B. unpublished data.
33. Bailey, F. E. (1976) Poly (ethylene oxide), Academic Press, New York.
34. Wong, S. F. and Fenn J. B. (1989) Proceedings of the 37th Annual Conference on Mass Spectrometry and Allied Topics p. 389.
35. Nohmi, T., Wong, S. F. and Fenn, J. B. (1990) Proceedings 38th Annual ASMS Conference on Mass Spectrometry and Allied Topics p. 10.
36. Whitehouse, C. M., Dreyer, R. N., Yamashita, M. and Fenn J. B. (1985) Anal. *Chem.* **57**, 675.
37. Ikonomou, M. G., Blades, A. T. and Kebarle, P. (1990) *Anal. Chem.* **62**, 957.
38. Alexandrov, M. L., Gall, L. N., Drasnov, M. V., Nikolaev, V. I. and Shkurov, V. A. (1985) *Zhurnal Analiticheskoi Khimii*, **40**, 1570.
39. Allen, M. H. and Lewis, I. A. S. (1989) *Rapid Commun. Mass Spectrom.* **3**,225; (b) Meng, C. K., McEwen, C. N. and Larsen, B. S. (1990) *Rapid Commun. Mass Spectrom.* **4**, 147; (c) Gallagher, R. T., Chapman, J. R. and Mann, M. (1990) Proceedings 38th ASMS Conference on Mass Spectrom. and Allied Topics p. 130, (will also appear in *Rapid Commun. Mass Spectrom*); (d) Lee, E. D., Shumate, C. B. and Radolovich, G. ibid. p. 437.
40. March, R. E. and Hughes, R. J. (1989) *Quadrupole Storage Mass Spectrometry*; Winefordner, J. D. ed. Chemical Analysis vol. 102, Wiley, New York (1989).
41. (a) Louris, J. N., Cooks, R. G., Kyka, J. E. P., Kelley, P. E., Stafford, G.C., Jr. and Todd, J. F. J. (1987); *Anal. Chem.* **59**, 1677; (b) Kaiser Jr., R. E., Cooks, R. G., Syka, J. E. P. and Stafford, G. C., Jr. (1990) *Rapid Commun. Mass Spectrom.* **4**, 30.
42. Van Berkel, G. J., Glish, G. L. and McLuckey, S. A. (1990) *Anal. Chem.* **62**, 1284.
43. Henry, K. D., Williams, E. R., Wang, B. H., McLafferty, F. W., Shabanowitz, J. and Hunt, D. F. (1989) Proc. natl. Acad. Sci. USA 86, 9075.
44. Mann, M., Meng, C. K. and Fenn, J. B. (1989) *Anal. Chem.* **61**, 1702.
45. Gallagher, R. T. and Chapman, J. R. Kratos Analytical Manchester, personal communication.
46. Smith, R. D., Loo, J. A., Barinaga, C. J., Edmonds, C. G. and Udseth, H. R. (1990) *J.Am Soc. Mass Spectrom.* **1**, 53.
47. Meng, C. K., McEwen, C. N. and Larsen, B. S. (1990) *Rapid Commun. Mass Spectrom.* **4**, 151.

# LASER DESORPTION MASS SPECTROMETRY PART I: BASIC MECHANISMS AND TECHNIQUES

F. HILLENKAMP AND H. EHRING
*Institut für Medizinische Physik der Universität Münster*
*Robert-Koch-Str. 31 D-4400 Münster, FRG*

ABSTRACT. The basic mechanisms und techniques of ultraviolet and infrared direct and matrix-assisted laser desorption/ionization of organic molecules are discussed. Emphasis is placed on results which elucidate energy transfer to the sample and ionization of molecules and their dependence on the laser parameters. Mass analysis by time-of-flight mass spectrometry is introduced.

## 1. Introduction

The first papers on the use of lasers for ion generation in mass spectrometric analysis appeared in the literature only a few years after the discovery of the laser principle. Early investigations on small organic molecules were conducted by Vastola et al. [1], for example. The rather poor quality of the results in those times was mostly attributable to the limited performance of the free run ruby lasers used in those studies. In the years to follow a large number of different lasers has been tested in conjunction with different types of mass spectrometers. An excellent account of the current state of laser mass spectrometry can be found in a recently published book [2]. For reasons discussed below, two instrumental combinations have been found most useful for a large variety of applications in organic mass spectrometry: Q-switched Nd-YAG lasers, mostly frequency quadrupled to a wavelength of 266 nm in the far ultraviolet (UV), combined with time-of-flight (TOF) mass spectrometers [3, 4] and $CO_2$-lasers at a wavelength of 10.6 $\mu$m in the far infrared (IR) attached to Fourier Transform Ion Cyclotron Resonance (FT-ICR) [5, 6, 7] or, much less frequently, to TOF [8] mass spectrometers. Many other combinations including particularly dye lasers and magnetic sector instruments for mass separation have been found useful for specific applications. New combinations are continuously being tested for matrix assisted laser desorption of large biomolecules such as the $N_2$-laser at 337 nm in the mid ultraviolet or the Er-YAG laser at 2.94 $\mu$m in the mid-infrared combined with with TOF mass spectrometers (see chapter 6). The current knowledge and understanding of the desorption and ionization mechanisms is far from being complete, but a few general principles, guiding most of the applications have evolved over the last 25 years. Some of the features seem to be general (*i. e.*, independent of the specific technique under consideration such as the relation between the laser wavelength as well as the sample absorption and the influence of the laser pulse width). Others, such as size of laser focus, geometry of the sample and the ion source, magnitude and time course of the ion extraction field, requirements for ion lifetime etc. vary greatly among the various combinations of lasers and mass spectrometers. As a consequence of this variation the results obtained are quite diverse and often difficult to compare to each other; the optimal instrumental choice depends considerably on the analytical problem to be solved.



*M. L. Gross (ed.), Mass Spectrometry in the Biological Sciences: A Tutorial*, 165–179.

## 2. The Process of Laser Desorption

The desorption of fragile, thermally labile molecules out of the condensed phase into the vacuum of the mass spectrometer, associated with their ionization is a complex process. Figure 1 is an attempt to describe at least the most important steps of this process to the extent that we understand it today. It is meant to ease the discussion of the different mechanisms and properties involved. The sequence of events as drawn may be correct for a single given ion, but should not be taken as real for the whole, irradiated sample. To avoid thermal decomposition of the sample molecules, the steps must in reality be considered to occur simultaneously in a very short time, typically on the nanosecond time scale or less and in a very small volume of only micrometers in dimension. In a first step the energy of the laser beam must be deposited into the sample. This may occur by absorption of the sample itself or via heat conduction after absorption in the substrate. The excitation of the solid, the details of which are very poorly understood so far, will then lead to the actual desorption. It is generally assumed that even under threshold conditions, this is a collective process of a limited sample volume rather than a desorption of individual molecules or ions. It is not unreasonable to assume that this step bears definite similarities to the equivalent steps in particle desorption such as SIMS or plasma desorption (it should be kept in mind that even though single desorption of preformed ions from the surface is postulated in these cases, this is induced by a collective excitation and disintegration of a given sample volume). For a short time during the disintegration the system will consist of molecules with high mobility at a high density. It is, therefore, reasonable to assume that in this "selvedge", a variety of chemical reactions are quite probable. It can, in fact be assumed, that the striking similarity of spectra obtained with the different desorption methods reflects the common chemical pathways and reactions in this step, leading to a selection of the most stable molecular and ionic species. The energy, deposited into the volume from the laser beam and, in the many cases of rather high charge density, also the Coulomb repulsion will then lead to a fast expansion of the "laser plume" into the vacuum. Some chemistry such as cationization can continue during this phase and metastable decay of the ions will continue to alter the ion spectrum all the way down through the mass spectrometer and to the detector.

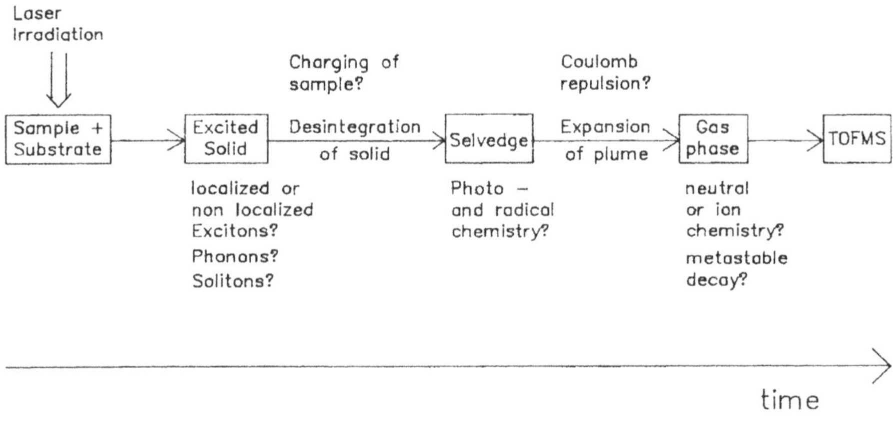


Figure 1. Schematic diagram of the different steps in laser desorption. For explanation see text.

## 2.1 ENERGY TRANSFER AND THE INFLUENCE OF THE LASER WAVELENGTH

In principle energy can be transferred from the laser beam to either the sample or the substrate. In the majority of cases, energy transfer proceeds exclusively or predominantly via linear resonant absorption. In this case the amount of energy transferred per unit volume of the absorber $E_a/V$ follows from Beer's law and is determined by its molar extinction coefficient $e_n$ and the molar concentration $c_n$.

$$E_a/V = 2.3 \times e_n \times c_n \times H \tag{1}$$

$$H = E \times \tau \tag{2}$$

For a sample with a given extinction coefficient $e_n$, it can be controlled by variation of the radiant exposure H, the total energy per unit area, incident on the sample per pulse. For a given pulse width $\tau$, the controlling parameter is the irradiance E (*i. e.* , the power, incident on the sample per unit area). The latter is often called laser intensity. For aromatic molecules or molecules with aromatic ligands $e_n$ typically is in the range of $10^3$-$10^4$ L $mol^{-1}cm^{-1}$. Irradiance and radiant exposure are usually given as mean values, averaged over the whole area and pulse width. Ion detection will set in at some threshold irradiance or radiant exposure, which depends on the sample/substrate combination in relation to the wavelength used, as well as on the instrumental configuration. Even though this is not a basic parameter, related directly and exclusively to the desorption process, it has turned out to be very useful to specify this parameter for a given experiment and scale all other irradiances to this threshold value. For the instrumental configurations discussed in this paper, threshold irradiance for ns-pulses in the whole wavelength range from 193 nm to 10.6 $\mu$m range from ca. $10^5$ to $10^7$ $Wcm^{-2}$. Particularly in molecular ion desorption, good results are obtained in a limited range above threshold irradiance only; typically this range spans only one order of magnitude or less. If sample and substrate do both not absorb at the laser wavelength used, energy can be transferred via laser induced, nonlinear absorption. This typically sets in at irradiances at or above ca. $10^7$ $Wcm^{-2}$. Control of the amount of deposited energy is much more difficult in nonlinear absorption as compared to the linear case, and one easily induces optical breakdown and plasma formation at the sample, which are highly undesirable for organic ion desorption. It is important to keep in mind that the irradiance is usually not uniform across the irradiated area and as a function of time. Poor results may be obtained for very nonuniform distribution, particularly if the patterns are not reproducible from laser shot to laser shot. Good mode control of the laser is, therefore, advisable.

The different possible cases of energy transfer are shown schematically in Figure 2. Metal substrates are used in the majority of cases to maintain a fixed electrical sample potential. Metals are more or less strong absorbers (and reflectors) of optical radiation throughout the whole wavelength range of usable lasers from the far ultraviolet (UV) at 193 nm to the far infrared (IR) at 10.6 $\mu$m. If the substrate is covered with only a thin layer (usually much less than 1 $\mu$m in thickness) of a non - or weakly absorbing sample, most of the energy will go into the substrate (Figure 2a). Sample heating will then occur with some delay and at a rate, limited by the -usually rather low - thermal conductivity of the sample. More often than not, the top volume of the substrate will get ablated along with the sample in this case. This mode will usually result in extensive fragmentation of organic molecules and complicated mass spectra. If substrate and sample are both nonabsorbing at the wavelength used, nonlinear absorption must be used for energy transfer (Figure 2b). The threshold irradiance for the onset of nonlinear absorption depends on the wavelength and also quite strongly on the presence of residual, resonant absorption. In UV desorption experiments, it usually sets in at ca. $10^7$ to $10^8$ $Wcm^2$.

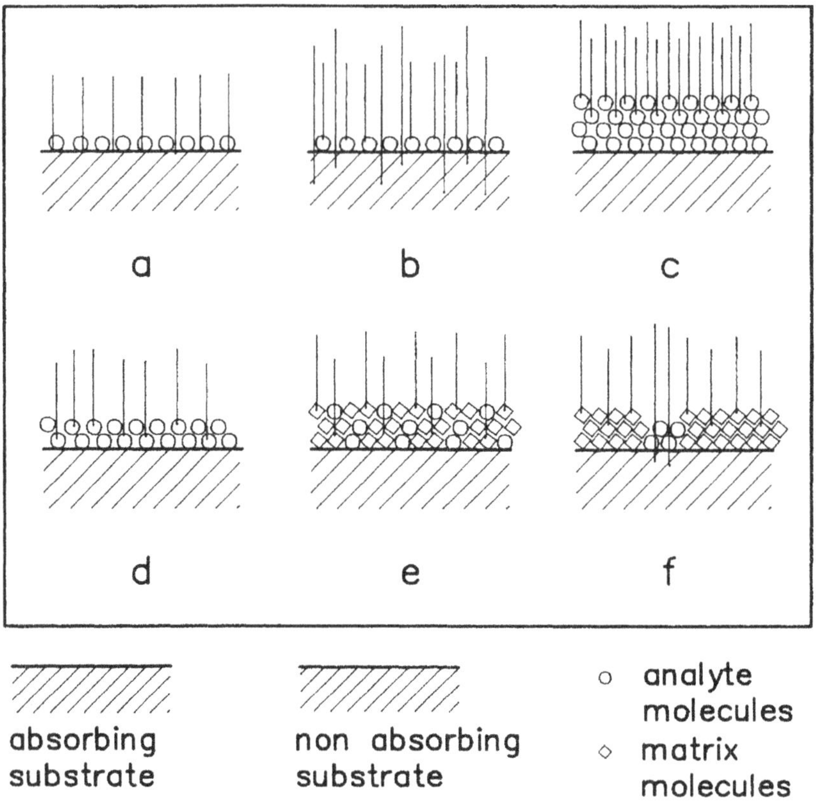


Figure 2. Energy transfer processes from the laser beam to absorbing or non absorbing analyte molecules on absorbing or non absorbing substrates. For explanation, see text.

Irradiances in the IR are expected to be somewhat higher. If suitable substrates are used, such as quartz in the UV and ZnSe or Ge in the IR, samples will usually exhibit a lower threshold for the onset of nonlinear absorption than the substrate, so substrate involvement will stay low or negligible. Very similar results can be obtained with "thick" layers of a nonabsorbing sample, typically several $\mu$m in thickness as shown in Figure 2c. In this case desorption of sample ions, based on nonlinear absorption can be induced without substrate involvement, even if the latter absorbs strongly [12]. As shown in Figure 3, quite good spectra can be obtained in this desorption mode, but it is difficult to get reproducible spectra because of the unavoidable energy fluctuation of the laser from shot to shot and the exceedingly strong irradiance dependence of the nonlinear process.

By far the best results have, so far, been obtained if a wavelength is chosen at which the sample has a strong, resonant absorption (Figure 2d). No substrate involvement is detectable in these cases. For UV-laser desorption this has been demonstrated in a series of systematic

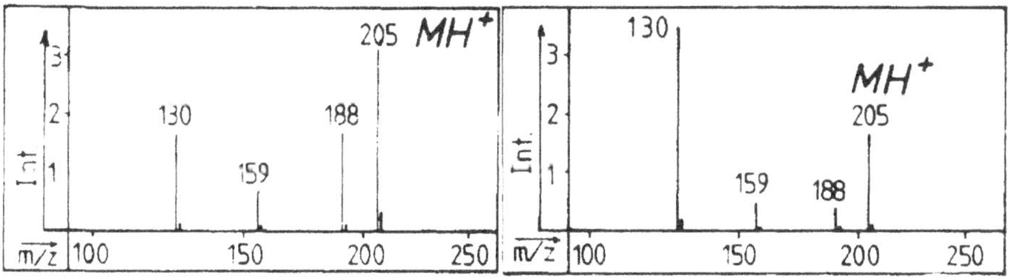


Figure 3. Direct laser desorption/ionization spectra of tryptophane with wavelengths of 310 nm (left) and 266 nm (right).

experiments [13, 14]. For IR-laser desorption, substrate involvement and its role in optimal desorption is less clear. In the UV, absorption results from electronic transitions in the molecules. In the wavelength range above 193 nm (where strong enough lasers, which can be oparated in atmosphere, are available), only aromatic molecules or those with aromatic residues such as the aromatic amino acids tryptophane and tyrosine and the nucleobases have a reasonable absorption. All organic molecules have vibrational and rotational absorption bands in the IR, but as a rule these are about one order of magnitude weaker than the electronic transitions in the UV. Figure 4 shows a series of UV-desorption/ionization spectra of four dipeptides, with varying degree of aromatic content, obtained with a wavelength of 266 nm [13]. All spectra were taken at the respective threshold irradiances. As expected, tryptophane, having the highest absorption at 266 nm, has the lowest threshold irradiance $E_0$. The less absorbing residues of tyrosine and phenylalanine show a 5-times higher threshold, and for the purely aliphatic peptide valylproline, nonlinear absorption must be induced at 12 times the threshold irradiance as compared to that of tryptophane. More important than the increasing threshold is the fact that the spectra get increasingly more complex with decreasing absorption, owing to increased fragmentation and the appearance of alkali ions and cationized organic ions in the spectra. Even though these spectra clearly demonstrate the importance of

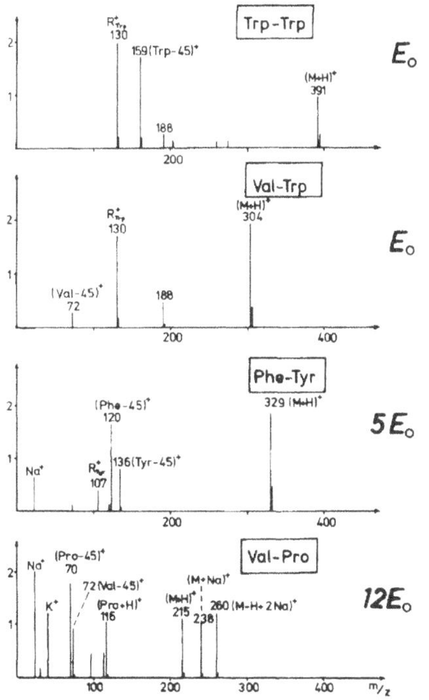


Figure 4. Direct laser desorption/ionization spectra of four dipeptides, recorded at their respective threshold irradiances (266 nm wavelength). Republished from ref. [13] with permission

linear sample absorption, some contribution by nonlinear absorption cannot be ruled out even in these cases. It was shown that even very thin absorbing tryptophane layers, too thin to shield effectively the absorbing substrate from the incoming laser radiation, will prevent substrate involvement, whereas nonabsorbing samples will not do so, even at much higher thicknesses [12].

Sometimes efficient photofragmentation competes with ion desorption after resonant energy absorption by chromophoric groups of the analyte molecules. In such cases, selected spectra, obtained with nonresonant wavelengths and taken close to desorption threshold, may show less fragmentation than those obtained with a resonant line as discussed below. Figure 3 shows this as an example for tryptophane.

The majority of organic molecules are aliphatic and lack absorption in the accessible UV wavelength range. They are, therefore, not accessible by direct resonant UV-laser desorption/-ionization. This has led to the development of matrix-assisted laser desorption/ionization [15, 16] as discussed in more detail in the following chapter 10. It had been observed that aliphatic amino acids such as alanine were codesorbed with aromatic ones such as tryptophane at the low threshold irradiance of the latter (Figure 5). Subsequent systematic investigations revealed that the principle of desorbing non absorbing molecules or ions out of an absorbing matrix can be generalized [17, 18, 19, 20]. It was found that small molecules with a strong resonant

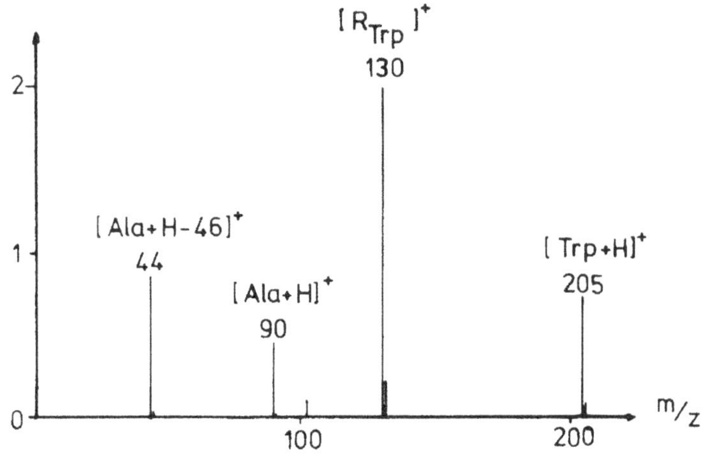


Figure 5. Spectrum of the aliphatic amino acid alanine codesorbed with the aromatic amino acid tryptophane at the irradiance threshold of the latter (266 nm wavelength). Republished from ref. [13] with permission

absorption and molar extinction coefficients at the wavelength of maximum absorption in the range of $10^3$ to $10^4$ L $mol^{-1}cm^{-1}$ work best as matrices. Typically they should be present at an about 100-1000 fold excess (V/V) relative to the analyte molecules. Out of a large number of potentially suitable matrix substances tested, only a few, however, gave good results. It can, therefore, be concluded that for this type of desorption to be successful the matrix must serve more than just the purpose of mediating the energy transfer. Matrix isolation of analyte molecules most probably is one of them. Many, particularly the biologically active molecules, have a strong tendency to form large aggregates, particularly if the concentration increases upon evaporation of the solvent before the sample can be introduced into the vacuum. Dilution of analyte molecules with the matrix will prevent them from aggregating to complexes too large to be desorbed and analyzed. The sample is in this way transformed from a dilute aqueous solution to a dilute "solid solution". This requirement for a homogeneous solution is shown in Figure 2e. Many matrix/sample systems will segregate upon evaporation of the solvent as shown in Figure 2f because of different physicochemical properties such as solubility or condensed phase energy. Such samples are not suited for desorption. The possible contribution of the matrix to an efficient ionization of the analyte molecules will be discussed later. Several groups have searched for and found suitable matrices for the different, routinely available laser wavelengths in the UV and IR [15, 16, 17, 21, 22, 23, 24, 25, 26]. Figure 6 shows the absorption spectra of two commonly used matrices along with some of the standard UV-wavelengths used for matrix assisted laser desorption/ionization. It has been this technique of matrix assisted laser desorption that has brought about the real breakthrough in laser desorption/-ionization. Ions with masses above 300,000 Da have been observed; sensitivity of the technique is in the low femtomol range. More detailed information is given in chapter 10.

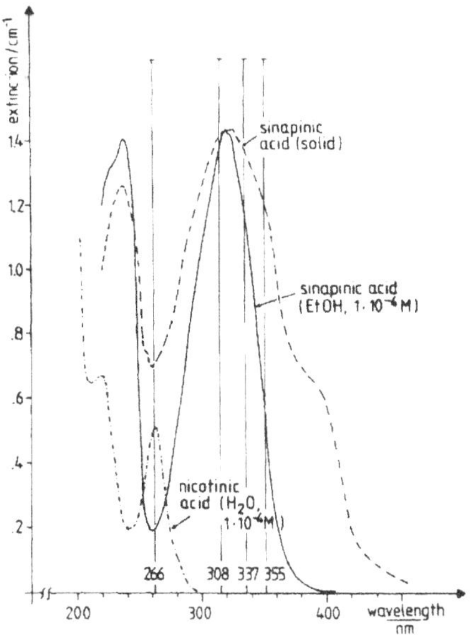


Figure 6. Optical absorption spectra of two widely used matrices. The two solution spectra were taken at equal concentrations of $10^{-4}$ L mol$^{-1}$cm$^{-1}$. The spectrum of solid sinapinic acid was taken by mixing a small amout of the sample into a $CaSO_4$ slurry and drying it onto a glass slide. Pure $CaSO_4$ was used as reference; this spectrum was scaled to the spectrum of the solution for equal maximum absorption at 315 nm. The vertical lines show four laser wavelengths commonly used in UV-laser desorption.

## 2.2 THE INFLUENCE OF THE LASER PULSE WIDTH

As was mentioned above, slow classical heating with energy equilibration among all available degrees of freedom in the sample and the individual molecules will result in pyrolysis (*i. e.*, a thermal decomposition of the sample). Consequently only short-pulsed lasers with pulse widths of ca. 200 ns or less have been used successfully for laser desorption of thermally labile, organic molecules. Within the two orders of magnitude from ca. 100 ns to 1 ns, results do not appear to depend critically on the pulse width. Very little is known so far about desorption of organic molecules with pulses in the picosecond or femtosecond time regime.

Some authors describe the desorption process as "fast heating" [9], others prefer the term "nonequilibrium process" [10], but all models have so far been very general, and too little is known about the details of the process to develop a model that would be really predictive in the details of the influence of the various parameters. It must, in fact, be kept in mind that nanoseconds are very long compared to equilibration times on the molecular level and in typical solids. It must, therefore, be assumed that the desorption of molecules, which are internally cool enough to be stable, is associated with the collective nature of the disintegration of the solid. This has led to the suggestion of a model that assumes a microexplosion of the excited sample volume [11].

## 3. Laser Ionization

Two schemes have been used to ionize the molecules for analysis and detection in mass spectrometers. Ionization, as an integral part of the desorption process itself, has historically been the first one used. It has the advantage of experimental simplicity, but the ion yield - - and thereby the sensitivity - - is rather low, at least for the desorption of small molecules without matrix. It is, in fact, comparable to that in particle-induced desorption. Also, intermolecular differences of the ion yield and its strong dependence on the molecular environment make quantification very difficult for this desorption scheme. Because desorption and ionization are one integral step in this scheme, it is termed "desorption/ionization". The second scheme involves a second laser for the postionization of the desorbed neutral molecules in vacuo.

### 3.1 LASER DESORPTION/IONIZATION

Models developed for the ionization step in particle desorption have been suggested to also apply in **direct** laser desorption/ionization. Particularly the desorption of preformed ions has been favored. This appears quite likely in certain cases such as the desorption of quarternary ammonium ions and possibly also of cationized molecules, the most common species in direct IR-laser desorption/ionization. For protonated or deprotonated molecules, usually the dominating species in UV-laser desorption/ionization spectra, this model is much less convincing because the molecules are not present in the dissociated state in the solid phase from which they are desorbed. Experiments to establish clear cut evidence for such a mechanism by systematic variation of the pH in the solution before evaporation of the solvent have also failed [27]. Ion pair desorption has been suggested as another mechanism and some evidence has been collected to support this model [28]. Hardly ever, however, are the signals of protonated molecules in the positive ion spectrum and those of the negative ions in the corresponding spectrum of a given substance reasonably equal in intensity. Based on

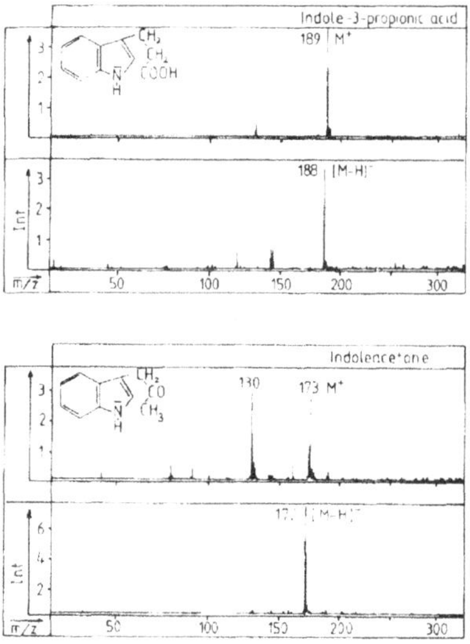


Figure 7. Direct laser desorption/ionization spectra of indole-3-proprionic acid and indoleacetone with radical parent molecular positive and even electron negative ions (266 nm wavelength). Republished from ref. [29] with permission.

this information, ion pair formation cannot be ruled out as a primary step, however, if a subsequent modulation of the yields of ions of the two polarities is assumed by the chemistry in the selvedge.

Considerable evidence has been collected for a model which postulates photoionization in the condensed state as the primary ionization step in direct UV-laser desorption/ionization. As most of these radical photoions are highly reactive, they are expected to react to give those more stable even electron ions usually observed in the spectra. A systematic investigation of different indole derivatives revealed that the postulated radical precursor ions can indeed be observed if they happen to be chemically stable, or if the subsequent chemical reactions are suppressed (*e. g.*, by sample cooling) [29]. Figure 7 shows the spectra of two such derivatives.

Ion formation in **matrix assisted** laser desorption/ionization is even less understood as yet. The ion yield has not been determined so far, but the high sensitivity of the technique at least suggests a much higher ion yield as compared to direct desorption. In contrast to direct desorption, all attempts to detect desorbed neutrals of the analyte molecules by postionization have also failed so far. As for direct desorption/ionization, photoion generation as the primary step appears quite likely for UV wavelengths. The fact, however, that very similar results are obtained in matrix assisted IR-laser desorption/ionization, where as many as 20 to 80 photons would be needed to account for the ionization potential, casts some doubt on this assumption or suggests that two different mechanisms are active in the two wavelength ranges. Contrary to direct IR-desorption/ionization, cations do not seem to be the dominating species in the matrix-assisted case [25]. Considerably more and more systematic work needs to be done to clarify the - - most probably several - - ion formation processes in laser desorption/ionization.

### 3.2 LASER POSTIONIZATION OF LASER-DESORBED NEUTRALS

This chapter concentrates on the desorption/ionization scheme; postionization is, therefore, discussed only superficially, and the interested reader is referred to the original literature. As the ion yield is typically ca. $10^{-3}$ for direct laser desorption, almost all of the desorbed particles are neutrals. It has been demonstrated by several groups that a substantial fraction of these neutrals can be postionized by a second laser beam that intercepts the neutrals somewhere off the sample [30, 31, 32]. Resonant postionization with one or several suitably tuned dye lasers is an excellent technique for molecular spectroscopy, particularly if the desorbed neutrals are rovibronically cooled in a supersonic jet of a noble gas prior to the ionization. Nonresonant postionization with short UV-wavelengths and high irradiances offers considerable potential for quantification of neutrals desorbed with lasers. If the cooling step is omitted, such experiments can also be used to obtain information on the desorption process itself [33].

## 4. Initial Energy Distribution

It was observed already early on that in direct UV-desorption/ionization, ions have a quite high initial kinetic energy of up to ca. 100 eV [34, 35]. Originally this energy was assumed to result directly from the desorption process. More recent measurements of the initial energy distribution of the desorbed neutrals revealed that they have much less initial kinetic energy typically in the meV-range, although mostly still somewhat above the expected values for a thermal process with temperatures of no more than a few hundred °C, at which one could still expect intact molecules to be desorbed [36]. Because it is somewhat unlikely that the ions and neutrals receive drastically different initial energies from the desorption process itself, it is now believed that most of the high initial energy of the ions results from Coulomb repulsion during the initially quite high charge density. Very little reliable information on initial energy distribution for direct IR-desorption/ionization is available in the literature. In matrix-assisted laser desorption/ionization of large molecules, initial energies seem to be lower than those in direct desorption, typically in the range of just a few eV, but again the experimental database is rather limited as yet.

## 5. Fragmentation and Metastable Decay

Considerable fragmentation of ions is observed in direct laser desorption/ionization at all wavelengths. Most of this fragmentation appears to be "prompt" (*i. e.*, taking place within the first few 10 ns of the desorption). This can be concluded from the fact, that these fragment ions are observed in the time-of-flight spectra at their correct (*i. e.*, daughter ion) mass with peak widths, usually comparable to those of the parent molecular ion. Results from postionization experiments suggest that similar prompt fragmentations also happen for the neutrals. Metastable decay with decay time constants in the $\mu$s-domain on the other hand appears to be rather low and limited mostly to special groups of molecules such as highly aromatic systems [37]. No prompt fragmentation has, in contrast, been observed in matrix-assisted laser desorption/ionization. Somewhat as a surprise, some metastable decay is, however, observed for large molecules in this case. It appears that peptide- and protein ions tend to lose small neutrals such as $H_2O$, HCOOH or a terminal amino acid. These observation support the notion that considerable differences exist between the processes in direct and matrix-assisted desorption.

## 6. Instrumentation and Sample Preparation

As an example for a time-of-flight laser desorption/ionization instrument, Figure 8 shows the schematic diagram of a set up as typically used in the authors laboratory. It consists of four main subunits: the laser, the sample chamber with the optics for laser focusing and sample observation as well as the ion optics for ion beam collimation, the time-of-flight mass spectrometer, consisting of a field free drift tube of typically ca. 1 - 2 m in length with a time focusing ion reflector, and finally the ion detector with the signal processing electronics.

The sample carrying substrate is mounted in the sample chamber on a xyz-movable stage, which permits selection of different sample locations under observation through an optical microscope with 50-250 magnification and an optical resolution of 3 to 10 micrometers. A three element immersion ion lens accelerates the desorbed ions to an energy of 3 to 20 keV and focuses them onto the detector at the end of the flight path. Six different lasers with a variety of wavelengths in the UV and IR have been used successfully in these instruments. Frequency tripled and quadrupled, Q-switched Nd-YAG lasers at wavelengths of 353 nm and 266 nm and a pulse duration of 3 to 15 ns have been used most extensively. Excimer lasers with wavelengths of 193 nm (ArF), 248 nm (KrF) and 308 nm (XeCl) and 15 ns pulse width have been applied for comparison. Particularly the wavelength of 193 nm with its quite high photon energy of 6.3 eV has provided some interesting results and insights into the desorption process [14]. An excimer pumped, frequency doubled dye laser has been used to scan the wavelength range from ca. 220 nm to 290 nm [11]. More recently, a small $N_2$-laser at 337 nm and 3 ns pulse width has been used successfully for matrix-assisted desorption/ionization [26]. It is by far the simplest and cheapest of all the laser. In the IR, a TEA-$CO_2$ laser at ca. 10.6 $\mu$m and 80 ns pulse width and a Er-YAG laser at 2.94 $\mu$m wavelength, Q-switched by a rotating cavity mirror to a pulse width of ca. 200 ns, have been combined with the mass spectrometer [24,25]. The beams of these lasers have been focused to spots of 3-500 $\mu$m in diameter on the sample. Spots down to 10 $\mu m$ in diameter in the UV and 50 $\mu m$ in the IR can quite easily be achieved for all of the above listed lasers by using simple planoconvex lenses of typically 100 mm focal length, positioned outside of the vacuum. Usually a straightforward telescopic intermediate optics is needed to adapt beam diameter and divergence of these lasers to those

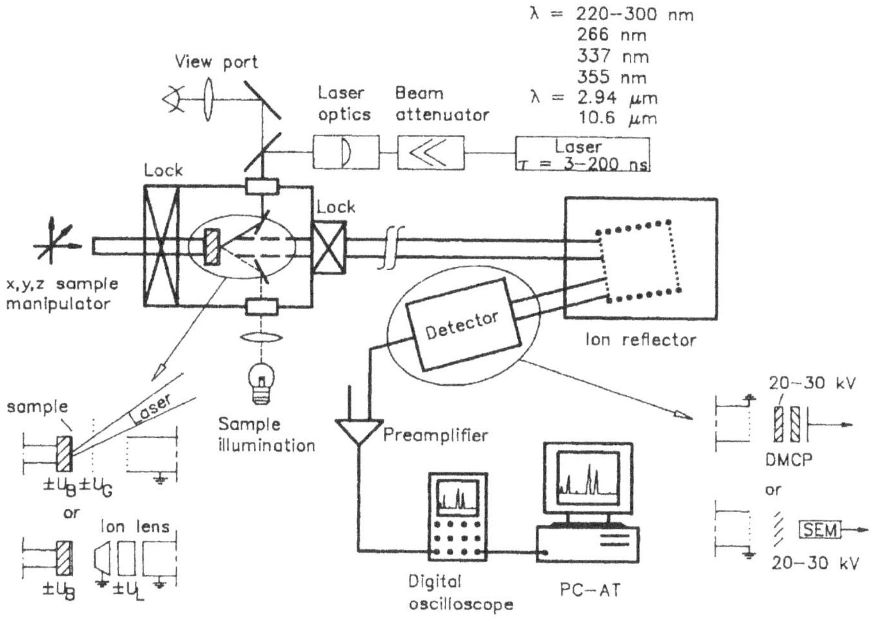


Figure 8. Schematic diagram of instruments used for direct - and matrix-assisted laser desorption/ionization

required for optimal focusing [33]. As was mentioned above, a uniform and reproducible beam pattern on the sample improves the results. For the solid state lasers, this is achieved by careful mode control. The gas lasers usually have a reasonably smooth, near Gaussian irradiance distribution in only the far field. This requires careful adjustment of the focusing optics to bring the sample exactly into the focal plane of the focusing lens (at the wavelength used!). A suitable attenuator is needed to control carefully the irradiance on the sample. For a maximum irradiance of ca. $10^7$ $Wcm^{-2}$, sufficient for the vast majority of desorptions of organic materials, a spot size of typically 100 $\mu m$ and a pulse width of 10 ns only 10 $\mu J$ of energy, incident on the sample, are needed.

Ions of different mass are separated by their different flight times in the mass spectrometer as described in a separate chapter in this book. First and second order ion reflectors as well as gridless focusing reflectors have been used in such instruments. Mass resolution varies from ca. 100 to 1000, depending on the instrument configuration. It appears that the mass resolution is usually limited by the ion formation process and ion stability rather than the theoretical resolution of the given instrument. Time errors, which can be compensated by the ion reflector only to a limited degree and at the expense of an additional energy dispersion as well as metastable ion decay, seem to be the major limiting factors. For macromolecules with masses above a few thousand Dalton, a resolution of 1000 to 2000 is actually quite sufficient, because it is equivalent to the unavoidable, natural $^{13}C$ isotope distribution.

Detection of relatively small ions of masses up to several thousand Dalton with secondary electron multipliers (SEM) or multichannel plates (DMCP) is straightforward. Larger ions require a higher acceleration potential to acquire enough momentum or impact speed at the detector for an efficient ion/electron or ion/ion conversion. Even at a kinetic energy of 20 keV an ion of mass 100,000 Da has a speed of only $6x10^3$ $ms^{-1}$. Until recently this speed was considered too low for an efficient conversion and detection. Results obtained for matrix assisted desorption/ionization apparently contradict this hypothesis, but the ion detection process for such big ions still awaits clarification. In some instruments, the ions are accelerated to the high kinetic energy of 10 to 35 keV in the ion source before entering the mass spectrometer, whereas in others they are postaccelerated from initially 3 to 5 keV to this high energy by a separate conversion dynode in front of the detector. Depending on the equivalent path length in the mass spectrometer as well as the mass range and - resolution, the signals of the ions of a given mass are as short as ca. 10 ns or may last as long as several microseconds. The total time length of a complete spectrum is up to ca. 100 $\mu$s for direct laser desorption/ionization; for matrix assisted laser desorption/ionization of large molecules this time can be as long as 1 ms. Storage and processing of such spectra requires a fast analog/digital conversion of the analog detector signal. Transient digitizers with a 10 ns time resolution, a dynamic range of 8 bit and a storage capacity of 32 kbytes or more are standard in up-to-date instruments. Special applications may require even more sophisticated electronics. Because of the collective nature of the laser desorption process, single ion generation and detection is very difficult, if not impossible. This limits the dynamic range to only two to at most three orders of magnitude, because the ion detectors as described above are suited for either single pulse or low current CW operation rather than intense, short pulsed (*i. e.*, analog) signals; the need for analog/digital conversion imposes similar limitations. PC-AT type computers are used for summation, averaging, mass calibration and similar operations. The mass calibration is done with the help of two or more ions of known mass. For small ions, this is rarely a problem; ubiquitous $Na^+$ and $K^+$ ions are typically used in these cases. Mass scale calibration is much more of a problem in matrix-assisted desorption/ionization of large ions. It is considerably more difficult to obtain calibration samples of exactly known mass and high purity in the high mass range. Moreover, formation of adducts between matrix molecules or their photoreaction products and the calibrant ion and/or cleavage of small parts off the calibrant molecule may falsify the calibration [21, 22]. Great care is needed in these cases to prevent systematic calibration errors.

Sample preparation for direct desorption/ionization of small molecules up to a mass of ca. 500 -1000 Dalton is usually straightforward. A volume of typically 1 $\mu$L of a solution in water or an organic solvent is dried onto an inert metal substrate such as silver or stainless steel either in air prior to or directly by introduction into the vacuum. For absorbing samples layer thicknesses of ca. 0.1 - 1 $\mu$m usually give best results; if substrate involvement shall be excluded in the analysis of non absorbing samples layers of ca. 10 $\mu$m thickness are more appropriate. The thickness can be controlled by the concentration of the solution and, to a limited extent, by the speed of the solvent evaporation. Surface tension of the solvent also influences the thickness via the spreading of the droplet on the substrate.

## 7. Conclusion

Direct and matrix-assisted laser desorption/ionization mass spectrometry of organic molecules have been developed into a useful and routinely applied analytical technique. Short pulsed lasers have evolved as most useful. Depending on the analytical problem at hand wavelengths in the ultraviolet or the infrared should be used for optimal results; the visible portion of the spectrum of optical radiation is less desirable because of lack of absorption of the analyte or matrix molecules. Time-of-flight mass separators are most commonly used with laser ion sources, but other mass spectrometers must be considered for special applications. Simple and fast sample preparation is another important feature of the technique. Applications of matrix-assisted laser desorption to the analysis of large organic molecules are discussed in detail in the following chapter.

## 8. References


1. Vastola, F. J., Mumma, R. O. and Pirone, A. J. (1970)
Analysis of Organic Salts by Laser Ionization
*Organic Mass Spectrom.* **3**, 101-104
2. Lubman, D. M. (1990) ed.
Lasers and Mass Spectrometry
Oxford Series in Optical Sciences, Oxford University Press,
Oxford, New York
3. Hillenkamp, F., Kaufmann, R., Nitsche, R. and Unsöld, E. (1975)
A High Sensitivity Laser Microprobe Mass Analyzer.
*Applied Physics* **8**, 341 - 348.
4. Hillenkamp, F. (1983)
Laser Induced Ion Formation from Organic Solids. In: A. Benninghoven ed.,
Ion Formation from Organic Solids, pg. 190 - 205,
Springer Series in Chemical Physics, 25, Springer Verlag,
5. Posthumus, M. A., Kistemaker, P. G., Meuzelaar, H. L. C. and
Ten Noever de Brauw, M. C. (1978), Laser Desorption Mass Spectrometry of Polar
Nonvolatile Bio-Organic Molecules, *Anal. Chem.* **50**, 985
6. Wilkins, Ch. L., Weil, D. A., Yang, C. L. C. and Ijames, C. F. (1985)
High Mass Analysis by Laser Desorption Fourier Transform Mass Spectrometry
*Anal. Chem.* **57**, 520-524,
7. Shomo, R. E. II., Marshall, A. G. and Lattimer, R. P. (1986)
Laser Desorption/Fourier Transform Ion Cyclotron Resonance vs. Field
Desorption/Magnetic Sector Mass Spectrometry. Erythromycin, Amoxicillin, Digoxin
and Daunorubicin, *Int. J. Mass Spectrom. Ion Proc.* **72**, 209-217
8. Cotter, R. J., Honovich, J., Qureshi, N. and Takayama, K. (1987)
Structural Determination of a Lipid A from Gram Negative Bacteria Using Laser
Desorption Mass Spectrometry,
*Biomed. Environm. Mass Spectrom.* **14**, 591 - 598
9. Lucchese, R. R. and Tully, J. C. (1984)
Laser Induced Thermal Desorptions from Surfaces.
*J. Chem. Phys.* **81**, 6313-6319
10. Hillenkamp, F. (1986)
Laser desorption mass spectrometry, A review. In: Benninghoven, A., Colton R. J.,
Simons D. S. and Werner, H. J., eds., Secondary Ion Mass Spectrometry SIMS V.
Springer Series in Chemical Physics, Vol. 44., Springer Verlag, Berlin

11. Hillenkamp, F., Karas, M., Bahr, U. and Ingendoh, A. (1990) On Popping Corn, Erupting Volcanos, Tasting Wine, Massive Proteins and Related Matters. In: A. Hedin, B. U. R. Sundquist and A. Benninghoven. eds.,Ion Formation from Organic Solids V, Wiley & Sons Inc., Chichester, pg. 111 - 118
12. Hillenkamp, F., Karas, M., Holtkamp, D., Klüsener, P. (1986) Energy deposition in ultraviolet laser desorption mass spectrometry of biomolecules. *Int. J. Mass Spectrom. Ion Proc.* **69,** 265 - 276.
13. Karas, M., Bachmann, D. and Hillenkamp, F. (1985) The Influence of the wavelength in High Irradiance Ultraviolet Laser Desorption Mass Spectrometry of Organic Molecules, *Anal. Chem.*, **57,** 2935 - 2939.
14. Spengler, B., Karas, M., Bahr, U., Hillenkamp, F. (1987) Excimer laser desorption mass spectrometry of biomolecules at 248 and 193nm. *J. Phys. Chem.* **91,** 6502 - 6506.
15. Karas, M., Bachmann, D., Bahr, U., Hillenkamp, F. (1987) Matrix assisted ultraviolet laser desorption of non-volatile compounds. *Int. J. Mass Spectrom. Ion Proc.* **78,** 53 - 68.
16. Karas, M. and Hillenkamp, F. (1988) Laser Desorption Ionization of Proteins with Molecular Mass Exceeding 10.000 Daltons *Anal. Chem.* **60,** 2299-2301.
17. Karas, M., Bahr, U. and Hillenkamp, F. (1989) UV-Laser Matrix Desorption Mass Spectrometry of Proteins in the 100,000 Dalton Range, *Int. J. Mass Spectrom. Ion Proc.*, **92,** 231 -242
18. Karas, M., Bahr, U., Ingendoh, A. and Hillenkamp, F. (1989) Laserdesorptions-Massenspektrometrie von Proteinen mit Massen zwischen 100.000 und 250.000 Dalton. (in English edition: Laser Desorption/Ionization Mass Spectrometry of Proteins of Mass 100.000 to 250.000 Dalton) *Angewandte Chemie,* **101,** 805 - 806; *Int. Ed. Engl.* **28,** 760
19. Karas, M., Ingendoh, A., Bahr, U. and Hillenkamp, F. (1989) UV-Laser Desorption/Ionization Mass Spectrometry of Femtomol Amounts of Large Proteins. *Biomed. Environm. Mass Spectrometry,* **18,** 841 - 843
20. Tanaka, K., Waki, H., Ido, Y., Akita, S., Yoshida, Y., and Yoshida, T. (1988) Protein and Polymer Analysis up to m/z 1000,000 by Laser Ionization Time-of Flight Mass Spectrometry. *Rapid Comm. Mass Spectrom.* **8(2),** 151
21. Beavis, R. C. and Chait, B. T. (1989) Cinnamic Acid Derivatives as Matrices for Ultraviolet Laser Desorption Mass Spectrometry of proteins, *Rapid. Commun. Mass Spectrom.* **3,** 432
22. Beavis, R. C., and Chait, B. T. (1989) Matrix assisted Laser-desorption Mass Spectrometry Using 355 nm Radiation. *Rapid. Commun. Mass Spectrom.* **3,** 436
23. Hillenkamp, F. (1990) (UV)-Laser Desorption/Ionization Mass Spectrometry of Large Biomolecules: Techniques, Results, Promises and Limitations. Proceedings of the 38th ASMS Conf. on Mass Spectrom. and Allied Topics, Tuscon, AZ, June 3 - 8, pg. 8-9 (extended abstract)
24. Overberg, A., Karas, M., Bahr, U., Kaufmann, R. and Hillenkamp, F. (1990) Matrix-Assisted Infrared (2.94 $\mu$m) Desorption/Ionization Mass Spectrometry of large Biomolecules, *Rapid Communications in Mass Spectrometry,* **4,** 293-296
25. Overberg, A., Karas, M. and Hillenkamp, F. (1991) Matrix-Assisted laser Desorption of Large Biomolecules with a TEA-$CO_2$-Laser, *Rapid Commun. Mass Spectrometry,* **5(3),** 128-131

26. Stahl, B., Steup, M., Karas M. and Hillenkamp, F. (1991) Mass Spectrometry of Oligosaccharides by Matrix-Assisted Laser Desorption/Ionization (LDI-MS), *Anal. Chem.*, in press
27. Bachmann, D. (1987), personal communication
28. Parker, C. D. and Hercules, D. M. (1986) Intermolecular Proton Transfer Reactions in the Laser Mass Spectrometry of Organic Acids. *Anal. Chem.* **58**, 25-34
29. Karas, M. and Hillenkamp, F. (1989) Photochemical Reaction Pathways and their Relevance to Ultraviolet Laser Desorption Mass Spectrometry. Proceedings of the Fourth Int. Conf. on Ion Formation from Organic Solids: IFOS IV, A. Benninghoven, ed., John Wiley & Sons, Chichester, pg. 103 - 107,
30. Lubman, David M. and Li, L. (1990) Resonant Two-Photon Ionization Spectroscopy of Biological Molecules in Supersonic Jets Volatlized by Pulsed Laser Desorption. in: Lasers and Mass Spectrometry, pgs 353 -382, D. M. Lubman, ed. Oxford Series in Optical Sciences, Oxford University Press, New York, Oxford
31. Grotemeyer, J. and Schlag, E. W. (1988) Peptides Investigated by Laser Desorption-Multiphoton Ionization Mass Spectrometry. *Organic Mass Spectrom.* **23**, 388-396
32. Becker, C. H., Jusinski, L. E. and Moro, L. (1990) Infrared Laser-Induced Desorption of Neutral Organic Compounds from Frozen Aqueous Solutions Followed by Single-Photon Ionization. *Int. J. Mass Spectrom. Ion Proc.* **95**, R1 -R4
33. Spengler, B., Bahr, U., Karas, M. and Hillenkamp, F. (1988) Postionization of Laser-Desorbed Organic and Inorganic Compounds in a Time of Flight Mass Spectrometer. *Analytical Instrumentation,* **17(1&2)**, 173 - 193.
34. Mauney, T. and Adams, F. (1984) Ion Kinetic Energy Measurements on Laser-Induced Plasmas in Laser Microprobe Mass Analysis (LAMMA). Part I. Methodology. *Int. J. Mass Spectrom. Ion Proc.* **59**, 103-119
35. Michiels, E., Mauney, T., Adams, F. and Gijbels, R. (1984) Ion Kinetic Energy Measurements on Laser-Induced Plasmas in Laser Microprobe Mass Analysis (LAMMA). Part II. Titanium Oxide *Int. J. Mass Spectrom. Ion Proc.* **61**, 231-246
36. Dreisewerd, K., Bahr, U. and Hillenkamp, F. (1990) UV-Laser Desorption of Neutral Organic Molecules: Speed Distribution and Fragmentations. Proceedings of the Conference on Resonance Ionization Spectroscopy and its Applications (RIS-90), Varese, Italy, Sept. 16. - 21.
37. Rosmarinowsky, J., Karas, M. and Hillenkamp, F. (1985) Metastable Decay of Laser Desorbed Ions from Aromatic Compounds. *Int. J. Mass Spectrom. Ion Proc.*, **67**, 109-119.

# LASER DESORPTION MASS SPECTROMETRY. PART II PERFORMANCE AND APPLICATIONS OF MATRIX-ASSISTED LASER DESORPTION/IONIZATION OF LARGE BIOMOLECULES

A. OVERBERG, A. HASSENBÜRGER AND F. HILLENKAMP
*Institut für Medizinische Physik der Universität Münster*
*Robert-Koch-Str. 31, D-4400 Münster, FRG*

ABSTRACT. Sample preparation and typical performance characteristics of ultraviolet and infrared matrix-assisted laser desorption/ionization mass spectrometry of large organic molecules are discussed. Examples of the application of the technique to the analysis of a variety different biopolymers are presented.

## 1. Introduction

The basic mechanisms and techniques of laser desorption mass spectrometry have been introduced in chapter 9 of this book. Some details of instrumentation and general considerations for sample preparation can also be found there. This chapter discusses the special aspects and applications of matrix-assisted laser desorption/ionization mass spectrometry (LDI/MS) of large biomolecules. With this technique, it is today possible to desorb intact molecular ions from biomolecules up to molecular weights of about 500,000 Da. The method promises to become a powerful tool for the analysis of molecules. This coincides with an increasing demand in fields like biochemistry, molecular biology, gen- and biotechnology. The method brings about considerable progress in the characterization of biomolecules by accurate molecular weight determination. In addition it has the advantage of a short analysis time. Sample preparation and mass spectrometric analysis usually take no more than 15 minutes. With the following spectra, the performance of this technique and some applications of laser desorption mass spectrometry in the field of analytical biochemistry are reported. Results obtained with ultraviolet (UV) lasers are reported along with first results obtained with infrared (IR) lasers; the advantages and shortcomings of the two wavelength ranges are discussed.

## 2. Sample Preparation

The basic considerations and procedures for sample preparation are discussed in chapter 9. The main special aspect of **matrix-assisted** LDI is the fact that optimum results depend strongly on the correct choice of the matrix for a given laser wavelength as well as on the optimum volume ratio of matrix to analyte of ca. $10^2$ - $10^3$ in the final sample as introduced into the mass spectrometer. Relative molar concentrations of matrix and analyte, therefore, depend on the matrix used and the size of the analyte molecules. The larger the molecules, the larger should



*M. L. Gross (ed.), Mass Spectrometry in the Biological Sciences: A Tutorial*, 181–197.

be the excess of matrix molecules by number. Table 1 lists typical values for protein analysis with three selected matrices and the respective wavelengths. Mass concentrations are listed rather than volume ratios because they can be used directly in sample preparation. Small admixtures of trifluoroacetic acid (TFA) are often used to improve protein solubility. It is worth noting that the requirement for optimum V/V excess of the matrix results in a molar protein concentration that decreases with increasing molecular mass of the protein. For 2,5-dihydroxybenzoic acid (DHB) a 1 $\mu$L sample typically contains 20 ng of protein, that is 200 fmol for a 10,000 Da protein and only 20 fmol for one of 100,000 Da molecular mass. In favorable cases, good spectra have been obtained from as little as 100 nL of a $10^{-8}$ mol/L protein solution (*i. e.*, from a total amount of protein for the preparation

| Matrix | | | | Protein | Solvent |
|---|---|---|---|---|---|
| Usable wavelengths nm | Name | Structure | Concentration g/l | Concentration g/l | |
| 220 – 300<br>248, 266<br>2.94 $\mu$m | Nicotinic acid | 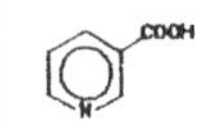 | 5 | 0.05 | $H_2O$ +<br>0.1% TFA +<br>10% EtOH |
| 337, 355<br>(266)<br>2.94, 10.6 $\mu$m | Dihydroxy-benzoic acid | 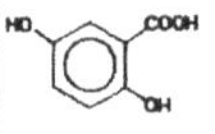 | 8 – (10) | 0.02–(0.005) | $H_2O$ +<br>0.1% TFA +<br>(10% EtOH) |
| 337, 355<br>(266)<br>2.94 $\mu$m | Sinapinic acid | 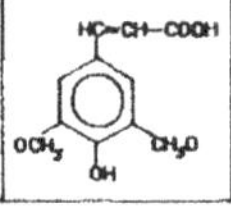 | 8 – (10) | 0.02–(0.005) | 0.1% TFA in acetonitrile/<br>$H_2O$: 1/2 |

Table 1. Sample preparation conditions for three selected UV- and IR-matrices

of only 1 fmol). The choice of the matrix is sometimes governed not only by the laser wavelength and relative sensitivity but also by the fact that different matrices exhibit drastically different sensitivities towards interference by sample impurities such as buffer salts or detergents. In the mass range above ca. 50,000 Da, best results are, so far, obtained with nicotinic acid as a matrix, but the signal is lost even for small amounts of such impurities. The cinnamic acid derivatives such as sinapinic acid and also DHB are remarkably insensitive and tolerate admixtures at concentrations up to 0.1 - 1 mol/L; DHB even tolerates up to 10 % of sodium dodecyl sulfate (SDS), which is advantageous in the analysis of samples separated by SDS gel electrophoresis. All matrices found so far for UV matrix-assisted LDI are rather acidic with pH-values below 4.5; upon solvent evaporation matrix concentration and acidity increase even further. Such a low pH sometimes interferes with the analysis because most proteins lose their quarternary structure in solution, whereas it appears that this structure is not affected by the desorption process if it survives the preparation procedure. Similar considerations apply for the investigation of enzyme kinetics. One advantage of IR-desorption/ionization is the availability of matrices with neutral or basic pH such as glycerol, urea or even water.

## 3. Performance Characteristics of Matrix-Assisted Laser Desorption/Ionization

Figure 1 shows a typical mass spectrum of matrix-assisted UV-LDI. The spectrum was obtained with a wavelength of 266 nm and nicotinic acid as matrix. It is the positive ion mass spectrum of an IgG monoclonal antibody of the mouse with a molecular mass of ca. 150,000 Da. Such spectra usually contain exclusively signals of the matrix in the low mass range up to ca. 500 Da; all signals in the high mass range can be attributed to the protein. The singly charged parent molecular ions typically give rise to the base peaks. Besides this base peak signals due to the multiply charged monomer (with up to three charges in this case) and to singly and multiply charged oligomers of the parent molecule (up to the trimere in this example) are seen. The extent of oligomerization depends on the protein structure and increases with the protein concentration in the sample. It is suspected that these oligomers are already formed in solution [1]. As explained later, the degree of multiple charging depends strongly on the matrix/wavelength combination. Typically the matrix signals have about the same intensity as those of the protein, even though matrix molecules outnumber the protein in the sample by about a factor of $10^3$. This is another indication for the participation of the matrix in the analyte ionization. The signal-to-noise (S/N) ratio is surprisingly good; summation of typically 10-30 spectra results in a further improvement of the S/N ratio. No fragment ions of the protein are usually observed. This is particularly surprising in this case of an antibody, where the two light and two heavy chains are linked only by disulfide bonds of rather small bond energy; the small signal around 23,000 Da most probably results from a sample impurity by light chain molecules. The lack of fragment ions testifies to the particular softness of the matrix-assisted desorption in contrast to the direct desorption without matrix. For most analytes, negative ion spectra of comparable intensity can be obtained (see for example Figure 8). In the following, a few special characteristics of the method are presented and discussed.

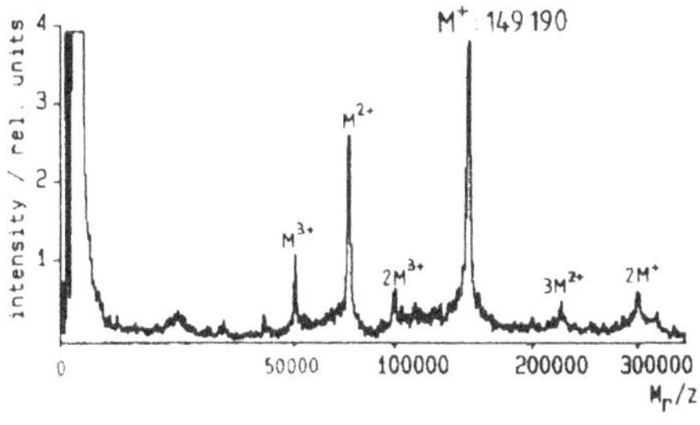


Figure 1. LDI-spectrum of an IgG monoclonal antibody. Matrix: nicotinic acid; wavelength 266 nm. Republished from ref. [16] with permission.

### 3.1. MASS RESOLUTION

Mass resolution for protein signals, defined here as $m/\delta m$, with m the ion mass and $\delta m$ the full width at half maximum (FWHM) of a given peak depends on instrument characteristics as well as on the ion formation process. For most of the spectra reported in this contribution, the mass resolution is only a rather low 50-100, whereas mass resolution for small ions and direct desorption in the same instrument is typically 300 to 500. Several processes contribute to these rather broad mass peaks. The main contribution to the peak width is a distribution of adducts by matrix ions or fragments thereof; cationization and multiple salt formation at acidic sites is a particular problem in nucleic acid analysis; in protein analysis cationization seems to be significant only for smaller peptides according to somewhat preliminary recent results obtained in the authors group. Even though the lack of fragment ions in the spectra signals that there is only negligible prompt fragmentation, recent experiments have demonstrated that there may be some metastable decay that also leads to a peak broadening [2]. All large ion signals in the spectra reported here were detected with a secondary electron multiplier (SEM) fitted with an additional conversion electrode ca. 30 mm in front of it (see Figure 8, chapter 9). The flight time dispersion of electrons and secondary ions from this conversion dynode to the SEM forms another contribution to the peak width. Most of these limitations in mass resolution can be circumvented. With an instrument comprising a straight drift tube, an initial ion acceleration to 20 kV and a double channel plate for ion detection, Beavis and Chait attained a mass resolution of 500 [3]; also essential in their experiments was the use of sinapinic acid as matrix, which exhibits a much smaller tendency for adduct formation as compared to nicotinic acid.

Figure 2 shows a spectrum of angiotensin I, obtained recently in the authors laboratory with nicotinic acid as matrix and a new instrument equipped with a first order ion reflector and a double channel plate as detector. The mass resolution of this spectrum is 3500 as demonstrated by the resolution of the isotopic peaks of the peptide. For biomolecules with higher molecular weight (*e.g.*, insulin or cytochrome c), a mass resolution of ca. 1000 has been achieved. With such a resolution adduct ions, cations and those resulting from metastable decay can be resolved from the protonated parent molecular ion up to protein masses of several 10,000 Da. Moreover, proteins of closely spaced mass can be detected in a single spectrum of a mixture.

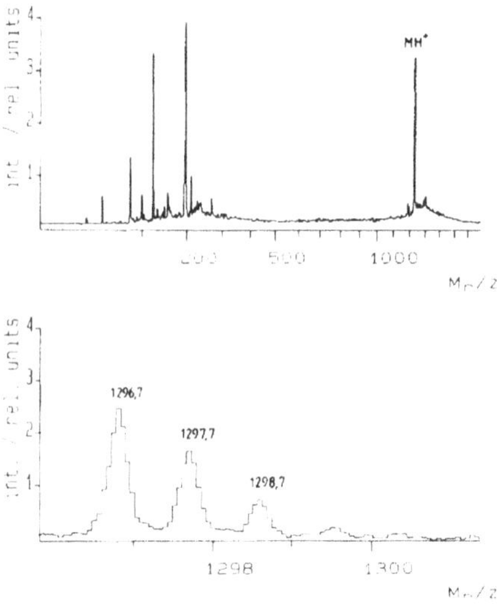


Figure 2. LDI-spectrum of angiotensin I. Matrix: nicotinic acid; wavelength: 266 nm.

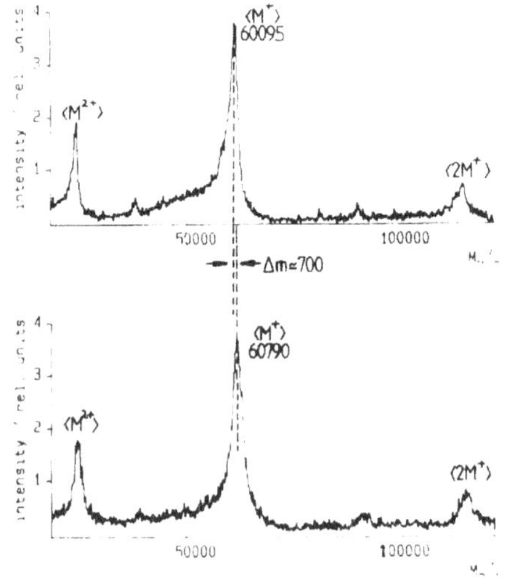


Figure 3. LDI-spectra of two isoenzymes: lipases OF 1 and OF 2 from *candida cylindracea*. Matrix: nicotinic acid; wavelength: 266 nm. Republished from ref. [16] with permission.

## 3.2. ACCURACY OF MASS DETERMINATION

The accuracy of the mass determination is given by the precision to which the center of a given (broad) peak can be determined and the absolute accuracy of the mass scale calibration. Mass peaks are centrioded with a semiautomatic computer algorithm. The operator can set the baseline for centroiding at the baseline of the spectrum for fully resolved peaks and above the bifurcation point for not fully resolved peaks. Initially, known matrix ion peaks were used for mass scale calibration. Adduct ion formation of the proteins will, however, lead to systematically too high mass values for the proteins with this procedure. More recently proteins of well-known mass such as cytochrome c or myoglobin have been used for mass scale calibration as internal or external standards. For spectra with a mass resolution of only less than 100, a mass accuracy of $5x10^{-4}$ - $2x10^{-3}$ has been obtained with this procedure. For proteins up to a molecular mass of 30,000 Da, observed with a mass resolution of 500, Beavis and Chait [4] have obtained an accuracy of $10^{-4}$.

Figure 3 serves as a demonstration that accuracy of mass determination can be quite high despite of rather poor mass resolution. It shows the spectra of two lipases from *candida cylindracea* with molecular masses of ca. 60,000 Da. The mass difference of 700 Da of these isoenzymes could be determined, but not in a single spectrum. The two lipases had to be separated chromatographically before the mass spectrometric analysis.

## 3.3. MASS RANGE

The upper mass limit of LDI technique is not known as yet. Figure 4 shows the positive ion spectrum of an IgG monoclonal antibody of molecular mass 150,000 Da. The trimer ion signal at mass 450,000 Da is clearly detected. The largest functional enzyme so far detected was catalase at mass 236,230 Da [5]. The upper mass limit is most probably not so much determined by the desorption and ion formation process but rather by the ion detection. Proteins with masses above ca. 100,000 Da also almost exclusively consist of several non covalently bound subunits that decompose in the acidic matrices used in matrix-assisted UV desorption. Transition to IR wavelengths and associated more physiological matrices may, therefore, extend the mass range somewhat.

## 3.4. SENSITIVITY

Sensitivity is already at a high level under standard preparation conditions. This results from the dilution of the sample in the matrix. Usually no more than 1 pmol of sample is used for a preparation. It was shown recently [6] that by reducing the sample concentration to a level of $10^{-8}$ mol/L and the total sample volume to a mere 100 nL, 1 fmol of cytochrome c was suffi-

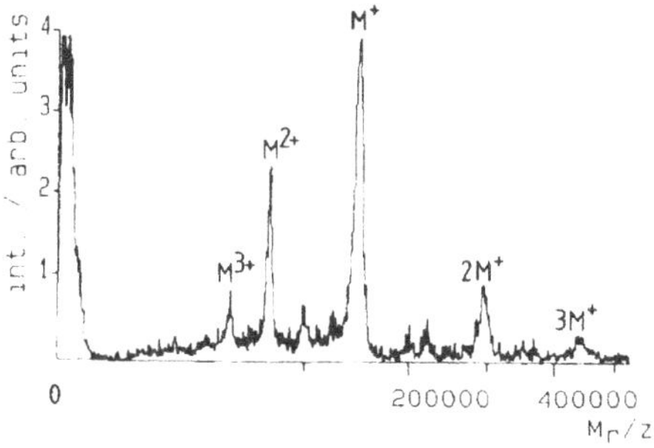


Figure 4. LDI-spectrum of an IgG monoclonal antibody. Matrix: sinapinic acid; wavelength: 2.94 $\mu$m

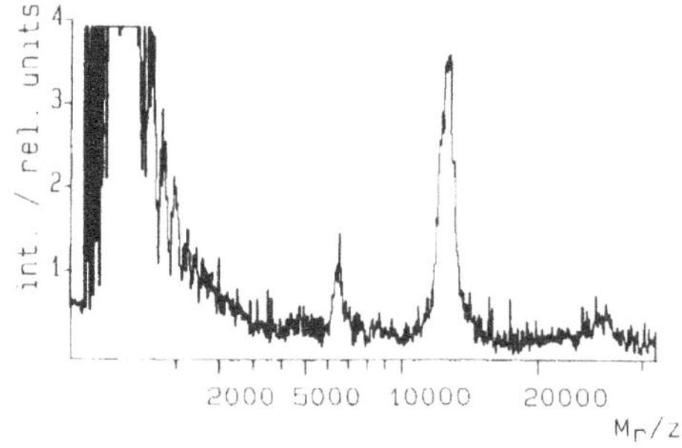


Figure 5. LDI-spectrum of a 1 fmol sample of cytochrome c. Matrix: DHB; wavelength 337 nm. The sample was radioactively labelled with KCNO; its heterogeneous distribution causes the broad peak.

cient to get the spectrum shown in Figure 5. At least 100 spectra could be obtained from this sample, setting the upper limit for the amount of sample consumed for a single spectrum to 10 attomole. Attempts to determine more accurately this amount failed so far. Estimates of the total ablated volume by electronmicroscopic inspection of cytochrome c containing DHB crystals after irradiation with several hundred laser shots and parallel registration of the protein spectra, however, suggest that the actual amount of consumed sample is at least 1-2 orders of magnitude smaller stlll. It, therefore, appears that the ultimate sensitivity of the method is strictly limited by the sample handling procedure. The fact that up to several hundred spectra can be obtained under favorable conditions from the same sample spot without degradation of the spectrum quality proves that the laser radiation does not cause any radiation damage; at least in principle, the non desorbed sample amount can, therefore, be retrieved for further use.

## 4. Matrix-Assisted Laser Desorption/Ionization with Infrared Wavelengths

### 4.1. CHARACTERISTICS OF MATRIX-ASSISTED IR-LDI

As described in chapter 9, the matrix should serve several different functions in matrix-assisted LDI. A suitable matrix isolates the large biomolecules and absorbs the laser energy. This means the matrix should have a strong absorption at the laser wavelength used. For the UV-LDI, it was suggested that matrix molecules also play a role in the ionization process of the analyte molecules, but this is far from fully established. For a better understanding of the ionization processes of large molecules, matrix-assisted LDI with infrared lasers was investigated [7, 8]. Three lasers were used for these experiments: a TEA (Transversely Excited Atmospheric pressure) $CO_2$-laser emitting at a wavelength of 10.6 $\mu$m and two Erbium-lasers, one emitting at a wavelength of 2.94 $\mu$m (Er-YAG) and the other at one of 2.79 $\mu$m (Er-YSSG). $CO_2$-lasers have been used for direct laser desorption for over 20 years while Q-switched Er-lasers have only recently become commercial available.

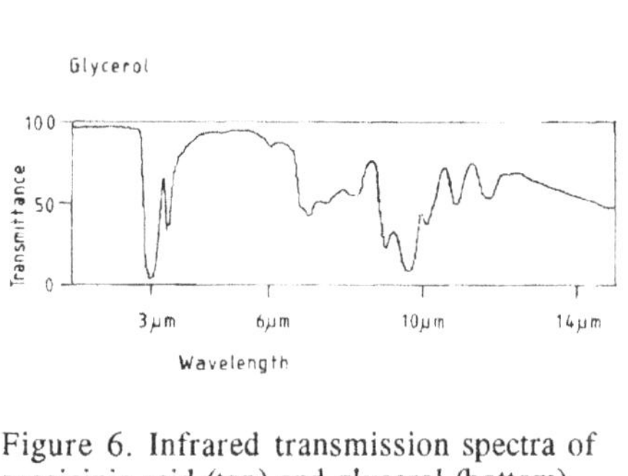


Figure 6. Infrared transmission spectra of succicinic acid (top) and glycerol (bottom).

Whereas photons in the UV-wavelength range can excite molecules electronically, radiation with wavelengths in the IR region will excite them vibrationally or rotationally. It is noteworthy that the absorption coefficients in the IR are weaker than those in the UV by one or two orders of magnitude, corresponding to a larger penetration depth of the radiation into the sample. At 2.94 $\mu$m in the IR, NH- and OH-stretching modes lead to a comparatively high absorption, the highest actually in the whole wavelength range for organic molecules. Typical examples for absorbing molecules are urea, glycerol, water and carboxylic acids. The IR transmission spectra of succinic acid, a di-carboxylic acid and of glycerol are shown in Figure 6 [9]. There is a sharply rising absorption band near 3 $\mu$m, caused by the OH-stretching mode. It is shifted slightly towards longer a wavelength for the carboxyl-group as compared to the hydroxyl-groups of the glycerol. This shift leads to a different behaviour for the two Er-lasers emitting at 2.79 and 2.94 μm respectively. The band is broadened by hydrogen bonds. There is also an absorption comparable in magnitude at the wavelength of the $CO_2$-laser caused by the OH-bending mode. Besides the two examples shown in Figure 6, a great variety of other matrices have been tested successfully for IR-LDI with the Er-lasers. Among them are all the matrices known to be useful for UV-LDI, a large variety of carboxylic acids, the liquid FAB matrices glycerol and triethanolamine, as well as urea. Most of these matrices worked also for IR-LDI at 10.6 μm with the exception of lactic acid and triethanolamine, most probably because of a too weak absorption at this wavelength. Because of the lack of quantitative data in IR-absorption spectroscopy, it is at this point not clear what the minimum requirements are for the absorption of the matrix to work at the chosen IR wavelength. Very recently good spectra were also obtained from frozen aqueous solutions of proteins as shown in Figure 7. This result opens the door for desorption of biomolecules out of a physiological environment, provided it can be demonstrated that the addition of a suitable buffer does not interfere with the desorption process, as is the case for nicotinic acid. Table 2 summarizes the features of the different IR-matrices tested so far. Basically the results in IR-LDI of proteins are very similar to those obtained with UV wavelengths. Some differences are exemplified in the following section.

| Matrix | Solid / Liquid | 266 nm | 337nm/355nm | 2.79μm/2.94μm | 10.6 μm |
|---|---|---|---|---|---|
| Nicotinic Acid | S | + | – | –/+ | + |
| Pyrazinecarboxylic Acid | S | + | – | not tried | not tried |
| Benzoic Acid Derivatives: | | | | | |
| 2,5–Dihydroxybenzoic Acid | S | + | + | +/+ | + |
| Aminobenzoic Acid | S | + | + | not tried/+ | not tried |
| Vanillic Acid | S | + | – | +/+ | + |
| Thymine | S | + | – | not tried | not tried |
| Cinnamic Acid Derivatives: | | | | | |
| e.g. Sinapinic Acid | S | + | + | +/+ | + |
| Caffeic Acid | S | + | + | +/+ | + |
| 3–Nitrobenzylalcohol | L | + | – | not tried | not tried |
| Phloroglucinol | S | not tried | not tried | +/+ | + |
| Carboxylic Acid Derivatives: | | | | | |
| e.g.: Succinic Acid | S | – | – | –/+ | + |
| Malic Acid | S | – | – | –/+ | + |
| Malonic Acid | S | – | – | not tried/+ | + |
| Lactic Acid | L | – | – | +/+ | – |
| Glycerol | L | – | – | +/+ | + |
| Triethanolamine | L | – | – | +/+ | – |
| Sucrose | S | – | – | not tried/+ | + |
| Urea | S | – | – | +/+ | + |
| Thiourea | S | + | – | +/+ | + |
| Bishydroxyethylglycine | S | – | – | +/+ | not tried |
| Ice (Frozen Aqueous Solution) | S | – | – | not tried | + |

Table 2. Properties and wavelength ranges of selected organic matrices for matrix-assisted laser desorption/ionization.

## 4.2. EXAMPLES FOR MATRIX-ASSISTED IR-LDI

As a typical spectrum of IR-LDI obtained with the Er:YAG-laser, the negative ion spectrum of cytochrome c with succinic acid as matrix is presented in Figure 8. It very much resembles matrix-assisted UV-LDI spectra of either polarity. The singly charged parent ion is responsible for the base peak and the doubly charged monomer and singly charged oligomers give rise to further intense peaks. No fragment ions are observed, and the low mass matrix signals are comparable in strength to those of the analyte. The mass spectra obtained with the TEA-$CO_2$-laser and the same matrix are also almost identical to those obtained with the Er:YAG-laser. Perhaps due to the weaker absorption coefficient and the higher penetration depth for IR radiation, respectively, the amount of removed material per laser shot is much larger compared to UV-LDI. Only a few spectra can be taken from the same sample area, corresponding to a depth of removed material of ca. 0.1 $\mu$m to 0.5 $\mu$m. This is much higher than for UV-LDI, where almost no surface change can be recognized after several hundred laser shots.

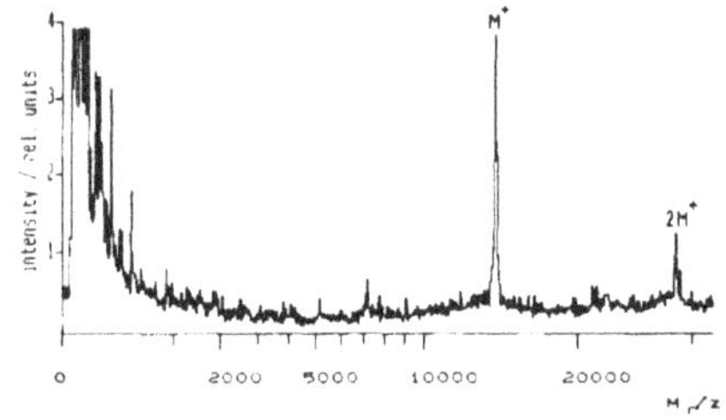


Figure 7. LDI-spectrum of lysozyme (MW 14,306). Matrix: $H_2O$ (ice); wavelength: 10.6 $\mu$m.

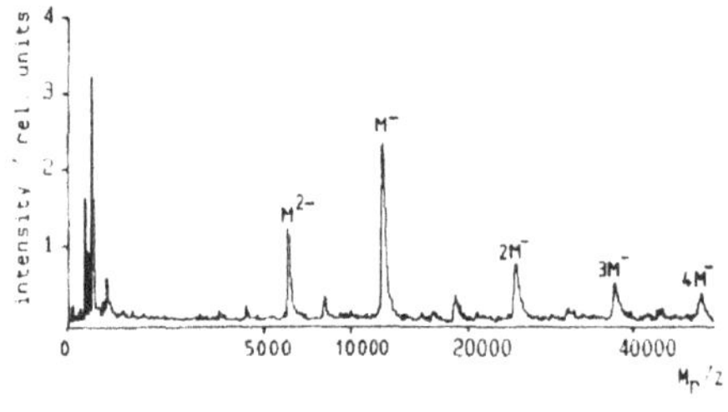


Figure 8. LDI-negative ion spectrum of cytochrome c (MW 12,360). Matrix: succinic acid; wavelength: 2.94 $\mu$m. Republished from ref. [7] with permission.

In IR-LDI, there is a general tendency toward the formation of ions with higher charge states. Irrespective of the wavelength, this tendency is particularly pronounced for the matrix caffeic acid as shown in Figure 9. The mass spectrum of an IgG monoclonal antibody of the mouse, obtained with nicotinic acid as matrix and a wavelength of 2.94 $\mu$m, resembles closely spectra obtained by UV-LDI. If caffeic acid is used as matrix, ions with up to 13 charges per ion can be observed in the spectrum; ions with up to 8 charges are observed with LDI at 10.6 $\mu$m [8]. This tendency was also observed for other analyte molecules and for some other matrices, but it is less pronounced for smaller proteins. This multiple charging, however, was neither observed for the other cinnamic acid derivatives nor for UV-LDI of smaller proteins. The possibility for multiple charging may allow the use of mass spectrometers with an upper mass limit below the mass of the analyte. It must be kept in mind that the peak intensities of multiple charged ions in the mass spectra do not represent the true abundance of ions. Because there is higher kinetic ion energy of multiply charged ions after post acceleration to a potential of 20 kV, the detection efficiency increases for increasing number of charges per ion. For most of the matrices examined so far, multiple charging is restricted to three to four charges depending on the size of the analyte molecule; multiple charging also appears to be less pronounced for $CO_2$-laser desorption.

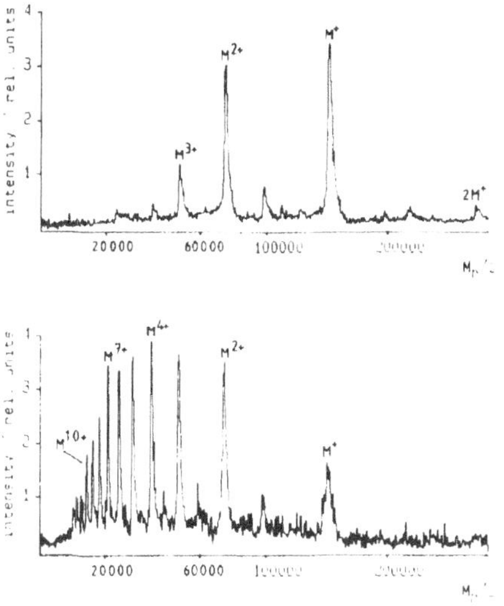


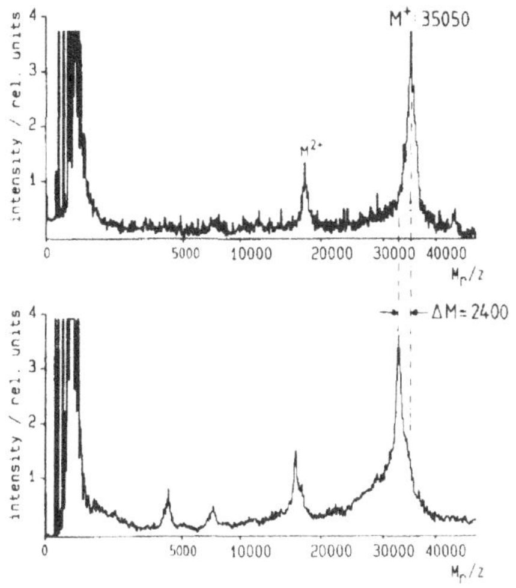


Figure 9. LDI-spectra of an IgG monoclonal antibody. wavelength: 2.94 $\mu$m. Top spectrum: matrix: nicotinic acid; Bottom spectrum: matrix: caffeic acid. Republished from ref. [7] with permission.

Figure 11. LDI-spectra of the enzyme violet phosphatase. Matrix: nicotinic acid; wavelength: 266 nm. Top spectrum: native enzyme; Bottom spectrum: deglycosylated enzyme. Republished from ref. [16] with permission.

## 5. Applications of Laser Desorption/Ionization Mass Spectrometry

In the following, some applications of matrix-assisted laser desorption to different analytical situations are presented.

### 5.1. GLYCOPROTEINS

More than one hundred different protein samples have been analyzed in the authors group; a number of them were glycoproteins with a considerable carbohydrate content. Whereas they often cause problems for accurate mass determination in other analytical techniques such as gel electrophoresis, they exhibit no special behavior in matrix-assisted laser desorption. As an example of a mass spectrum of a glycoprotein, the protein endoglucanase I from *Trichoderma reesei* with an ca. 15% carbohydrate content is shown in Figure 10. Another example of a glycoprotein is violet phosphatase. Figure 11 shows the spectrum of the intact protein and that of the molecule, deglycosylated with the enzyme endoglycosidase H. As for the two lipases, it would have been difficult to resolve the two ions in a single spectrum, but the mass difference, representing the carbohydrate fraction of 2400 Da, can easily be found from the mass difference of the base peaks in the two spectra.

### 5.2. HYDROPHILIC AND HYDROPHOBIC PROTEINS

Not only is LDI mass spectrometry well suited for the analysis of large proteins, but also its sensitivity, speed and ease of application make it well-suited for the analysis of peptides and smaller proteins in the mass range between several hundred to several thousand Da. In this mass range, matrix-assisted LDI seems to be quite independent of protein properties such as hydrophilicity or hydrophobicity. This feature is notable with respect to other desorption methods such as FAB-MS and PD-MS. The spectrum of a highly hydrophilic Zn-binding protein with a molecular weight of 11,365 Da is shown in Figure 12. For a total of 100 amino acids, the protein contains 37 glutamic acids and 11 aspartic acids (the two amino acids with acidic residues).

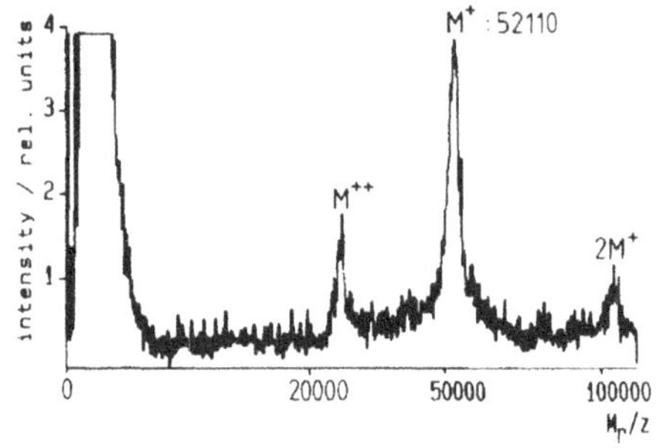


Figure 10. LDI-spectrum of the glycoprotein endoclucanase I from *T. reesei*. Matrix:nicotinic acid; Wavelength:266 nm. Republished from ref. [17] with permission.

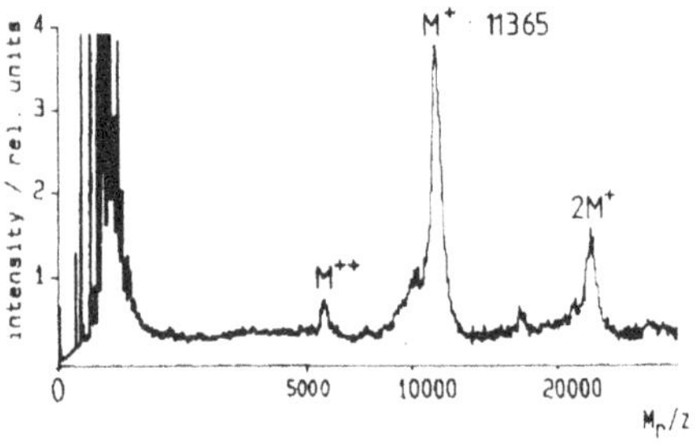


Figure 12. LDI-spectrum of a zinc-binding protein. Matrix: nicotinic acid; wavelength: 266 nm. Republished from ref. [13] with permission.

Figure 13 shows a spectrum of porin from *Rhodobacter Capsulatus* as an example of a hydrophobic membrane protein. It consists of three non covalently bound subunits. To bring the hydrophobic molecule into an aqueous solution, it was necessary to use a suitable detergent. The successful desorption of the protein demonstrates that the method is insensitive to such additives at rather high concentration levels. The capability to get mass spectra from both hydrophilic and hydrophobic proteins is an important feature of the matrix-assisted technique.

## 5.3. MIXTURE ANALYSIS

It is also possible to obtain spectra from protein mixtures. This is important if the proteins cannot be separated chromatographically or if separation would take too much time. Figure 14 shows the spectrum of a protein mixture prepared for mass calibration experiments. The relative concentrations were adjusted to yield comparable signals for all of them. With a maximum concentration ratio of only ca. 6 (between insulin on one hand and cytochrome c and trypsin on the other) this is a quite remarkable result, and would be difficult to reproduce with other desorption techniques such as FAB-MS or PD-MS.

## 5.4 REQUIREMENTS FOR SAMPLE PURITY

Requirements for sample purity appeared rather critical in the early experiments. In the meantime, closer examinations revealed that certain matrices tolerate contaminations or additives to protein samples sometimes even at relatively high concentrations. In nicotinic acid samples, for example, detergents may be present up to a concentration of $10^{-3}$ M, inducing only slight degradation in quality of the spectra. Furthermore, it was observed that citrate buffers at a 50 mM concentration, as used in enzymatic reactions of carboyxpeptidases, have no negative effects on the quality of the LDI-spectra out of a ferulic acid matrix. This observation seems to be generally true for cinnamic acid derivatives as matrices [10, 11]. Physiological salt concentrations (*e.g.*, of NaCl) in the protein solution are tolerated by these matrices without any deterioration of the spectrum quality. Figure 15 shows a UV-LDI mass spectrum of porcine trypsin (MW 23,298 Da) taken from a 1 pmol protein sample in 150 mM sodium chloride solution by using a sinapinic acid matrix. Even a 1 M salt concentration is tolerated by this matrix, possibly because salt and protein crystallize in different areas of the sample or on top of each other upon solvent evaporation. This is reflected in the mass spectra by a gradual decrease of the sodium ion signal if the same sample area is repeatedly irradiated. Concerning practical applications, this behavior is highly desirable because proteins can be examined under physiological conditions whereas problems due to denaturation and limited solubility can be avoided.

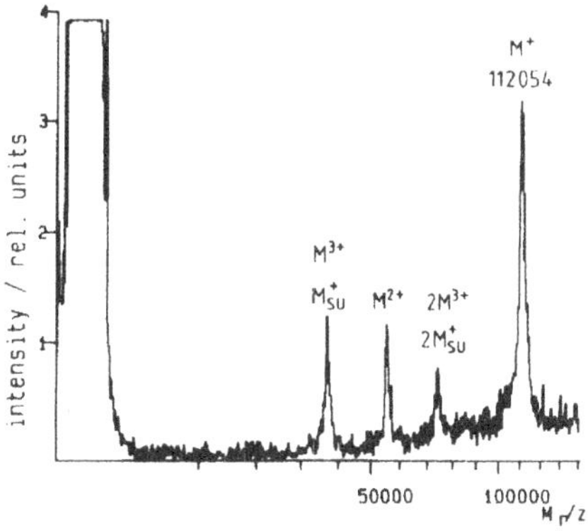


Figure 13. LDI-spectrum of the protein porin from *Rhodobacter capsulatus*. Matrix: nicotinic acid; wavelength: 266 nm.

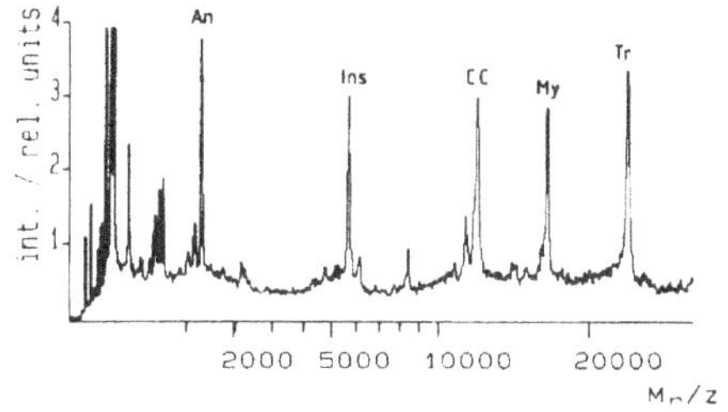


Figure 14. LDI-spectrum of a mixture of the peptide angiotensin I ($1.5 \times 10^{-6}$M) and the proteins insulin ($0.3 \times 10^{-6}$M), cytochrome c ($2 \times 10^{-6}$M), myoglobin ($2 \times 10^{-6}$M) and trypsin ($1 \times 10^{-6}$M). Matrix: DHB; wavelength:337 nm.

## 5.5. THE INFLUENCE OF MATRIX pH

The advantage of the availability of matrices with a neutral pH-value for IR-LDI was mentioned earlier. Glycerol and urea can be taken as examples. With these matrices, it may be possible, in principle, to desorb intact molecules that are pH sensitive in their quarternary structure. Figure 16 shows the positive ion spectrum of bovine trypsin obtained with the $CO_2$-laser at 10.6 $\mu$m and urea as matrix. Another interesting feature of the $CO_2$-LDI with urea as matrix can be recognized in this spectrum. In contrast to UV-LDI of the same molecule and with the same instrument, the parent peak is only slightly over 100 Da wide, corresponding to a mass resolution of ca. 200. Such a small peak width for trypsin could not be obtained with any of the other matrices. The reason for this observation is not yet clear.

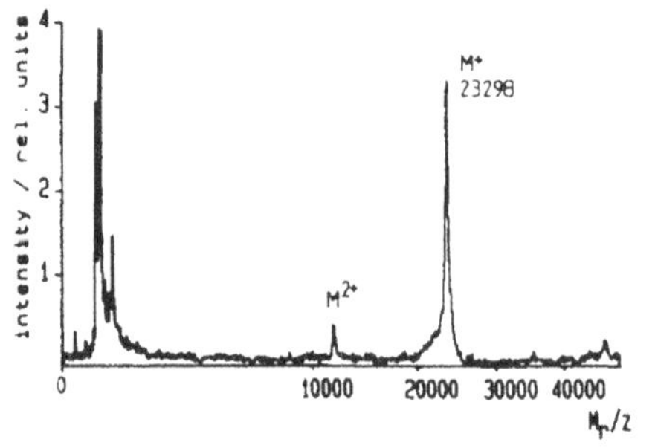


Figure 15. LDI-spectrum of the protein bovine trypsin. Matrix: sinapinic acid + 150 mM NaCl. Wavelength: 337 nm. Republished from ref. [13] with permission.

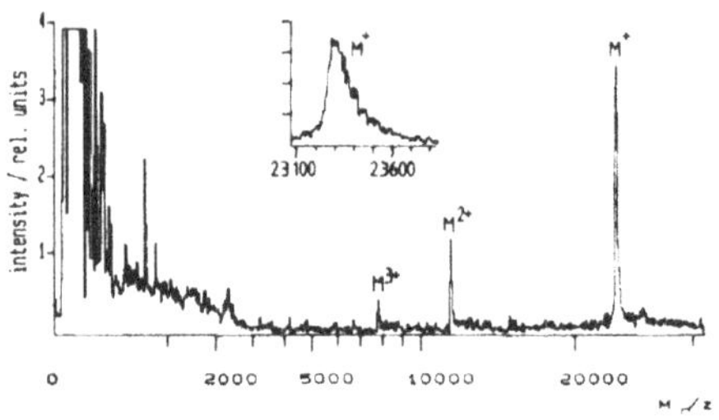


Figure 16. LDI-spectrum of the protein porcine trypsin. Matrix: urea; wavelength: 10.6 $\mu$m. Inset with extended mass scale in the parent ion region. Republished from ref. [8] with permission.

## 5.6. LIQUID MATRICES

Matrix-LDI with the liquid matrix glycerol is shown for lysozyme from chicken egg white (MW = 14,300 Da) in Figure 17. The top spectrum was obtained with the Er:YAG- laser at 2.94 $\mu$m, the bottom one with the $CO_2$-laser at 10.6 $\mu$m. Both are single laser shot spectra. The peak broadening was observed for other analytes as well when desorbed with glycerol as a matrix. It is suspected that it originates from adduct formation with matrix molecules or fragments thereof. For Er:YAG-LDI, two further liquid matrices, lactic acid and triethanolamine, were tested. Figure 18 shows a positive ion mass spectrum of the enzyme glucose isomerase desorbed with a lactic acid matrix. This enzyme consists of four equal, noncovalently bound subunits of 43,100 Da each. The enzyme was adsorbed at the surface of silica gel beads taken without further preparation from a glucose/fructose bioreactor. It is expected that the enzyme divides into its subunits already upon adsorption. The beads were soaked in lactic acid and then introduced into the mass spectrometer. The singly charged subunit and singly charged multimers up to the pentamer were detected. Nitrobenzyl alcohol, a liquid aromatic compound and thus a strong absorber at 266 nm, is the only liquid matrix that has been successfully used with UV lasers [12]. Although nitrobenzyl alcohol is not strictly miscible with water or aqueous solutions, good results can be obtained by simply spreading the aqueous analyte solutions on top of a droplet of nitrobenzyl alcohol, or immersing a small crystal of the protein in the matrix. Some of these liquid matrices are established as matrices in Fast Atom Bombardement (FAB). As in FAB, they may permit a direct coupling of liquid chromatography with LDI.

## 5.7. COMBINATION OF BIOCHEMICAL TECHNIQUES WITH MATRIX-ASSISTED LASER DESORPTION/IONIZATION

The lack of any significant fragmentation in LDI is on the one hand advantageous for the determination of the molecular mass of large ions, but on the other it strongly limits the potential of the method for the collection of structural information. Added collision induced - or photofragmentation are difficult to implement in a pulsed time-of-flight system. It is, therefore, a natural approach to combine biochemical techniques of sample digestion or alteration with LDI-MS.

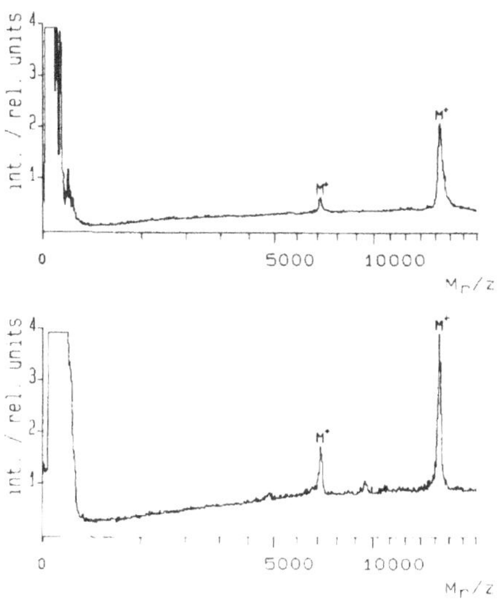


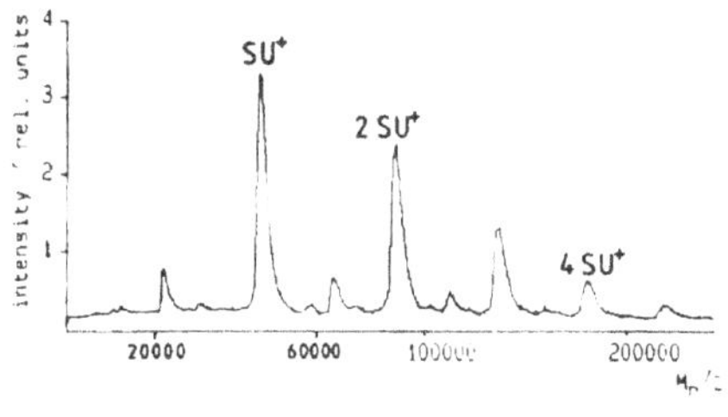


Figure 17. LDI-spectra of lysozyme from chicken egg white. Matrix: glycerol; wavelength: 2.94 $\mu$m (top) and 10.6 $\mu$m (bottom). Republished from ref. [7] and [8] with permission.

Figure 18. LDI-spectrum of the enzyme glucose isomerase from *streptomyces albus* desorbed from silicagel beads. Matrix: lactic acid; wavelength: 2.94 $\mu$m.

As was explained before, IgG monoclonal antibodies (MAB) consist of two light and two heavy chains linked by disulfide bonds. In conjugation studies, it is of interest to which part of the MAB the conjugates bind. Cleavage of the disulfide bonds can be achieved by treatment of the protein by dithiothreitol (DTT) [13] as shown in Figure 19. The top spectrum is typical of the intact MAB, as presented before. The bottom spectrum was obtained after treatment of the MAB with DTT. Even though the light chain is ca. 23,000 Da and the heavy chain is ca. 50,000 Da and are necessarily present in the sample at equal concentration, the signal of the light chain exceeds that of the heavy chain by a factor of ca. 10. The reason for this suppression of the heavy chain signal is not yet known. A control experiment with a deglycosylated MAB revealed that the suppression is not caused by the carbohydrate fraction, which is associated almost exclusively with the heavy chain. The signal at ca. 73,000 Da represents the combination of a light and a heavy chain due to incomplete cleavage or noncovalent association.

The kinetics of enzymatic reactions in (partial) protein sequencing can also be investigated by UV-LDI technique. An example of a small peptide (porcine renin substrate) that is enzymatically digested from the carboxy terminal by the enzyme carboxypeptidase Y is shown Figure 20. In this experiment, several separate droplets of protein solution were mixed with corresponding amounts of enzyme solution on the sample substrate. After preselected times, suitable amounts of matrix solution were added to a given droplet. The low pH of the matrix solution abruptly terminates the enzymatic reaction. After the reaction in the last droplet had been stopped the solvent was dried off in the usual manner and the substrate introduced into the mass spectrometer. By proper choice of the times for matrix addition to the individual droplets, the kinetics of the process can then be followed via analysis of the individual

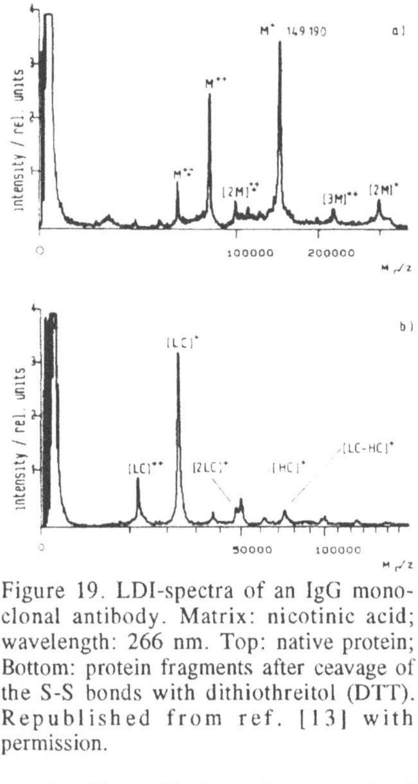


Figure 19. LDI-spectra of an IgG monoclonal antibody. Matrix: nicotinic acid; wavelength: 266 nm. Top: native protein; Bottom: protein fragments after ceavage of the S-S bonds with dithiothreitol (DTT). Republished from ref. [13] with permission.

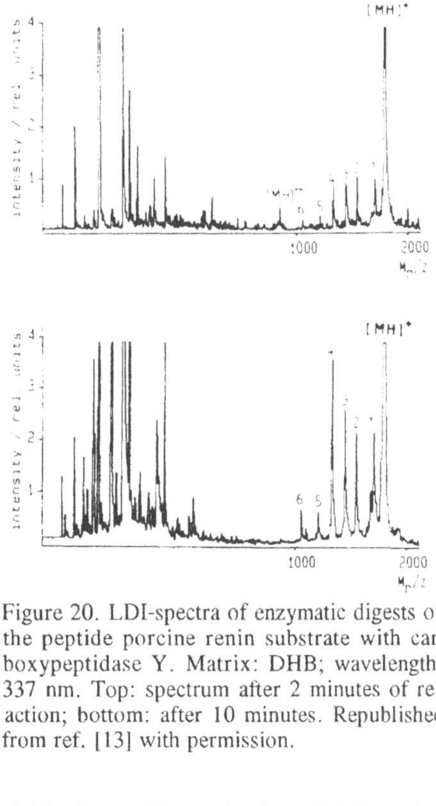


Figure 20. LDI-spectra of enzymatic digests of the peptide porcine renin substrate with carboxypeptidase Y. Matrix: DHB; wavelength: 337 nm. Top: spectrum after 2 minutes of reaction; bottom: after 10 minutes. Republished from ref. [13] with permission.

samples. Figure 20 shows the states after 2 and 10 minutes. The peaks denoted 1 through 6 represent the peptide minus the respective number of amino acids. The distribution of the signal intensities relative to each other represents the enzyme kinetics *i. e.*, different cleavage rates, depending on the specific amino acid cleaved, as well as a partially different sensitivity of the LDI for the different reaction products. The details of these contributions can be elucidated by repeating the experiment with different types of carboxy peptidases. Best results have been obtained with a cocktail of different enzymes [14]. The highly desirable extension of this technique into the mass range of 10,000 to 20,000 Da requires an improvement of the mass resolution to values between 500 or 1000, in order to separate the single digestion products in a given spectrum and determine their mass with sufficient accuracy.

## 5.8. ANALYSIS OF OLIGO- AND POLYNUCLEOTIDES

The potential of matrix-assisted LDI for the analysis of oligonucleotides is shown in Figure 21. This figure presents a negative ion spectrum of a mixture of the six synthetic oligonucleotides $d(pT)_{12}$ to $d(pT)_{18}$. The spectrum was taken with a nitrogen laser and aminopyrazine carboxylic acid as matrix. A distribution of seven oligonucleotides is resolved. Multiple salt formation by Na and K as well as cationization lead to relatively broad, structured peaks and thereby loss in mass resolution. A further example is the positive ion spectrum of a 5S ribosomal RNA from *Escherichia coli* with a molecular weight of ca. 38,500 Da. The Er:YAG-laser and nicotinic acid were used to obtain this sum spectrum presented in Figure 22. Several t-RNAs with molecular weights of ca. 25,000 Da were analyzed successfully as well. All had relatively broad peaks, pointing to the need for better preparation techniques to get more homogeneous samples.

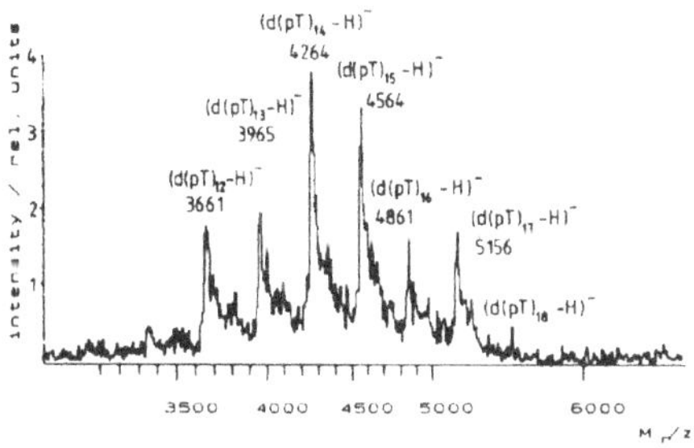


Figure 21. LDI-spectrum of a mixture of oligonucleotides $(d(pT))_{12}$ - $(d(pT))_{18}$ (negative ions). Matrix: aminopyrazine carboxylic acid; wavelength: 337 nm.

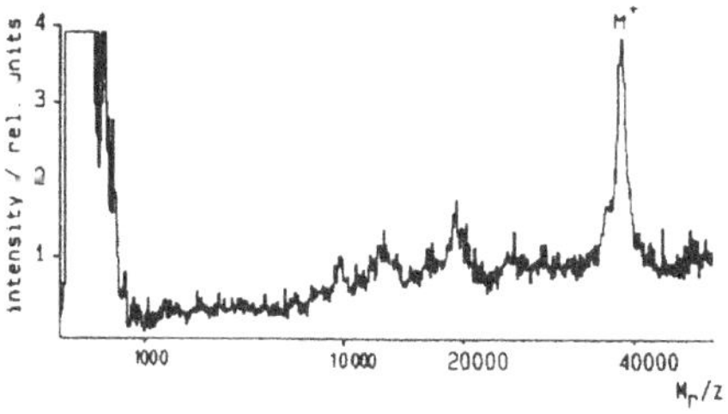


Figure 22. LDI-spectrum of a 5S ribosomal RNA from *Eschericia coli* (MW 38,500). Matrix: nicotinic acid; wavelength: 2.94 $\mu$m.

## 5.9. ANALYSIS OF PURE CARBOHYDRATES

A number of oligosaccharides have been investigated using both UV- and IR-lasers.[15] Figure 23 shows the positive ion spectra of mixtures of underivatized oligosaccharides, taken with a nitrogen laser at 337 nm wavelength and DHB as matrix. Spectrum A represents a complex mixture of maltodextrins ($\alpha$-1,4-linked oligoglucans) with a mass difference of one glucose unit of 162 Da and a maximum mass of ca. 3000 Da. B shows a spectrum of a dextran ($\alpha$-1,6-linked oligoglucans) with masses up to ca 7000 Da. The distribution peaks at 2311 Da. As expected, all ions of pure carbohydrates contain metal cations rather than an additional proton.

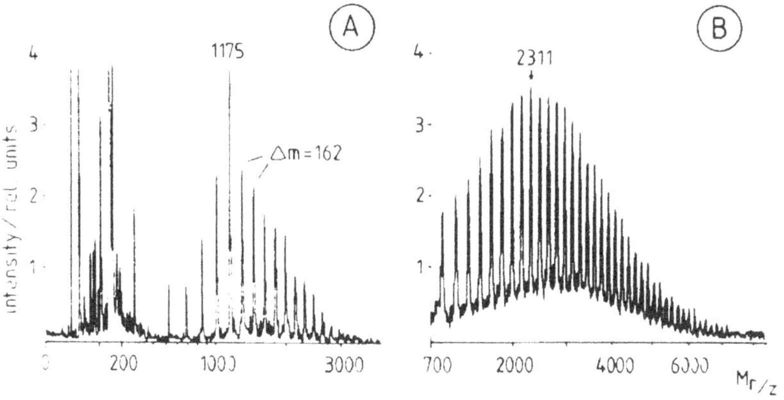


Figure 23. LDI-spectra of underivatized carbohydrates. Matrix: DHB; Wavelength: 337 nm. A: mixture of maltodextrins; B: mixture of dextrans. Republished from ref. [15] with permission.

## 6. CONCLUSIONS

The results reported here show the high potential of matrix-assisted LDI for molecular mass determination of large molecules. Because there are numeous advantages and features, this techniques should become a valuable tool in the field of biochemical analysis. Improvements in mass resolution and mass accuracy as well as in optimization of sample preparation techniques with respect to reproducibility and sensitivity will further the use of this technique. Different wavelength/matrix combination for particular applications will undoubtedly be found in the future and will increase the versatility of the method.

## 7. REFERENCES


1. Karas, M., Ingendoh, A., Bahr, U. and Hillenkamp, F. (1989) UV-Laser Desorption/Ionization Mass Spectrometry of Femtomol Amounts of Large Proteins. *Biomed. Environm. Mass Spectrometry,* **18,** 841 - 843
2. Spengler, B., Kirsch, D., Kaufmann, R. (1991) Metastable Decay of Peptides and Proteins in Matrix-Assisted Laser Desorption Mass Spectrometry. *Rapid Commun. Mass Spectrom.,* **5**, 198-202
3. Beavis, R., Chait, B. T. (1989) Factors Affecting the Ultraviolet Desorption of Proteins *Rapid Commun, Mass Spectrom* **3,** 233
4. Beavis, R., Chait, B. T. (1990) High-Accuracy Molecular Mass Determination of Proteins Using Matrix-Assisted Laser Desorption Mass Spectrometry. *Anal. Chem.* **62,** 1836-1840
5. Karas, M, Bahr, U., Ingendoh A. and Hillenkamp, F. (1989) Laserdesorptions-Massenspektrometrie von Proteinen mit Massen zwischen 100.000 und 250.000 Dalton. (in English edition: Laser Desorption/Ionization Mass Spectrometry of Proteins of Mass 100.000 to 250.000 Dalton) *Angewandte Chemie* **101,** 805 - 806; Int. Ed. Engl. **28,** 760
6. Strupat, K. (1991) Untersuchungen zur Matrix-unterstützten ultravioletten Laser Desorptions/Ionisations Massenspektrometrie großer Moleküle: Einfluß der Probenpräparation bei der Benutzung unterschiedlicher Matrix-Wellenlängen Kombinationen Diploma Thesis, Universität Münster
7. Overberg, A., Karas, M., Bahr, U., Kaufmann R. and Hillenkamp, F. (1990) Matrix-Assisted Infrared (2.94 $\mu$m) Desorption/Ionization Mass Spectrometry of Large Biomolecules. *Rapid Commun. Mass Spectrometry* **4,** 293-296
8. Overberg, A., Karas M. and Hillenkamp, F. (1991) Matrix-Assisted laser Desorption of Large Biomolecules with a TEA-$CO_2$-Laser. *Rapid Commun. Mass Spectrometry.* **5**, 128-131
9. Mecke, R. and Langenbucher, F. (1965) Infrared Spectra Heydon & Son Ltd., London,
10. Beavis, R. C. and Chait, B. T. (1989) Cinnamic Acid Derivatives as matrices for Ultraviolet Laser Desorption Mass Spectrometry of Proteins. *Rapid. Commun. Mass Spectrom.* **3,** 432
11. Beavis, R. and Chait, B. T. (1990) Rapid, Sensitive Analysis of Protein Mixtures by Mass Spectrometry *Proc. Nat. Acad. Sc.,* **87,** 6873

12. Karas, M. and Hillenkamp, F. (1988)
Laser Desorption Ionization of Proteins with Molecular Mass Exceeding 10.000 Daltons
*Anal. Chem.* **60,** 2299-2301.
13. Karas, M., Bahr, U., Ingendoh, A., Nordhoff, E., Stahl, B., Strupat, K. and Hillenkamp, F. (1990)
Principles and Applications of Matrix-Assisted UV-Laser Desorption/Ionization Mass Spectrometry
*Analytica Chimica Acta* **241,** 175-185
14. Nordhoff, E.
Untersuchungen zur Partiellen Sequenzanalyse von Proteinen und Polypeptiden durch enzymatischen Abbau und Matrix UV-Lasersdesorptions/Ionisations Massenspektrometrie. Diploma thesis, Universität Münster, 1991
15. Stahl, B., Steup, M., Karas, M. and Hillenkamp, F. (1991)
Mass Spectrometry of Oligosaccharides by Matrix-Assisted Laser Desorption/Ionization (LDI-MS), *Anal. Chem.*, in press
16. Hillenkamp, F., Karas, M. (1990)
Mass Spectrometry of peptides and Proteins by Matrix-Assisted Ultraviolet Laser Desorption/Ionization. In J. A. McCloskey (ed.), Methods in Enzymology, Vol. 193, Academic Press, Inc., New York, pg. 280-295.
17. Allmaier, G., Schmid, E. R., Hagspiel, K., Kubicek, C. P., Karas, M. and Hillenkamp,F. (1990)
Strategy for the Characterization of the Glycoprotein Endoglucanase I, isolated from *Trichoderma reesei*: Combination of Plasma Desorption and UV Laser Dersorption Time-of-Flight Mass Spectrometry.
*Analytica Chimica Acta* **241**, 321-327.

# LASER DESORPTION, CHEMICAL IONIZATION, AND LASER DESORPTION/CHEMICAL IONIZATION APPLICATIONS WITH FOURIER TRANSFORM MASS SPECTROMETRY

J. Paul Speir, Greg S. Gorman, I. Jonathan Amster
*Department of Chemistry*
*University of Georgia*
*Athens, Georgia 30602*
*USA*

ABSTRACT. Laser desorption coupled with Fourier transform mass spectrometry has been applied to the analysis of several classes of nonvolatile molecules. Tandem mass spectrometric and high resolution mass measurements have been made of ions produced by laser desorption. Chemical ionization applications to Fourier transform mass spectrometry require specialized experimental conditions. Chemical ionization of laser desorbed neutrals in a Fourier transform mass spectrometer is described.

## 1. Laser Desorption/FTMS

### 1.1 BACKGROUND

During the 1980s, many practitioners of mass spectrometry were attracted to Fourier transform mass spectrometry (FTMS) by its high performance capabilities for high resolution measurements and tandem mass spectrometry. These attributes, along with the ability to trap ions and detect all ions produced during a pulsed ionization event, have stimulated interest in using FTMS for the analysis of large molecules. New desorption and ionization methods for large molecules were applied to FTMS, paralleling their rapid development in other areas of mass spectrometry. Examples of laser desorption [1, 2], fast atom bombardment [3], $^{252}$Cf plasma desorption [4, 5], secondary ion mass spectrometry [6, 7], and electrospray ionization [8] to FTMS appeared in the literature over the last decade. The first demonstration of laser desorption in combination with FTMS was for forming transition metal ions, used for ion-molecule studies, in 1980 [9]. Laser desorption/FTMS for the analysis of organic molecules was first reported by Gross and coworkers in 1982 [1]. Laser desorption/FTMS has since been applied to a broad spectrum of nonvolatile compounds including peptides [10-13], saccharides [14-18], nucleotides and nucleosides [19-21], porphyrins [22-24], steroids and steroid glycosides [25-27], pharmaceuticals [28], polymers [29-36], and surfactants [37]. Laser desorption predominated over other ionization methods in the early stages of the development of FTMS for the analysis of nonvolatile compounds for two reasons. Laser desorption is compatible with the low pressure requirement of FTMS (background pressure of $10^{-8}$ torr or less), and the desorption particles (photons) are not influenced by magnetic and electric fields, facilitating their introduction into the analyzer region. The application of fast atom bombardment and of electrospray ionization to FTMS evolved only after the development of the external ion source [3, 38].



*M. L. Gross (ed.), Mass Spectrometry in the Biological Sciences: A Tutorial*, 199–212.

Laser desorption produces ions in a burst having a short duration and a low duty cycle. Mass spectrometers capable of detecting all the ions produced in a single pulse, such as Fourier transform mass spectrometers or time-of-flight mass spectrometers, couple more effectively with laser desorption ionization sources than do scanning mass analyzers, such as sector or quadrupole instruments. FTMS has the advantages of higher mass resolution and a facility for tandem mass spectrometry in comparison to time-of-flight mass spectrometry, although recent developments of the latter method suggest that these advantages are diminishing [39, 40].

## 1.2 EXPERIMENTAL CONSIDERATIONS

In a laser desorption/FTMS experiment, a solid sample is deposited on a substrate, introduced into the vacuum system, positioned adjacent to the analyzer cell, and irradiated by a pulsed laser. Desorbed ions leaving the surface of the sample enter the analyzer cell and are trapped and mass analyzed. Once trapped, ions may also be subjected to further fragmentation by collisionally activated dissociation [9] or photodissociation [22, 41]. Figure 1 illustrates the relative positions of the sample and the analyzer cell for a laser desorption experiment. Typically, the sample is introduced into the vacuum system on the end of a direct insertion probe. Such probes and their required vacuum locks must be designed to preserve the near ultra-high vacuum required for FTMS analysis.

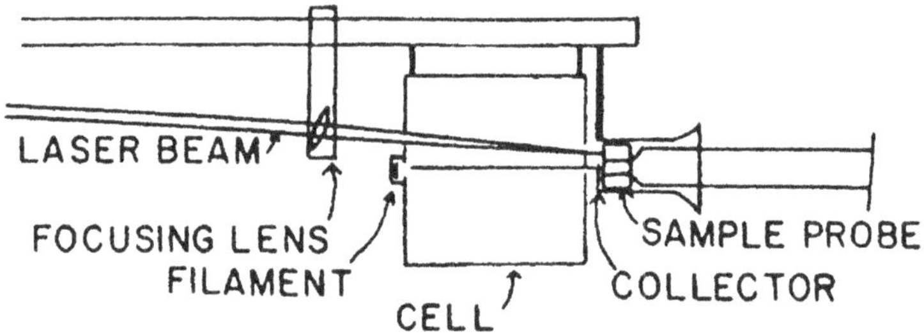


Figure 1. Experimental arrangement for a LD/FTMS experiment, typical of systems using a solenoidal magnet. The magnetic field lines in the region of the cell are parallel to the horizontal axis of the figure. Republished from ref. 2 with permission.

Samples are applied to the probe substrate by one of a variety of methods. A solution of a sample may be deposited onto the probe tip, followed by evaporation of the solvent to leave a precipitate of the sample. Solutions can be electrosprayed onto the probe tip to produce a more uniform thin film [42]. Insoluble polymers have been pressed onto the probe tip [32]. Samples may be applied neat, or mixed with additives to enhance desorption or ion formation. Alkali salts have been used extensively as an additive to form cation- attached molecular ions [29]. Matrix assisted laser desorption coupled with FTMS is being investigated by several groups. The impressive enhancement in forming molecular species of large peptides and proteins that have been observed using time-of-flight mass spectrometers has not been reported yet for FTMS.

Wavelengths from the infrared to the ultraviolet have been used for laser desorption/FTMS experiments. Pulsed $CO_2$ lasers (10.6 μm radiation) were used in most of the early work. Nd:YAG lasers (1.06 μm) and excimer lasers (193 nm, 248 nm, 308 nm, or 351 nm) have also been successfully applied. The irradiance (laser power density) is an important parameter in laser desorption experiments. Typical values of irradiance used in laser desorption/FTMS experiments range from $10^6$ W/cm$^2$ to $10^9$ W/cm$^2$. Lower power densities ($10^6$-$10^7$ W/cm$^2$) have been reported to favor formation of the molecular ion species, while a higher irradiance ($10^8$ W/cm$^2$) causes an increase in the formation of fragment ions [43].

Desorbed ions encounter magnetic and electric fields along their paths into the analyzer cell. Such fields would ordinarily prevent the entrance of an isolated ion into the cell. For example, the first laser desorption/FTMS experiments used an instrument with an electromagnet [1], in which the direct probe and the laser beam were brought into the analyzer region perpendicular to the magnetic field lines. Ions produced at the probe tip and moving toward the analyzer cell travel transverse to the magnetic field lines. This is expected to exert a force on an ion perpendicular to its motion that should be sufficiently strong to prevent its entry into the cell. Despite this, ions desorbed by this configuration are observed to be trapped in the cell. Collisions with the neutrals that are also desorbed by the laser are postulated to cause the diffusion of ions into the analyzer cell. FTMS instruments that utilize superconducting, solenoidal magnets are designed to bring in the sample probe and laser beam parallel to the magnetic field lines, as shown in Figure 1. With this configuration, desorbed ions formed at the probe tip do not cross magnetic field lines en route to the analyzer cell, allowing them to move freely toward the cell. The ions, however, must pass a trapping plate with its applied potential to enter the analyzer cell. Ions having sufficient kinetic energy to pass the first trapping plate should collide with the opposite trapping plate and be neutralized. Nevertheless, laser desorbed ions are found to be trapped by the analyzer cell. The mechanism for trapping ions made externally to the cell has not been determined, although collisions of the ions with desorbed neutrals may be responsible in part.

### 1.3 LASER DESORPTION/FTMS EXAMPLES

Laser desorption/FTMS has been successfully applied to the analysis of peptides and peptide mixtures. Mixing a peptide sample with KBr has been utilized to enhance the formation of a $K^+$-attached molecular ion species. With such sample preparation, spectra obtained with moderate values of laser irradiance typically contain the cationized molecule and few fragment ions [43]. The abundance of the cationized molecule formed in this manner is sufficiently large to obtain tandem mass spectrometry data [10]. Higher values of laser irradiance produce peptide spectra containing both fragment ions and a cationized molecule [2]. High resolution mass spectra have been recorded for peptide ions formed by laser desorption. A mass resolution of 150,000 is demonstrated at mass 1180 for the $K^+$-attached molecule of the cyclic decapeptide, gramicidin S, shown in Figure 2 [43].

Laser desorption/FTMS has been particularly useful for determining molecular weight distributions for polymers. The mass spectra of polymers contain abundant molecular ion species, and show little fragmentation. Figure 3 show the molecular ion region of the negative ion mass spectrum of Krytox, a polymeric diffusion pump oil [2]. The average molecular weight and the molecular weight distribution of the polymer can be determined from this mass spectrum.

Porphyrin spectra obtained with laser desorption/FTMS are also characterized by abundant molecular ion species and few fragment ions. Figure 4 shows the positive ion mass spectrum of tetrakis[4-(dimethylamino)phenyl]porphyrin [22], in which potassium, sodium, and proton attached molecular ions are observed with excellent signal to noise.

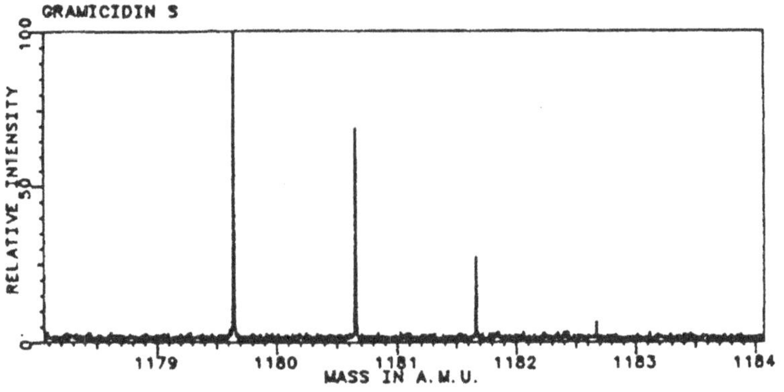


Figure 2. The peaks corresponding to (M+K)$^+$ and its isotopes in the LD/FTMS high resolution mass spectrum of gramicidin S. Republished from ref. 43 with permission.

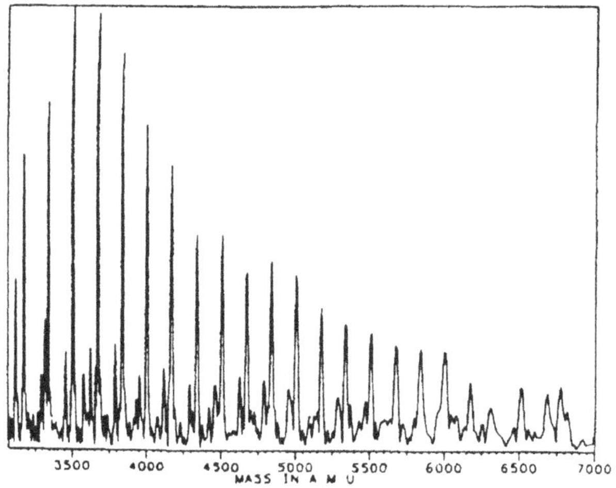


Figure 3. LD/FTMS negative ion mass spectrum of Krytox, a polymeric oil. The distribution of chain lengths in the polymer produces a corresponding distribution of peaks corresponding to (M+K)$^+$ in the mass spectrum. Republished from ref. 2 with permission.

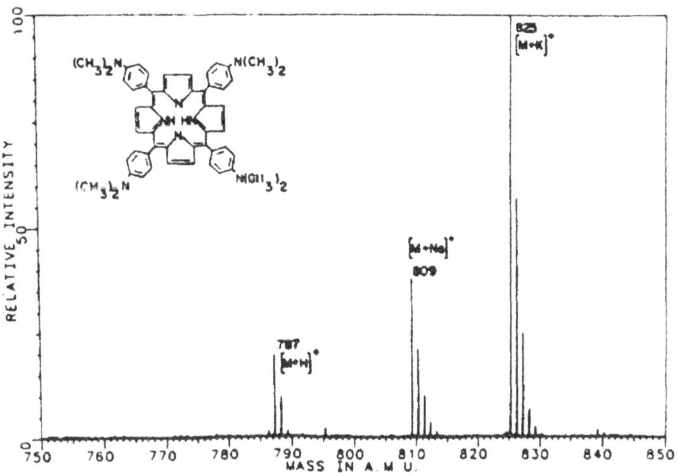


Figure 4. LD/FTMS positive ion mass spectrum of tetrakis[4-(dimethylamino) phenyl]porphyrin. Republished from ref. 22 with permission.

## 2. Chemical Ionization/FTMS

### 2.1 CHEMICAL IONIZATION - BACKGROUND AND THEORY

Chemical ionization (CI), introduced by Field and Munson in 1966, was the first example of a "soft" ionization method in mass spectrometry, that is, an ionization method in which formation of a molecular ion or molecular ion species is enhanced relative to the formation of fragment ions [44, 45]. Sample molecules are ionized by a reaction with a reagent ion, hence the name chemical ionization. Proton transfer and charge exchange are both examples of CI reactions [46]. This section will focus on proton transfer reactions, as they have been utilized more frequently for the analysis of biomolecules.

CI occurs through a set of sequential reactions, in which reagent ions are formed by electron bombardment of a reagent gas, and sample molecules are ionized by a reaction with the reagent ions. The chain of reactions for chemical ionization of a generic molecule, M, using methane as a reagent, serves to illustrate the general scheme.

$$CH_4 + e^- \text{---}> CH_4^{+\cdot}, CH_3^+, CH_2^{+\cdot}, CH^+$$

$$CH_4^{+\cdot} + CH_4 \text{---}> CH_5^+ + CH_3^\cdot$$

$$CH_3^+ + CH_4 \text{---}> C_2H_5^+ + H_2$$

$$CH_5^+ + M \text{---}> (M+H)^+ + CH_4$$

$$C_2H_5^+ + M \text{---}> (M+H)^+ + C_2H_4$$

The formation of the reagent ions $CH_5^+$ and $C_2H_5^+$ results from bimolecular reactions which occur between methane molecules and the ions formed directly by electron ionization of methane. In sector and quadrupole mass spectrometers, an ion source pressure of 1 torr is required for this ion-molecule reaction to occur, a much higher pressure than the $10^{-6}$ torr ion source pressure typically used for electron ionization. The transfer of a proton from the reagent ion to the sample molecule completes the ionization sequence. The internal energy of the protonated molecule increases by a quantity equal in magnitude to the enthalpy of the proton transfer reaction:

$$RH^+ + M \dashrightarrow (M+H)^+ + R$$

$$\Delta H = P.A.(R) - P.A.(M)$$

where $RH^+$ is a reagent ion and P.A. is a proton affinity. The protonation of the sample is observed to a significant extent only if the reaction is exothermic. If the enthalpy of the reaction is small (and negative), a protonated molecule is formed, accompanied by little or no fragmentation of the molecular species. If the reaction is sufficiently exothermic, the protonated molecule will undergo unimolecular dissociation to form fragment ions.

An important application of chemical ionization is determination of molecular weight, particularly for those compounds that produce only fragment ions by electron ionization. For such studies, the formation of the protonated molecule is enhanced by selecting a reagent ion with a proton affinity that is slightly smaller than that of the sample. Common reagent ions and their proton affinities are listed in Table I. With this wide range of protonating species, chemical ionization is extremely flexible in matching the proton affinity of the reagent ion to that of the sample. Compounds having a proton affinity less than that of the reagent ion will not be protonated. Targeted compounds present in complex mixtures can be selectively ionized by choosing a reagent ion with a proton affinity value smaller than that of the sample of interest and larger than the proton affinity values of the other components of the mixture.

**TABLE I**

Proton Affinities of Selected Reagent Ions

| Reagent Ion | Conjugate Base | Proton Affinity[a] (eV) |
|---|---|---|
| $Me_3NH^+$ | $Me_3N$ | 9.7 |
| $EtNH_3^+$ | $EtNH_2$ | 9.4 |
| $NH_4^+$ | $NH_3$ | 8.9 |
| $t\text{-}C_4H_9^+$ | $i\text{-}C_4H_8$ | 8.5 |
| $C_3H_7^+$ | $C_3H_6$ | 8.0 |
| $H_3O^+$ | $H_2O$ | 7.5 |
| $C_2H_5^+$ | $C_2H_4$ | 7.1 |
| $COH^+$ | $CO$ | 6.1 |
| $N_2OH^+$ | $N_2O$ | 5.9 |
| $CH_5^+$ | $CH_4$ | 5.7 |
| $N_2H^+$ | $N_2$ | 5.1 |
| $H_3^+$ | $H_2$ | 4.4 |

[a]Proton affinity values are from Reference 46 and are for the conjugate base of the reagent ion. Proton affinity is defined as $-\Delta H$ for the reaction $B + H^+ \dashrightarrow BH^+$.

## 2.2 CHEMICAL IONIZATION IN FOURIER TRANSFORM MASS SPECTROMETRY

The implementation of CI using a Fourier transform mass spectrometer requires different experimental conditions from those used in sector or quadrupole instruments, because there are fundamental differences in their operating principles. The main consideration is the length of time available for the ion-molecule reactions leading to protonation. The ion residence time in the source of a conventional mass spectrometer is on the order of microseconds. A high reagent gas pressure (0.1-1 torr) is required so that the sequence of ion-molecule reactions that produce a protonated molecule will occur in less than 1 μs. This high pressure is incompatible with ion detection by FTMS. The pressure in the FTMS analyzer cell should be less than $10^{-7}$ torr for the optimal detection of ions. With FTMS, however, ions can be trapped in the analyzer cell for seconds or longer, allowing the ion-molecule reactions required for chemical ionization to occur at a much lower pressure [47]. For example, at a pressure of $1 \times 10^{-7}$ torr, methane and methyl ions formed by electron ionization will experience a collision with a methane neutral to form a reagent ion within 1 s. Chemical ionization at this low pressure is observed simply by waiting for the necessary reactions to occur.

The experimental sequence used for chemical ionization in an FTMS instrument incorporates a delay between the ionization and detection events. Consider the chemical ionization observed with a reagent pressure of $1 \times 10^{-7}$ torr and a sample pressure of $1 \times 10^{-8}$ torr. Electron ionization of the mixture produces ions derived from both the reagent and the sample, with the reagent products being ten times more abundant. 1 s following the electron ionization event, the ions that were initially formed ($CH_4^{+\cdot}$ and $CH_3^+$, for the example of methane) have reacted with the reagent gas to form the reagent ions ($CH_5^+$ and $C_2H_5^+$). Within 10 s, all the reagent ions have reacted with the sample. The mass spectrum produced by this experiment is a mixture of chemical ionization and electron ionization of the sample, although the CI products will be 10 times more abundant.

The ratio of CI to EI products can be raised by increasing the ratio of the methane to sample pressure. This can be accomplished by increasing the methane pressure or decreasing the sample pressure. The former will cause a decrease in mass resolution, whereas the latter will require a longer reaction time for the sample to be protonated. An alternative solution utilizes pulsed valve reagent addition [48]. Reagent gas admitted into the vacuum chamber through a pulsed valve causes the pressure to increase to $10^{-5}$ torr within 100 ms, followed by an exponential decay to the background pressure, as the reagent gas is pumped away. A time profile of the pressure achieved by using a pulsed valve for reagent addition is shown in Figure 5, and is compared with the timing sequence for a pulsed valve CI experiment. Electron ionization occurs during the peak of the pressure pulse. At this high pressure, the sequence of reactions to form reagent ions occur in approximately 10 ms. A suitable delay period follows, during which the sample, admitted through a leak valve at a constant low pressure, is protonated, and the excess reagent gas is pumped away. The detection of the ions occurs after the pressure of the reagent has fallen below $10^{-7}$ torr. In this manner, a large reagent to sample ratio ($10^2$-$10^3$) is maintained during the electron ionization event, and the system pressure is kept low during ion detection.

Chemical ionization of a sample in a Fourier transform mass spectrometer can be observed without the addition of a reagent. The sample itself can serve as the source of reagent ions in a "self-chemical ionization" experiment [49]. Electron ionization of the sample produces fragment ions which, for some molecules, may serve as reagent ions. Compounds containing alkyl chains readily exhibit this behavior. If the detection event is delayed after electron ionization, reactions of the sample with the initially formed ions can

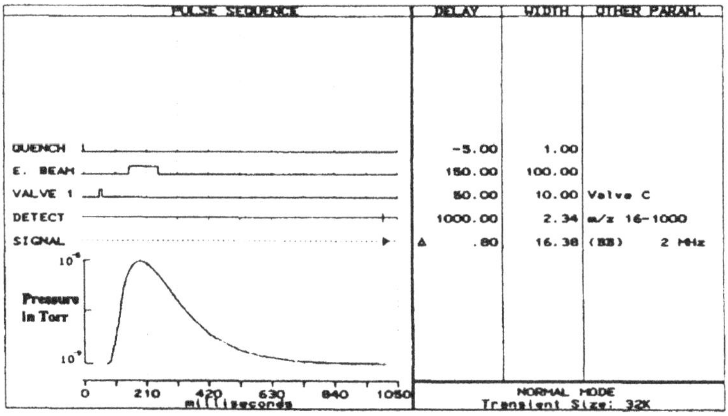


Figure 5. Time profile of reagent gas pressure in a pulsed valve CI experiment compared with the time profile of the experimental sequence.

lead to the formation of a protonated molecule. Such reactions are observed in sector mass spectrometers if the sample pressure is high enough to allow ion-molecule reactions to occur in the ion source [50]. This process is observed at low pressure with FTMS if the reaction period prior to ion detection is suitably long. As an example, the self-CI mass spectrum of methyl stearate is presented in Figure 6. The protonated molecule gives rise to the base peak of the mass spectrum. Electron ionization of this compound produces some unreactive fragment ions, which are found at lower mass than the protonated molecule. The protonated molecule itself can react with neutral methyl stearate to form a proton bound dimer. This ion has a relative abundance of 10% in the mass spectrum displayed, but with longer reaction periods, this ion becomes most abundant in the mass spectrum.

## 3. Laser Desorption/Chemical Ionization/FTMS

### 3.1 POSTIONIZATION METHODS IN LASER DESORPTION

Irradiating an organic sample with a pulsed laser can lead to the desorption of odd electron ions, cationized molecules, ionic fragments, adduct ions and neutral species. The type of desorption products that will be observed depends on several experimental factors, including laser wavelength, laser power density (irradiance), sample thickness, the nature of the matrix (if present), substrate composition, and the presence of alkali salts. At low laser irradiance (less than $10^6$ W/cm$^2$), neutral desorption has been observed to occur without the formation of any ions. Even at higher values of irradiance, the yield of neutrals is often much larger than the yield of ions [51]. Although most research in the area of laser desorption mass spectrometry has focused on the ions that are produced, several experiments have been developed to utilize desorbed neutrals. To be analyzed by mass spectrometry, the neutral species are ionized in a separate step, called postionization, which emphasizes that ionization occurs in a second process, after laser desorption. Laser

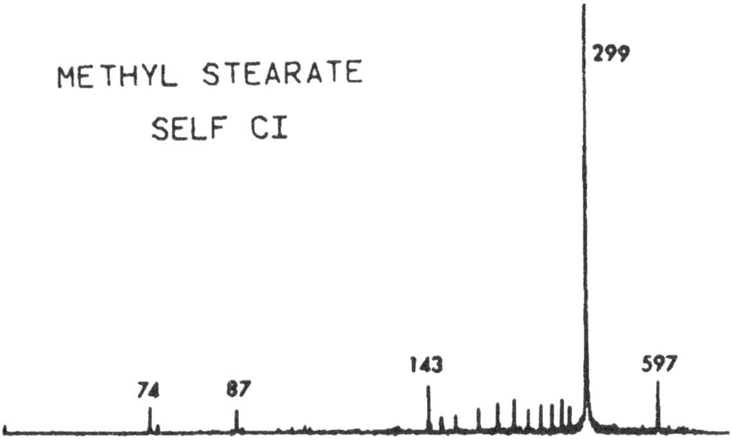


Figure 6. Self-chemical ionization mass spectrum of methyl stearate. Republished from ref. 49 with permission.

multiphoton ionization [52-55], electron ionization [56, 57] and chemical ionization [58-60] have been used as postionization methods for laser desorbed neutrals. This section focuses on the application of chemical ionization to laser desorbed molecules.

Chemical postionization, or laser desorption chemical ionization (LD/CI) was first demonstrated by Cotter [58] in 1980, to utilize the neutrals which were observed to desorb from a sample for several hundred microseconds after the termination of the laser pulse. Cotter observed that the protonated molecules for two steroid glucuronides, that had previously resisted analysis by chemical ionization, were formed by LD/CI, in a double focusing mass spectrometer. The combination of a pulsed ionization method with a scanning mass analyzer resulted in poor sensitivity for these experiments. Gross and co-workers [42] suggested the use of FTMS for LD/CI experiments, because of the capability of this mass spectrometer to couple with pulsed ionization methods. Amster et al. [59, 60] performed the first LD/CI experiments using FTMS, and were able to form protonated molecules for several low volatility compounds, including pentadecylacridine and thymidine.

The LD/CI/FTMS experiment is illustrated in Figure 7, which portrays the sequence of events occurring inside the analyzer cell. Reagent gas enters the analyzer region through a pulsed valve, and is ionized by an electron beam. Ion-molecule reactions occurring at the peak of the pressure pulse rapidly lead to the formation of reagent ions. These are trapped in the analyzer cell while the reagent gas is pumped away. The laser is then fired at the sample, which causes neutrals to desorb in a short burst. As the neutrals pass through the trapped ions, a fraction of the molecules undergo proton transfer. The protonated sample molecules, and the fragment ions derived from these, are trapped in the analyzer cell, and are subsequently mass analyzed.

### 3.2 APPLICATIONS TO PEPTIDE ANALYSIS

Current research in this laboratory is directed toward developing LD/CI/FTMS for the determination of peptide structures [61]. Experimental conditions have been determined that allow the desorption of intact peptide molecules without interference from ions formed

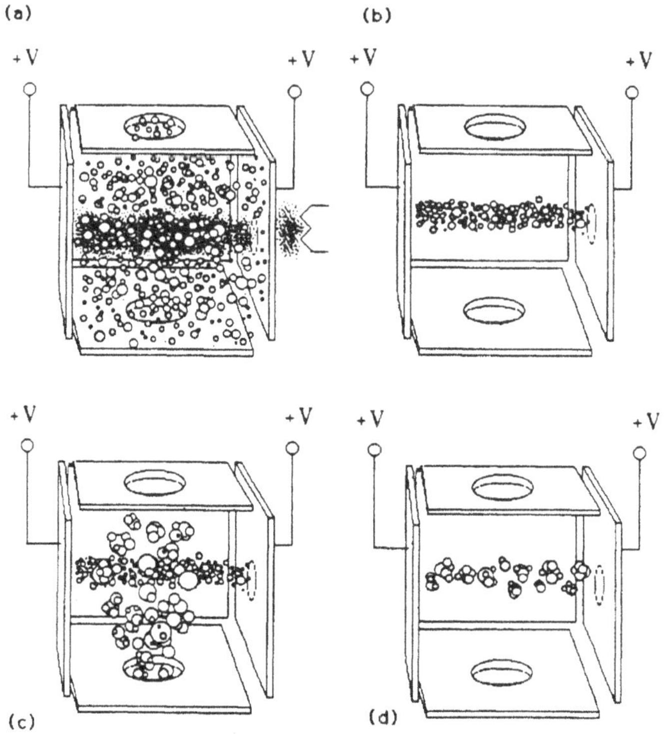


Figure 7. A graphical representation of the events occurring in a FTMS analyzer cell during a LD/CI experiment: (a) pulsed valve admits a reagent gas into the cell while the electron beam is on; (b) after the pulsed valve is closed, the excess reagent gas pumps away and reagent ions are stored in the cell; (c) the laser is fired and desorbed neutrals are ionized by the reagent ions; (d) protonated sample molecules are trapped in the cell and mass analyzed by FTMS. Republished from ref. 59 with permission.

directly by the laser or from fragmented neutral species. The key experimental parameters are laser wavelength (193 nm), laser irradiance ($10^6$ W/cm$^2$), substrate composition (75 nm gold/palladium film on a Macor stub), sample thickness (1000 monolayers), and sample purity (peptides are desalted by using HPLC).

The desorption of an intact molecular neutral can be demonstrated by chemical ionization with a reagent ion having a proton affinity value close to, but smaller than that of the peptide. Peptide proton affinities should be close to those of their constituent amino acids, to a first approximation. Amino acid proton affinities range from 9 eV for those having aliphatic hydrocarbon side chains, to 10 eV for those containing basic functional groups. The ammonium ion, $NH_4^+$, with a proton affinity of 8.9 eV, should protonate peptides comprised of aliphatic amino acids without causing fragmentation. The LD/CI mass spectrum of the tripeptide val-pro-leu, using $NH_4^+$ as the reagent ion, is shown in

Figure 8a. The only ion formed is the protonated molecule, demonstrating that the tripeptide was desorbed as an intact species without contamination from neutral fragments. Fragmentation of the peptide can be induced by selecting a reagent ion with a lower proton affinity, making the ionization process more exothermic. This is demonstrated in Figure 8b and 8c, in which val-pro-leu is protonated by $C_2H_5^+$ and $N_2OH^+$, respectively. LD/CI with these ions is more exothermic by 1.8 eV and 3.0 eV than when using $NH_4^+$. The extent of fragmentation of the peptide is controlled in a predictable manner by the enthalpy of protonation. The relative abundance of the two principal fragment ions, of *m/z* 229 ($y_2$, pro-leu + 2H) and of *m/z* 197 ($b_2$, val-pro) undergoes a reversal in the latter two spectra. The data suggest a tighter transition state for the formation of the "b" ion than for the "y" ion [61], implying that the mechanism of the formation of the "b" ion involves a rearrangement reaction, and not the previously suggested direct bond cleavage.

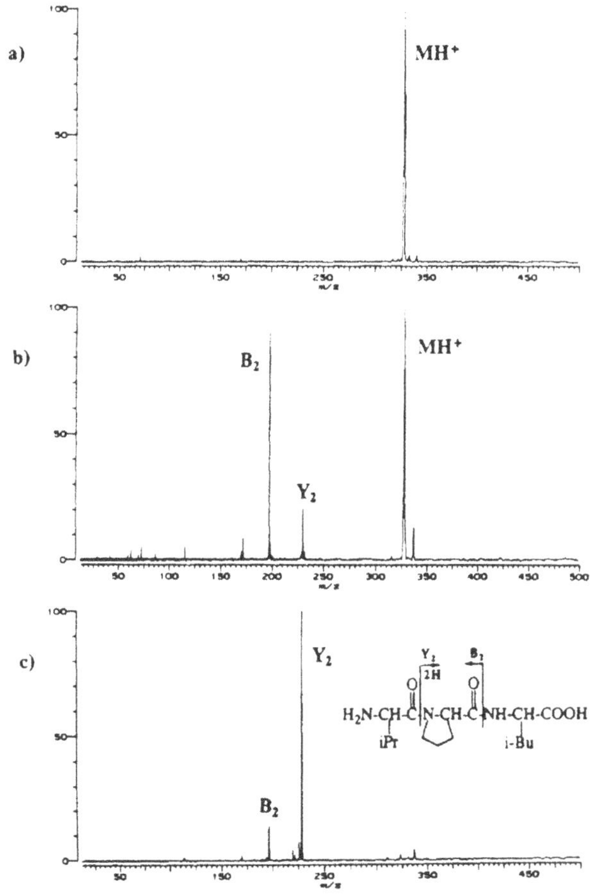


Figure 8. LD/CI mass spectra of the tripeptide val-pro-leu obtained with (a) $NH_4^+$, (b) $C_2H_5^+$, and (c) $N_2OH^+$ as reagent ions. Reprinted from ref. 61 with permission.

Other ion-molecule reactions of peptide neutrals can be investigated by using this technique. For example, metal ions can be made and stored in the analyzer cell and then reacted with laser desorbed peptides. Such a reaction can be thought of as chemical ionization with a metal ion as the reagent [62, 63]. Studies of gas-phase metal ions with volatile molecules suggests that a broad range of reactivities are available through the selection of different elements. We have studied the reactions of transition metal ions with peptides to search for analytically useful reactions. The metal ions are formed by laser ablation of a metal target near the analyzer cell. This leads to the formation of metal ions with a broad distribution of excited electronic states. Some of these excited states are sufficiently long-lived to be observed, and have different reactivities than those of the ground state ions. The mass spectrum in Figure 9 shows the products of the reaction of val-ala with $Fe^+$. The formation of $(M+Fe)^+$ is evident, as are fragment ions containing iron. The protonated molecule is hypothesized to result from a charge exchange reaction between an excited state metal ion and a peptide molecule, followed by self-chemical ionization in a reaction with a second peptide neutral. Although the reaction of $Fe^+$ with this peptide does not provide product ions that are structurally significant, the results demonstrate the feasibility of this technique, which can be used to study the reactions of any metal ion with peptide molecules. The reactions of the first row transition metal ions copper and nickel are currently being investigated.

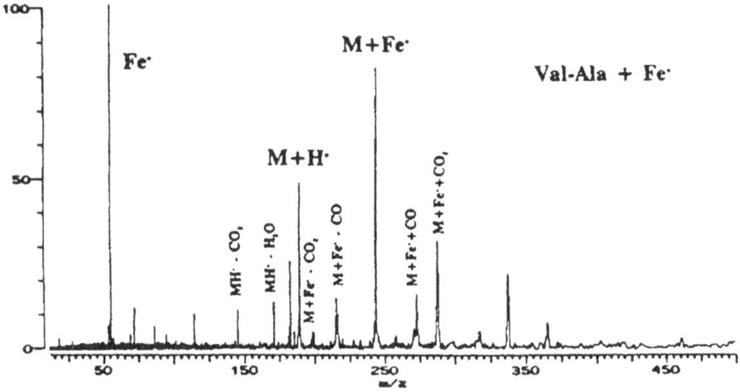


Figure 9. Mass spectrum of the products of the reaction of $Fe^+$ with the dipeptide val-ala.

## 4. Acknowledgements

This work was supported by a research award from the American Society for Mass Spectrometry, sponsored by Finnigan MAT. Acknowledgement is made to the donors of the Petroleum Research Fund, administered by the American Chemical Society, for partial support of this work.

**5. References**


1. McCrery, D. A.; Ledford, E. B., Jr.; Gross, M. L. (1982) *Anal. Chem.* **54**, 1435-1437.
2. Wilkins, C. L.; Weil, D. A.; Yang, C. L. C.; Ijames, C. F. (1985) *Anal. Chem.* **57**, 520-524.
3. Hunt, D. F.; Shabanowitz, J.; McIver, R. T.; Hunter, R. L.; Syka, J. E. P. (1985) *Anal. Chem.* **57**, 765-768.
4. Tabet, J. C.; Rapin, J.; Poretti, M.; Gaumann, T. (1986) *Chimia* **40**, 169-171.
5. Loo, J. A.; Williams, E. R.; Amster, I. J.; Furlong, J. J. P.; Wang, B. H.; McLafferty, F. W. (1987) *Anal. Chem.* **59**, 1880-1882.
6. Castro, M. E.; Russell, D. H. (1984) *Anal. Chem.* **56**, 578-581.
7. Amster, I. J.; Loo, J. A.; Furlong, J. J. P.; McLafferty, F. W. (1987) *Anal. Chem.* **59**, 313-317.
8. Henry, K. D.; Williams, E. R.; Wang, B. H.; McLafferty, F. W.; Shabanowitz, J.; Hunt, D. F. (1989) *Proc. Natl. Acad. Sci. U.S.A.* **86**, 9075-9078.
9. Cody, R. B., Jr.; Burnier, R. C.; Reents, W. D., Jr.; Carlin, T. J.; McCrery, D. A.; Lengal, R. K.; Freiser, B. S. (1980) *Int. J. Mass Spectrom. Ion Phys.* **33**, 37-43.
10. Cody, R. B., Jr.; Amster, I. J.; McLafferty, F. W. (1985) *Proc. Natl. Acad. Sci, USA* **82**, 6367-6370.
11. Chiarelli, M. P.; Gross, M. L. (1989) *Anal. Chem.* **61**, 1895-1900.
12. Wilkins, C. L.; Yang, C. L. C. (1986) *Int. J. Mass Spectrom. Ion Proc.* **72**, 195-208.
13. Yang, L. C.; Wilkins, C. L. (1989) *Org. Mass Spectrom.* **24**, 409-414.
14. Coates, M. L.; Wilkins, C. L. (1987) *Anal. Chem.* **59**, 197-200.
15. Coates, M. L.; Wilkins, C. L. (1985) *Biomed. Mass Spectrom.* **12**, 424-428.
16. Lam, Z.; Comisarow, M. B.; Dutton, G. G. S.; Weil, D. A.; Bjarnason, A. (1987) *Rapid Commun. Mass Spectrom.* **1**, 83-86.
17. Lam. Z.; Comisarow, M. B.; Dutton, G. G. S. (1988) *Anal. Chem.* **60**, 2304-2306.
18. Lam, Z; Dutton, G. G. S.; Comisarow, M. C.; Weil, D. A.; Bjarnason, A. (1988) *Carbohydr. Res.* **180**, C1-C7.
19. McCrery, D. A.; Gross, M. L. (1985) *Anal. Chim. Acta* **178**, 91-103.
20. Chiarelli, M. P.; Gross, M. L. (1989) *J. Phys. Chem.* **93**, 3595-3599.
21. Giam, C. S.; Holliday, T. L.; Williams, J. L.; Weller, R. R.; Mayernik, J. A.; Hinton, D. E. (1989) *Mar. Environ. Res.* **28**, 309-312.
22. Brown, R. S.; Wilkins, C. L. (1986) *Anal. Chem.* **58**, 3196-3199.
23. Nuwaysir, L. M.; Wilkins, C. L. (1989) *Anal. Chem.* **61**, 689-694.
24. Forest, E. (1989) *Org. Mass Spectrom.* **24**, 197-200.
25. Fung, E. T.; Wilkins, C. L. (1988) *Biomed. Environ. Mass Spectrom.* **15**, 609-613.
26. Coates, M. L.; Wilkins, C. L. (1986) *Biomed. Environ. Mass Spectrom.* **13**, 199-204.
27. Shomo, R. E., II; Chandrasekaran, A.; Marshall, A. G.; Reuning, R. H.; Robertson, L. W. (1988) *Biomed. Environ. Mass Spectrom* **15**, 295-302.
28. Shomo, R. E., II; Marshall, A. G.; Lattimer, R. P. (1986) *Int. J. Mass Spectrom. Ion Proc.* **72**, 209-217.
29. Wilkins, C. L.; Weil, D. A.; Yang, C. L. C.; Ijames, C. F. (1985) *Anal. Chem.* **57**, 520-524.
30. Brown, R. S.; Weil, D. A.; Wilkins, C. L. (1986) *Macromolecules* **19**, 1255-1260.
31. Miller, L. L.; Thomas, A. D.; Wilkins, C. L.; Weil, D. A. (1986) *J. Chem. Soc., Chem. Commun. 661-663.*
32. *Nuwaysir, L. M.; Wilkins, C. L.* (1988) Anal. Chem. **60**, 279-282.
33. Brown, C. E.; Kovacic, P.; Wilkie, C. A.; Cody, R. B., Jr.; Kinsinger, J. A. (1985) *J. Polym. Sci., Polym. Lett. Ed.* **23**, 453-463.

34. Brown, C. E.; Kovacic, P.; Wilkie, C. A.; Kinsinger, J. A.; Hein, R. E.; Yaniger, S. I.; Cody, R. B., Jr. (1986) *J. Polym. Sci., Part A: Polym. Chem.* **24**, 255-267.
35. Brown, C. E.; Kovacic, P.; Wilkie, C. A.; Cody, R. B.; Hein, R. W.; Kinsinger, J. A. (1986) *Synth. Met.* **15**, 265-279.
36. Brown, C. E.; Kovacic, P.; Welch, K. J.; Cody, R. B.; Hein, R. E.; Kinsinger, J. A. (1988) *J. Polym. Sci., Part A: Polym. Chem.* **26**, 131-148.
37. Chiarelli, M. P.; Gross, M. L.; Peake, D. A. (1990) *Anal. Chim. Acta,* **228**, 169-176.
38. Hunt, D. F.; Shabanowitz, J.; McIver, R. T.; Hunter, R. L.; Syka, J. E. P. (1985) *Anal. Chem.* **57**, 2728-2733.
39. Bergmann, T.; Martin, T. P.; Schaber, H. (1989) *Rev. Sci. Instrum.* **60**, 792-793.
40. LaiHing, K.; Cheng, P. Y.; Taylor, T. G.; Willey, K. F.; Peschke, M.; Duncan, M. A. (1989) *Anal. Chem.* **61**, 1458-1460.
41. Watson, C. H.; Baykut, G.; Eyler, J. R. (1987) *Anal. Chem.* **59**, 1133-1138.
42. McCrery, D. A.; Gross, M. L. (1985) *Anal. Chim. Acta* **178**, 105-116.
43. Cody, R. B., Kinsinger, J. A.; Ghaderi, S.; Hein, R. E. Proceedings of the 33rd Annual Conference on Mass Spectrometry and Allied Topics, San Diego, California, May 26-31, 1985, pp. 186-187.
44. Munson, M. S. B.; Field, F. J. (1966) *J. Am. Chem. Soc.* **88**, 2621-2630.
45. Field, F. H. (1990) *J. Am. Soc. Mass Spectrom.* **1**, 277-283.
46. Harrison, A. G. (1983) Chemical Ionization Mass Spectrometry, CRC Press, Boca Raton.
47. Hunter, R. L.; McIver, R. T. (1979) *Anal. Chem.* **51**, 699-704.
48. Johlman, C. L.; Laude, D. A., Jr.; Brown, R. S.; Wilkins, C. L. (1985) *Anal. Chem.* **57**, 2728-2733.
49. Ghaderi, S.; Kulkarni, P. S.; Ledford, E. B., Jr.; Wilkins, C. L.; Gross, M. L. (1981) *Anal. Chem.* **53**, 428-437.
50. McLafferty, F. W. (1980) Interpretation of Mass Spectra, 3rd ed., University Science Books, Mill Valley, Chapter 6.
51. van Breemen, R. B.; Snow, M.; Cotter, R. (1983) *Int. J. Mass Spectrom. Ion Proc.* **49**, 35-50.
52. Tembreull, R.; Lubman, D. M. (1987) *Anal. Chem.* **59**, 1082-1088.
53. Boesl, U; Grotemeyer, J.; Walter, K.; Schlag, E. W. (1987) *Anal. Instrum.* **16**, 151-171.
54. Grotemeyer, J.; Schlag, E. W. (1987) *Org. Mass Spectrom.* **22**, 758-760.
55. Engelke, F.; Hahn, J. H.; Henke, W.; Zare, R. N. (1987) *Anal. Chem.* **59**, 909-912.
56. Sherman, M. G.; Land, D. P.; Hemminger, J. C.; McIver, R. T., Jr. (1987) *Chem. Phys. Let.* **137**, 298-300.
57. Land, D. P.; Tai, T.-L.; Lindquist, J. M.; Hemminger, J. C.; McIver, R. T., Jr. (1987) *Anal. Chem.* **59**, 2924-2927.
58. Cotter, R. J. (1980) *Anal. Chem.* **52**, 1767-1770.
59. Amster, I. J.; Land, D. P.; Hemminger, J. C.; McIver, R. T., Jr. (1989) *Anal. Chem.* **61**, 184-186.
60. Amster, I. J.; Land, D. P.; Hemminger, J. C.; McIver, R. T., Jr. (1989) *Adv. Mass Spectrom.* **11**, 680-681.
61. Speir, J. P.; Gorman, G. S.; Cornett, D. S.; Amster, I. J. (1991) *Anal. Chem.* **63**, 65-69.
62. Burnier, R. C.; Byrd, G. D.; Freiser, B. S. (1980) *Anal. Chem.* **52**, 1641-1650.
63. Peake, D. A.; Gross, M. L. (1985) *Anal. Chem.* **57**, 115-120.Spectra, 3rd ed., University Science Books, Mill Valley, Chapter 6.

# PLASMA DESORPTION MASS SPECTROMETRY: PRINCIPLES AND APPLICATIONS TO PROTEIN STUDIES

PETER ROEPSTORFF
*Department of Molecular Biology, Odense University*
*DK-5230 Odense M*
*Denmark*

ABSTRACT. The introduction of plasma desorption mass spectrometry (PDMS) has had a major impact on the practical application of mass spectrometry for analysis of biological macromolecules. The principles of the method and its application in protein studies are described.

## 1. Introduction

Mass spectrometric analysis of involatile molecules with molecular weight above 1000 has, in the last decade, dramatically changed the use of mass spectrometric analysis in the biological sciences. The first method that was demonstrated to yield molecular ions of involatile biomolecules in this high mass range was plasma desorption mass spectrometry (PDMS) invented by Macfarlane and coworkers in 1974 [1]. Although this first paper only described low molecular weight involatile compounds such as sodium acetate, it was predicted that the method might be used for the analysis of much larger molecules. This was soon after proven by the analysis of peptides with molecular weights of 1881 [2] and 3464 [3] and of complex biological compounds such as the marine toxin palytoxin [3]. Eight years would pass before the dream of many protein chemists (i.e., the observation of the molecular ion of a protein) became a fact. This achievement was obtained in the group of B. Sundquist at the Uppsala University by bombardment of a solid sample of insulin with highly charged I-127 ions produced in the tandem accelerator and mass analysis of the desorbed molecular ions by time-of-flight [4]. Nearly simultaneously M. Barber and coworkers [5, 6] described another desorption/ionization method, fast atom bombardment (FAB) which also could produce molecular ions of insulin.

Both methods rapidly expanded the molecular weight range accessible to mass spectrometry. Thus, spectra obtained by PDMS of proteins like snake toxins and phospholipase A in the 13 kDa range [7, 8] and enzymes such as trypsin [9] and pepsin [10] in the 20 to 35 kDa range were soon published. The practical applications of PDMS were delayed by the fact that commercial instruments were not available until a small company, BioIon, created by the Uppsala physicists commercialized a prototype instrument by using the original Macfarlane design where desorption was effected by bombardment with fission fragments from a source of Cf-252. This instrument was based on the concept that it, being fully computerized, should be simple to use in the daily routine without the need for a highly qualified operator. We purchased the first commercial plasma desorption mass spectrometer from BioIon in 1985, and it has since then been used as a routine tool in our protein studies and operated on a daily basis by up to 20 different users. By the end of 1990, approximately 40 BioIon-instruments have been installed, many in laboratories without prior experience in mass spectrometry. Furthermore, about



*M. L. Gross (ed.), Mass Spectrometry in the Biological Sciences: A Tutorial*, 213–228.

15 custom-built instruments are operative in laboratories mainly in USA, Germany and in France. Recently a company, Selmi, in Sumi, Ukraine USSR announced a commercial instrument.

In this paper, I will briefly describe the principles of plasma desorption mass spectrometry and illustrate its use as a routine tool in the protein chemistry laboratory by using examples from our own studies.

## 2. Principles of Plasma Desorption Mass Spectrometry

### 2.1. THE PLASMA DESORPTION TIME-OF-FLIGHT MASS SPECTROMETER

The principle of the plasma desorption mass spectrometer is illustrated in figure 1. The sample is deposited on a 0.5-1 $\mu$m-thick aluminized polyester foil. A 10 $\mu$Ci Cf-252 source is placed behind the sample foil. In each fission event, two collinear fission fragments with an energy of 80 to 120 MeV are created. One fragment hits the start detector and triggers the time measurement, whereas the other penetrates the sample and causes desorption of a number of secondary ions. These ions are accelerated by a 10-20 kV potential between the sample foil and the grounded acceleration grid, allowed to drift through the field free flight tube to the stop detector, and their flight time recorded by the time-to-digital converter (TDC). Each fission event results in formation of only a few ions. It is, therefore, necessary to accumulate the data from a large number of events ($10^5$-$10^7$) corresponding to between a few minutes and some hours of recording time to obtain sufficient ion statistics.

The flight time T is correlated to the mass to charge ratio of the ion (m/z) by the equation $T=k_1\sqrt{m/z}+k_2$, where $k_1$ and $k_2$ are constants. Thus, calibration of a spectrum can be performed if the masses and the corresponding flight times for two ions are known. Normally, the ions for $H^+$ and $NO^+$ are used because these two peaks are always abundant in the spectra when the nitrocellulose support is used (see below).

### 2.2. SAMPLE PREPARATION FOR PLASMA DESORPTION MASS SPECTROMETRY

The quality of the sample and with this the method of sample preparation is of utmost importance for success in all types of secondary ion mass spectrometry, but even more so in PDMS because the simplicity of the instrument does not give any other possibilities for improvement of the performance. The californium source is placed behind the sample in most PD mass spectrometers. Therefore, the combined backing and sample layer must be thin enough to allow the fission fragments to penetrate both. This creates certain restraints for the use of matrices and the thickness of the sample layer.

The original, and until recently, most popular method for sample deposition was the electrospray method [11]. With this method, molecular ions have been observed for a number of peptides and proteins (reviewed in ref 12) with trypsin (MW 23463) being the largest. Many samples, however, failed with the electrospray deposition technique. The major reason seemed to be the presence of low molecular weight contaminants, most especially alkali metal ions, that often reduced the spectrum quality and sometimes entirely quenched the spectrum. Another drawback of this sample preparation technique is that aqueous solvents, which are the most appropriate solvents for proteins, are difficult or impossible to electrospray.

Complete removal of the alkali metal salts in samples of biological origin proved to be very difficult because numerous sources - like glassware, water and organic solvents - liberate or contain small amounts of sodium salts. The introduction of nitrocellulose as

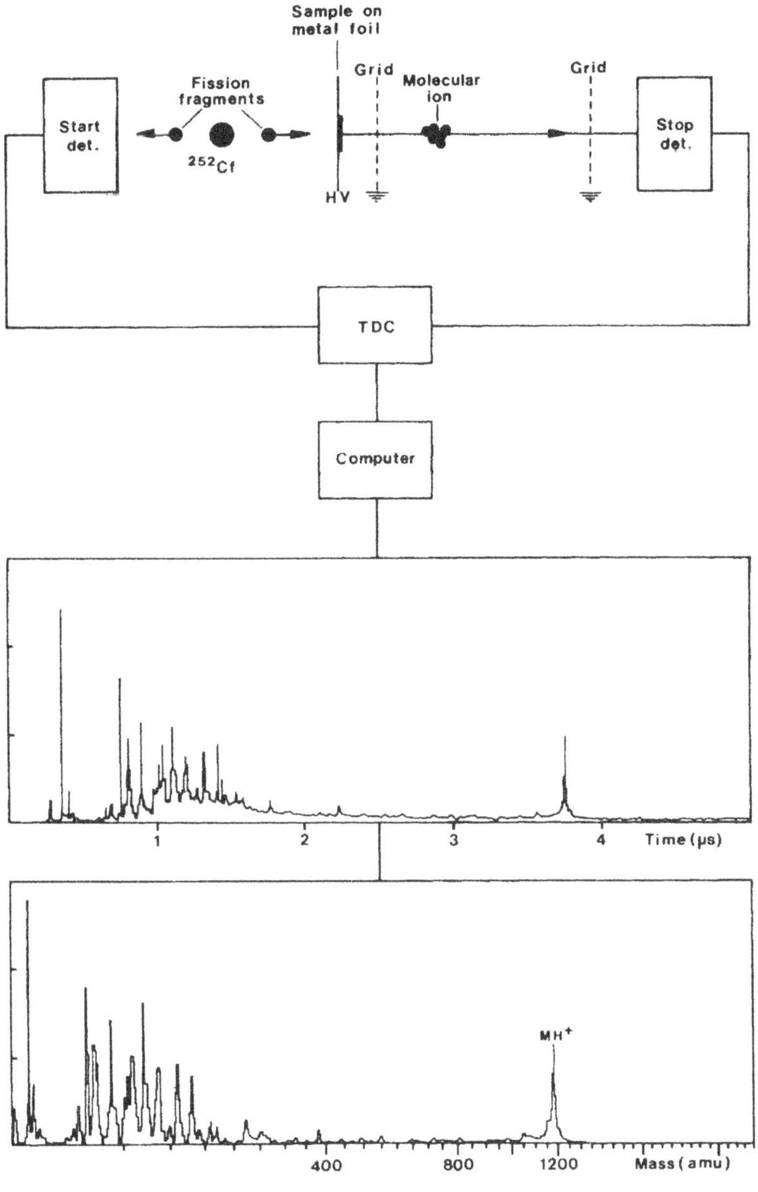


**Figure 1.** Principle of the BioIon plasma desorption mass spectrometer (reproduced by courtsey of Applied Biosystems Inc).

a matrix for PDMS solved this problem [13]. The principle is that a thin nitrocellulose layer is deposited on the aluminium foil and the sample adsorbed, or simply deposited, on the nitrocellulose from an aqueous solution. Finally, the salts are removed by washing with ultra pure solvents. Removal of the salts by this method is very successful and, unexpectedly, it dramatically improves molecular ion yields and increases the abundance of multibly charged ions [13]. Similar, but not as dramatic, improved results were obtained by adding an excess of reduced glutathione to the sample solution prior to sample application by electrospray [14] and, recently, the use of aracadonic acid as matrix has been demonstrated to have a similar effect [15].

A further improvement of the nitrocellulose technique was obtained by application of the sample on a spinning nitrocellulose target [16]. The spin-drying method seems to combine the adsorption and washing in a single step because of simultaneous drying and migration of the sample solution to the periphery of the target. As a result a thin sample layer is deposited in the central area (the area exposed to fission fragments) whereas excess sample and the more soluble salts are deposited at the periphery. The spin-drying method has been found very useful, not only for its original purpose, namely application of small peptides and for other low molecular weight compounds that bind poorly to nitrocellulose, but also for large peptides and proteins because better sensitivity was obtained. The nitrocellulose techniques are superior to the spray techniques by about a factor of five with regard to molecular ion yields (i.e., the number of molecular ions formed per 100 fission fragments) and by several orders of magnitude with regard to sensitivity [16]. The reasons for the improved results are poorly understood, probably because several factors are involved, and also because the fundamental mechanism for desorption and ionization of large molecules is unclear. It seems, however, that the sample layer thickness - as well as the content of metal ions - have an influence because the application of larger amounts of sample by the spin technique as well as the addition of sodium salts result in a decrease of molecular ion yields to the level obtained with the spray techniques. Simultaneously, the spectra change their qualitative aspects to that of a sprayed sample (i.e., the intensity of the multiply charged ions is reduced) and the peaks become broader [16, 17]. These results suggest that protein molecules that are in contact with only the nitrocellulose matrix are easier to desorb and form more stable molecular ions after desorption than protein molecules that interact with other sample molecules.

Nitrocellulose is now the most widely used matrix for PDMS of peptides, proteins and many other types of compounds. In the author's laboratory, the sample is normally applied by the spin-drying method from a 0.1% trifluoroacetic acid solution containing 15% ethanol or acetonitrile to decrease the surface tension. Other solvents such as isopropanol, concentrated trifluoroacetic acid, salt-containing buffers or 8 M urea solutions may also be used with the latter two requiring extensive washing. If the spectra are of poor quality, the most likely reasons are poor sample quality, a sample that desorbs or ionizes poorly, too little sample, too much sample or too high a content of alkali metal ions. The alkali metal content can be judged by comparing the abundance of the peaks for $H^+$ and $Na^+$ (or $K^+$). It is our general experience that the alkali metal content is too high if the peaks for the latter are more than 50% of the intensity of the $H^+$-peak.

As PDMS is practically non-destructive because there is low flux of primary ions, it is often possible to improve a poor result by removing the sample from the mass spectrometer, washing it with water or dilute acid to remove excess sample or alkali metal ions followed by re-analysis. The nitrocellulose matrix is an ideal interface between the mass spectrometer and the wet procedures in protein chemistry because it is possible to apply the sample directly from chromatographic fractions or from the reaction mixture from chemical or enzymatic reactions, or to carry out such reactions directly on the nitrocellulose bound sample. It is also possible to fish out the peptides or proteins from extremely dilute solutions [13] and to transfer protein samples from electrophoretic gels to the nitrocellulose covered target by using blotting procedures [18].

### 2.3. MASS PRECISION, ACCURACY AND MOLECULAR WEIGHT RANGE

In a discussion of precision in mass determination, it is important to separate systematic errors and the accuracy that can be obtained in determination of the centroid of a peak of interest.

The systematic errors are most frequently related to calibration and, especially, to the use of wrong empirical values for the mass of the ions used for calibration. These values may shift with time and also seem to show a dependency of the sample layer thickness and the method for preparing the sample layer. Determination of the time centroid of the calibration ions is also critical because even a slight error when extrapolated to high mass becomes significant. Both types of errors usually give rise to a shift of all the determined masses in the same direction; sometimes, however, the determined masses are too low in the low mass range, correct in the mid-range and too high in the high mass range or *vice versa*. The best way to avoid systematic errors of both types is for all operators to analyse frequently a number of standard samples to check their centroiding and to verify the calibration constants.

The accuracy with which the centroid of a given peak can be determined depends on the peak shape and abundance. Both of these features are related to the mass, the purity, the amount of the sample and to the stability of the ion. Broad peaks dominated by a metastable component gives a rather poor accuracy simply because it is difficult to establish reproducibly which area should be centroided. The width of the peaks increase with increasing mass probably because the higher mass ions are less stable. It also seems that protein molecules that are cross-linked by many disulfide bridges are more stable and, therefore, give narrower peaks than proteins without or with only a few disulfide bridges. The influence of impurities seems to be related to two different phenomena. First, many contaminants, especially alkali metal ions, tend to form adduct ions with the sample molecules. The resolution of the time-of-flight mass spectrometer is not sufficient to resolve the adduct ions from $(M + H)^+$ at high mass, and the determined centroid will thus represent an average of a mixture of molecular ion species. Second, the presence of impurities, in many cases, seems to increase the internal energy of the desorbed ions, thereby causing an increase of the metastable, and a decrease of the stable component of the peak.

As a general experience the mass precision and accuracy obtained with the commercial plasma desorption mass spectrometer is around or better than 0.1%.

The time-of-flight mass spectrometer does not have a theoretical mass limit like, for example, a magnetic sector or quadrupole instrument. The accessible mass range seems to be related to the desorption process or, in other words, is limited only to how large a molecule can be lifted off the surface upon impact of a single fission fragment. The largest protein molecules observed until now have been porcine pepsin (MW 34600) [10] and ovalbumin (MW 45000) [19]. That this does not represent the upper limit at which protein molecules can be brought in the gas phase has been demonstrated by laser desorption time-of-flight mass spectrometry where molecular ions of proteins between 100 and 200 kDa have been observed [20].

In practice, most peptides and proteins with molecular weights below 10,000 and many proteins between 10,000 and 20,000 are readily analysed by PDMS, whereas the success rate beyond 20,000 is limited.

## 3. Applications of PDMS in protein studies

The most widespread use of PDMS has been in the area of peptide and protein chemistry. This, however, does not indicate that PDMS is not useful for other types of compounds. On the contrary, PDMS has been found well-suited for the analysis of numerous different compound classes such as porphyrins [21, 22], crown ethers [23], organophosphorous compounds [24], and other types of macrocyclic heterocyclic and metal organic compounds [25]. The limited application of PDMS to other compound classes merely reflects a historical situation where several of the laboratories first engaged in PDMS were involved in protein studies - a reflection of a tremendous need by protein chemists for routine mass spectrometric analysis - a niche that PDMS immediately filled.
In the following, some examples from studies in the authors laboratory will be used to illustrate the usefulness of PDMS in protein studies.

### 3.1. PROTEIN ANALYSIS BY PDMS COMBINED WITH DNA SEQUENCE DATA

A substantial amount of protein sequence information has recently been derived from cDNA sequence data. This information, however, does not necessarily reflect the correct primary structure of the actual protein because the protein may have been processed or modified after translation, the reading frame may have been incorrectly assigned, or the actual protein studied may be a variant originating from a gene other than the one cloned. The following examples illustrate the combined use of DNA sequence information and molecular weight determination by PDMS for identification of protein primary structures.

In studies of the proteins in photosystem I of barley, two polypeptides were identified by SDS-gel electrophoresis and, based on their apparent molecular weights, termed the 1.5 and 16 kDa polypeptides, respectively. Both polypeptides were subjected to analysis by PDMS. For the first polypeptide, the molecular weight was determined to be 4066. The polypeptide was N-terminally blocked, and, because it was chloroplast coded, suspected to be blocked with an N-formylmethionine. This was supported by partial removal of the blocking group by incubation with formic acid and sequencing of the first 15 residues by automatic Edman degradation. Only partial sequence data could be obtained on the remainder of the peptide because it was very resistant to enzymatic and chemical cleavage with cyanogen bromide, in spite of the presence of one additional methionine residue.

Based on the DNA-sequence for the photosystem I genes and the few pieces of protein sequence information, a polypeptide with molecular weight 4008 could be predicted. If a formyl group is added, there is still a mass difference of 30 u. between the calculated mass and the measured mass. Re-examination of the amino acid analysis revealed that 50% of the methionine content was present as methionine sulfone, whereas methionine sulfoxide was nearly absent. As this, within experimental error, could explain the observed mass difference and the resistance to cyanogen bromide cleavage, it was concluded that one of the methionine residues was present as the sulfone [27].

The molecular weight of the second polypeptide was determined to be 10,817 by PDMS. Based on this molecular weight, and the sequence of a few amino acids from the N- terminus, the polypeptide could be located relative to the DNA-sequence as the processing product of a 15,457 Da precursor polypeptide. The agreement between the calculated molecular weight (10,821) and the measured showed that no post-translational modifications were present [30].

Processing of large precursor proteins is an important pathway in the production of biologically active peptides. Such processes are often tissue-specific, and the processing may consist of both enzymatic cleavage of the precursor protein and post-translational modification (e.g., acylation, amidation or glycosylation). The sequence of many precursor proteins is known from the *c*DNA, but this information does not allow prediction of the final products and the processing pathway.

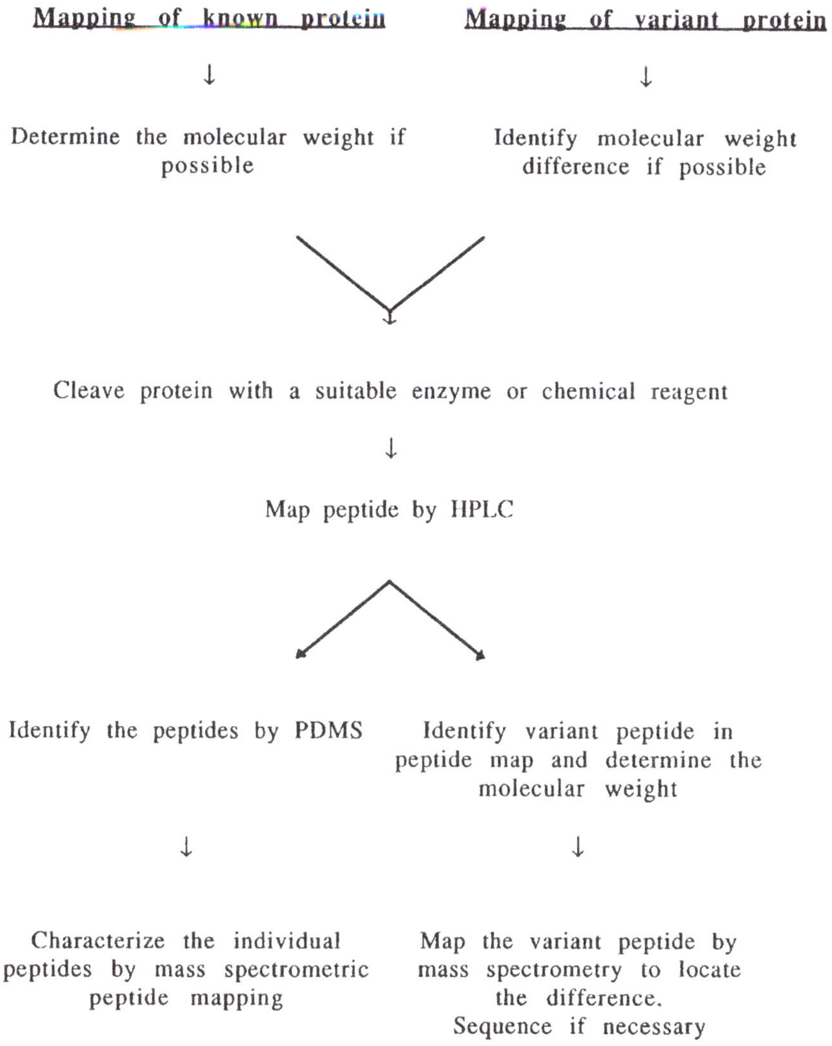


**Figure 2**. Strategy for analysis of a variant protein relative to the known protein.

We have studied the processing products of the precursor proteins known to be active in pituitary tissue. A simple acid extract of pituitary tissue was fractionated by HPLC, and the molecular weight of the components of each fraction was determined by PDMS. The *c*DNA sequences for three precursor proteins active in rat and bovine pituitary tissue are known. Any given molecular weight can, therefore, be matched with that of any peptide corresponding to a contiguous partial sequence of the precursor proteins by using a computer program [28] developed for this purpose. If a match is obtained, the identity of the peptide can be further confirmed by a few steps of Edman degradation, or in situ reactions with exo- or endopeptidases. If a match is not obtained, the search can be repeated, taking into account possible post-translational modifications.

By this strategy, peptides accounting for the total sequence of the three known precursor molecules in bovine [29] and rat [30] pituitary were identified. Several of the peptides were post-translationally modified or found in both the modified and non-modified version. The strategy has been termed peptide charting because it gives a total picture of the peptides present in a given tissue.

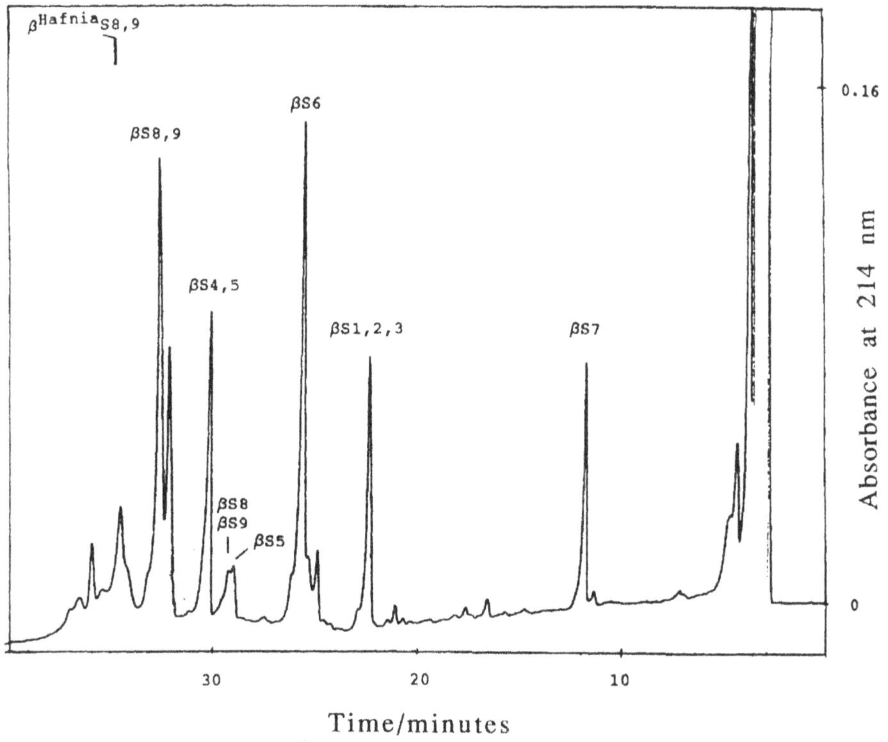


**Figure 3.** RP-HPLC peptide map of the staphylococcus aureus V8 proteinase generated peptides of a hemoglobin s-pyridylethylated ß-chain. The peaks are identified relative to the nomenclature in Table 1. The elution position of the variant peptide from hemoglobin Hafnia (ß S-Hafnia8,9) is also indicated.

3.2. STUDIES OF VARIANT PROTEINS

Mass spectrometric methods in combination with chromatographic or electrophoretic mapping techniques have their particular strength for structure determination of variants of proteins for which the primary structure is known. This includes naturally mutated proteins, isoforms of proteins, and proteins produced by protein engineering techniques. The general strategy for such studies is illustrated in Figure 2. The important initial step is to determine the appropriate peptide mapping conditions for the normal proteins and to determine the molecular weight for each of the peptides. Once this is accomplished variant proteins are easily analyzed because further mass spectrometric analysis and if necessary, sequencing are only needed for the peptides for which deviating retention times are observed in the peptide map.

We have applied this scheme to locate the sites of mutation in natural hemoglobin variants [31] as also discussed in the chapter by Marino. Hemoglobin consists of four subunits, two $\alpha$- and two ß-chains, each containing a heme group. The first step is to separate the chains and, from them, the heme group. This can be achieved in a single HPLC-run, which also normally shows which of the chains carry the variation. For mapping of the ß-chains, cleavage at glutamic acid residues with *Staphylococcus aureus* protease has been found best-suited as this cleavage results in a reasonably simple peptide map (Figure 3) and peptides of a suitable size (i.e., 1000 to 4000 Da) (Table 1).

| Peptide name | Position | Sequence | MW (average) | MW (measured) | Difference |
|---|---|---|---|---|---|
| βS1,2,3 | 1-22 | VHLTPEEKSAV TALWGKVNVD E | 2422.7 | 2423.0 | +0.3 |
| βS5 | 27-43 | ALGRLLVVYP WTQRFFE | 2095.5 | 2097.5 | +2.0 |
| βS4,5 | 23-43 | VGGEALGRLLV VYPWTQRFFE | 2437.8 | 2438.8 | +1.0 |
| βS6 | 44-90 | SFGDLSTPDA VMGNPKVKAH GKKVLGAFSD GLAHLDNLKG TFATLSE | 4843.5 | 4843.7 | +0.2 |
| βS7[a)] | 91-101 | LHCDKLHVDP E | 1410.6 | 1411.0 | +0.4 |
| βS8[a)] | 102-121 | NFRLLGNVLV CVLAHHFGKE | 2371.8 | 2369.9 | -1.9 |
| βS9 | 122-146 | FTPPVQAAYQ KVVAGVANAL AHKYH | 2681.1 | 2681.7 | +0.6 |
| βS8,9[a)] | 102-146 | NFRLLGNVLV CVLAHHFGKE FTPPVQAAYQ KVVAGVANAL AHKYH | 5034.9 | 5032.9 | -2.0 |

**Table 1**. PDMS analysis of hemoglobin $\beta$ chain *Staphylococcus aureus* proeinase generated fragments.
a) Cysteine is *S*-pyridylethylated.

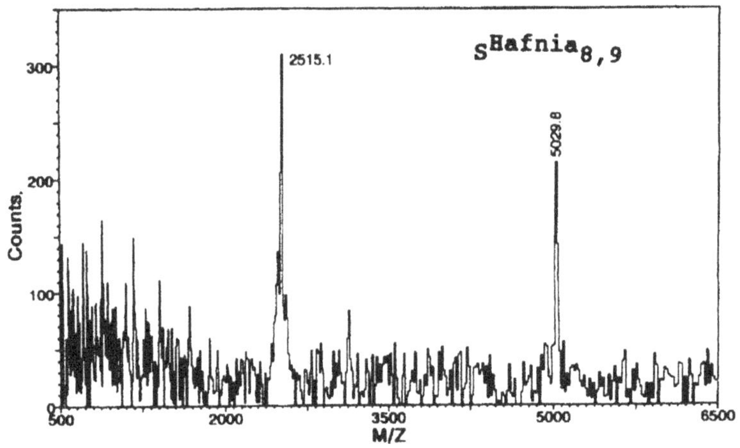


**Figure 4.** Plasma desorption mass spectrum of the variant peptide (ß S-Hafnia8,9) from hemoglobin Hafnia. Its molecular weight is measured to 5028,5 3 based on the single and double charged ions corresponding to a decrease of 6.4 u. relative to the normal peptide ß S8,9.

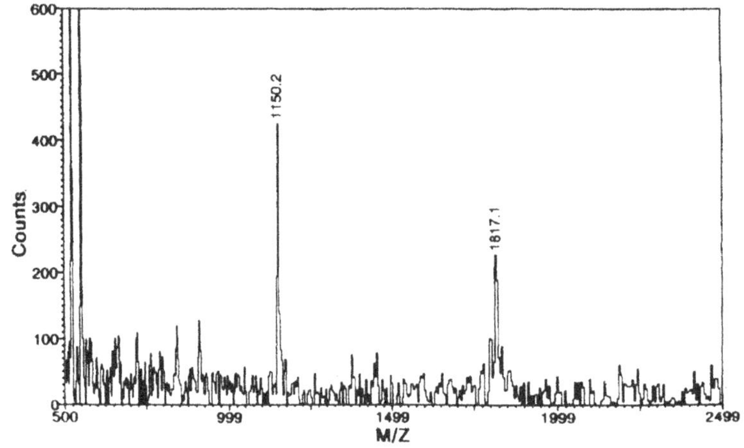


**Figure 5.** Plasma desorption peptide map of a tryptic digest of ß s-Hafnia8,9. In the mass range shown 2 out of 3 expected peptides are observed. The peptide at m/z 1150.2 correspond to normal ß133- 143, whereas the peptide at m/z 1817.1 is 9.1 u. below that of normal ß106-130, assuming a single base mutation and considering the sequence this can only be due to a His-Gln exchange. As the sequence contain two adjacent His residues in positions 116 and 117, sequence determination was necessary to locate unambiguously the mutant to a ß116 His-Gln.

The $\alpha$-chain, for similar reasons, is best analysed after cleavage at Lys and Arg residues with trypsin.

Once the variant peptide is located from the peptide map and its molecular weight determined by PDMS (Figure 4), possible mutations leading to the observed mass difference between the normal and the variant peptide can be postulated. If more possibilities are present, the correct one or ones can be established by mass spectrometric peptide mapping of the peptide after digestion with an alternative enzyme (Figure 5) or by a few steps of Edman degradation.

Structure determination of a hemoglobin variant by the procedure described here can be performed on as little as 15 $\mu$l of blood and completed within two days. Although direct mass spectrometric mapping of the cleaved proteins may appear faster, we have found the described procedure safer and in practice not more time- and sample-consuming because suppression effects often observed in direct mass spectrometric mapping are eliminated. Moreover, the combined chromatographic and mass spectrometric approach has the advantage that mass spectrometric "silent" mutations (i.e., amino acid exchanges that within the mass accuracy cannot be located, will be observed in the chromatographic procedures, whereas chromatographically silent mutations (e.g., exchange of one hydrophobic amino acid residue with another) will be revealed by mass spectrometry.

An example of the former was observed in studies of isoforms of fatty acid binding proteins. A variant form contained asparagine in a given position compared to aspartic acid in the normal protein, resulting in identical amino acid analysis data, only one u mass difference, but a significant difference in hydrophobicity. Therefore, the variant peptide could be located by reversed phase HPLC, but the nature of the change not directly revealed by mass spectrometry. After performing a methyl esterification reaction on the suspected peptides, the difference became clear because one methyl group less than expected was introduced in the variant peptide compared to the normal peptide [32].

### 3.3. PDMS IN PROTEIN SEQUENCE DETERMINATION

Plasma desorption mass spectra of peptides contain sequence specific fragment ions [33, 34], but in our experience these are rarely sufficient to sequence unknown peptides. Enhancement of the sequence specific ions can be obtained by N-acetylation and N,O-permethylation of the peptide [35], but this reaction is time and sample consuming and, therefore, only advantageous in specific cases, for example for sequence determination of N-terminally blocked peptides. The main advantage of PDMS in protein sequencing is its use in combination with conventional sequence determination by Edman degradation following the strategy shown in Figure 6 [36]. This strategy, of course, may be modified or supplemented to meet the needs of the specific protein under investigation, but the general outline has been used for the last five years in all our protein sequencing studies. The major advantages are that the overall sample consumption and time required for sequencing an unknown protein is reduced, the sequenator use is approximately halved and the costs correspondingly reduced. Most important, however, is that inclusion of mass spectrometric analysis increases the confidence, provides an independent confirmation of the results, and discloses the presence of post-translationally modified residues - as illustrated recently by the structure determination of a heavily modified structural protein [37].

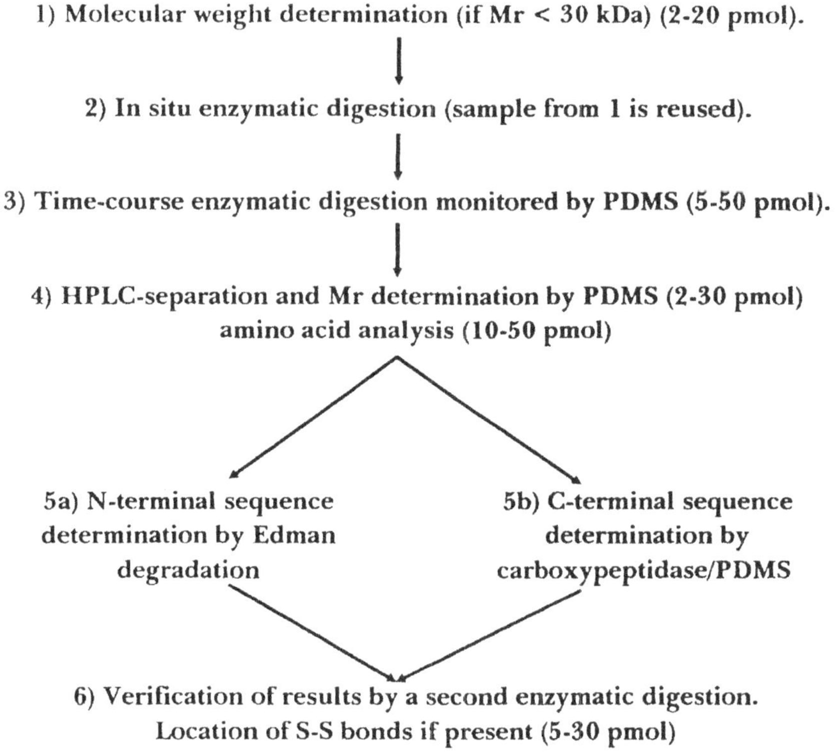


**Figure 6.** Strategy for the incorporation of PDMS in protein sequence determination. The sample amounts needed for PDMS are indicated in parenthesis.

## 4. Conclusion

Plasma desorption mass spectrometry has in the past decade played an important, innovative role in protein chemistry. The ease of sample preparation, mainly due to the nitrocellulose matrix, and the simplicity of operation of the instrument make it ideal for integration in the daily routine of a protein chemistry laboratory. There are limitations in the mass range that the recently developed laser desorption and electrospray mass spectrometry can overcome (cf. Chapters 8-10 of this volume), but it is still unclear if these methods will be as easy to use on a routine basis.

The sensitivity of PDMS is appropriate for most practical problems, although here the new methods also have some advantages. The coming decade will show if PDMS can survive and expand or if it will be replaced by other methods. The main function of PDMS as "ice breaker" in the field of mass spectrometry of biological macromolecules, however, has been fulfilled.

## 5. References

1. Thorgerson,D.F., Skowronski, R.P., and Macfarlane R.D. (1974) 'New approach to the mass spectroscopy of non-volatile compounds', *Biochim. Biophys. Res. Commun.*, **60**, 616-621.
2. Macfarlane , R.D. and Thorgerson, D.F. (1976) 'Californium-252 plasma desorption mass spectrometry', *Science*, **191**, 920-925.
3. Macfarlane, R.D., Uemura, D., Ueda, K., and Hirata, Y., (1980) '252-Cf plasma desorption mass spectrometry of palytoxin', *J. Am. Chem. Soc.*, **102**, 875-876.
4. Håkansson, P., Kamensky, I., Sundqvist, B., Fohlman, J., Peterson, P., McNeal, C.J. and Macfarlane, R.D. (1982) '127-I plasma desorption mass spectrometry of insulin', *J. Am. Chem. Soc.*, **104**, 2948-2949.
5. Dell, A., and Morris, H.R. (1982) 'Fast atom bombardment - highfield magnetic mass spectrometry of 6000 dalton polypeptides', *Biochem. Biophys. Res. Commun.*, **106**, 1456-1461.
6. Barber, M., Bordoli, R.S., Elliot, G.J., Sedgwick, R.D., -Tyler, A.N. and Green, B.N. (1982) 'Fast atom bombardment mass spectrometry of bovine insulin and other large peptides', *J. Chem. Soc. Chem. Commun.*, 1982, 936-938.
7. Kamensky, I., Håkansson, P., Kjellberg, J., Sundqvist, B., Fohlman, J., and Peterson, P. (1983) 'The observation of quasi molecular ions from a tiger snake venom component (MW 13309) using 252-Cf plasma desorption mass spectrometry', *FEBS Lettters*, **155**, 113-116.
8. Sundqvist, B., Håkansson, B., Kamensky, I., Kjellberg, J., Sahlepour, M., Widdiyasekera, S., Fohlmann, J., Peterson, P., and Roepstorff, P. (1984) 'Californium-252 plasma desorption time-of-flight mass spectrometry of proteins', *Biomed. Mass Spectrom.*, **11**, 242-257.
9. Sundqvist, B., Roepstorff, P., Fohlmann,J., Hedin, A., Håkansson, P., Kamensky, I., Lindberg, M., Sahlepour M., and Säve, G. (1984) 'Molecular weight determination of proteins by californium plasma desorption mass spectrometry', *Science*, **226**, 696-698.
10. Craig, A.G., Engström, Å., Bennich, H., and Kamensky I., (1987) 'Enhancement of molecular ion yields in plasma desorption mass spectrometry', presented at the 35th. ASMS conference on Mass Spectrometry and Allied Topics, Denver Colorado, May 1987.
11. McNeal, C.J., Macfarlane, R.D., and Thurston, E.L., (1979) 'Thin film deposition by the electrospray method for californium-252 plasma desorption studies of involatile molecules' *Anal. Chem.*, **51**, 2036-2039.
12. Sundqvist, B., and Macfarlane, R.D. (1985) '252-Cf plasma desorption mass spectrometry' *Mass Spectrom. Rev.*, **4**, 421-460.
13. Jonsson, G.P., Hedin, A.B., Håkansson, P.L., Sundqvist, B.U.R., Säve, G.S., Nielsen, P.F., Roepstorff, P., Johansson, K.E., Kamensky, I., and Lindberg, M.S.L. (1986) Plasma desorption mass spectrometry of peptides and proteins absorbed on nitrocellulose', *Anal. Chem.*, **58**, 1084-1087.
14. Alai, M., Demirev, P., Fenselau. C., and Cotter, R.J. (1986) 'Glutathione as matrix for plasma desorption mass spectrometry of large peptides', *Anal. Chem.*, **58**, 1303-1307.
15. Wolf, B. and Macfarlane, R.D. (1991) 'Small molecules as substrates for adsorption/desorption in 252-Cf plasma desorption mass spectrometry', *J. Am. Soc. Mass Spectrom.*, **2**, 29-32.
16. Nielsen, P.F., Klarskov, K., Høјrup, P., and Roepstorff, P. (1988) 'Optimization of sample preparation for plasma desorption mass spectrometry of peptides and proteins', *Biomed. Environ. Mass Spectrom.*, **17**, 355-362.

17. Nielsen, P.F. and Roepstorff, P. (1988) ' Sample-preparation dependent fragmentation in plasma desorption mass spectrometry of nisin', *Biomed. Environ. Mass Spectrom.*, **17**, 137-141.
18. Roepstorff, P., Talbo, G., Klarskov, K., and Højrup, P., (1990) 'Nitrocellulose, the interface between plasma desorption mass spectrometry and protein chemistry', in A.L. Burlingame and J.A. McCloskey (eds.), Biological Mass Spectrometry, Elsevier, Amsterdam, pp. 25-48.
19. Jonsson, G., Hedin, A., Håkansson, P., Sundqvist, B.U.R., Bennich, H., and Roepstorff, P. (1989) 'Compensation for non-normal ejection of large molecular ions in plasma desorption mass spectrometry', *Rapid Commun. Mass Spectrom.*, **3**, 190- 191.
20. Karras, M., Ingendoh, A., Bahr, U. and Hillenkamp, F. (1989) 'Ultraviolet-laser desorption/ionization mass spectrometry of femtomolar amounts of large proteins', *Biomed. Environ. Mass Spectrom.*, **18**, 841-843.
21. Chait, B.T. and Field, F.H. (1982) '252-Cf fission fragment ionization mass spectrometry of chlorophyll A', *J. Am. Chem. Soc.*, **104**, 5519-5521.
22. Tuszinski, W., Angermann, R., Hillmann, F. and Maier-Schwartz, K. (1990) 'PDMS in an institute of marine research, chlorophyll and other pigments in photoactive and buried marine biological mats' in E.R. Hilf and W. Tuszynski (eds.), Mass spectrometry of large non-volatile molecules for marine organic chemistry, World Scientific, Singapore, pp. 131-146.
23. Malhotra, N., Roepstorff, P., Hansen T.K., and Becher, J. (1990) 'Alkali metal ion complexation of crown ethers and related ligands studied by 252-californium plasma desorption mass spectrometry' *J. Am. Chem. Soc.*, **112**, 2709-3710.
24. Allmaier, G., Schmid, E., and Roepstorff, P. (1990) 'Positive and negative ion 252-californium plasma desorption mass spectrometry of polar agrochemical metabolites', *Biomed. Environ. Mass Spectrom.*, **19**, 75-79.
25. Jørgensen, T., Becker, J., Hansen, T.K., Kristiansen, K. Roepstorff, P., Larsen, S., Nygaard, A. (1991) 'Structural characterization and properties of a new tetrathiafulvalene-based cage-molecule derived from an azacrown', *Advanced Materials*, in press.
26. Scheller, H.V., Okkels, J.S., Høj, P.B., Svendsen I., Roepstorff, P., and Møller B.L. (1989) 'The primary structure of a 4.0 kDa photosystem I polypeptide encoded by the chloroplast psaI gene', *J. Biol. Chem.*, **264,** 18402-18406.
27. Scheller, H.V., Okkels, J.S., Roepstorff, P., Jepsen, L.B., and Møller, B.L., (1990) 'Characterization of a cDNA clone for the psaI gene from barley and plasma desorption mass spectrometry of the corresponding photosystem I polypeptide PSI-E' in M. Baltscheffsky (ed.), Proceedings of the VIIIth International Congress on Photosysthesis, Stockholm august 1989, Klüwer Academic Publishers, Dortrecht, Holland, pp. 609-612.
28. Højrup, P., (1990) 'General protein mass analysis (GMPA), a convenient program in studies of proteins by mass spectrometry' in A.Hedin, B.U.R. Sundqvist, and A. Benninghoven (eds.), Ion Formation from Organic Solids (IFOS V), J. Wiley and Sons, Chichester, England, pp. 61-66.
29. Feistner, G.J., Højrup, P., Evans, C.J., Barofsky, D.F., Faull, K.F,. and Roepstorff, P. (1989) 'Mass spectrometric charting of bovine posterior/interior pituitary peptides', *Proc. Natl. Acad. Sci. (USA)*, **86**, 6013-6017.
30. Feistner, G.J., Faull,K.F., Evans, C.J., Roepstorff, P., Højrup, P., and Barofsky, D.F. (1989) 'PDMS and FABMS for the charting of neuropeptides' in P. Longevialle (ed.), Advantages in Mass Spectrometry vol 11, J. Wiley and Sons, Chichester, England, pp. 1380-1381.

31. Jensen, O.N., Højrup, P., and Roepstorff, P. (1991) 'Plasma desorption mass spectrometry as a tool in characterization of abnormal proteins. Application to variant human hemoglobins.' *Anal. Biochem.*, in press.
32. Unterberg, C., Börchers, T., Højrup, P., Roepstorff, P., Knudsen J., and Spener, F. (1990) 'Cardiac fatty acid binding protein. Isolation and characterization of the mitochondrial fatty acid binding protein and its structural relationship with the cytosolic isoforms' *J. Biol. Chem.* **265**, 16255-16261.
33. Fohlman, J., Peterson, P., Roepstorff, P., Højrup, P., Kamensky, I., Säve, G., Håkansson, P., and Sundqvist, B. (1985) 'Comparison of 252-Californium plasma desorption and fast atom bombardment mass spectrometry for the analysis of small peptides', *Biomed. Mass Spectrom.*, **12**, 380-387.
34. Vorst, H.J., Van Tilborg, M.W.E.M., Van Veelen, P.A., Tjaden, U.R. and Van der Greef, J. (1990) 'Sequence-informative fragmentation of peptides up to a molecular weight of 4.6 kDa in plasma desorption mass spectrometry', *Rapid Commun. Mass Spectrom.*, **4**, 202-205.
35. Talbo, G. and Roepstorff, P. (1990) 'Sequence determination of N-acetylatedN,O-permethylated peptides by plasma desorption mass spectrometry', *Biomed. Environ. Mass Spectrom.*, **19**, 589-596.
36. Roepstorff, P., Klarskov, K., and Højrup, P. (1989) 'Strategy for the use of plasma desorption mass spectrometry in protein sequence analysis' in B. Wittman-Liebold (ed.), Methods in Protein Sequence Analysis 1988, Springer Verlag, Berlin. BRD, pp.191-198.
37. Talbo, G., Højrup, P., Rahbek-Nielsen, H., Andersen, S.O. and Roepstorff, P. (1991) 'Determination of the covalent structure of an N- and C-terminally blocked glycoprotein from endocuticle of Locusta migratotia. Combined use of plasma desorption mass spectrometry and Edman degradation to study post-translationally modified proteins', *Eur. J. Biochem.*, **195**, 495-504.

# CONTINUOUS-FLOW FAST ATOM BOMBARDMENT MASS SPECTROMETRY IN ANALYTICAL BIOCHEMISTRY

WILLIAM T. MOORE and RICHARD M. CAPRIOLI
*The Analytical Chemistry Center and the*
*Department of Biochemistry and Molecular Biology*
*The University of Texas Medical School*
*P. O. Box 20708*
*Houston, Texas 77225*

ABSTRACT. This chapter describes the operation of a mass spectrometric technique termed continuous-flow fast atom bombardment mass spectrometry (CF-FABMS) and presents some applications concerning the analysis of proteins and peptides. A description of the operational aspects of the technique is presented along with a discussion of the features and criteria of stable operation. Analytical advantages are presented by using problems encountered in protein chemistry. The use of the method to evaluate the automated chemical synthesis of peptides is presented to demonstrate the facility of the CF-FAB method in an analytical biochemistry laboratory.

## 1. Introduction

Fast atom bombardment (FAB) [1] is one of several desorption-ionization techniques that have permitted the direct mass spectrometric analysis of non-volatile polar polymers having moderately high molecular weights (<6000). The energy for desorption arises from a collision cascade arising from the interaction of analyte molecules with highly energetic "fast" (kinetically activated, 6-10 keV) neutral atoms (Xe, Ar) focussed into a beam by a device termed a FAB gun. This technique overcomes the analyte volatility requirement that generally had been a prerequisite for mass spectrometric analysis and has extended mass-specific analysis to molecules of biological interest without prior purification and derivatization [2]. The method has proven itself to be an extremely important tool in analytical biochemistry.

The FAB technique is considered a "soft" ionization process that results generally in the detection of an intact molecular ion in the form of an adduct with a single proton if the analyte is dissolved in a low concentration alkali-salt acidic liquid matrix and the instrument is operated in the positive ion mode. The sample matrix for standard FAB usually consists of a 1-3 $\mu$L droplet of a viscous liquid mixture consisting of glycerol, water, and an acid. The droplet of matrix provides a nonvolatile polar environment for the analyte on a probe tip placed in the high vacuum environment of the mass spectrometer ion source. An advantage of the liquid matrix is that analyte molecules at the surface are continually replenished from



*M. L. Gross (ed.), Mass Spectrometry in the Biological Sciences: A Tutorial*, 229–252.

the interior of the sample droplet by diffusion. Molecules are desorbed or "sputtered" by the atom bombardment to form ions, which are subsequently extracted by the ion optics of the source and focussed for delivery to the mass analyzer portion of the mass spectrometer. The mass spectrum consists of ions derived from the analyte and also those derived from the glycerol matrix as fragments and clusters. The mechanism of ionization by the FAB process is highly complex and involves both liquid- and gas-phase reactions [3]. Due to "ion suppression" effects [4, 5], the mass spectrum of an analyte mixture may not be a reliable reflection of the actual quantitative relationships existing in the sample. The clear advantage of FAB analysis, however, has been that a broad class of compounds of biochemical interest, once thought of as poor candidates for direct mass spectrometric analysis (e.g., peptides, sugars, nucleotides, antibiotics, and drug metabolites) have been successfully analyzed. A disadvantage of FAB is the high background arising from the matrix liquid. The matrix constituents essentially contaminate the analyte and significantly decrease the limit of detection. Another disadvantage is that the viscous matrix enhances ion suppression effects (i.e. more hydrophilic compounds tend to remain in the droplet interior whereas hydrophobic compounds tend to migrate to the surface of the droplet, suppressing the ionization of the more hydrophilic compounds).

Continuous-flow (CF) FAB was devised by Caprioli and co-workers [6] to diminish problems associated with static FAB analysis while retaining and extending the advantages of the overall FAB process. CF-FAB MS utilizes a sample introduction probe that permits the continuous flow of a carrier liquid containing only low concentrations of the organic matrix onto the target end of the probe while it is in the ion source of the mass spectrometer. The pumping capacities of commercial instruments limits the flow rate of liquid introduction into the source to a maximum of about 20 $\mu$L/min. Higher flow rates result in unstable operation and to high voltage arcing in magnetic instruments that are operated at high accelerating voltages. Early experimentation [5, 6, 7] indicated that the requirement for a viscous organic matrix component (glycerol) could be markedly reduced to 1-5%. The major component of the CF-FAB carrier solvent is water, giving several operational advantages: the direct introduction of aqueous samples by a flow-injection method, higher relative sensitivity, reduction in matrix-derived chemical noise, decreased ion suppression effects, temporal relationship between the background carrier signal and the injected sample peak, and the ability to do quantitative analysis by comparing flow-injection peak profiles. In addition, the lower glycerol content allows the inclusion of other volatile solvents such as acetonitrile and methanol, permitting CF-FAB to be used as an interface for the coupling of separation methods such as liquid chromatography (LC) and capillary electrophoresis (CE) with mass spectrometry. The technique of CF-FAB has also opened up the possibility of on-line biochemical reaction monitoring *in vitro* by direct infusion of a reaction mixture [8]. In this way, product-reactant relationships can be directly assessed [9]. Use of a microdialysis probe in conjunction with CF-FAB has permitted the *in vivo* monitoring of drug metabolism [10]. For a recent and extensive review of the CF-FAB technique, see the monograph edited by Caprioli [11] containing contributions from several workers who have successfully used the technique to address the solution of a wide variety of analytical problems.

This chapter will cover some of the instrumental and methodological aspects of CF-FAB MS operation and will review several applications that demonstrate its unique use in analytical biochemistry. Many examples will involve peptide analyses because they encompass several different modes of operation of CF-FAB and address real analytical problems encountered in the laboratory.

## 2. Methodology and Operational Features

### 2.1 CF-FAB PROBE DESCRIPTION

Basically, the CF-FAB probe is a direct insertion probe comprised of a hollow stainless steel shaft accommodating a high voltage insulating fused silica capillary, the end of which emerges at the sample target stage (Figure 1A). A syringe pump delivers an aqueous sample carrier solution through the narrow bore (75 $\mu$m) capillary at a flow rate of about 5 $\mu$L/min so that a thin film of liquid sample matrix is steadily maintained over the stainless steel target surface of the sample stage. To maintain the continuous flow of a thin film of aqueous sample matrix, the rates of delivery and removal of liquid need to be balanced and dead volume spaces at the tip need to be eliminated. The commercially available flow probe shown in Figure 1 is designed to fit the Finnigan MAT 90 high performance magnetic mass spectrometer. Some of the hardware features that contribute to stable operation are: (1) the inclusion of a septum placed near the target tip; (2) the placement of a filter pad "wick" near the target; and (3) a mechanical means to adjust capillary extrusion at the target. The target tip septum prevents backflow of emerging matrix solution down the outside surface of the capillary into the shaft of the probe. Without the septum in place, collection of liquid in the probe shaft space occurs and disruptive evaporative bursting of these liquid pools lead to sporadic high source pressures and unstable operation. The filter pad "wick" acts as an absorbent sink for excess liquid that may collect on the target surface. Figure 1B shows the sample matrix target surface deposition and flow condition with and without the filter pad "wick" in place. The filter pad insures a thin film of matrix flowing over the target surface. Capillary tip adjustment permits an even transfer of matrix eluate to the surface of the target. If the capillary protrudes too much, either beading and bursting of matrix may occur. The lack of a good thermal contact may lead to the formation of a mixture of water and ice crystals because of the high rate of evaporative cooling of the solvent. Heating the probe tip region by setting the source block temperature to about 45 °C and optimally adjusting the protrusion of the capillary so that it is almost flush, barely extending beyond the target surface, prevents the icing condition. Another factor that contributes to stable probe operation is the composition of the liquid matrix carrier solvent. Although several solvent systems ranging from mostly aqueous to organic liquids have been used for various analyses, the most generally useful for samples of biological interest are those that are primarily aqueous, although often containing some organic constituents. We have found that a matrix consisting of 92% water, 5% glycerol, and 3% acetonitrile gives excellent performance. We believe the low level of glycerol functions to aid the FAB process by acting as a viscous seal that coats and fills in any surface irregularities that are present on the target, creating a surface conducive to smooth flow of a hydrophilic liquid. Any small dead volume spaces are thought to be filled preferentially by the more viscous glycerol component because its rate of evaporation is very low. The acetonitrile concentration, although not critical, was

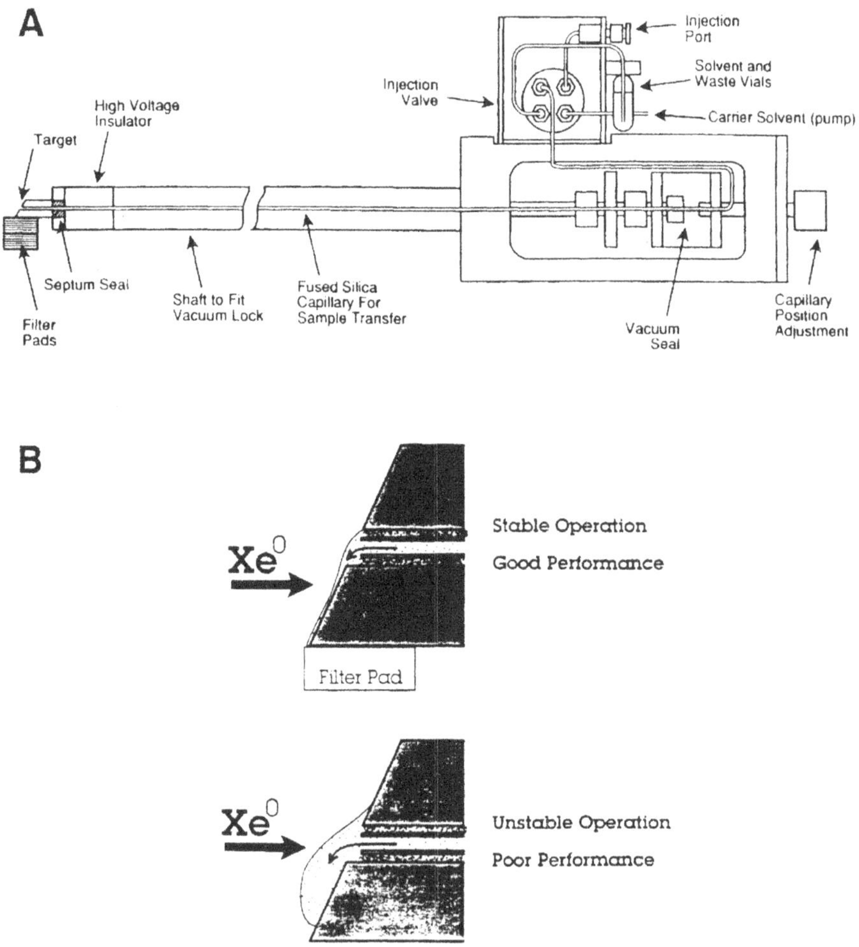


Figure 1. Continuous flow-FAB probe features. A: Schematic of the Finnigan MAT 90 probe, and B: The target condition for stable operation (top) and the condition that leads to unstable operation and poor performance (bottom) (see text for explanation).

empirically derived and is believed to increase the wetness of the liquid matrix by lowering the surface tension of the solvent.

### 2.2. CRITERIA FOR STABLE OPERATION

Initial flow rate adjustments, capillary tip placement, and establishment of the thin film flow of matrix over the target are all essential to stable operation and can be conveniently assessed while the probe is outside the source of the mass spectrometer. Once these external assessments have been made, the CF-FAB probe is inserted through the direct insertion interlock. Stable operation conditions are then usually achieved within 15 min with some minor adjustments of capillary position, flow rate, or source temperature. Once these minor adjustments have been made for a defined sample carrier, the probe may be removed and re-inserted with a stabilization period of 5 min. Stable operation is easily maintained for 5-6 hours at a flow rate of 5 $\mu$L/min. The instability that may result after prolonged use is usually caused either by a saturated absorbent filter pad wick or a dirty target surface. Replacing filter pads and cleaning the target surface with a brief exposure to concentrated acid are remedies.

The criteria that we use as indications of stable probe operation are: (1) less than ± 10% variation in the intensity of carrier solvent ions such as acetonitrile and glycerol (see Figure 2A) over a prolonged period of time; (2) sharp and smooth injection profiles (Figure 2B) and good reproducibility (less than ± 10%) of injection profile peak areas; (3) maintenance of a steady source pressure as indicated by a nonwaivering needle of the ionization gauge output. Operationally, a steady source ionization gauge reading is initially the most informative indication of stable probe operation. Using the Finnigan MAT 90 equipped with a liquid nitrogen cold trap and running the flow probe at 5 $\mu$L/min, we find that stability is indicated by a steady source vacuum gauge reading of 2 X $10^{-4}$ torr.

### 2.3. OPERATIONAL AND ANALYTICAL ADVANTAGES

*2.3.1. Direct Analysis of Aqueous Sample Solutions.* The most common use of CF-FAB MS is to employ it in batch sample processing in the flow injection mode of operation. Figure 3A schematically shows the experimental set-up for this mode of operation. The essential feature is the use of an injection valve having a sample loop capable of accepting 0.5 $\mu$L to 5 $\mu$L volumes. This permits mass spectrometric analysis by direct sampling of an ionic analyte in an aqueous environment without extraction and prior derivatization. The injection of a volume of 1 $\mu$L or less results in an injection profile resembling that of a chromatographic peak as shown in Figure 3B. The mass spectrum taken at the apex of this peak is shown in Figure 3C. The profile may be output as either a total ion chromatogram or a selected ion chromatogram by the data acquisition software of the mass spectrometer. The mass spectra of the matrix background contribution immediately before and after the peak permits effective and valid background subtraction, if necessary. Typically, the analyte is bombarded for a period of 30 to 60 seconds, defined by the injection peak width. Figure 2B shows the flow injection profiles of increasing amounts of an aqueous sample of the

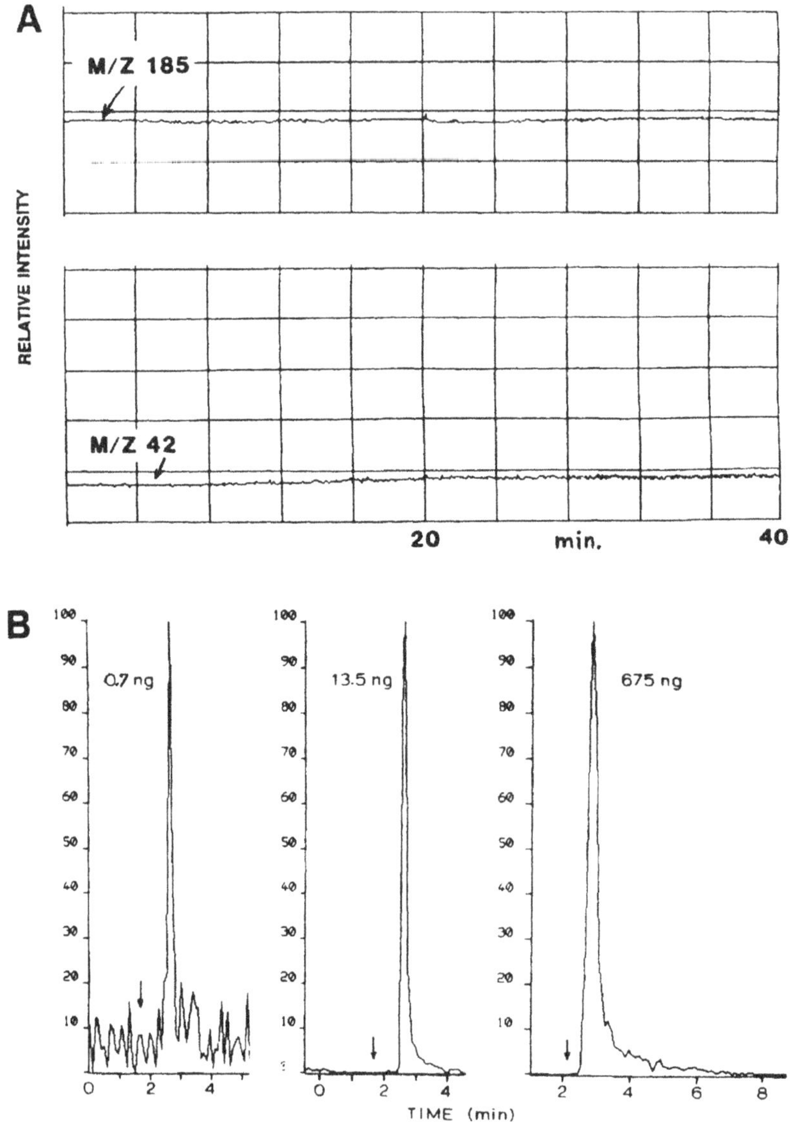


Figure 2. Response criteria of stable CF-FAB probe operation. A: Steady signal for selected ion monitoring of matrix solvent (82% water, 15% glycerol, 3% acetonitrile) components: glycerol dimer (of $m/z$ = 185) and acetonitrile molecular ion (of $m/z$ = 42). Components were monitored for 40 min. B: Narrow injection profiles with minimal trailing. Selected ion chromatograms for amounts of substance P injected in 0.5 $\mu$L volumes.

peptide substance P (*m/z* 1348). Excessive peak trailing can lead to potential memory effects and, thus, it is important to eliminate all dead-volumes in the plumbing fittings. The results of an experiment designed to evaluate the degree of peak trailing and the concomitant problem of memory effects that may occur upon closely spaced repetitive injections is shown in Figure 4. In this experiment we found that 100 pmoles each of substance P and tetradecapeptide renin substrate may be repetitively and alternately injected 2.3 minutes apart without a significant amount of carry-over. Figure 4, B and C, show the mass spectra for the last two separate injections of these two peptides after one hour of a closely

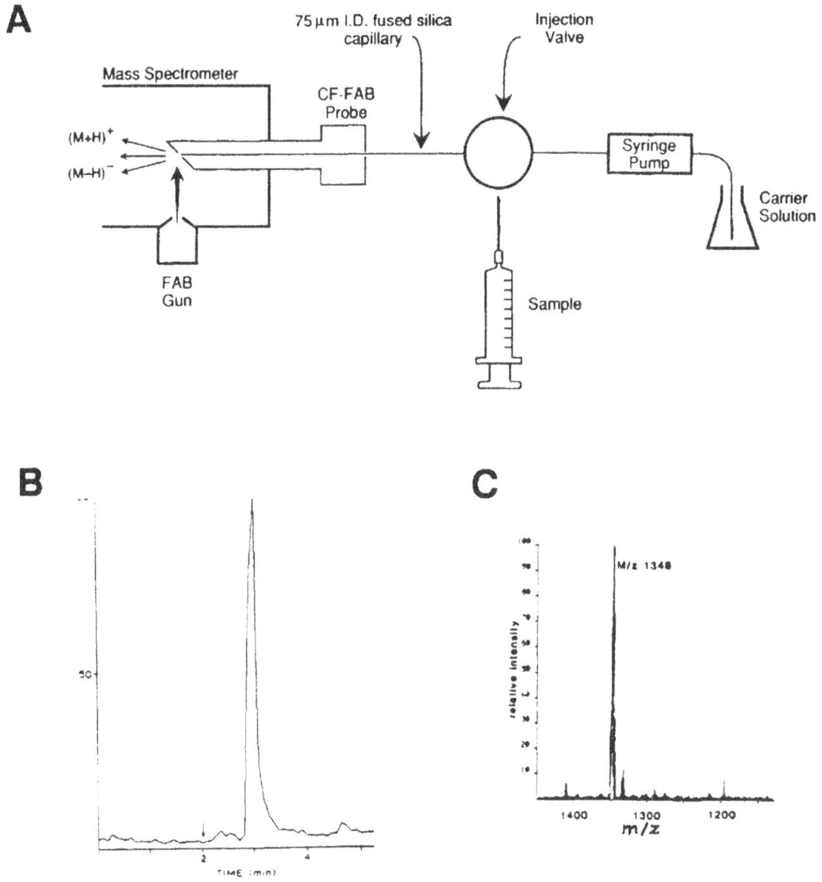


Figure 3. Flow injection mode of operation. A: Experimental set-up. B: 0.5 $\mu$L injection of 13.5 ng of substance P, and C: Mass spectrum derived from a single scan.

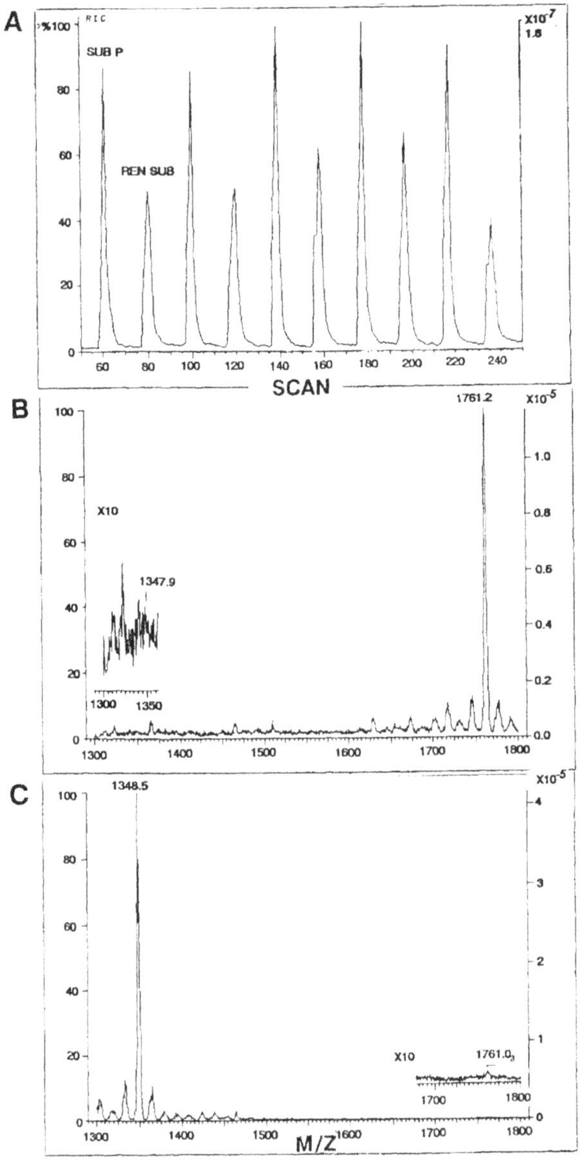


Figure 4. Alternate and repetitive injections of 200 pmoles substance P and Renin substrate. A: Total ion chromatogram injection profiles for first five injections of each peptide. B: The mass spectrum for the 20th injection of renin substrate, and C: The mass spectrum for the 20th injection of substance P. (Reprinted with permission from reference 11.)

spaced series of repetitive and alternating injections. Only a negligible carryover was observed (see expanded intensity regions of the spectra, Figure 4B and C). This result also demonstrates the capability of CF-FAB for a high rate of sample throughput with this approach.

*2.3.2. Increased Sensitivity and Background Reduction.* One of the major advantages of CF-FABMS over that of standard FAB is the substantial increase of apparent sensitivity [7]. A sensitivity of 150-fold greater for CF-FABMS over standard FAB has been observed under some conditions. This increase in apparent sensitivity arises from the combination of a decrease in background and an increased ion production from the sample. The increase in apparent sensitivity is more significant when very low levels of analyte are assayed. Figure 5 shows the results of a comparison of the analyses of two levels of substance P by both standard FAB and CF-FAB. In Figure 5A, the sensitivity enhancement is approximately 30 to 50-fold for 5 picomoles of substance P, while in Figure 5B, the comparison for 100 femtomoles gives an enhancement of 150-fold. The limit of detection for this particular peptide was about 10 femtomoles.

*2.3.3. Modification of Ion Suppression Effects.* One of the disadvantages of standard FABMS analysis is the ion suppression effects that are observed upon the analysis of complex mixtures

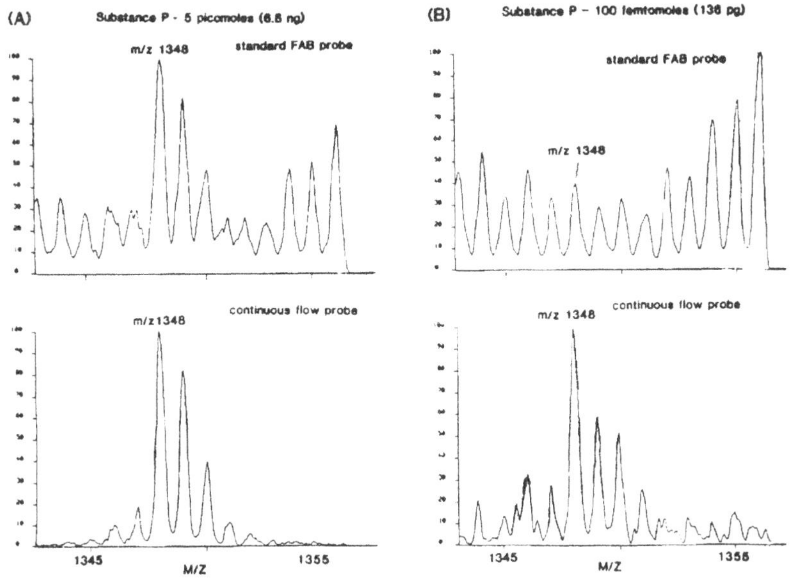


Figure 5. Enhanced sensitivity of CF-FAB. Comparisons of mass spectra with standard FAB (matrix - 90% glycerol/10% water) and CF-FAB (matrix - 80% water/20% glycerol). A: (top and bottom) 5 picomoles of substance P and B: 100 fmoles of substance P. (Reprinted with permission from reference 7).

of analytes. This suppression effect has been attributed to a competition between hydrophobic and hydrophilic species for occupancy of surface layers in the sample-matrix droplet [4]. Often quantitative measurements are unreliable for a complex mixture of peptides and, moreover, some peptides in the mixture may go undetected [4, 5, 11]. The CF-FAB technique diminishes the ion suppression effect, as shown in Figures 6 and 7. Figure 6A is a spectrum encompassing two peptides derived from a tryptic digest of myoglobin. The tryptic peptide having an $(M + H)^+$ at *m/z* 1515 is completely suppressed in the standard FAB analysis while it is clearly detectable by CF-FAB analysis. Another example shown in Figure 6B for the standard FAB and CF-FAB spectra obtained for a tryptic digest of glucagon. In this example, hydrophobic peptides in the digest (e.g., that corresponding to amino acid residues 19-29 having an $(M + H)^+$ of *m/z* 1352) completely suppress the more hydrophilic peptide which should appear at *m/z* 1357 for fragment 1-29 in the standard FAB analysis. However, CF-FABMS analysis of the same mixture shows the presence of both peptides in the mass spectrum. A more complex example is presented in Figure 7. Seven closely related synthetic hydrophilic heptapeptides were analyzed as an equimolar mixture by

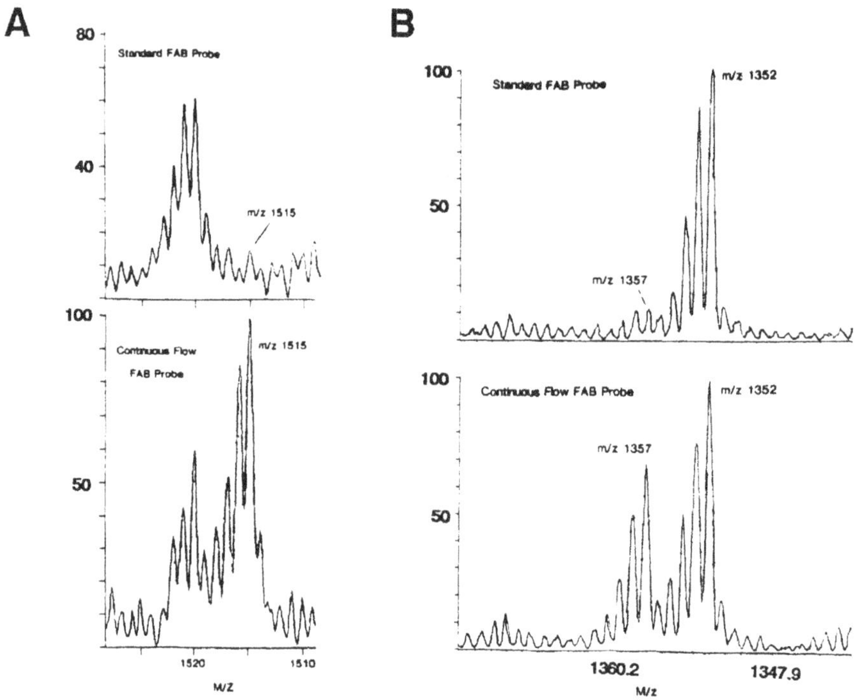


Figure 6. Diminished ion suppression effects using CF-FAB. A: Myoglobin digest; comparing standard FAB (upper) and CF-FAB (lower) B: Glucagon digest with similar comparison.

both standard FAB and CF-FAB. In the upper panel of Figure 7, ion signals for peptides I, III, IV, and V were significantly suppressed in the standard FAB analysis. Upon analysis of an aliquot of the identical sample solution by CF-FAB MS (lower panel), the same peptides gave signals of markedly higher intensity. These four peptides contained specific amino acid substitutions that made them multiply-charged at acidic pH and are very hydrophilic. The ion abundances for the other three peptides, interestingly, show the same

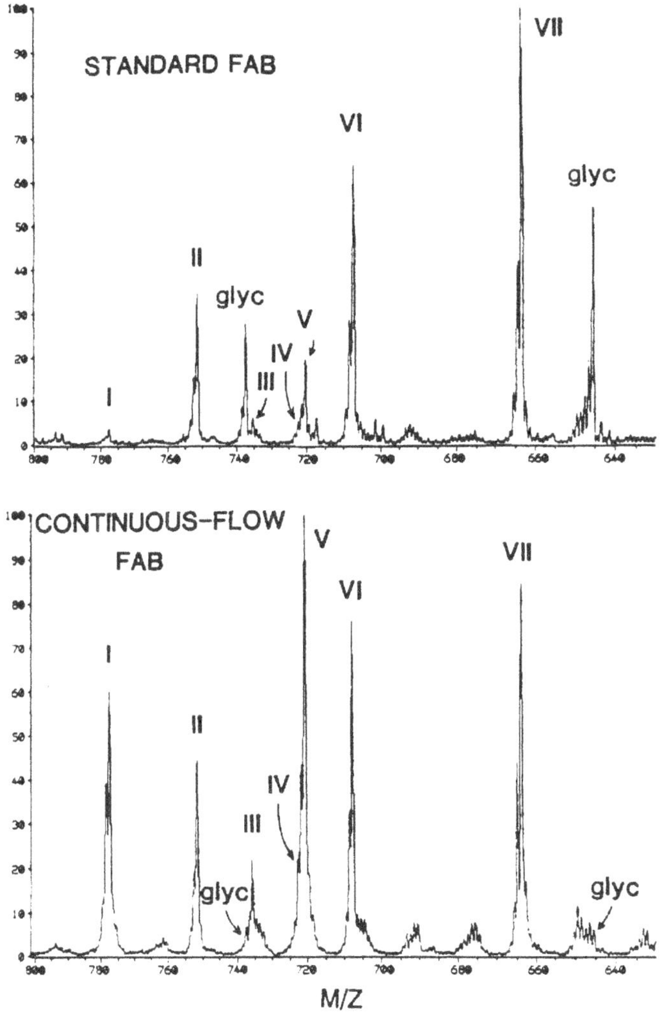


Figure 7. Comparative analysis of seven hydrophilic peptides by standard FAB (upper) and CF-FAB (lower). Glyc refers to peaks due to glycerol. See text for explanation. (Reprinted with permission from reference 12.)

relative abundances in both spectra. We attribute this decrease in the ion suppression effect seen with CF-FAB to the formation of a thin film over the target and the mechanical mixing occurring with the flow of the thin film over the target surface. This condition may simply permit a more homogeneous solution of a chemically diverse analyte population. It is noted, however, that CF-FAB does not necessarily eliminate the problem of ion suppression; small hydrophilic peptides still tend to markedly exhibit suppression effects.

*2.3.4. On-line Reaction Monitoring.* The general techniques employed in CF-FAB enable the continuous monitoring of a reaction mixture. A reaction mixture may be directly infused into the high vacuum of the ion source by atmospheric pressure. The experimental setup can be as simple as placing the end of the fused silica capillary directly into a reaction vial if the reaction solution can serve as an appropriate carrier (Figure 8A). The progress of a reaction may then be followed by the simultaneous monitoring of product and reactants. In an experiment involving the pepsin catalyzed digestion of $\alpha$MSH (data not shown), reaction progress profiles for both substrates and products were obtained by plotting selected ion chromatograms for the $(M + H)^+$ ions of the intact peptide and pepsin-generated fragments. Initial rate data can be obtained for this reaction from the decrease in abundance of the quasi-molecular ion of the substrate and the parallel increase of product ions. In another example, amino acid sequence data were obtained for a peptide subjected to C-terminal exopeptidase digestion and a time-based mass spectral analysis of the reaction progress (Figure 8B, C). In this particular case, glycerol was not present in the reaction mixture and was provided by the incorporation of a make-up tee in-line with the direct infusion set-up. The make-up tee provided a 1:1 mixture of reaction solution and a 30% glycerol solution containing 0.3% TFA. The peptide digest reaction consisted of 100 pmol/$\mu$L of S peptide from Ribonuclease dissolved in 50 mM Tris pH 6 buffer and a mixture of carboxypeptidases Y and P. The results presented in Figure 8B indicate the temporal nature of the signal intensities for some of the substrates and products for this reaction. M1 is the signal for the intact peptide and M2, M3, and M7 are for the C-terminal truncated forms of the peptide minus two, three, and seven C-terminal residues, respectively. The spectrum for the mixture representing a snapshot of the reaction at 20 min into the reaction indicates the presence of nine truncated forms of the peptide (Figure 8C). The mass differences between these forms allow the correct deduction of the C-terminal sequence of this peptide. Any ambiguity in interpretation can be clarified by the temporal information provided by the selected ion chromatograms for these truncated forms. It is important to note that the reactant analyzed is the C-terminal truncated intact peptide and not the individual amino acids released, the usual end point of C-terminal sequencing using carboxypeptidases. The mass differences between the truncated forms provide the identification of the individual amino acid residues involved.

*2.3.5. Coupling Separation Methods to Mass Spectrometry.* Some of the advantageous features of CF-FAB such as high sensitivity, good peak shapes for injected samples, compatibility with organic solvents, and continuous introduction capability have allowed the successful coupling of the separation techniques of liquid chromatography [13] and capillary

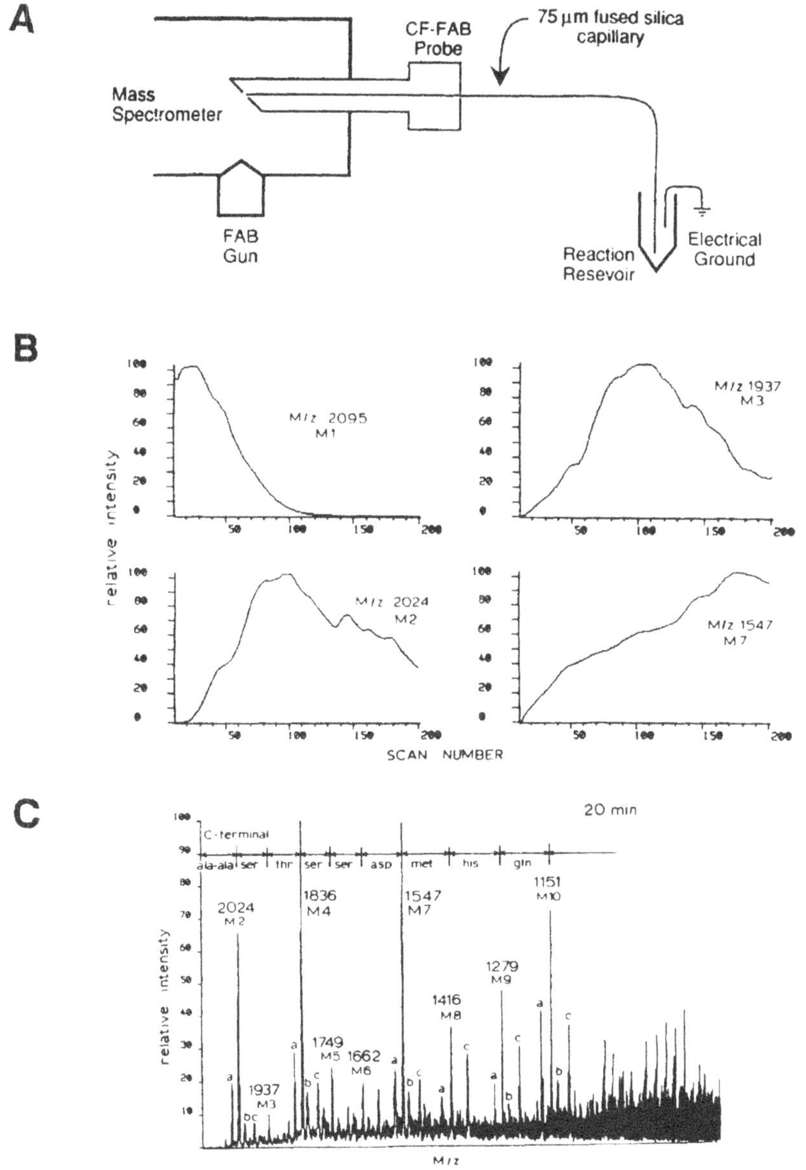


Figure 8. Constant flow reaction monitoring. A: Experimental set-up for constant reaction monitoring with CF-FAB. B: Carboxypeptidase (Y and P) treatment of ribonuclease S peptide; selected ion chromatograms for substrate and products. C: Mass spectrum for scan taken 20 min into the reaction. See text for explanation. (Reprinted with permission from reference 11.)

electrophoresis [14] with mass spectrometry. A technique similar to CF-FAB, known as frit-FAB, was first reported and developed by Ito et al. [15] for the specific purpose of coupling a mass spectrometer to an LC for the separation and analysis of bile acids. The designation 'frit'-FAB refers to the use of a stainless steel mesh or frit at the target end of the probe to disperse the liquid. Prior to this work, the only mechanism for coupling LC and FABMS was through a moving belt interface [16, 17], a technique that lacked good sensitivity and was mechanically complex. Other ionization methods, such as thermospray, electrospray (ES), particle beam and others have been successfully employed as ionization interfaces. We will only cover specific applications involving LC/CF-FAB MS.

The basic instrumental setup that we used for the LC/MS analysis of peptide mixtures is shown in Figure 9A. Microbore HPLC (1 mm i.d. X 50 mm C-18 column) is performed using a system that has syringe pumps capable of accurately delivering mobile phase at low flow rates without pulsing. A flow rate of 25 $\mu$L/min is adequate for efficient chromatography with a reverse-phase column and a glycerol-containing mobile phase. The effluent from the column is split 1:4 so that a 5 $\mu$L/min flow is conducted through the CF-FAB probe capillary into the source. To establish a conveniently rapid gradient, the gradient is formed at 500 $\mu$L/min and split 1:19 so that the column receives the gradient at 25 $\mu$L/min. An example of the results obtained with this system are shown in Figure 9B and 9C. This result is the fractionation of the tryptic peptides of a 200 pmol digest of $\beta$-lactoglobulin A. The total ion chromatogram is shown in Figure 9B and the specific ion chromatograms for 19 definable peptide fragments are shown in Figure 9C. The 19 fragments constitute mostly tryptic, partial tryptic, and some other fragments, the latter being derived from contaminating chymotryptic activity that was presumably in the trypsin preparation. This example, in addition to one presented in the next section, shows the clear advantage of combining mass spectrometry with liquid chromatography as an analytical tool.

Capillary electrophoresis is another micro separation technique that is dimensionally suited to coupling with CF-FABMS. Two interfacing approaches have been successfully employed, both having operational advantages and disadvantages. The two approaches are the liquid junction interface [14, 18, 19] and the coaxial interface [20]. A specific application involving the analysis of synthetic peptides will be presented in the next section.

## 3. Evaluation of the Chemical Synthesis of Peptides

Automated solid phase peptide synthesis entails the instrumentally controlled assembly of a protected peptide chain [21]. The C-terminal amino acid is attached to a solid support and each additional amino acid residue is then added stepwise by a coordinated series of reaction steps described by a programmed reaction cycle. Any automated peptide synthesis may run into difficulty for various reasons such as instrument malfunction, reagent quality, sequence specific difficulties, and operator error. The need to document a chain assembly is important so that synthetic strategy for a particular peptide can be fine tuned and the final product structure validated. The desired synthetic product may be contaminated with by-products that are the results of modification events such as deletion, insertion, termination, and

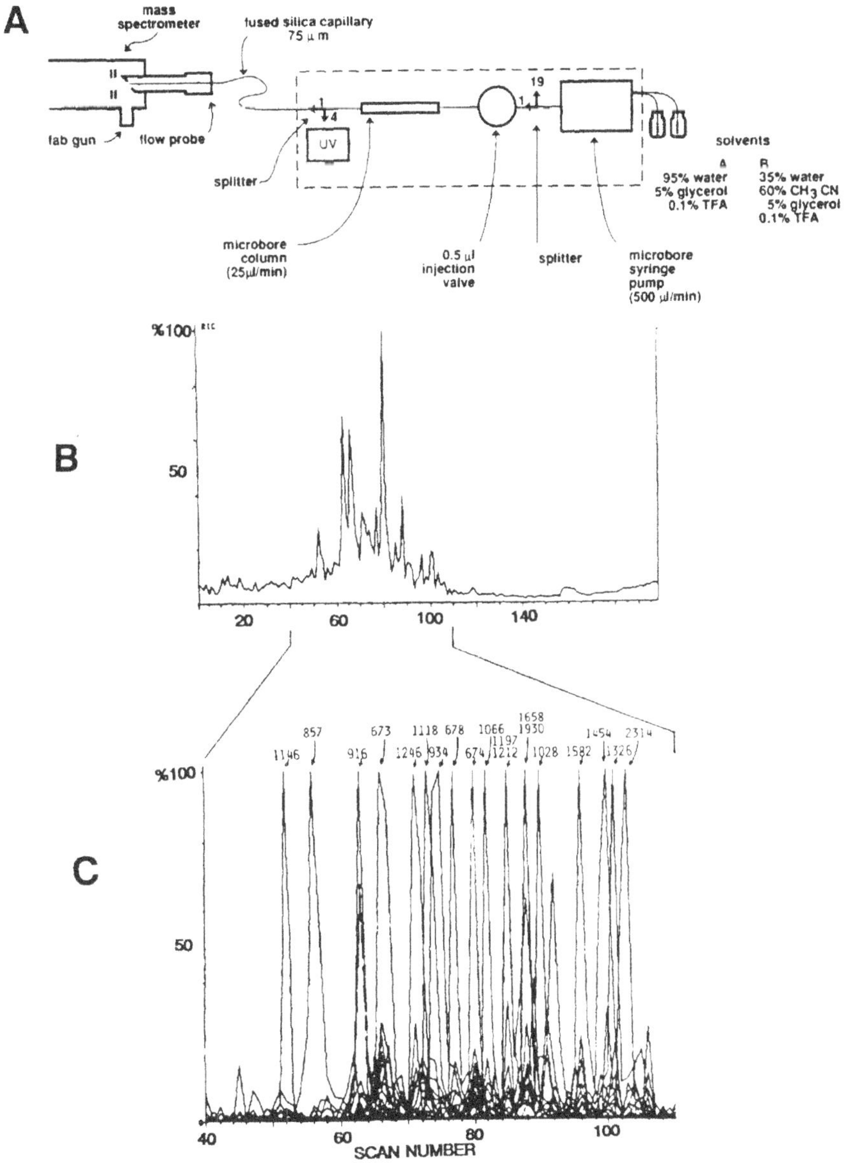


Figure 9. Peptide mapping by LC/MS. A: Experimental set-up for microbore LC/MS. B: LC/CF-FABMS analysis of a tryptic digest of $\beta$-lactoglobulin A (reverse phase - $C_{18}$). Total ion chromatogram. C: "Peptide chromatogram" generated by overlaying the normalized selected ion chromatograms for identifiable peptides.

branching reactions [22, 23]. There are several existing rapid monitoring methods that give a partial analysis of the situation involving reagent uptake and spectrophotometry. The quantity of unreacted amino component is usually the only property assessed and may give an incomplete picture of a particular stage of synthesis. A slower but more thorough method, preview sequencing or quantitative Edman degradation of the protected peptide-resin product, has been developed to provide a more direct analysis [24, 25]. This method, however, is lengthy and does not lend itself as a "real time" assay system.

We have used CF-FABMS to perform a stepwise assessment of automated peptide syntheses [26]. In addition, we have taken advantage of a computer controlled resin-sampling feature of the automatic peptide synthesizer that removes an aliquot of peptide-resin product after each amino acid addition cycle. Analyte peptide was obtained from a 1 mg aliquot of peptide-resin with a micro-acidolytic cleaving and deblocking method based on the treatment of peptide-resin with trifluoromethanesulfonic acid (TFMSA). Figure 10 shows the results of a stepwise assessment of a 14 residue synthetic peptide by CF-FABMS. Figure 10A shows the sequence of the desired peptide (the N-terminus being methionine (M) and the C-terminus being alanine (A), the synthetic reaction cycle number for each amino acid addition (synthesis proceeds from the C-terminus to the N-terminus), and the expected $(M + H)^+$ values for the resulting peptide at each synthetic step. Figure 10B shows the respective mass spectra obtained from samples containing micro-TFMSA released peptide obtained from peptide-resin products that were removed under instrument control after each amino acid addition cycle (only cycles 9, 11, and 13 are presented). The mass spectra indicate that the synthesis went well. The major ion signals are at *m/z*-values corresponding to those expected (Figure 10A). The mass spectrum for cycle 9 also has a major signal 17 u less than the expected. Because the peptide product at this stage has an N-terminal glutamine (Q), this ion was assumed to arise from a deamidation and cyclization to pyroglutamic acid for that residue. The presence of this ion solely in the cycle 9 spectrum indicates that this reaction took place as a consequence of the micro-TFMSA cleavage procedure and did not occur in the reaction vessel of the automated peptide synthesizer.

Retrospectively, a complete synthetic history of a peptide can be recorded by a slightly different procedure requiring just a few CF-FABMS analyses. In the case of the synthesis described above, the reaction product at every amino acid addition step can be determined by analyzing the products released from only three "peptide-resin pools". The three peptide-resin pools were prepared by mixing the four or five consecutive peptide-resins representing the early, middle, and final stages of assembly (Figure 10A). As shown in Figure 10C, the mass spectra for the micro-TFMSA treated peptide-resin pools contain all the expected ion signals from every step of synthesis except cycle 1 (the product was outside the mass range scanned). Cycle 2 product is present as a TFMSA adduct (+150 u). This adduct formation was limited to the Pool 1 derived sample because the peptide contained arginine (R) and TFMSA was present at a high level because the peptides were too short and not precipitable with ethyl ether. Ether precipitation of the released peptides and ether extraction of the pellet, as was possible for the longer peptides in resin pools 2 and 3, removes excess TFMSA to a degree where adduct formation is not detected. Ether precipitation and extraction of the peptide sample is the last step of the micro-TFMSA cleavage procedure. Dissolution of

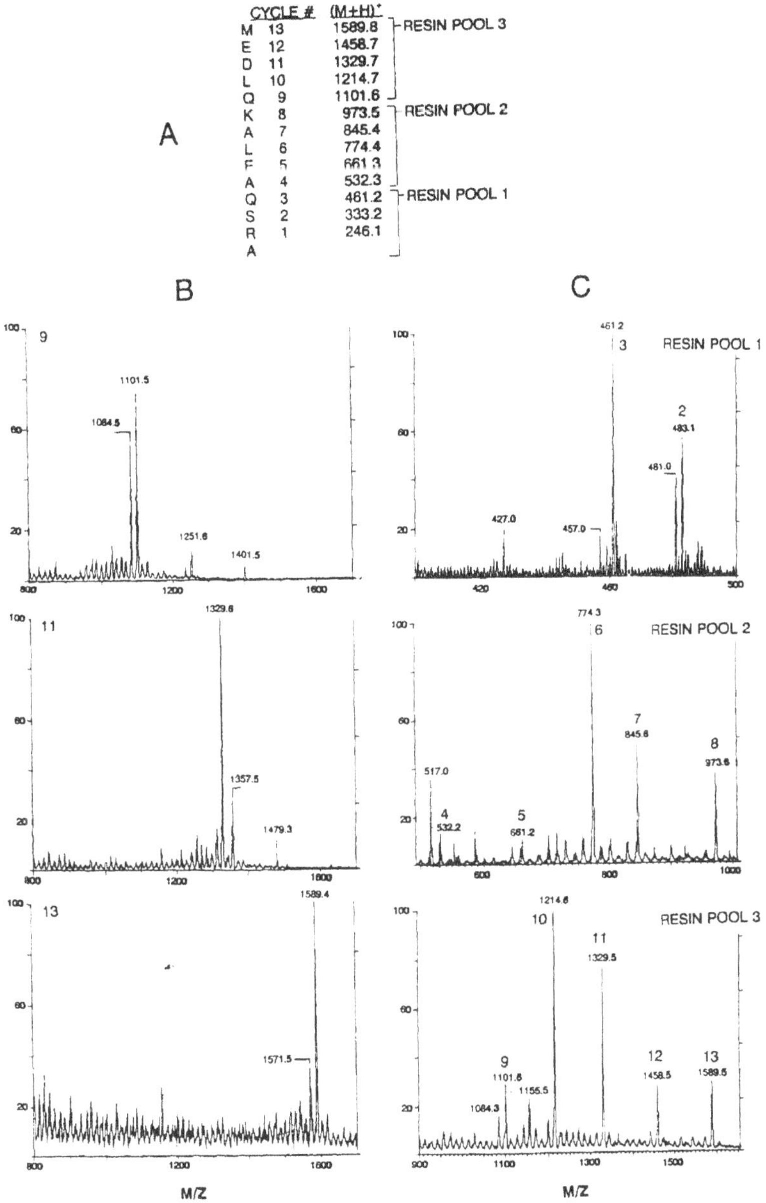


Figure 10. Stepwise CF-FABMS analysis of automated solid phase synthesis of a 14 residue peptide. A: Peptide sequence, synthetic cycle numbers, and the $(M + H)^+$ values for expected products. B: Single cycle product analysis. C: Pooled cycle product analysis. (Reprinted with permission from reference 26.)

the peptide pellet in 0.1% TFA is performed prior to CF-FAB MS analysis. Although the results presented indicate that this particular synthesis was successful, a modification at any particular step, if it had occurred, would have been easily tracked. Because the peptide-resin pool approach is stepwise, covering groups of steps, and the synthesis is "time-based", any modification could have been unambiguously and quickly located. Such modifications would appear as mass shifts in products following the cycle where the modification event occurred. This type of localization of a structural feature could not be obtained from a single MS analysis of a final product.

Success at using CF-FAB MS and the micro-TFMSA cleavage procedure as a stepwise evaluation method for straightforward peptide syntheses as described above suggested that we might be able to apply these methods to more challenging peptide synthesis problems. To generate multisubstrates for use in protease substrate structure function relationship studies [27], we attempted simultaneous multipeptide syntheses, or the synthesis of several closely related peptides in a single automated synthetic run [26]. One example will be presented that addresses a complex and challenging peptide synthesis problem and involves use of CF-FAB with flow-injection analysis and LC and CE to define the products and help clarify quantitative relationships.

Six derivatives of angiotensin II (Figure 11) were prepared by mixing equimolar amounts of six BOC-amino acids (Phe, His, Met, Glu, Asp, and Ile) in the single vial scheduled to be present in the fifth amino acid addition cycle (Figure 11A). This particular group of six BOC-amino acids was chosen on the basis that they all had identically or similarly written instrument vessel programs. The CF-FAB mass spectrum for the acid cleaved (anhydrous hydrogen fluoride) and deprotected product mixture is shown in Figure 11 B. Signals were present for each of the desired products; however, the intensities for the respective peptide signals suggest considerable variability in individual yields. Because the problem of ion suppression effects exist for standard FAB, and to a lesser extent also for CF-FAB, the ion intensities in Figure 11B do not necessarily reflect the true quantitative relationships. To assess more realistically the quantitative aspects, we utilized LC and CE coupled to CF-FAB. Figure 12 compares UV absorbance profiles for a conventional HPLC analysis, a comparable microbore HPLC analysis, and the "peptide chromatogram" described by overlapping the six mass spectral selected ion chromatograms for the six component angiotensin II analog mixture. Five major peaks are noted in the conventional run. The presence of shoulders on the first two major peaks of the microbore HPLC run indicate that the microbore analysis is not as well-resolved. As shown in Figure 12C, the mass spectral selected ion chromatograms for each analog derived from the LC/MS analysis of the microbore column clearly indicate the peak identities for the UV traces for the microbore and conventional chromatograms. As expected, the hydrophilic analogs elute before the more hydrophobic analogs. Because the peptide analogs are all of the same length, and therefore contain the same number of peptide bonds, the UV trace data (absorbance at 234 nm) reflect quantitative relationships whereas the mass spectral data permit peak identity assignments.

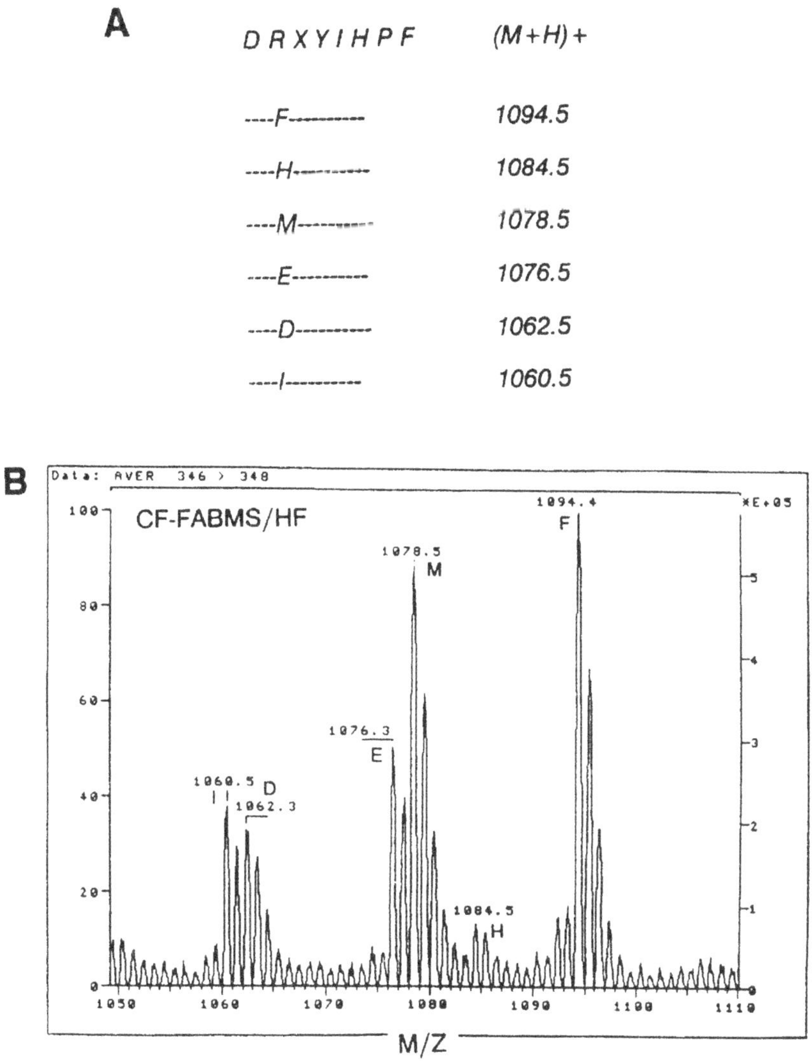


Figure 11. CF-FABMS analysis of an Angiotensin II analog mixture containing six products. A: Sequences and expected masses of the product mixture. B: CF-FAB spectrum derived from flow-injection analysis of the six component mixture.

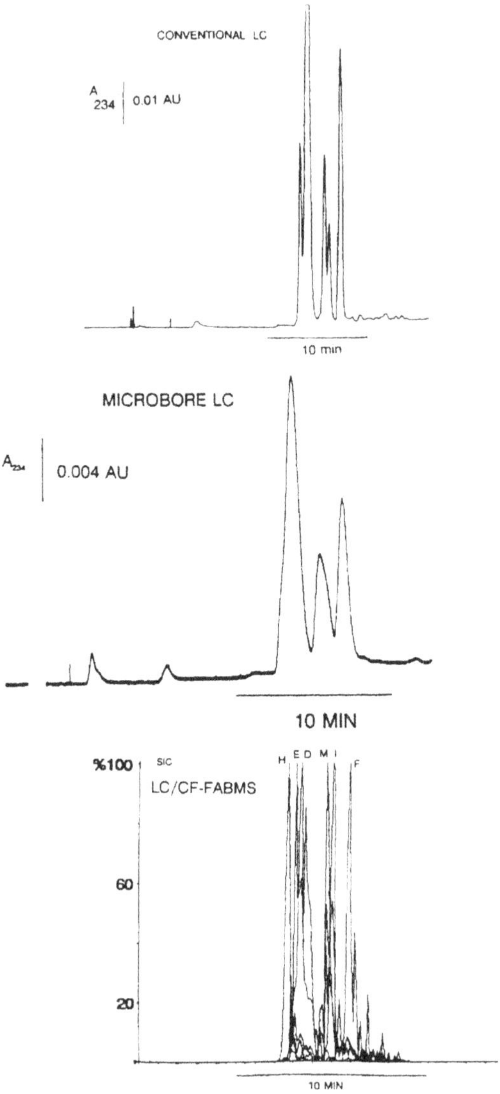


Figure 12. HPLC and LC/CF-FABMS analysis of the six component Angiotensin II analog mixture. A: The UV detector trace for a "conventional" LC run. B: The UV detector trace for a microbore LC/CF-FABMS run. C: The overlapping mass spectral specific ion chromatograms for the individual peptides in the mixture. The letters represent the amino acid substitutions present in the six Angiotensin II analogs.

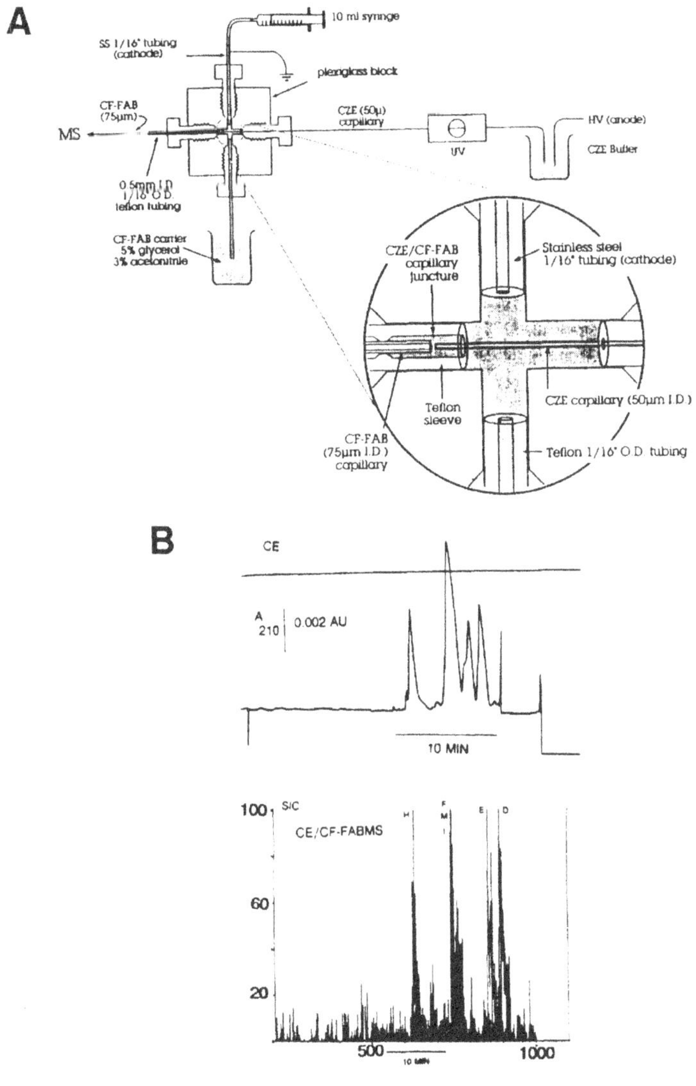


Figure 13. CE/CF-FABMS analysis of the Angiotensin II analog mixture containing six products all derived from a single automated run. A: Instrumental arrangement for CE/CF-FABMS and enlargement of the liquid junction interface between the CE capillary and the CF-FAB probe capillary. B: Upper panel - UV detector trace of a CE run performed on the peptide mixture. Lower panel - the overlapping mass spectral specific ion chromatograms for each peptide in the mixture. The single letters represent the amino acid substitutions present in the six Angiotensin II analogs.

To gain an additional perspective on the variable yields of the individual components of the multipeptide mixture, an aliquot of the same sample of the octapeptide mixture was analyzed by CE/CF-FABMS. The liquid junction interface previously described [14] is shown in Figure 13A. The interface shown in the magnified area allows 5 $\mu$L/min of solution from the CF-FAB reservoir to transfer efficiently CE effluent (< 0.1 $\mu$L/min) into the mass spectrometer via the CF-FAB probe. The results of the CE/CF-FAB MS analysis of the six component octapeptide mixture are shown in Figure 13. Figure 13B shows the UV trace of the electrophoretic separation of the peptides at a point 2/3 the length of the CE capillary. The selected ion chromatograms shown in the lower portion of Figure 13B indicate the electrophoretic pattern of the peptides as detected in the ion source of the mass spectrometer following additional electrophoretic separation in the last third of the CE capillary and transfer through the liquid junction interface. As expected, the overlapped selected ion chromatograms indicate that the most basic substituted analog emerged first; then the unresolved neutral residue substituted group emerged followed by the resolved acidic residue substituted analogs. Again the UV trace data suggest quantitative yield estimations whereas the mass spectral data confer peak identity.

## 4. Conclusion

The unique features of CF-FAB permit the direct analysis of aqueous samples enabling the direct examination of dynamic biochemical processes and systems. The molecular weight information gained by this technique permits an accurate and direct characterization of any chemical modifications that occur. The organic matrix reduction afforded by CF-FAB significantly increases the apparent sensitivity of FAB, and the dynamic flow condition at the target modifies and lessens the sample ion suppression effect that occurs with FAB. CF-FAB has been shown to be a valuable tool for structure evaluation of peptides derived either from natural or synthetic sources. We have found the technique to be of great use in the assessment of automated peptide syntheses, and it can be applied in either a stepwise fashion during actual synthesis or retroactively to define the complete synthetic history of particular synthesis. The capacity to couple CF-FABMS to separation techniques such as microbore LC and capillary electrophoresis has allowed us to deduce quickly the outcome of multipeptide synthetic reactions. The chromatographic mass identification information derived from the LC/MS analyses reveals preparative strategies to obtain each peptide in pure form. CF-FABMS-related techniques provide the peptide chemist with tools that can be used to design, improve, and validate synthetic peptide strategies.

## 5. References

1. Barber, M., Bordoli, R.S., Sedwick, R.D. and Tyler, A.N. (1981) 'Fast atom bombardment of solids (F.A.B.): a new ion source for mass spectrometry', *J. Chem. Soc., Chem. Commun.*, 325.
2. Lehmann, W.D., Kessler, M. and Koenig, W.A. (1984) 'Investigations on basic aspects of fast atom bombardment mass spectrometry: Matrix effects, sample effects, sensitivity and quantification', *Biomed. Mass Spectrom.*, **11**, 217.
3. Sunner, J., Morales, A. and Kebarle, P. (1988) 'Mechanism of formation of FAB spectra', *Intl. J. Mass Spectrom. and Ion Phys.* **86**, 169.
4. Naylor, S., Findeis, F., Gibson, B.W. and Williams, D.H. (1986) 'An approach toward the complete FAB analysis of enzymic digests of peptides and proteins'. *J. Am. Chem. Soc.* **108**, 6359.
5. Caprioli, R.M., Moore, W.T. and Fan, T. (1987) 'Improved detection of "suppressed" peptides in enzymic digests analyzed by FAB mass spectrometry', *Rapid Commun. Mass Spectrom.*, **1**, 15.
6. Caprioli, R.M., Fan, T. and Cottrell, J.S. (1986) 'Continuous-flow sample probe for fast atom bombardment mass spectrometry', *Anal. Chem.*, **58**, 2949.
7. Caprioli, R.M. and Fan, T. (1986) 'High sensitivity mass spectrometric determination of peptides: Direct analysis of aqueous solutions', *Biochem. Biophys. Res. Commun.*, **141**, 1058.
8. Caprioli, R.M. (1987) 'Enzymes and mass spectrometry: a dynamic combination', *Mass Spectrom. Rev.*, **6**, 237.
9. Moore, W.T., Fan, T. and Caprioli, R.M. (May 24-29, 1987) 'Determination of enzymic reaction parameters by on-line FABMS', Proc. 35th ASMS Conf. on Mass Spectrom. and Allied Topics, Denver, CO, p. 566.
10. Caprioli, R.M. and Lin, S-N. (1990) 'On-line analysis of penicillin blood levels in the live rat by combined microdialysis/fast atom bombardment mass spectrometry', Proc. Natl. Acad. Sci., **87**, 240.
11. Caprioli, R.M. (1990) (ed.) Continuous-Flow Fast Atom Bombardment Mass Spectrometry, John Wiley & Sons Ltd., England.
12. Caprioli, R.M., Moore, W.T., Petrie, G. and Wilson, K. (1988) 'Analysis of mixtures of hydrophilic peptides by continuous-flow fast atom bombardment mass spectrometry', *Int. J. Mass Spectrom. Ion Proc.*, **86**, 187.
13. Caprioli, R.M., DaGue, B., Fan, T. and Moore, W.T. (1987) 'Microbore HPLC/mass spectrometry for the analysis of peptide mixtures using a continuous flow interface', *Biochem. Biophys. Res. Comm.*, **146**, 291.
14. Caprioli, R.M., Moore, W.T., Wilson, K.B. and Moring, S. (1989) 'Coupling capillary zone electrophoresis and continuous-flow FAB mass spectrometry for the analysis of peptide mixtures', *J. of Chromatogr.*, **480**, 247.
15. Ito, Y., Takeuchi, T., Ishi, D. and Goto, M. (1985) 'Direct coupling of micro high-performance liquid chromatography with fast atom bombardment mass spectrometry', *J. Chromatogr.*, **346**, 161.

16. Stroh, J.G., Cook, J.C., Milberg, R.M., Brayton, L., Kihara, T., Huang, Z., Rinehart, K.L. and Lewis, I.A.S. (1985) 'On-line liquid chromatography/fast atom bombardment mass spectrometry', *Anal. Chem.*, **57**, 985.
17. Dobberstein, P., Korte, E., Meyerhoff, G. and Pesch, R. (1983) 'Investigation of an LC/MS interface for EI-, CI- and FAB-ionization', *Int. J. Mass Spectrom. Ion Phys.*, **46**, 185.
18. Minard, R.D., Chin-Fatt, D., Curry, Jr., P. and Ewing (June 5-10, 1988) 'Capillary electrophoresis/flow FAB MS', Proc. 36th ASMS Conf. on Mass Spectrom. and Allied Topics, San Francisco, CA, 5-10, 905.
19. Reinhoud, N.J., Niessen, W.M.A., Tjaden, U.R., Gramberg, L.G., Verheij, E.R. and van der Greef, J. (1989) 'Performance of a liquid-junction interface for capillary electrophoresis mass spectrometry using continuous-flow fast-atom bombardment', *Rapid Commun. Mass Spectrom.*, **3**, 348.
20. Moseley, M.A., Deterding, L.J., Tomer, K.B. and Jorgenson, J.W. (1988) 'Capillary-zone electrophoresis/fast-atom bombardment mass spectrometry: Design of an on-line coaxial continuous-flow interface', *Rapid Commun. Mass Spectrom.*, **3**, 100.
21. Merrifield, R.B. (1965) 'Automated synthesis of peptides', *Science,* **150**, 178.
22. Barany, G. and Merrifield, R.B. (1980) 'Solid phase peptide synthesis', in E. Gross and J. Meienhofer (eds.), The Peptides, Analysis, Synthesis, Biology, Academic Press, Inc., New York, pp. 3-253.
23. Kent, S.B.H. (1988) 'Chemical synthesis of peptides and proteins', *Ann. Rev. Biochem.*, **57**, 957.
24. Niall, H.D., Treageer, G.W. and Jacobs, J. (1972) 'Automated Edman degradation monitoring of solid-phase peptide synthesis', in Meinhofer, J. (ed.), Chem. Biol. Pept., Proc. Amer. Pept. Symp., 3rd, Ann Arbor, MI: Ann Arbor Sci., pp. 695.
25. Applied Biosystems, Inc.: Sequence analyses of synthetic, side-chain protected, resin-bound peptides. Applied Biosystems Peptide Synthesizer User Bulletin No. 4.
26. Moore, W.T. and Caprioli, R.M. (1991) 'Monitoring peptide synthesis stepwise by mass spectrometry', in Villafranca, J. (ed.), Techniques in Protein Chemistry II, Academic Press, San Diego, CA, 511.
27. Seifert, W.E., Moore, W.T., Ballatore, A. and Caprioli, R.M. (June 3-8, 1990) 'Evaluation of multi-substrate enzyme assays by continuous-flow FAB MS', Proc. 38th ASMS Conf. on Mass Spectrom. and Allied Topics, Tucson, AZ, p. 1329.

# COMBINED LIQUID CHROMATOGRAPHY/MASS SPECTROMETRY (LC/MS): A Review

P. J. ARPINO
*Institut National Agronomique*
*Laboratoire de Chimie Analytique*
*16, rue Claude Bernard*
*75231 Paris 05*
*France*

ABSTRACT. This chapter reviews the various recent techniques for the on-line coupling of high performance liquid chromatography with mass spectrometry.

## 1. Introduction

Combined liquid chromatography/mass spectrometry (LC/MS) has been the subject of intensive research since its introduction in the early 70s and is well-documented. In particular, publications describing early LC/MS methods have been reviewed by Games [1]. The method referred to as "Thermospray" (TS) has developed rapidly and is the most widely used, while other methods have either not changed appreciably or have declined in importance [2].

Active research programs on LC/MS by the end of the 70s were soon followed by the introduction on the market of different types of LC/MS interfaces, adaptable to magnetic sector or quadrupole instruments. As a consequence, it was hoped that the utility of the technique in various fields of application would increase considerably. This has not yet occurred to the expected extent. A possible reason for this situation is that early LC/MS interfaces, such as the original moving belt interface [3, 4], or first models of Direct Liquid Introduction interfaces (DLI), were rather expensive, with a limited range of application, and were sometime difficult to operate routinely in a service laboratory. They consisted of interfaces to standard mass spectrometers that also were equipped with other conventional inlets, (e.g., a GC interface and a direct insertion probe).

Running LC/MS analyses often required long installation and tuning time, and results were sometime unsatisfactory. Many laboratories put their LC/MS interface aside, preferring to use the mass spectrometer for more well-established analysis methods such as GC/MS, or off-line LC/MS; for example, the combinations of preparative high performance liquid chromatography with field desorption [5, 6], or fast atom bombardment (FAB) [7, 8] were found to be particularly useful. In one specific field, the application of off-line LC/MS to environmental studies was reviewed [9], and the advantages and the disadvantages are the same today. In the following, the different methods for on-line coupling are presented because they tend to be used more often than off-line methods.



*M. L. Gross (ed.), Mass Spectrometry in the Biological Sciences: A Tutorial*, 253–267.

## 2. Transport interfaces

This approach to LC/MS aims at the complete removal of the solvent from the column eluent prior to introduction of the solutes into the mass spectrometer. Solutions are fed continuously to a moving Kapton belt. Commercial systems are available from Finnigan-MAT and VG Analytical, and a comparison between different models was reported [10]. The influence of construction details, such as extension of the belt into the ionization chamber of the mass spectrometer was emphasized. Consequently, the quality of results in some reports varies depending on the instrument that was available to the analyst.

An old Finnigan moving belt was used by Krost [11] for the analysis of polynuclear aromatic mixtures, including anthracene, pyrene, chrysene and benzo[*a*]pyrene. As already reported in the past, the sensitivity was found to be a direct function of compound volatility, and the detection limit was approximately 10 ng.

Levsen et al. [12] used the new Finnigan-MAT interface and a quadrupole mass spectrometer for the analysis of nonionic surfactants of the alkylphenolethoxylate and oxoalcoholethoxylate classes. Information on the distribution of homologs and the presence of impurities was obtained. Biodegradation of non-ionic surfactants in surface water was investigated. After adding 0.3 g of nonylphenolethoxylate to 4 L of river water and stirring the solution for 3 days, the LC/MS analysis revealed that biodegradation to the homolog with two ethylene oxide units had occurred.

In another report [13], it is noted that deactivation of the Kapton belt in the same interface with a 50-ppm Carbowax 20M solution reduced memory effects during recycle of the belt, and in general increased the detection limit for various compounds. For instance, a detection limit of 35 pg of bromazepam was obtained when using CI. Various isomeric tocopherols and $\alpha$-tocatrienol were identified in a maize germ oil.

These techniques have been applied to substances of environmental concern other than those listed in ref. 11 and 12; for example, in the analysis of linuron, diuron and monuron, which belong to two groups of thermally labile pesticides: carbamates and ureas [14]. The method was satisfactory for regulatory analysis of crop residues at the ppm level. Application to other classes of substances may be of interest for evaluation of instrument performance. Such studies include the identification of antioxidant and ultraviolet light stabilizing additives in plastics [15]; sequence analysis of derivatized peptides [16] and identification of C-methylation artifacts from the permethylation reaction of peptides [17]. There are many applications to analyses of natural substances, such as a series of naturally occurring phospholipids [18] and sphingoid base [19]; gossypol and some of its derivatives [20]; pepper and capsicum oleoresins [21]; vanilla extracts [22]; dexmethasone and betamethasone.

Three major improvements have been made to the moving belt in the recent years and have significantly increased the overall performance.

### 2.1 SPRAY DEPOSITION ONTO THE BELT

Special attention was paid to the solution deposition onto the moving plastic film. Initially, the solution was simply fed onto the belt by a continuous stream via a narrow tube located close to the belt. Liquid deposition was uneven, especially for aqueous solutions. The liquid film would break into small beads that degraded the chromatographic resolution, producing typical christmas-tree shaped peaks, and caused vacuum instabilities inside the mass spectrometer source housing.

Vouros et al. [24, 25] showed that spray deposition from a nebulizer located 4-8 mm above the belt and at an angle of 60° with the belt, eliminated this problem. Fine spraying is obtained by means of a coaxial pneumatic nebulizer. In recent models, the gas is preheated, or the solution is directly heated electrically [26] in a manner similar to that used in thermospray nebulizer (*vide supra*). The heating improves the spray deposition and, in addition, removes part of the solvent; thus, a concentrated, sometime dry, mist reaches the belt.

### 2.2 MICRO-HPLC AND POST COLUMN CHEMISTRY

Micro-packed columns are feasible for HPLC, whereas open tubular columns are not. The former offer advantages when the amount of sample is limited or when the solvent flow rate must be reduced. Consequently, narrow-bore columns (*ca.* 1 mm i.d.) are now used in conjunction with moving-belt interfaces because they dramatically reduce the amount of solvent that must be handled by the interface [25, 27]. Instrumentation compatible with 1-2 mm i.d. packed tubes, including pumps delivering pulseless flow rates in the range 50-200 μL/min, have become available, and was the subject of several reviews [28, 29]. As a consequence, the infrared heater used in early moving-belt models, known to be difficult to adjust and to cause severe problems when running aqueous solutions, and various splitting devices, used to remove solvent excess, are no longer necessary with narrow-bore PC.

Karger and Vouros [27] believe that even smaller columns should be investigated for optimum chromatographic and mass spectrometric conditions; for example, 200-μm fused silica capillary columns packed with small diameter (3 μm) particles. They emphasized the utility in conjunction with moving-belt LC/MS of a post-column solvent/solvent extraction and on-line synthesis of volatile derivatives. Although they have pioneered post-column chemistry since 1979 and have demonstrated its possibilities [30], the number of groups that have followed the same approach remains small [31, 110].

### 2.3. IMPROVED IONIZATION METHODS

Conventional ionization methods following thermal desorption of the sample from the belt are electron impact or chemical ionization using methane or isobutane as the reactant gas. Investigation on the use of soft ionization techniques applied to samples still adsorbed onto the belt, however, has led to some interesting results.

Several research programs, combining HPLC and Secondary Ion Mass Spectrometry (SIMS) were stopped (see [32] and [33] for early work). The same situation occurred in the case of laser desorption [34] (LDMS). In one paper [35], the performance of these two ionization methods was compared by using a moving steel belt and a quadrupole mass spectrometer, based on analyses of 14 amino acids and 4 nucleosides. Laser desorption provided more reliable molecular weight information; however, because the laser energy absorbed by the sample varies uncontrollably, the mass spectra as well as the LC response areas are less reproducible. On the other hand, chromatographic peak shapes, mass spectral reproducibilities and resolution were much better with SIMS than with LDMS.

Future progress on the on-line combination with SIMS or LDMS appear uncertain. Another effective soft ionization technique, $^{252}Cf$ plasma desorption, has so far been exclusively applied in a discontinuous mode. Sample solution were sprayed onto a mechanized support, and solutes were ionized after complete solvent evaporation [36].

The combination with Fast Atom Bombardment (FAB) was also explored. FAB implies bombardment by high kinetic energy atoms of samples in solution with liquid matrices such as glycerol or thioglycerol. Initially, the matrix was spread separately onto the belt by using a reservoir and an applicator [37]; however, this methods is incompatible with the maintenance of a high vacuum in the mass spectrometer. Sometimes, the matrix is dissolved into the HPLC eluent at very low concentration, typically 0.005%. This may result, however, in an important modification of the LC retention parameters. The clean-up heater used in conventional belt systems for regenerating the belt surface was replaced by a wash bath containing methanol for removing nonionized samples and the liquid matrix.

FAB-LC/MS was applied to the molecular weight determination and sequencing of series of peptides including nonderivatized hexadecapeptide antiamoebin I (M = 1669 u) and a mixture of tetra- to nonapeptides obtained after its partial hydrolysis. Total ion current profiles similar to traces from a conventional UV detector were produced. One weakness of the method is poor sensitivity; for example, 125 ng of an N-acylpentapeptide gave an $(M+Na)^+$ ion with a signal-to-noise ratio of 3 [26]. The recent emergence of continuous flow-FAB probes has contributed to the decline of FAB-belt studies, because they are more simple to use and produce good spectra with a low detection limit.

It has often been said that a conventional moving belt is a sensitive and reliable LC/MS interface, but that a limited range of compounds can be handled. This may explain why the number of application reports has not increased significantly over the recent years. Conventional moving-belts can be used in conjunction with SFC, thus extending the use of this often expensive mass spectrometer accessory.

## 3. Direct Liquid Introduction (DLI)

In this method, a portion of the eluent from the liquid chromatograph is fed directly into the ion source of a mass spectrometer configured for CI; solvent induced CI spectra are produced from solute molecules [1, 38].

### 3.1. INSTRUMENTATION

DLI probes are generally modified solid probes and can be made in the laboratory or purchased from accessory dealers (Vacumetrics in the US, Hositrad in Holland), and can be fitted to nearly any model of mass spectrometer. Quadrupoles are generally used, but work using sector instruments was also reported [39, 41]. A diaphragm is no longer the only method of vacuum nebulization. Other devices including fused-silica tubes [42, 45] and coaxial pneumatic nebulizers [46, 47] were described.

Classical DLI models comprise a thin metal diaphragm pierced by a 2-5 μm pinhole for vacuum nebulization of chromatographic effluents. Modifications of a standard mass spectrometer for DLI/LCMS operations include the addition to the CI source block of a desolvation chamber for better droplet evaporation [49] and a cryotrap for a better vacuum pumping [50]. The trap, however, imposes some practical constraints for routine work; it can be advantageously replaced by a rough pump directly attached to the ion source [43], in a manner first described for thermospray experiments.

DLI interfaces are simple to design and to install, rather easy to operate, and have often produced good analytical results when applied to difficult samples [49-53]. Optimization of the operating condition, modification of standard mass spectrometers for DLI coupling, designs of desolvation chambers and laboratory-built DLI probes and source are well-described [51, 52]. The life time of the heated metallic ribbon, which emits the electron beam for inducing solvent CI reactions, can be short if it is surrounded by a high pressure of solvent vapors; replacement of the filament assembly by a discharge tube has been proposed [44].

Nevertheless, all DLI interfaces suffer more or less from severe plugging problems and irregular responses, especially those models that utilize a small diaphragm [54]. Another inherent limitation is the poor heat transfer to rapid liquid droplets when attempting to evaporate them under a vacuum, even when they travel through a very hot (>300 °C) desolvation chamber [55]. The situation can be improved by preheating the solution while it flows through the DLI probe, a method sometime called "hot DLI" [43], the difference with a micro-thermospray interface being probably small. The plugging problem remains so serious that the use of this interfacing method has declined in the recent years. The origin of the plugging problem was elucidated: silica from the HPLC packing can dissolve in aqueous or methanolic eluents, even when chemically bonded supports are used, although less silica dissolves in that case. Note that combined SFC/MS also uses narrow restrictors. They do not plug as long as pure $CO_2$ is used as the eluent; however, addition of polar modifiers, such as methanol, increases the risk of restrictor plugging as minute amount of silica from fused-silica column walls are dissolved and carried to the column end.

### 3.2. COMBINATION WITH MICRO-HPLC

Because the liquid flow rate sampled by the DLI interface into the ion source is usually very small, ca 10-80 μL/min, combination with narrow-bore columns has become increasingly popular among DLI users [56-62]. Splitting the column effluent, an operation that negatively affects the detection limit and the overall sensitivity, is avoided when utilizing narrow-bore (0.5-2 mm id) PC, since the full column effluent can be fed into the ion source [42, 44, 56, 61].

One attempt to use 5 and 10 μm i.d. fused silica open tubular columns, in combination with a 4 μm diaphragm nebulizer was reported by Niessen and Poppe [63]. Although the detection limits in the low picogram range appear promising, no application to difficult samples was presented; thus, the potential for routine work was not demonstrated.

Use of solvent addition and post-column extraction, as done with moving-belt interfaces, to exchange solvents that are difficult to vaporize, were investigated [31, 64], but did not find a general application.

### 3.3. ANALYTICAL APPLICATIONS

In environmental studies, thermally labile herbicides and pesticides were investigated, including the active ingredients of two herbicides, chlorosulfon and sulfometuron [65], 25 carbamates, chlorinated carboxylic acids and methylureas [52, 66, 67], as well as several organophosphorous pesticides [68]. Of related interest is the analysis of two mycotoxins, nivalenol and deoxynivalenol [69].

Another group of application reports address the determination of biological substances, including ranitidine and its metabolites in urine [70], nonderivatized sialic acids [71], and several optically active drugs [72, 73]. A series of corticosteroids were determined in equine biological fluids [74]; because the same compounds were also analysed by another group using a moving-belt interface [23], a comparison of the two methods can be made. Other biological applications include quantification of biotin and dethiobiotin in biological fluids [75]; sequencing of nonderivatized peptides [53]; liposoluble vitamins [76]; glycerolipids and glycerophospholipids including diacylglycerol moieties of natural phosphatidylcholines [77, 78].

DLI instrumentation was found particularly useful for the analysis of nitroaromatic explosives, including trinitrotoluene, which surely belong to a class of thermally sensitive compounds [39, 79]. Reduction of nitro- to amino groups, however, may occur. When studying nitropolycyclic aromatic compounds by several analytical methods, Quilliam et al. [80] observed that reduction was most extensive when using an LC/MS with a DLI interface. They estimated that both qualitative and quantitative analyses at trace levels can be impaired by the irreproducibility of the mass spectra.

Although the number of DLI application reports is large, its use has declined over the recent years [2]. Many users have shifted to a thermospray (TS) interface because the change from a DLI set-up only requires minor modifications. Several comparison studies have been published [65, 67, 81]. Although the sensitivities for given compounds were often comparable, and molecular information sometime better in DLI than in TS, the later was found to be easier to use routinely and less prone to, although not totally free of, plugging problems.

A remaining interest in DLI systems is for the combination with micro-HPLC. For the moment, TS appears to accommodate a low liquid flow rate with some difficulties, unless the column effluent is diluted to 1 mL/min before entering the interface, which is a chromatographic nonsense.

Continuous flow FAB may be regarded as an example of DLI, and it is discussed fully in another chapter in this volume.

## 4. Thermospray

### 4.1. PRINCIPLES OF NEBULIZATION AND SOLUTE IONIZATION BY THERMOSPRAY

The method called "thermospray" has evolved from large crossed-beam machines assembled in Salt Lake City by Vestal and co-workers in the early 70s' (see [82-84] for reviews). In its present form, it is mechanically simple and simultaneously achieves two different processes: firstly, a fine nebulization and a nearly complete desolvation of the HPLC solvent prior to entering the vacuum of the mass spectrometer; secondly, a soft chemical ionization of nonvolatile and thermally labile solute samples.

Typically, a solution of the sample in an aqueous solvent containing *ca.* 0.1 M of a volatile salt, for instance ammonium acetate, is heated by an appropriate energy source. Part of the solvent evaporates and the resulting vapor rapidly expands to sonic velocities. The vapor induces pneumatic nebulization of the remaining liquid solution into a fine mist.

The exact ionization process acting under the experimental conditions has been the subject of a controversy. Originally, it was assumed that the resulting droplets are electrically charged as a result of liquid film ruptures and random partitioning of positive and negative moieties of the ionized salt between individual liquid droplets. The high electric field caused by an excess of either positive or negative ions in a small droplet induces desorption of preformed ions from the liquid solution into the gas phase. The preformed ions result from the solute molecule by protonation, cationization with an alkaline cation or addition of solvent cluster ions [84]. Direct comparison with other MS ionization techniques including field desorption [90], FAB [91], and SIMS [92] have been made to evaluate the possibility that a thermospray ionization process can ionize any nonvolatile molecule as long as a solute-derived ion can be formed in liquid solution.

The current explanation is rather different. The fortunate selection of ammonium acetate was because of the salt volatility compared to other possible choices; thus, no accumulation of ammonium acetate is found inside the ion source after a TS experiment. It turned out that heating an aqueous solution of ammonium acetate produces large concentrations of gas phase ions, including $NH_4^+$, $H_3O^+$, $CH_3COO^-$, and various cluster ions derived from these precursor ions.

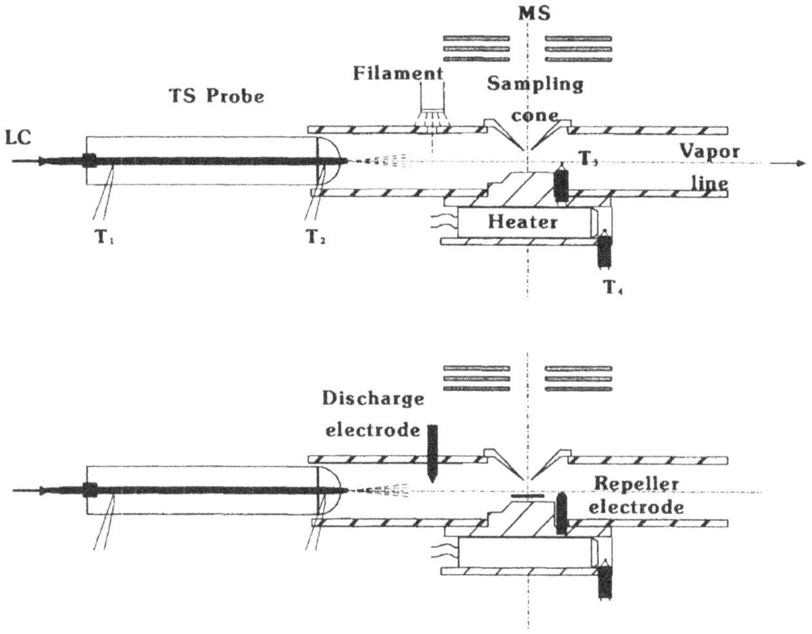


**Figure 1.** Schematic diagrams of thermospray interfaces. The upper diagram shows the design with a filament for electron ionization. $T_1$, $T_2$, $T_3$, and $T_4$ are thermocouples. The lower diagram shows the substitution of a discharge electrode for the filament and the inclusion of a repeller.

The abundances of the solvent ions can be predicted with accuracy from well-known gas phase equilibria. Consequently, thermospray ionization is now viewed to operate almost entirely from the gas phase and cannot ionize solute molecule with no vapor pressure in the vacuum conditions of the mass spectrometer. Short lived thermally labile molecules can, however, be ionized before they decompose and produce intact structure related ions.

### 4.2. SPECIFIC ADVANTAGES

Three major features have contributed to the success of this interfacing method:

(i) The design of the interface was continuously simplified to achieve a simple device applicable to different models of mass spectrometers. For example, nebulization was achieved successively by a high powered laser, an oxy-hydrogen torch, several electrical cartridges [85], and finally by a tube directly heated by an electrical current [86]. The original cost was reduced by at least four orders of magnitude. Direct electrical heating is now used in all commercial versions of this method, and this version is available from all major instrument manufacturers. In general, they are used with quadrupole instruments, but coupling to magnetic sector instruments has been reported, although less frequently. Both instrument types show equal performance at masses below 1000 u [87-89]. The high kinetic energy of ions from the source of a magnetic instrument, however, can induce ion dissociation upon collision with residual solvent molecules that are always abundant in a TS experiment. Thus, molecular mass information for high mass molecules can be obscured.

(ii) The experimental production of solute ions by simply spraying an electrolytic solution, in the absence of other ionization devices [85], the so-called "filament-off mode" is still one of the most spectacular features. A minimum percentage of water (*ca.* 20%) and a high salt concentration (0.1-0.5 M), however, must be present. Thus the method applies mainly to hydrophilic polar compounds and to some stable preformed ions. This explains why TS is often combined with reversed phase HPLC. For nonpolar samples that dissolve only in low polarity solvents (for instance, hydrocarbons, pure acetonitrile, pure methanol), adding a salt such as ammonium acetate has no effect, probably because the salt does not get ionized by the solvent. In this case, electrons from a conventional filament, or from a discharge electrode, are necessary to induce the series of chemical ionization reactions that produce sample-derived ions. This "abnormal" mode is referred to as "filament-on", or "discharge" mode depending on the source of electrons, and leads to solvent-derived CI. It is entirely equivalent to the ionization conditions that prevail in DLI. Which of these two ionization mode, filament on, or off, is sometimes difficult to predict and is often determined by trial and error attempts.

iii) TS interfaces are more reliable with respect to plugging than previous DLI interfaces, at least when the TS nebulizer is a tube with a diameter *ca.* 100 um. Many DLI users have turned to thermospray for this reason.

iv) Mass spectra often exhibit very few fragment ions; mainly cluster ions derived from the intact molecule are formed, and these allow molecular weight confirmations. Structure elucidation is more difficult owing to the absence of fragments. High energy collision induced dissociation can fragment such clusters, but it requires the use of multisector instruments that are costly. Some fragmentation control of the ions produced by thermospray, however, is possible by introducing a repeller electrode into the source block.

4.3. OPTIMIZATION OF INTERFACE AND SOURCE PARAMETERS

As for DLI, sensitivity is compound dependent. The detection limit is comparable in some instances, or worse in others [67, 71, 81]. There is a direct relationship between the diameter of the heated tube and the optimum liquid flow rate that can be fed to the ion source. For 100-150 µm i.d. tubes, the optimum flow rate is 1-1.5 mL/min. Lower flow rates, as delivered by micro-HPLC columns require narrower tubes, and the risks of plugging become as severe as with a DLI. The need for a very high flow rate of solvent vapor through the ion source has some practical consequences:

The mass spectrometer must be customized to high vapor throughputs by attaching a rough pump directly to the ion source [85, 86]. Unfortunately, many of the solute ions are dragged to this pump, instead of being extracted by the ion optics towards the analyser. This accounts for the poor sensitivity of some thermospray determinations.

A special ion source block must be used, which is not suitable for other modes of MS recording, such as GC/MS or direct insertion probe analyses. To minimize this inconvenience, instrument manufacturers have designed plug-in source modules for an easy and rapid exchange.

The salt required for filament-off mode should be added at the HPLC column outlet [89, 93], because it may modify the retention parameters if added to the eluting HPLC mobile phase.

Small-bore columns cannot be connected directly. Attempts to use narrow capillary tubes, with 10-25 µm i.d., as thermospray nebulizers have failed because of rapid plugging. Thus, it becomes necessary to dilute the micro-column effluent and to use standard larger nebulizing tubes.

Thermal vaporization of a constant fraction of the HPLC solvent is required for stable operating conditions, even if this fraction is close to 99% [85, 86]. As for DLI, the liquid flow rate entering the interface must be rigorously constant. Flow pulsations produced by reciprocating piston pumps can pose difficulties. A syringe pump that delivers a pulseless flow of mobile phase avoids this particular problem.

The critical influence of the temperature and the voltage at different locations in the interface and ion source assembly was observed [85, 94]. Small changes in these parameters often produce strong variations in the abundances of the ions and modify the recorded mass spectra. The nature and abundances of cluster ions, multiply-charged ions, cationized molecules are often difficult to predict and make mass assignments difficult if some prior knowledge of the sample is not available before the LC/MS analysis. Optimization of experimental conditions and studies of samples containing unknowns can present difficulties. No round robin test has yet been conducted to evaluate the reproducibility of mass spectra produced by TS instruments, and it remains to be done if, for example, official analytical methods for environmental analyses are to be established.

Compared to other mass spectrometric methods, the background level at the low mass end is rather high. In addition, the method introduces contamination to the ion optic; thus, the ion source block must be cleaned frequently. Fortunately, replacement of the source block has been simplified in most recent instruments.

4.4. RECENT APPLICATIONS.

As for the previous LC/MS methods, application of thermospray during the recent years generally involves determination of biological substances, and a large number of studies describe real case problems and analyses of *a priori* unknown mixtures.

Peptide sequencing has been achieved by combination of sequential enzymic hydrolysis on a column of immobilized enzymes, followed by TS-LC/MS [95-97]. Thermospray LC/MS has also permitted the study of patients suffering carnitine deficiency; carnitine and its derivatives [98], and a drug metabolite, valproylcarnitine, in patient urine [99] were determined at trace levels. In addition, several other examples of analyses of drugs and their metabolites in biological fluids were reported [67, 89, 100]. In many cases, a rapid clean-up extraction from an aqueous solution was followed by TS-LC/MS investigation; the operating protocol could be adapted to aqueous solutions from environmental samples.

Other applications include the investigation of diquaternary ammonium salts [101], and constituents of two synthetic dyes: Basic Red 14 and Basic Orange 21 [102].

### 4.5. FUTURE PROSPECTS

Since the recognition of its nature as a gas-phase ionization process, thermospray seems to have reached a plateau, with little hope to extend its application range to species having high masses (above 5000 u). Nevertheless, the method has proven its capability to produce good mass spectral information from samples that would be difficult to analyse by other mass spectrometric methods.

## 5. Miscellaneous LC/MS coupling methods

None of the previous method is perfect. Work in several laboratories is still under way to find new concepts that would permit improved performance, especially for samples of high mass.

### 5.1. ION EVAPORATION AND ELECTROSPRAY

The two names in fact cover the same physical process. The HPLC effluent is nebulized into a spray of small droplets which are effectively charged (and not "presumably" charged as in thermospray) by direct contact with, or induction from, metallic surfaces held at high voltages (typically 10-50 kV). Stripping the solvent in a nitrogen bath at atmospheric pressure sets free solute ions, which subsequently are drawn into the vacuum of the mass spectrometer through a narrow orifice, and finally mass analysed. Compared to thermospray, nebulization and ionization are achieved at nearly ambient temperature and pressure, and the electric charge carried by the droplets is much higher.

Ion evaporation was the name given by Thomson and co-workers [103], and an instrument was available from Sciex (Canada) since 1980. Dole et al. [104] had called the method electrospray in the late 60s, and the method was reinvestigated 10 years later by Fenn and co-workers [105, 106]. Mass spectra from many nonvolatile substances have been reported, and the authors estimate that from the standpoint of flexibility, convenience, sensitivity, cleanliness of the mass spectrometer, and ease of maintenance, the method may comprise an effective and practical LC/MS interface. Electrospray is described more fully in another chapter of this volume.

### 5.2. MAGIC OR PARTICLE BEAM INTERFACE

Magic is the acronym for Monodisperse Aerosol Generation Interface for Combining liquid chromatography with mass spectrometry [107]. It employs a sophisticated cross-flow pneumatic nebulizer for producing a monodisperse (uniform droplet sizes) aerosol from HPLC solutions. The solvent is removed in a molecular beam separator using a two-stage jet system similar to the old jet separator once used for GC interfacing. The sample forms a cloud of dry particles that are directed towards the ion source. The mist is thermally vaporized and ionized by an electron beam in a classical manner. This elegant method provide EI spectra and detection limits sometime in the nanogram range. It applies, however, only to moderately volatile molecules, and a large part of the sample, up to 70%, can be lost in the molecular separator, being pumped away with the solvent.

### 5.3. PERVAPORATION THROUGH SEMI-PERMEABLE MEMBRANES

In a typical set-up, one side of a methylsilicone membrane is in physical contact with a liquid solution, the other side is under a vacuum and directly connected to the MS source. Many volatile nonpolar solutes in polar solvents (*e.g.*, water) can selectively be dissolved into the apolar polymeric membrane, migrate through it by diffusion, cross the barrier and be mass analyzed. Polar solvents are not soluble into the membrane and cannot cross it. Application to the determination of chloro-alkanes in water was reported [108].

A variation of the method uses a semipermeable silicon capillary tubing and a triple quadrupole MS [109]. In this work, no chromatographic equipment is used. The solution from a flask is continuously circulated through the capillary tube. A number of mono- and diaromatic hydrocarbons gave detection limits in the range $10^{-3}$-$10^{-6}$ M, depending on the sample. The authors suggest that the membrane device be used in conjunction with a robot-controlled organic reactor as an industrial process monitor.

## 6. Conclusion

It should be reemphasized that today there is still no universal LC/MS system. The potential of current methods for organic analyses, however, is high, especially for polar, thermally labile molecules in the molecular weight of range 300-500 u. These methods, especially the thermospray, extend the range of molecules amenable to the combination of an on-line chromatographic separation and a mass spectrometric identification. For marginally volatile molecules present in several samples, for example, river and drinking water, which cannot be analysed by GC, this is a possible alternative. The results obtained in the case of several pesticide investigations are good examples. For these substances, it is possible to reconstruct the chromatographic signal and to produce mass spectral data at trace levels, without any sacrifice of the chromatographic separation.

On the other hand, it must be stated that for higher molecular weight sustances, for example, oligosaccharides, oligopeptides, porphyrins, polyaromatic compounds, etc, with masses above 500 u, the molecular weight information can often be ascertained by using one of the available LC/MS systems, but at the expense of some more or less serious damages to the chromatographic data.

In the LC/MS litterature, there are some cases of misuses and abuses of the term "LC/MS" to cover work reporting only mass spectral data: for example, direct solution introduction of reference molecules with no chromatographic separation. Even in thermospray, nonvolatile samples showing mass spectra may become adsorbed on metallic surfaces and be thermally decomposed. The reconstructed chromatographic profiles then exhibit prominent tailing shapes, and sensitivity is strongly reduced.

The investigation of apolar or weakly polar molecules that have low volatility, (for example, plastic oligomers, polyaromatic mixtures from crude oils and oil shales, etc), in the molecular range 500-2000 u, are difficult to investigate by LC/MS. Thus, they may be more amenable to an SFC separation and a combined SFC/MS identification.

## 7. References

1. Games, D.E., McDowall, M.A., Foster, M.G. and Merez O. (1984) in G. Angeletti and A. Bjørseth (eds), Analysis of organic micropollutants in water, D. Reidel Publishing Co, Dordrecht, p. 68.
2. Arpino, P.J. *(1985) J. Chromatogr.*, **323**, 3.
3. McFadden, W.H., Schwartz, H.L. and Evans, S. (1976) *J. Chromatogr.*, **122**, 389.
4. Millington, D.S., Yorke, D.A. and Burns P. (1980) *Adv. Mass Spectrom.*, **8B**, 1819.
5. Schulten, H.R. (1982) *J. Chromatogr.*, **251**, 105.
6. Schulten, H.R., Bahr, U. and Monkhouse, P.B. (1983) *J. Biochem. Biophys. Methods*, **8**, 239.
7. Brueckner, H. and Przybylski, M. (1984) *J. Chromatogr.*, **296**, 263.
8. Crawthorne, B., Fielding, M., Steel, C.P. and Watts, C.D. (1984) *Environ. Sci. Technol.*, **18**, 797.
9. Watts, C.D., Crawthorne, B., Fielding, M. and Steel, C.P. (1984) in G.Angeletti and A. Bjørseth (eds), Analysis of organic micropollutants in water, D. Reidel Publishing Co, Dordrecht, p. 120.
10. Games, D.E., McDowall, M.A., Levsen, K., Schafer, K.H., Dobberstein, P. and Gower, J.L. (1984) *Biomed. Mass Spectrom.*, **11**, 87.
11. Krost, K.J. (1985) *Anal. Chem.*, **57**, 763.
12. Levsen, K., Wagner-Redeker, W., Schäfer, K.H. and Dobberstein, P. (1985) *J. Chromatogr.*, **323**, 135.
13. Van der Greef, J., Tas, A.C., Ten Noever de Brauw, M.C., Höhn, G., Meijerhoff, G. and Rapp, U. (1985) *J. Chromatogr.*, **323**, 81.
14. Cairns, T., Siegmund, E.G. and Doose, G.M. (1983) *Biomed. Mass Spectrom.*, **10**, 24.
15. Vargo, J.D. and Olson, K. (1985) *Anal. Chem.*, **57**, 672.
16. Yu, T.J., Schwartz, H.A., Cohen, S.A., Vouros, P. and Karger, B.L. (1984) *J. Chromatogr.*, **301**, 425.
17. Yu, T.J., Karger, B.L., Vouros, P. (1983) *Biomed. Mass Spectrom.*, **10**, 633.
18. Jungalwala, F.B., Evans J.E. and McCluer, R.H. (1984) *J. Lipid Res.*, **25**, 738.
19. Jungalwala, F.B., Evans, J.E., Kadowaki,H. and McCluer, R.H.(1984) *J. Lipid Res.*, **25**, 209.
20. Martin, S.A., Zhou, R.H., Games, D.E., Jones, A. and Ramsey, E.D. (1984) *HRC & CC, J. High Resolut. Chromatogr. Chromatogr. Commun.*, **7**, 196.
21. Games, D.E., Alcock, N.J., Van der Greef, J., Nyssen, L.M., Maarse, H. and Ten Noever de Brauw, M.C. (1984) *J. Chromatogr.*, **294**, 269.
22. Fraisse, D., Maquin, F., Stahl, D., Suon, K. and Tabet, J. C. (1984) *Analusis*, **12**, 63.
23. Cairns, T., Siegmund, E.G., Stamp, J. and Skelly, J.P. (1983) *Biomed. Mass Spectrom.*, **10**, 203.

24. Hayes, M.J., Lankmayer, E.P., Vouros, P., Karger, B.L. and McGuire, J.M. (1983) *Anal. Chem.*, **55**, 1745.
25. Hayes, M.J., Schwartz, H.E., Vouros, P., Karger, B.L., Thruston, A.D.Jr. and McGuire, J.M. (1984) *Anal. Chem.*, **56**, 1229.
26. Stroh, J.G., Cook, J.C., Milberg, R.M., Brayton, L., Kihara, T., Huang, Z.,Rinehart K.L.,Jr. and Lewis, I.A.S. (1985) *Anal. Chem.*, **57**, 985.
27. Karger, B.L. and Vouros, P. (1985) *J. Chromatogr.*, **323**, 13
28. Scott, R.P.W. (1985) *J. Chromatogr. Sci.*, **23**, 233.
29. Rabel, F.M. (1985) *J. Chromatogr. Sci.*, **23**, 247.
30. Vouros, P., Lankmayr, E.P., Hayes, M.J., Karger, B.L. and McGuire, J.M. (1982) *J. Chromatogr.*, **251**, 175.
31. Maris, F.A., Geerdink, R.B., Frei, R.W. and Brinkman, U.A.Th. (1985) *J. Chromatogr.*, **323**, 113.
32. Benninghoven, A., Eicke, A., Junack, M., Sichterman, W., Kriezek, J. and Peters, H. (1980) *Org. Mass Spectrom.*, **15**, 459.
33. Smith, R.D., Burger, J.E. and Johnson, A.L. (1981) *Anal. Chem.*, **53**, 1603.
34. Huber, J.F.K., Dzido, T. and Heresch,, F. (1983) *J. Chromatogr.*, **271**, 27.
35. Fan, T.P., Hardin, E.D.and Vestal, M.L. (1984) *Anal. Chem.*, **56**, 1870.
36. Jungclas, H., Danigel, H. and Schmidt, L. (1983) *Int. J. Mass Spectrom. Ion Phys.*, **46**, 197.
37. Dobberstein, P., Korte, E., Meyerhoff, G. and Pesch, R. (1983) *Int. J. Mass Spectrom. Ion Phys.*, **46**, 185.
38. Arpino, P.J. (1983) in T.M. Vickrey (ed), Liquid Chromatography Detectors, Marcel Dekker, New York, N.Y., p. 243.
39. Yinon, J. and Hwang, D.G. (1983) *J. Chromatogr.*, **268**, 45.
40. Chapman, J.R., Hardin, E.H., Evans, S. and Moore, L.E.(1983) *Int. J. Mass Spectrom. Ion Phys.*, **46**, 201.
41. Alborn. H. and Stenhagen, G. (1985) *J. Chromatogr.*, **323**, 47.
42. Bruins, A.P. and Drenth, B.F.H. (1983) *J. Chromatogr.*, **271**, 71.
43. Arpino, P.J. and Beaugrand, C. (1985) *Int. J. Mass Spectrom. Ion Proc.*, **64**, 275.
44. Hirter, P., Walther, H.J. and Dätwyler, P. (1985) *J. Chromatogr.*, **323**, 89.
45. Tiebach, R., Blaas, W., Kellert, M., Steinmeyer, S. and Weber, R. (1985) *J. Chromatogr.*, **318**, 103.
46. Apffel, J.A., Brinkman, U.A.T., Frei, R.W. and Evers, E.A.I.M. (1983) *Anal. Chem.*, **55**, 2280.
47. Yoshida, H., Matsumoto, K., Itoh, K., Tsuge, S., Hirata, Y., Mochizuki, K., Kokobun, N. and Yoshida, Y. (1982) *Fresenius'Z. Anal. Chem.*, **311**, 674.
48. Christensen, R.G., Hertz, H.S., Meiselman, S. and White, E. (1981) *Anal. Chem.*, **53**, 171.
49. Dedieu, M., Juin, C., Arpino, P.J. and Guiochon, G. (1982) *Anal. Chem.*, **54**, 2372.
50. Eckers, C., Skrabalak, D.S. and Henion, J.D. (1982) *Clin. Chem.*, **28**, 1882.
51. Sugnaux, F.R., Skrabalak, D.S. and Henion, J.D. (1983) *J. Chromatogr.*, **264**, 357.
52. Voyksner, R.D., Bursey, J.T. and Pellizzari, E.D. (1984) *Anal. Chem.*, **56**, 1507.
53. Kenyon, C.N. (1983) *Biomed. Mass Spectrom.*, **10**, 535.
54. Mauchamp, B. and Krien, P. (1982) *J. Chromatogr.*, **236**, 17.
55. Arpino. P.J., Bounine, J.P., Dedieu, M. and Guiochon, G. (1983) *J. Chromatogr.*, **271**, 43.
56. Henion, J. (1984) *J. Chromatogr. Libr.*, **28**, 260.
57. Lee, D.E. and Henion, J.D. (1985) *J. Chromatogr. Sci.*, **23**, 253.

58. Tsuge, S. (1985) *J. Chromatogr. Libr.*, **30**, 217.
59. Westwood, S.A. (1984) *Anal. Proc.*, **21**, 418.
60. Henion, J.D. (1985) *J. Chromatographic Libr.*, **323**, 37.
61. Esmans, E.L., Geboes, P., Luyten, Y. and Alderweireldt, F.C. (1985) *Biomed. Mass Spectrom.*, **12**, 241.
62. Tiebach, R., Blaas, W. and Kellert, M. (1985) *J. Chromatogr.*, **323**, 121.
63. Niessen, W.M.A., and Poppe, H. (1985) *J. Chromatogr.*, **323**, 37.
64. Apffel, J., Brinkman, U.A. and Frei, R.W. (1984) *J. Chromatogr.*, **312**, 153.
65. Shalaby, L.M. (1985) *Biomed. Mass Spectrom.*, **12**, 261.
66. Voyksner, R.D., Bursey, J.T. and Pellizari, E.D.(1984) *J. Chromatogr.*, **312**, 221.
67. Voyksner, R.D., Bursey, J.T., Hines, J.W. and Pellizzari, E.D. (1984) *Biomed. Mass Spectrom.*, **11**, 616.
68. Parker, C.E., Yamaguchi, K., Harwan, D.J., Smith, R.W. and Hass, J.R. (1985) *J. Chromatogr.*, **319**, 273.
69. Tiebach, R., Blaas, W., Kellert, M., Steinmeyer, S. and Weber, R. (1985) *J. Chromatogr.*, **318**, 103.
70. Lant, M.S., Martin, L.E. and Oxford, J. (1985) *J. Chromatogr.*, **323**, 143.
71. Shukla, A.K., Schauer, R., Schade, U., Moll, H. and Rietschel, E. T. (1985) *J. Chromatogr.*, **337**, 231.
72. Crowther, J.B., Covey, T.R., Dewey, E.A. and Henion, J.D. (1984) *Anal. Chem.*, **56**, 2921.
73. Eckers, C. and Henion, J.D. (1985) *J. Chromatogr. Sci.*, **32**, 115.
74. Skrabalak, D.S., Covey, T.R. and Henion, J.D. (1984) *J. Chromatogr.*, **315**, 359.
75. Azoulay, H., Desbene, P.L., Frappier, F. and Georges, Y. (1984) *J. Chromatogr.*, **303**, 272.
76. Milon, H. and Bur, H. (1983) *J. Chromatogr.*, **271**, 83.
77. Pind, S., Kuksis, A., Myher, J.J. and Marai, L.(1984) *Can. J. Biochem. Cell Biol.*, **62**, 301.
78. Kuksis, A., Myher, J.J. and Marai, L. (1984) *JAOCS, J. Am. Oil Chem. Soc.*, **61**, 1582.
79. Yinon, J. and Hwang, D.G. (1985) *J. Chromatogr.*, **339**, 127.
80. Quilliam, M.A., Messier, F., D'Agostino, P.A., McCarry, B.E. and Lant, M.S. (1984) *Spectrosc.: Int. J.*, **3**, 33.
81. Crowther, J.B., Covey, T.R., Silvestre, D. and Henion, J.D. (1985) *LC Mag.*, **3**, 240.
82. Vestal, M.L. (1983) *Mass Spectrom. Rev.*, **2**, 447.
83. Vestal, M.L. (1984) *Science*, **226**, 275.
84. Vestal, M.L. (1983) in A. Benninghoven (ed), Ion formation from organic solids, Springer-Verlag, Heidelberg, p. 246.
85. Blakley, C.R. and Vestal, M.L. (1983) *Anal. Chem.*, **55**, 750.
86. Garteiz, D.A. and Vestal, M.L. (1985) *LC Mag.*, **3**, 334.
87. Vestal, M.L. (1984) *Anal. Chem.*, **56**, 2590.
88. Chapman, J.R. (1985) *J. Chromatogr.*, **323**, 153.
89. Catlow, D.A. (1985) *J. Chromatogr.*, **323**, 163.
90. Schmelzeisen-Redeker, G., Giessman, U. and Röllgen, F.W. (1984) *J. Phys. Colloq.*, p. 297.
91. Fenselau, C., Liberato, D.J., Yergey, J.A., Cotter, R.J. and Yergey, A.L. (1984) *Anal. Chem.*, **56**, 2759.
92. Schmelzeisen-Redeker, G., Wong, S.S., Giessman, U. and Röllgen, F.W. (1985) *Z. Naturforsch. A: Phys. Phys. Chem. Kosmophys.*, **40A**, 430.
93. Voyksner, R.D., Bursey, J.T. and Pellizzari, E.D. (1984) *Anal. Chem.*, **56**, 1507.

94. Schmelzeisen-Redeker, G., McDowall, M.A., Giessman, U., Levsen, K. and Röllgen, F.W. (1985) *J. Chromatogr.*, **323**, 127.
95. Kim, H.Y., Pilosof, D., Dyckes, D.F. and Vestal, M.L. (1984) *J. Am. Chem. Soc.*, **106**, 7304.
96. Pilosof, D., Kim, H.Y., Dyckes, D.F. and Vestal, M.L. (1984) *Anal. Chem.*, **56**, 1236.
97. Pilosof, D., Kim, H.Y., Vestal, M.L. and Dyckes, D.F.(1984) *Biomed. Mass Spectrom.*, **11**, 403.
98. Yergey, A.L., Liberato, D.J., and Millington, D.S. (1984) *Anal. Biochem.*, **139**, 278.
99. Millington,D.S., Bohan, T.P., RoeR, C.R. and Yergey, A.L. (1985) *Clin. Chim. Acta*, **145**, 69.
100. Covey, T., Crowther, J.; Dewey, E.A. and Henion, J.D. (1985) *Anal. Chem.*, **57**, 474.
101. Schmelzeisen-Redeker, G., Giessman, U. and Röllgen, F.W. (1984) *Angew. Chem.*, **96**, 889.
102. Betowski, L.D. and Ballard, J.M. (1984) *Anal. Chem.*, **56**, 2604.
103. Thomson, B.A., Iribarne, J.V. and Dziedic, P.J. (1982) *Anal. Chem.*, **54**, 2219.
104. Gieniec, J., Mack, L.L., Nakamae, K., Gupta, C; Kumar, V., Dole, M. (1984) *Mass Spectrom.*, **11**, 259.
105. Yamashita, M. and Fenn, J.B. (1984) *J. Phys. Chem.*, **88**, 4451.
106. Whitehouse, C.M., Dreyer, R.N., Yamashita, M. and Fenn, J.B. (1985) *Anal. Chem.*, **57**, 675.
107. Willoughby, R.C. and Browner, R.F. (1984) *Anal. Chem.*, **56**, 2625.
108. Eustache, H. and Histi, G. (1981) *J. Membrane Sci.*, **8**, 105.
109. Brodbelt, J.S. and Cooks, R.G. (1985) *Anal. Chem.*, **57**, 1153.
110. Bigley, F., Grob, R.L. and Brenner, G. (1984) *J. Chromatogr.*, **288**, 293.

# COMBINED SUPERCRITICAL FLUID CHROMATOGRAPHY/MASS SPECTROMETRY (SFC/MS)

P. J. ARPINO
*Institut National Agronomique*
*Laboratoire de Chimie Analytique*
*16, rue Claude Bernard*
*75231 Paris 05*
*France*

ABSTRACT: Supercritical fluid chromatography coupled to mass spectrometry is an alternative method to LC/MS for the analysis of weakly polar, high boiling solutes. This chapter reviews possible methods permitting on-line SFC/MS coupling.

## 1. Introduction

It is well-known that the on-line coupling of a chromatographic method to mass spectrometry provides a powerful analytical tool for chemists; therefore, it is not surprising that, in the past, the emergence of new chromatographic separation techniques has always been rapidly followed by attempts to develop an efficient coupling method to a mass spectrometer. Such has been the successful case for GC and combined GC/MS during the 1960s and 1970s, then, and with greater difficulties, for LC and combined LC/MS during the next decade. Probably, one of the landmarks of the 1980s is the emergence of chromatographic separation methods utilizing supercritical fluids (SFC). Following the same general trend, the coupling of this last separation technique to mass spectrometry (SFC/MS) is being seriously investigated by several research groups, and commercial instruments are ready to be introduced on the market. This specific interest in SFC/MS accounts for the increasing number of published papers on methods and applications, and several reviews on SFC/MS have already appeared [2, 4, 9, 12, 13, 25, 37, 49].

The present situation of SFC/MS can be seen as the result of two separate approaches followed rather independently by chromatographers and mass spectrometrists. The first group of scientists had first explored GC, then LC, before becoming interested in SFC during recent years. The exponential growth of published articles on SFC, starting in 1982, which followed nearly 20 years of exploratory studies by a few research groups, was delineated recently [38]. Because SFC is a chromatographic technique that can be considered a hybrid of GC and LC, and because the application domains of the two techniques do not overlap extensively, chromatographers are today primarily interested in separation problems not easily amenable to GC or LC methods.

On the other hand, mass spectrometrists first considered sampling supercritical solutions, without prior separations, in an attempt to obtain gas-phase ions of nonvolatile molecules [1, 34, 35]. Consideration of SFC/MS came later from groups who had previously experimented with LC/MS and who felt that SFC/MS could be



*M. L. Gross (ed.), Mass Spectrometry in the Biological Sciences: A Tutorial*, 269–280.

an easier alternative than LC/MS. Most of the pioneering work that was published since 1982 [20-22, 41-50, 53, 54] was achieved by R.D. Smith and his co-workers, and remarkably was started before the establishment of SFC as an accepted chromatographic separation method. Today, mass spectrometrists view the situation of SFC/MS quite differently than do chromatographers. They estimate that, given comparable separation conditions using LC or SFC for nonvolatile or thermally labile molecules, then coupling the mass spectrometer to SFC should be less difficult than coupling to LC.

## 2. Gas flow to the mass spectrometer

When considering the interfacing difficulties for connecting a chromatographic technique to mass spectrometry, the first problem to be addressed is the compatibility between the pumping capacity of the vacuum equipment connected to the mass spectrometer envelope, and the volume flow rate of mobile-phase vapor delivered by the chromatographic column.

The situation occupied by SFC is intermediate between GC and LC, and this is reflected by the fact that either packed columns, similar to those used in LC, or open tubular columns, similar to capillary GC columns, can be equally considered for SFC. The reason for this situation arises because the values of mobile phase viscosities and solute diffusivities for the three techniques are different. Diffusion coefficients directly affect the plate height of the column, as predicted by the Van Deemter equation or the more modern Knox equation for packed columns, and the Taylor-Golay equation for open tubular columns.

Both the low viscosities of gases and the rapid diffusion of solutes in gases combine to favor the use of capillary columns with internal diameters between 0.2 and 0.5 mm for GC. On the other hand, the high liquid viscosities and the low solute diffusivities dictate the choice of short packed columns, filled with 3-10 µm particles, for LC. In SFC, the intermediate values of the two factors allow one to consider either the same packed column used in LC, but operated at higher mobile phase linear velocities (resulting in faster separations), or the GC-type capillary columns with narrower internal diameters (typically 50-100 µm).

The choice of the column type is important in SFC because the chromatographic equipment and the operating parameters for the separation are somewhat different. In addition, the mobile phase flow rate delivered at the column output differs widely, and different approaches must be considered for the coupling to the mass spectrometer.

Mass spectrometers are not all the same type and vary widely in terms of analysers and vacuum pumping equipment. Given the same simplified approach as above, mass spectrometers can be divided into three broad categories.

Group 1 includes old instruments and modern low-cost benchtop instruments designed for use under electron ionization exclusively. They generally are equipped with a single line of vacuum pumps. In the same group are also mass spectrometers that require a very low pressure in the analyser, such as an FTMS cell with an operating pressure 2 or 3 orders of magnitude lower than that in a conventional magnetic or quadrupole analyser (being typically $10^{-7}$ to $10^{-9}$ torr in a FTMS). All these mass spectrometers can accept the direct introduction of 1 mL/min STP of gas or less.

Group 2 includes standard mass spectrometers designed for both electron ionization (EI) and chemical ionization (CI). They generally include two separated vacuum pump lines for differential pumping of the source and the analyser and can accept the direct introduction of *ca.* 10 mL/min of gas into the source housing.

Group 3 includes mass spectrometers equipped with an interface that selectively eliminates part or all of the mobile phase prior to sample introduction into the ion source (*e.g.*, a moving belt, a jet separator). In the same group are also mass spectrometers fitted with an ion source operating under a high flow rate of gas, such as that from the thermospray interface (TS), originally designed for LC/MS coupling with packed columns and liquid flow rates up to 2 mL/min of water. An atmospheric pressure ion source is another example. The gas flow tolerance of instruments in this group can reach 2000 mL/min STP.

## 3. Packed column SFC/MS

It is not within the scope of this text to discuss extensively the relative merits of packed vs capillary columns for SFC. To the mass spectrometrist, packed columns mainly offer the advantage of rapid analyses under easy injection conditions. Single ion monitoring and full scan recording are not limited by the amount of sample tolerated by the column. The disadvantage that the mass spectrometer must absorb (or the interface eliminate) large volumes of mobile phase is counterbalanced by the lack of serious restrictor problems. In addition to the general method of density programming to control solute retention, packed columns are often used under gradient elution, as in LC, by changing the mobile phase composition: the most common mode is to add small amounts (1-5%) of polar modifiers (water, methanol, etc.) to the $CO_2$. This may limit the ionization mode in the mass spectrometer to proton exchange chemical ionization (see below) with no operator control of the reactant ions. Other advantages and disadvantages of packed columns are summarized in table 2 and are based on a more detailed paper written by Schoenmakers [39].

Although packed columns may seem inappropriate to the beginner in SFC/MS, because the volume of the mobile phase is large and because the opinion of some chromatographers is that packed columns have become obsolete, complete utility and specific advantages when linked to a mass spectrometer do exist and have been emphasized recently by Smith [22, 50], on the basis of long experience using packed and capillary columns for SFC/MS.

## 4. Capillary column SFC/MS

A 50-100-µm capillary column can be directly connected to a large number of existing mass spectrometers, thus explaining why they already have been utilized in the majority of the published work on SFC/MS. A 25-µm column is not commercially available and currently represents the ultimate possible miniaturization. Again the more general article of Schoenmakers [39] is recommended for complete details, and only the important features for MS interfacing are emphasized here.

The low flow rate delivered by capillary columns places a minimum load on the vacuum pumping system of the mass spectrometer, but it also negatively affects the dimensions of the necessary restrictor. This tiny part becomes the fragile link in the instrument chain assembled as a SFC/MS/Computer system.

With capillary SFC, the choice of possible mobile phases is more restricted than with packed SFC as polar modifiers do not significantly modify solute retentions; thus, pure $CO_2$ is even more widely used in capillary SFC. Consequently, the analysis time devoted to column equilibration is eliminated, and separation conditions can be optimized prior to a SFC/MS run by a preliminary analysis using a flame ionization detector (FID). This detector does not respond to $CO_2$ but is nearly universal for solutes. The situation is comparable to GC with FID detection for a preliminary screening of the complexity of an unknown mixture, followed by a more complete GC/MS investigation for identification purposes. In addition, pure $CO_2$ under goes ionization to give $CO_2^{+\cdot}$, which is a charge exchange reagent and produces electron ionization-like spectra that are library searchable.

The small volume of stationary phase available in a capillary column is responsible for certain limitations on sample size that affect quantification and detection of trace constituents in complex mixtures. Although reproducible injection techniques are now available, they often require the availability of relatively large amounts of sample beforehand, even if only a small fraction of it is finally transferred into the column.

The most common injection techniques use an LC-type high pressure injection valve with an internal sample loop of less than 200 nL capacity and include:

i) split injection (a flow splitter discards 50-80% of the injected solution)

ii) timed-split injection (the careful timing of the movement of the injection valve controls the transfer of a fraction of the inner loop content into the column)

iii) direct injection into a "retention gap" (a length of bare capillary tubing, installed between the injection valve and the analytical column, that causes solute focusing).

## 5. Flow restrictors

The primary purpose of the restrictor is to maintain a high mobile phase density along the analytical column. For capillary column SFC/MS, a 5-20 m capillary column with 50-100 μm internal diameter is connected, by means of a zero-dead volume union, to a restrictor metering the flow rate of mobile phase tolerable by the mass spectrometer. The head fluid pressure at the column entrance is generally increased from 100 to *ca.* 400 bars during a separation, and the pressure drop along the capillary column is only 3-4% (the flow regime into the column is quasi linear and the mobile phase velocity is constant to a good approximation). The fluid is nearly entirely decompressed during its passage through the restrictor. As this dramatically reduces the fluid density, the risk exists that solutes precipitate within the restrictor if the pressure barrier is too wide.

Because the restrictor is important for SFC/MS coupling, dimensions, operating parameters, transmission efficiency for nonvolatile solutes of different flow restrictor types were carefully studied by several groups [3, 8, 14, 47].

If total plugging is the ultimate risk, partial erratic plugging is more frequent and is also a serious nuisance because it reduces the mobile phase velocity (thus affecting both the retention parameters and the column plate height) and also reduces the gas pressure in the ion source. The drop in pressure in turn causes a change in the sample current produced upon ionization. Partial plugging also gives rise to flow instability because an increase in density results from partial plugging, increasing the fluid solvating power that may unplug the restrictor for a short time, and so on.

Solute precipitates due to a poorly controlled fluid decompression are responsible for spikes seen on the total ion current traces, especially on top of large chromatographic peaks, producing "christmas-tree" shaped peaks, and at the end of the chromatographic separation when nonvolatile substances elutes (baseline spikes). From this point of view, a burning flame ionization detector is more tolerant to microcrystal precipitates than is a mass spectrometer that requires a gas-phase ionization process to produce a signal. Thus, FID traces corresponding to the elution of molecules with molecular weight up to 9000 Da have been obtained, whereas, to date, only a limited number of SFC/MS studies of high mass samples, such as persilylated sugars having MWs up to 6000 Da [33, 36], have been reported. This may be a temporary situation, however, owing to the infrequent availability of suitable high mass instruments.

Different restrictor types were described, and their utility results from several contradictory factors including availability or ease of making suitable dimensions, transmission efficiency for nonvolatile solutes, long term stability, ease of replacement, and cost. Many of these factors had been investigated previously for LC/MS coupling with Direct Liquid Introduction (DLI), and severely discouraging plugging problems were frequently encountered. The situation is more favorable in SFC/MS because most plugging problems in DLI LC/MS are caused by silica from the chromatographic packing that was dissolved by aqueous mobile phases. SFC columns either have no packing material or are percolated by a fluid that does not dissolve silica.

The restrictor is generally inserted into the mass spectrometer through a transfer line either utilizing the GC interface or the solid probe inlet. A removable assembly that does not require venting the mass spectrometer is advantageous for rapid interchange in case of plugging.

### 5.1. LINEAR RESTRICTORS

Linear restrictors, consisting of a length of straight capillary tubing of uniform diameter and generally made of fused silica with internal diameter of 5-25 μm, are cheap and readily available. They are often quite satisfactory for samples with molecular weights less than 1000 Da, and mass spectrometers with limited mass ranges do not need more complex and fragile restrictor types. They have found a wide use in a number of laboratory for preliminary SFC/MS studies [15-17, 55].

### 5.2. THIN METAL DIAPHRAGMS

In theory, thin metal diaphragms should provide optimum results for transmitting nonvolatile molecules, although the pinhole diameter must be less than 2 μm diameter if placed at the end of a 50-100 μm capillary column. Larger diameter are only permitted with columns delivering higher flow rates of mobile phase (*e.g.*, packed columns). Frequent plugging, large dead volumes, and alignment difficulties (due to the difficulty to center a 2 μm pinhole in front of a 50 μm i.d. capillary tube), and pressure induced mechanical distortion causing bulging of the thin metal disc were reported. Smith et al. attempted to use diaphragms in early work; Crowther and Henion [11] used unmodified DLI probes originally designed for LC/MS and fitted with a 3 μm metal diaphragm in conjunction with packed columns. Recently, a new set-up that eliminates the alignment problem and is capable of transmitting solutes with molecular weights greater than 6000 Da was described [36].

### 5.3. FLAME-PULLED, FLAME-POLISHED, AND MANUALLY ERODED FUSED SILICA TUBINGS

These tubes constitute a group of tapered restrictors that work well for most difficult samples [8, 14] although some of them are mechanically fragile. Because they are not commercially available, they must be prepared in the laboratory, which can be a tedious task until skill and practice are acquired.

### 5.4. FRIT RESTRICTORS

Frit (or "ceramic") restrictors consist of an array of submicron chanels that provide multiple paths for the solute; these restrictors perform well for the transmission of many samples. The frit restrictor is commercially available and is relatively rugged with minimal plugging problems. The frit must be heated up to a rather high temperature (400-600 °C) to provide enough heat to the expanding gas [6, 7]: poor thermal conductivity of fused silica and high gas speed through the restrictor are responsible for this required temperature. The fluid itself is estimated to reach a much lower temperature, below 200 °C; therefore, thermally labile solutes are not degraded. Nevertheless, if the multiple paths offered to solutes tend to limit the possibility of a complete blockage, truly nonvolatile materials are unlikely to be transmitted through this restrictor.

Several other restrictor types were described, although they are used less frequently, and were reviewed elsewhere [49].

Common to all restrictors is the need to heat the supercritical fluid entering the restrictor, to compensate for the cooling resulting from fluid expansion; the Joule-Thomson coefficient of $CO_2$ is positive in the interval covered by SFC applications. Heating is also required for the transfer of less volatile solutes; thus, the restrictor housing is usually kept at 300-400 °C, or even higher in the case of frit restrictors. Heating the fluid in front of the restrictor also affects the mass flow rate of supercritical fluid through the restrictor and provides a limited degree of control of this parameter [47, 51]. Adjusting the input of mobile phase into the mass spectrometer, however, is generally done in rather an inelegant manner by changing the restrictor itself, using a variable length of tube in the case of a linear restrictor, or a different aperture size for tapered restrictors.

Because pressure or density programming produce variable output of mobile phase to the mass spectrometer, some kind of variable restrictor would be quite useful. Unfortunately, such a device is difficult to conceive; the only known system for mass spectrometry interfacing is the pulsed nozzle that was used in conjunction with laser multiphoton ionization and a time of flight mass spectrometer [23, 40]; its pressure operating range, however, is limited to 150 bars, and the coupling to SFC was not demonstrated.

## 6. Ionization modes

The molecular $CO_2^{+\cdot}$ ion is the major ion obtained upon primary ionization, caused either by an electron beam from a heated filament or an high voltage DC discharge of $CO_2$ if used as a pure supercritical mobile phase. When primary ionization is made within too short a distance from the restrictor, before the Mach disk, high abundances of ion clusters $(CO_2)_nCO_2^{+\cdot}$ are produced and significantly contribute to the total ion current. Series of clusters with n up to 7 were observed [17], the abundances are a function of ion source pressure, temperature, and restrictor to electron beam distance. If enough distance before entering the ion source is provided to the expanding jet, only the dimer at *m/z* 88 is of noticeable abundance and amounts to *ca.* 2-5% of that for the *m/z* 44 ion.

Because of the high recombination energy of $CO_2^{+\cdot}$ ions with electrons, 13.8 eV, charge exchange with most solutes occurs and produces odd-electron molecular ions,

$M^{+\cdot}$, with comparable, although somewhat lower, internal energies as those resulting from conventional electron ionization. The resulting mass spectra usually compare well to those in standard MS data bases.

Nevertheless, high $CO_2^{+\cdot}$ abundances must be reduced in the case of mass spectrometers that store ions for extended periods of time, such as an ion trap detector (ITD), or a Fourier transform mass spectrometer (FTMS). Although direct coupling of SFC to either an ITD [51], or an FTMS [26, 28] were reported, selective ejection of $CO_2$-cluster ions is required before scanning mass spectra. For conventional quadrupole or sector mass spectrometers, a high background due to these mobile-phase-derived ions can significantly decrease the sensitivity [55].

When polar modifiers are added to $CO_2$, or when a proton-containing gas is used as the supercritical mobile phase (*e.g.*, pentane or ammonia), solvent-mediated CI spectra are produced. For example, direct introduction of a supercritical mobile phase containing $CO_2$ with 10 % methanol, and submitted to an ionizing electron beam [6], predominantly produces methanol-derived reactant ions, $(CH_3OH)_nH^+$ with n= 2, 3. A moving belt (see below) is the only interface that may completely remove these proton donor fluids prior to MS introduction. For the other cases, part or all of the column eluent enters the ion source and induces proton exchange CI reactions. Adding a strong gas-phase base into the source (*e.g.*, $NH_3$), can level CI reactions to constant conditions for varying levels of polar modifiers. Competition between charge exchange and proton exchange reactions may result because ammonia is usually bled directly at a constant rate into the ion source, whereas the partial pressure of $CO_2$ continuously increases under pressure programmed SFC [10]. It remains that SFC in combination with chemical ionization has permitted one to record remarkably good spectra of series of nonvolatile molecules with molecular weights in excess of 4000 Da [33, 36].

## 7. Mobile phase separators

If interfacing SFC and MS is more difficult compared to GC and MS, because larger gas volumes must be discarded, the more restricted choice of supercritical mobile phases renders the coupling easier than in LC/MS. No new physical principle, however, has been found to lead to a specific mobile phase separator; the basic devices that in the past had been used for GC/MS or LC/MS interfacing have been tailored to SFC/MS requirements.

### 7.1. THE MOVING-BELT

This device was first used by Games and co-workers [5] and is basically the same one as originally designed for LC/MS. A simple metal tube, crimped at one end and directly heated by a low voltage current supply, and placed at the SFC column outlet, is the only required modification. The system is fully compatible with packed columns eluted with high flow rates of supercritical mobile phase (up to 4 mL/min of the equivalent liquid eluent), and containing high percentages of polar modifiers. Because all of the mobile phase can be eliminated prior to solute transfer into the mass spectrometer, any conventional ionization mode (EI, CI using a selected reagent gas) is left to the operator's convenience. The only serious limit comes from the belt itself, because it is now well-known that belts are inappropriate in the the case of truly nonvolatile or thermally labile molecules. Given this upper limit in compound volatility, the moving belt performs well, and at least two groups [5, 24] reported interesting results for a large set of natural product mixtures.

### 7.2. JET SEPARATORS

Skimming part of the expanding jet at the restrictor outlet is the basis for several interfacing devices that have been proposed for SFC/MS. They can be related to the many models of jet separators frequently used in GC/MS or LC/MS.

Early attempts to eliminate completely the mobile phase in a complex two-stage system were not convincing owing to poor sensitivity and lack of transmission of nonvolatile molecules [34, 35].

Although it was not combined with SFC instrumentation, the mass spectrometric analyser for individual aerosol particles built by Allen and Gould [1] could be adapted to SFC/MS operations. A system of aerodynamic nozzles and differential pumping can selectively separate the particulate matter from carrier gases. The collimated beam of solid particles is transported into a high vacuum region, flash vaporized by impacting on a hot filament, ionized by electron impact and finally analysed by a quadrupole mass spectrometer. The instrument was tested with several inorganic and organic aerosols.

Matsumoto et al. [29, 30] have adapted a vacuum nebulizing system, previously used for LCMS, that combines both the addition of helium for nebulizing the chromatographic column effluent and one stage of pumping by a rotary pump to remove excess of vapor prior to introduction into the mass spectrometer. Coupling to micropacked columns eluted with various supercritical mobile phases, including ethanol/hexane mixtures, was applied to the SFC/MS investigation of mixtures of fat-soluble and water-soluble vitamins [30]. Ethanol-mediated CI spectra were recorded as a consequence of the presence of large volumes of this solvent in the mobile phase. Another SFC/MS application by the same group was the analysis of nonionic surface active agents [31].

One stage of a mechanically pumped expansion region and a flame-pulled restrictor were adopted by R.D. Smith and co-workers [22, 50] for the latest version of their SFC/MS coupling system, referred to as "the High Flow Rate (HFR) interface". It operates under conditions of pressure programmed separations with narrow bore (1 mm i.d.) packed HPLC-type columns and variable solvating power of supercritical fluids by using increasing concentrations of polar fluid modifiers. Nonionic surfactant polymers, some herbicides, and thermally labile, high molecular weight, organophosphorus insecticides were determined, under ammonia-CI conditions; detection limits in the part-per-billion range were reported.

## 8. Direct coupling

Mass spectrometers with limited vacuum pumping capacity that only accept a small amount of mobile phase, less than 1 mL/min STP of $CO_2$, including simple benchtop machines [27, 51] or complex FTMS with a differentially pumped dual cell [26, 28], have been directly interfaced to capillary column SFC using restrictors of very low permeability. Because of the restrictor limitations, however, excessively long analysis time for moderately volatile samples and frequent plugging were often encountered. In the case of the FTMS experiments, residual $CO_2$ molecules in the analyser cell limited the mass resolution to *ca.* 6000 [26]. Although the coupling to these different mass spectrometers was reported to be possible, the feasibility for routine use in a control laboratory was not fully established.

Direct coupling of capillary SFC to differentially pumped conventional EI/CI instruments, with either magnetic sector [18] or quadrupole analysers as in most reports, is easier to achieve and was successfully realized by several groups, using laboratory-built interfaces.

Various designs utilizing different restrictor types were described by R.D. Smith and co-workers [20-22, 41-46, 53, 54] who applied them to the determination of several types of compounds, including polynuclear aromatic hydrocarbons from petroleum-derived samples [53], and trichotecene mycotoxins [46]. A heated integral frit restrictor interface was adopted by Games et al. [7] and used for the analysis of high boiling point alkanes.

Maybe the easiest way to interface SFC to a mass spectrometer is by using a standard LC/MS thermospray ion source (TS), originally conceived by Vestal [52], and equipped with both an electron beam ionizer and a discharge ionization facility. A TS source can accept the full effluent, up to 4 mL/min of equivalent liquid mobile phase, from large bore packed columns and allow the use of large diameter restrictors that do not plug easily. In theory, rapid sample transit through the source should reduce the sensitivity, but low detection limits for some model molecules have been reported. This approach was found to be simple to construct [7] and useful for the study of a wide range of compound types, including drugs, sulphonamides, pesticides, indole alkaloids, PTH amino-acids [7] and insect ecdysteroids [32]. Use of a repeller voltage allowed stable ions from CI reactions to be fragmented, leading to more significant structural information. Collision induced dissociation of selected precursor ions, using a tandem quadrupole instrument, was also reported [19]. It is probable that existing TS sources designed for LC/MS operations could be better optimized to match the less stringent requirements of SFC/MS coupling (*e.g.*, the fixed position of the source off-axis electron beam or the discharge ionizer may not be optimized).

## 9. Conclusion

Until recently, the major limitation to the development of SFC/MS was the chromatographic part of the combined system, owing to the frequent need to custom design the chromatographic apparatus. This is no more valid today because satisfactory commercial equipment is now available from several manufacturers. The separating conditions of SFC are better-understood and efficient and stable chromatographic columns can be obtained.

The coupling to the mass spectrometer is relatively easy, although the choice of interface depends on the sample structure to be investigated: for compounds up to 1000 Da, only minor modifications to existing EI/CI instruments are required, and useful work can be obtained by using instruments that are available in many laboratories. On the other hand, improved designs remain to be found for the investigation of high molecular weight, nonvolatile compounds, based on the encouraging preliminary results that have been reported recently.

The high background, affecting the sensitivity is due to the high mobile phase pressure into the ion source and is still a problem in many systems. Lowering the amount of fluid by reducing the SFC column diameter is not totally satisfactory as it complicates injection of complex mixtures and detection of trace substances. Currently, the solution to this problem is being investigated following two different directions:

i) by selectively eliminating part of the mobile phase prior to introduction into the MS ion source [22, 50], with the risk of discarding a sample fraction.

ii) by introducing the full column eluent into a high flow-through MS source, such as a TS source [6], with the risk of a low sensitivity due to the short sample residence time in the ionizer region.

These two approaches are fundamentally different and, for the present time, not enough data are available to make a clear decision, although the second approach is more simple and can already be adapted to many existing mass spectrometers. In addition, work should be made to investigate the possibilities of other interface designs, such as the atmospheric pressure ion source.

Based on reported results and on rapid progress accomplished in a short period of time, the future of SFC/MS is particularly promising. Because, from the start until the end, SFC/MS operates entirely with gas phase materials, with sometimes comparable or better results than when using LC/MS or a high energy process desorption from solid surfaces, this new mass spectrometric technique will probably contribute to elucidating some of the fundamental ionizing mechanisms of large biological molecules.

## 10. References


1. Allen, J. and Gould, R.K. (1981) *Rev. Sci. Instrum.*, **52**, 804.
2. Arpino, P.J., Cousin, J. and Higgins, J. (1987) *TrAC, Trends. Anal. Chem.*, **6**, 69.
3. Bally, R.W. and Cramers, C.A. (1986) *HRC&CC J. High. Resolut. Chromatogr. Chromatogr. Commun.*, **9**, 626.
4. Berry, A.J., Games, D.E. and Perkins, J.R. (1986) *Anal. Proc. (London)*, **23**, 451.
5. Berry, A. J., Games, D. E. and Perkins, J. R. (1986) *J. Chromatogr.*, **363**, 147.
6. Berry, A. J., Games, D. E., Mylchreest, I. C., Perkins, J. R. and Pleasance, S. (1988) *Biomed. Environ. Mass Spectrom.*, **15**, 105.
7. Berry, A. J., Games, D. E., Mylchreest, I. C., Perkins, J. R. and Pleasance, S. (1988) *HRC&CC J. High Resolut. Resolut. Chromatogr. Chromatogr. Commun.*, **11**, 61.
8. Chester, T.L., Innis, D.P. and Owens, G.D. (1985) *Anal. Chem.*, **57**, 2243
9. Cousin, J, and Arpino, P.J. (1986) *Analusis*, **14**, 215.
10. Cousin, J and Arpino, P.J. (1987) *J. Chromatogr.*, **398**, 125.
11. Crowther, J.B. and Henion, J.D. (1985) *Anal. Chem.*, **57**, 2711.
12. Games, D.E., Berry, A.J., Mylchreest, I.C., Perkins, J.R. and Pleasance, S. (1987) *Anal. Proc. (London)*, **24**, 371.
13. Games, D.E., Berry, A.J., Mylchreest, I.C., Perkins, J.R. and Pleasance, S. (1987) *Lab. Pract.*, **36**, 45.
14. Guthrie, E.J. and Schwartz, H.E. (1986) *J. Chromatogr. Sci.*, **24**, 236.
15. Hawthorne, S.B. and Miller, D.J. (1987) *J. Chromatogr.*, **388**, 397.
16. Holzer, G., Deluca, S. and Voorhees, K.J. (1985) *Proc. Int. Symp. Capillary. Chromatogr. 6th*, Huethig, Heidelberg, FRG, p. 919.
17. Holzer, G., Deluca, S. and Voorhees, K.J. (1985) *HRC&CC J. High. Resolut. Chromatogr. Chromatogr. Commun.*, **8**, 528.
18. Huang, E.C., Jackson, B.J., Markides, K.E. and Lee, M.L. (1988) *Chromatographia*, **25**, 51.
19. Hurst, J. (1988) *TS Reports, Finnigan Mat application report*, San Jose, CA, USA, **2**, 1.
20. Kalinoski, H.T., Udseth, H.R., Wright, B.W. and Smith, R.D. (1987) *J. Chromatogr.*, **400**, 307.

21. Kalinoski, H.T., Udseth, H.R., Chess, E.K. and Smith, R.D. (1987) *J. Chromatogr.*, **394**, 3.
22. Kalinoski, H.T. and Smith, R.D. (1988) *Anal. Chem.*, **60**, 529.
23. Köster, C., Grotemeyer, J., Rohwer, E.R., Lindner, J. and Schlag, E.W. (1989) *Advances Mass Spectrom.* **11B**, 1192.
24. Lane, S.J. (1988) In R.M. Smith (ed), Super Critical Fluid Chromatography, Royal Society of Chemistry, London, p. 175.
25. Later, D.W., Bornhop, D.J., Lee, E.D., Henion, J.D. and Wieboldt, R.C. (1987) *LC-GC.*, **5**, 804.
26. Laude, D.A.J., Pentoney, S.L., Griffiths, P.R. and Wilkins C.L. (1987) *Anal. Chem.*, **59**, 2283.
27. Lee, E.D. and Henion, J.D. (1986) *HRC&CC J. High. Resolut. Chromatogr. Chromatogr. Commun.*, **9**, 172.
28. Lee, E.D., Henion, J.D., Cody, R.B. and Kinsinger, J.A. (1987) *Anal. Chem.*, **59**, 1309.
29. Matsumoto, K., Tsuge, S. and Hirata, Y. (1986) *Chromatographia*, **21**, 617.
30. Matsumoto, K., Tsuge, S. and Hirata, Y. (1986) *Anal. Sci.*, **2**, 3.
31. Matsumoto, K., Tsuge, S., and Hirata, Y. (1987) *Bunseki*, **35**, 15.
32. Morgan, E. D., Murphy, S. J., Games, D.E., and Mylchreest, I. C. (1988) *J. Chromatogr.*, **41**, 165.
33. Pinkston, J.D., Owens, G.D., Burkes, L.J., Delaney, T.E., Millington, D.S. and Maltby, D.A. (1988) *Anal Chem.*, **60**, 962.
34. Randall, L.G. and Wahrhaftig. A.L. (1978) *Anal. Chem.*, **50**, 1705.
35. Randall, L.G. and Wahrhaftig. A.L. (1981) *Rev. Sci. Instrum.*, **52**, 1283.
36. Reinhold, V.N., Sheeley, D.M., Kuei, J.C.H., and Her, G.R. (1989) *Advances Mass Spectrom.*, **11B**, 1196.
37. Rose, M.E. (1987) *Mass. Spectrom.*, **9**, 264.
38. Sanagi, M.M. and Smith, R.M. In R.M. Smith (ed), Super Critical Fluid Chromatography, Royal Society of Chemistry, London, p. 29.
39. Schoenmakers, P.J. In R.M. Smith (ed), Super Critical Fluid Chromatography, Royal Society of Chemistry, London, p. 102.
40. Sin, C.H., Pang, H.M., Lubman, D.M. and Zorn, J. (1986) *Anal. Chem.*, **58**, 490.
41. Smith, R.D., Fjeldsted, J.C. and Lee, M.L. (1982) *J. Chromatogr.*, **247**, 231.
42. Smith, R.D., Felix, W.D., Fjeldsted, J.C. and Lee, M.L. (1982) *Anal. Chem.*, **54**, 1883.
43. Smith, R.D., Fjeldsted, J. and Lee, M.L. (1983) *Int. J. Mass. Spectrom. Ion. Phys.*, **46**, 217.
44. Smith, R.D., Udseth, H.R. and Kalinoski, H.T. (1984) *Anal. Chem.*, **56**, 2971.
45. Smith, R.D., Kalinoski, H.T., Udseth, H.R. and Wright, B.W. (1984) *Anal. Chem.*, **56**, 2476.
46. Smith, R.D., Udseth, H.R. and Wright, B.W. (1985) *J. Chromatogr. Sci.*, **23**, 192.
47. Smith, R.D., Fulton, J.L., Petersen, R.C., Kopriva, A.J. and Wright, B.W. (1986) *Anal. Chem.*, **58**, 2057.
48. Smith, R.D., Wright, B.W. and Udseth, H.R. (1986) *ACS. Symp. Ser.*, **297**, 260.
49. Smith, R.D., Kalinoski, H.T. and Udseth, H.R. (1987) *Mass. Spectrom. Rev.*, **6**, 445.
50. Smith, R.D. and Udseth, H.R. (1987) *Anal. Chem.*, **59**, 13.
51. Todd, J.F.J., Mylchreest, I. C., Berry, A. J., Games, D. E. and Smith, R.D. (1988) *Rapid Commun. Mass Spectrom.*, **2**, 55.
52. Vestal, M.L. and Fergusson, G.J. (1985) *Anal. Chem.*, **57**, 2373.

53. Wright, B.W., Udseth, H.R., Smith, R.D. and Hazlett, R.N. (1984) *J. Chromatogr.*, **314**, 253.
54. Wright, B.W., Kalinoski, H.T., Udseth, H.R. and Smith, R.D. (1986) *HRC&CC J. High. Resolut. Chromatogr. Chromatogr. Commun.*, **9**, 145.
55. Zaugg, S.D., Deluca, S.J., Holzer, G.U. and Voorhees, K.J. (1987) *HRC&CC J. High. Resolut. Chromatogr. Chromatogr. Commun.*, **10**, 100.

## DETERMINATION OF BIOMOLECULES BY USING LIQUID CHROMATOGRAPHY AND THERMOSPRAY MASS SPECTROMETRY: FROM PENICILLINS TO PEPTIDES

SOMCHAI SUWANRUMPHA,[1a] MATTHEW A. MCLEAN,[1b] SAUL W. FINK,[1c] CHERYL WILDER,[1d] KRISTINA STACHOWIAK,[1e] DOUGLAS F. DYCKES,[1f] and ROYAL B. FREAS
*Department of Chemistry*
*University of Houston*
*Houston, Texas 77204-5641*

ABSTRACT. One of the major advantages of using mass spectrometry to analyze biomolecules is the structural information obtained from characteristic fragment ions, in addition to molecular weight information. This structural information facilitates the rapid determination of known compounds and the structural elucidation of unknown compounds. such as previously undetermined metabolites. The coupling of mass spectrometry to other separation techniques, such as liquid chromatography or tandem mass spectrometry, provides an additional dimension to the analysis. In this work, we review our combination of liquid chromatography and thermospray ionization mass spectrometry to (I) the analysis of a class of antibiotics, the penicillins and their metabolites, and (II) the analysis of peptides by using an on-line enzymatic digestion.

### 1. Introduction

Mass spectrometry (MS) has long been recognized as a powerful detection technique for chromatography. The combination of mass spectrometry and gas or liquid chromatography provides a unique capability to separate and identify a component in a complex mixture. The on-line coupling of gas chromatography (GC) and mass spectrometry has successfully provided qualitative and quantitative data for biomedical applications [2]. Analytical techniques such as GC/MS, however, often require derivatization to increase the volatility and stability of the analyte. In comparison, the on-line coupling between liquid chromatography (LC) and mass spectrometry is often difficult to achieve because the basic operating conditions for conventional liquid chromatography and for mass spectrometry are quite different [3]. Liquid chromatography normally requires conditions that include liquid-phase operation, high pressure (between 10 to 300 bar), high flow rate (between 0.5 to 2 mL/min), and relatively low temperature (normally at room temperature). The optimal conditions for mass spectrometry require gas-phase operation, high vacuum (between $10^{-4}$ to $10^{-6}$ mbar), low flow rate, and higher temperatures (100-300 °C). To overcome this incompatibility, several research groups have developed different types of interfaces to combine liquid chromatography and mass spectrometry (LC/MS) [4, 5, 6]. At the present time, these include moving belt or transport LC/MS [7, 8], direct liquid introduction (DLI) [9, 10], thermospray [11, 12], a heated nebulizer interface with atmospheric-pressure chemical ionization [13], electrospray [14, 15], a monodisperse aerosol generation interface [16], and liquid chromatography/fast-atom bombardment [17, 18]. These developments are discussed more fully in other chapters of this volume. They make feasible direct on-line separation and identification of nonvolatile and



*M. L. Gross (ed.), Mass Spectrometry in the Biological Sciences: A Tutorial*, 281–301.

thermally labile components without prior derivatization.

### 1.1. THERMOSPRAY LIQUID CHROMATOGRAPHY/MASS SPECTROMETRY

The thermospray interface [19, 20] has proven to be useful for the coupling of liquid chromatography with mass spectrometry [21]. The utility of thermospray MS as an analytical tool has been demonstrated by increasing usage in the fields of environmental analysis and biomedical research [22, 23].

Thermospray is defined as "... the production of supersonic jet vapor with entrained particles as droplets by application of sufficient heat to a capillary to effect controlled partial vaporization of a liquid as its passes through the capillary" [3]. The liquid from a liquid chromatographic column is vaporized by direct electrical resistive heating of a capillary tube. This direct electrical heating can result in stable and reproducible conditions for the vaporization of solvent and solutes as well as a source of ionization for reproducible mass spectral data.

The entire mass flow of a conventional liquid chromatograph (0.5 to 2 mL/min) enters the thermospray interface in the mass spectrometer. This is accomplished by additional vacuum pumping of the ion source and by the controlled partial vaporization of the effluent as it traverses the conductively heated capillary tube. Free expansion of the heated effluent creates a supersonic jet, which contains particles and liquid droplets. Heat is supplied to the ion source to compensate for cooling during the free jet expansions and to allow for continued vaporization of the solvent.

The most commonly used mode of ionization is thermospray ionization (TI). No supplemental source of electrons is needed in this mode. Instead, a volatile buffer is added to the mobile phase. The volatile buffer acts as a source of charge in the droplets that emerge from the heated thermospray vaporizer. The partial vaporization of the solvent causes charge to concentrate in the unevaporated solvent droplets. Coulombic repulsive forces become greater than cohesive forces and the droplets break apart (Rayleigh instability). Evaporation of solvent molecules from these smaller droplets continues due to heat transferred from the walls of the source block. A high electric field is created on the droplets as the volumes of the droplets decrease. Ion evaporation starts to occur when the field strength on the droplets exceeds about $10^9$ V/m [24]. At this point, ions are ejected directly from the condensed phase into the gas phase. Hence, intact molecular ions of nonvolatile and thermally labile compounds can be obtained.

Although ion evaporation predominates for compounds that are multiply charged in solution, other compounds were shown to undergo ionization by gas-phase ion/molecule reactions in the thermospray ion source [25, 26, 27, 28]. Dual-beam thermospray experiments have shown that a number of compounds are ionized via chemical ionization (CI) [25]. In this case, neutral molecules are evaporated from thermospray droplets and are ionized via collisions with solvent and buffer ions in the gas phase. Compounds that undergo ionization via a CI-type mechanism will show a direct correlation between analyte volatility and thermospray sensitivity. For many compounds this appears to be the case and good sensitivity is achieved with compounds having some volatility. Compounds that exhibit poor sensitivity with thermospray are often compounds that are extremely nonvolatile and exist as pre-formed ions. A review and applications of thermospray mass spectrometry are covered in a previous chapter.

Thermospray ionization has become a practical method for routine LC/MS operation. Many applications of thermospray LC/MS and thermospray LC/MS/MS have appeared in the literature in recent years. Environmental applications using thermospray ionization were reported in the identification of dyes [29, 30] and quantification of herbicides and pesticides [31, 32]. Biomedical and pharmaceutical applications of LC/MS have been reported in several major areas. A major application of LC/MS is in the identification and quantitative analysis of drugs and their metabolites in biological materials. Another major application is in the identification of by-products and degradation products in bulk drugs and in dosage forms. Thermospray LC/MS was used to confirm the identity of several anticancer drugs such as triazinone, tetrahydrouridine, azolastone and sulfamic acid derivatives [33]. Confirmation of these drugs was unable to be obtained by using a UV or refractive index detector for HPLC analysis. Ribose, deoxyribose, and deoxytetrahydrouridine were identified as impurities by analyzing thermospray mass spectra. The detection limit for identification of these impurities was 50 ng. These impurities were by-products in the synthesis of the parent drugs. Thermospray ionization in conjunction with liquid chromatography was used to analyze the structures of pyocyanin and 1-hydroxyphenazine [34] The molecular weight of pyocyanin was determined with thermospray LC/MS and it was confirmed with desorption ion-impact mass spectrometry. Mass spectrometric data of these compounds could not be obtained with fast-atom bombardment or desorption chemical ionization. Because of the thermal lability of pyocyanin, this compound could not be analyzed by using GC/MS without prior derivatization. Chen et al. [35] used LC/MS and LC/MS/MS with a thermospray interface in analysis of degradation products of piroximone, a cardiotonic agent. The identification of piroximone degradation products were determined by comparison of retention times and collisionally activated dissociation spectra with known compounds. Liquid chromatography with a chiral stationary phase and mass spectrometry with thermospray interface was used to detect and confirm the identity of enantiomeric contaminants in (S)-methamphetamine [36] and to detect terbutaline enantiomers in human plasma [37].

### 1.2. Penicillins and Their Metabolites

Penicillins are a group of β-lactam antibiotics. They are widely used because they have broad antimicrobial activity against both gram-positive and gram-negative micro-organisms and low toxicity. The antimicrobial activity of penicillins is inhibition of the synthesis of the bacterial cell wall. The structure of the penicillins consist of bulky side chains attached to 6-aminopenicillanic acid. The structures of several penicillins are shown in Figure 1.

Penicillins can be hydrolyzed in the body by the enzyme penicillinase to form penicilloic acid, a major metabolite (structure shown in Figure 2). Analyses of penicillins and their metabolites were extensively reviewed [38, 39]. The techniques include microbiological, spectrophotometric, colorimetric, iodometric titration and spectrofluorometric methods. These techniques, however, often suffer from poor precision, specificity, sensitivity, and interference. For example, a poor correlation exists between the microbiological assay vs. the fluorometric assay for ampicillin [40]. The fluorometric assay is non-specific and involved chemical oxidation of active parent drugs as well as the inactive metabolites. Thus, the concentration of ampicillin in plasma determined by fluorometric methods yielded higher values than by microbial assay, which is specific only for the active parent drug.

In recent years, HPLC methods have been reported for the analysis of penicillins. Most of the chromatographic procedures were run by using a reverse-phase column. The types of samples were pharmaceutical preparations, urine, bile, lymph, plasma or blood. Protein in biological samples was removed by precipitation or ultrafiltration subsequently followed by organic phase extraction or liquid/solid cartridges extraction. UV adsorption and fluorescence were normally used for detection. These HPLC procedures were reviewed by Miners [38] and Rouan [39].

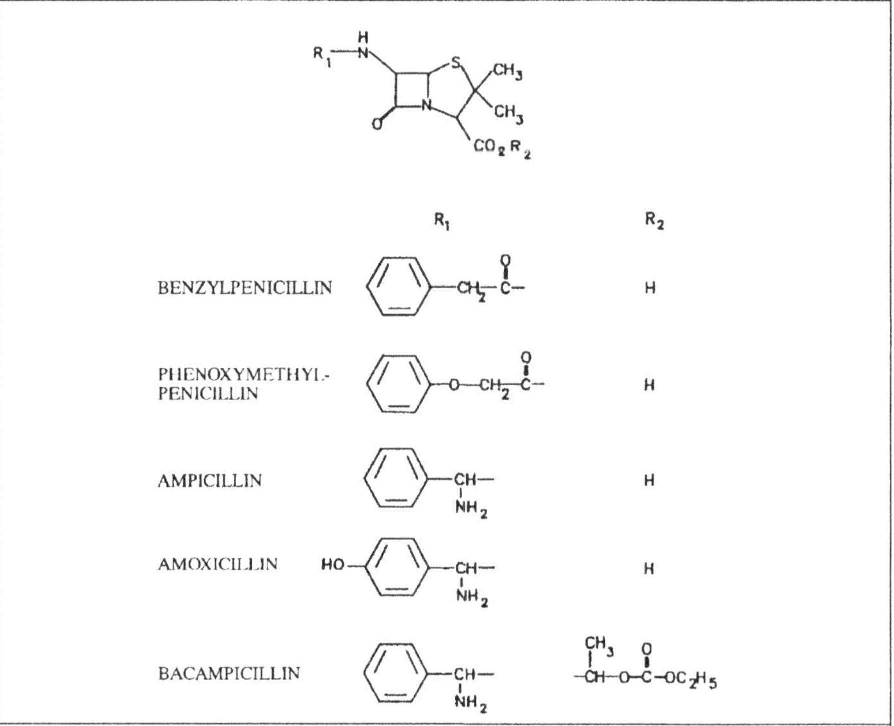


Figure 1: Structures of some penicillins.

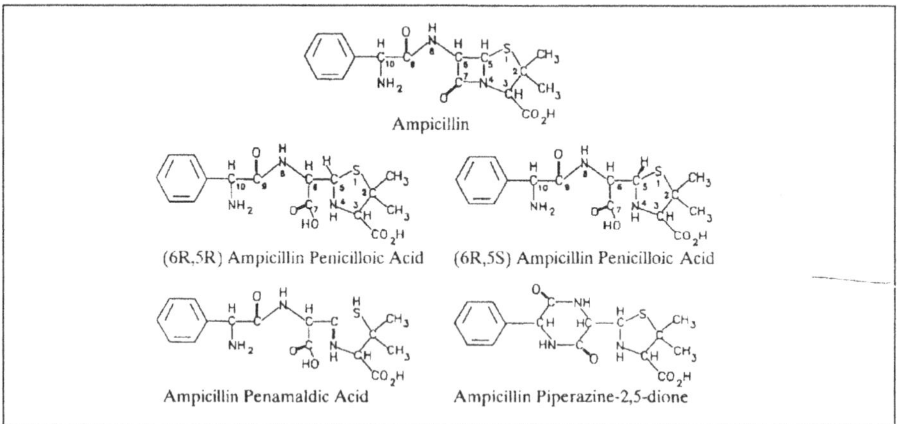


Figure 2: The structures of ampicillin and its known metabolites.

Penicillins do not have specific chromophores. With a UV detector, penicillins must be monitored at wavelengths where many endogenous compounds can interfere. Penicillins are unstable. They are sensitive to nucleophilic and electrophilic attack catalyzed by bases and acids, respectively. They can undergo degradation during the analysis leading to questionable results. For example, a new metabolite of ampicillin, penamaldic acid (Figure 2), was reported and consequently quantified from human urine samples by using an HPLC method with a UV detector [41, 42, 43]. Bird and co-workers [44], however, showed that penamaldic acid was not a new metabolite of ampicillin but that it was a 5S epimer of the well-known major metabolite, penicilloic acid. In a recent investigation of the metabolism of ampicillin by means of spin-echo $^1H$ nuclear magnetic resonance spectroscopy, piperazine-2,5-dione (Figure 2) was demonstrated as a new metabolite of ampicillin in rat urine [45]. The quantitative analysis of this metabolite in human urine by an HPLC method was also described [46].

The applications of mass spectrometry in the determination of penicillins and other antibiotics were reviewed by Border et al. [47]. The β-lactam antibiotics are thermally labile compounds with low vapor pressure. Barber [48] used fast-atom bombardment mass spectrometry to investigate the fragmentation of both positive and negative ions of some penicillins. The characteristic in-beam electron ionization mass spectra of 6-aminopenicillanic acids and several penicillins were reported by Ohasi et al. [49]. Barbalas et al. [50] used tandem mass spectrometry in the targeted class analysis of β-lactam antibiotics.

Several applications of the combination of chromatographic and mass spectrometric techniques have been reported. Pyrolysis gas/liquid chromatography and mass spectrometry were used to characterize some penicillins and cephalosporins [51, 52] from pharmaceutical preparations. The mass spectra obtained by GC/MS yield no detectable molecular ion peaks. McDowall et al. [53] have combined microbore LC/MS with a moving belt interface to analyze mixtures of penicillins.

## 2. Determination of Penicillins and Metabolites

Qualitative and quantitative analyses of drugs in body fluids are important in both clinical and pharmaceutical studies. In clinical studies, a high concentration of a drug may have toxic side effects. On the other hand, therapeutic activity may not be achieved if a certain dosage level is not maintained. In a pharmacokinetic study, the monitoring of a drug in the body is maintained over a period of time to determine drug absorption, distribution, metabolism and excretion. For both clinical and pharmacokinetic studies, the analytical chemist is required to provide techniques that are sensitive and specific. The combination of liquid chromatography and mass spectrometry can provide such a tool.

Thermally labile and nonvolatile compounds such as penicillins can be analyzed by using thermospray LC/MS without derivatization. Compounds do not always need to be completely separated chromatographically when using a mass spectrometer as a detector. The separation of these components is possible by constructing ion chromatograms of characteristic selected ions. In a recent study in our laboratory [54], ampicillin was only partially separated from other components in a urine sample within a reasonable retention time. The known ampicillin metabolites were not able to be identified by using a UV absorbance detector. In addition, the total ion chromatogram of the same urine sample containing ampicillin and its metabolites is less complex than the chromatogram using a UV-absorbance detector, because of the selectivity of the mass spectrometer. Utilizing selected ion chromatograms, we observed five chromatographic peaks corresponding to ampicillin and its metabolites. Thus, a chromatographic procedure can be simplified by using mass spectrometry. In addition, on-line sample clean-up using a solid phase extraction column can be employed instead of an extensive conventional sample clean-up procedure.

We now discuss in some detail the mass spectrometric determination of ampicillin and its metabolites. The reader is referred to the literature [54] for the details of sample preparation and experimental procedure.

Ampicillin and its known metabolites ampicillin penicilloate and ampicillin piperazine-2,5-dione were unambiguously determined in human urine samples by using thermospray LC/MS [54]. The identification was based on a comparison of retention times and on the thermospray mass spectra of ampicillin and its metabolites. A selected ion chromatogram in Figure 3 of a urine sample shows four peaks, two of which could be identified as the metabolite ampicillin penicilloic acid (peak 1) and unmetabolized ampicillin (peak 2). Two additional compounds were detected (peaks 3 and 4) and they were believed to be unreported ampicillin metabolites [54]. Ampicillin and its known metabolites, however, are relatively unstable and thermally labile. Thus, the additional unknown metabolites might also be (and are) relatively unstable. The thermospray mass spectra (Figure 5 in reference 54) of these unknown metabolites do not yield a significant abundance of the protonated molecule. Identification of fragment ions in the thermospray mass spectra were hampered by solvent ions and ions from other endogenous compounds in the urine samples. Although thermospray mass spectra provided the information used to detect the additional ampicillin metabolites in the urine samples, the thermospray spectra do not provide sufficient information to deduce the identity of these new metabolites.

Tandem mass spectrometry (MS/MS) was used to provide additional information leading to the structural elucidation of the two unreported metabolites [54]. These two compounds were identified as ampicillin penilloic acid and its epimer. The identities of these two possible metabolites were confirmed by comparing the retention times and mass spectra of synthesized ampicillin penilloic acids [54]. The structures of these two ampicillin penilloic acids differ by the absolute stereochemical configuration at $C_5$, (3S,5R) versus (3S,5S). The collisionally activated dissociation (CAD) spectra of the protonated ampicillin penilloic acid epimers do not show any significant difference in their fragmentation patterns. Therefore, without the observation of two separated peaks in the selected ion chromatograms (Figure 3), ampicillin penilloic acid cannot be distinguished from its epimer. The CAD spectrum of protonated ampicillin penicilloic acid (Figure 4) shows also a fragment ion at $m/z$ = 324, which has the same $m/z$ ratio as protonated ampicillin penilloic acid. Without the information obtained from separation with LC/MS, the ion at $m/z$ = 324 could be misinterpreted as a fragment ion from penicilloic acid. Thus, LC/MS and MS/MS are both needed and provide complementary information for the determination of these two epimers.

The interconversion of the two epimers was also demonstrated. By heating a solution of penilloic acid **3** (corresponding to the epimer that gives rise to chromatographic peak 3), an increasing amount of penilloic acid **4** (corresponding to the epimer of chromatographic peak 4) was observed and vice versa as shown in Figures 5 and 6. Sufficient information is not provided by using LC/MS and MS/MS to determine the absolute configuration of these epimers (either (3S,5R) or (3S,5S) penilloic acid). Only chromatographic peak 3, which is identified as ampicillin penilloic acid **3**, was observed in freshly taken human urine samples. Both ampicillin penilloic acid **3** and **4** were detected in human urine samples after storage at -5 °C for an extended period of time. This evidence suggests that ampicillin penilloic acid **3** has the same configuration (3S,5R) as the pure ampicillin. However, no other confirmation of the stereochemical configuration was obtained.

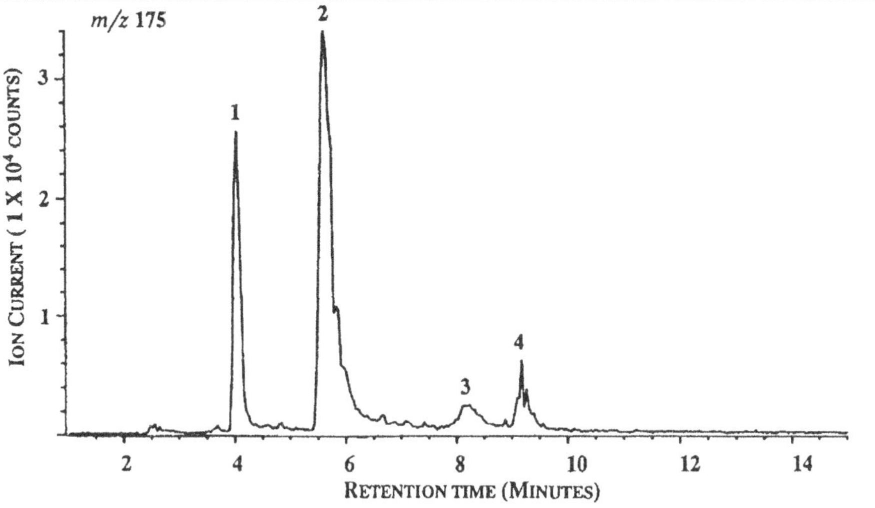


Figure 3: A selected ion chromatogram (for $m/z$ = 175) of a human urine sample collected at 3.5 hours after orally administering 500 mg of ampicillin.

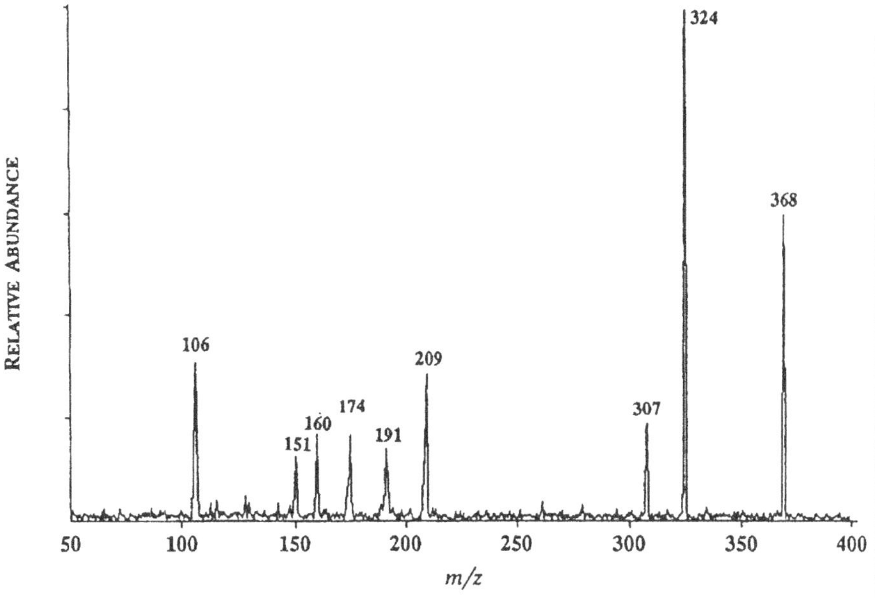


Figure 4: The low-energy CAD spectrum of protonated ampicillin penicilloic acid.

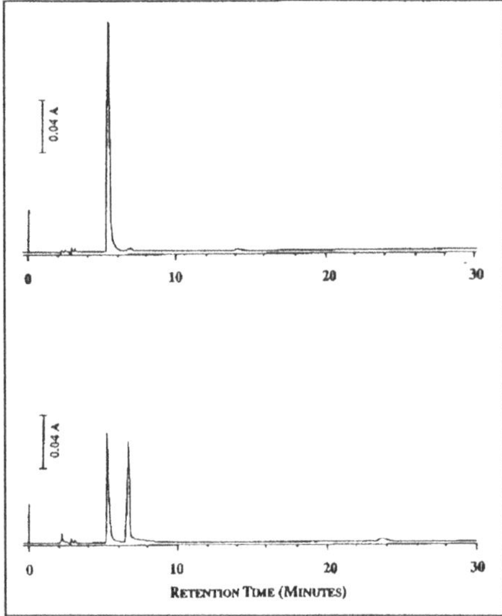


Figure 5: (a) The UV absorbance chromatograms of (a) 20 $\mu$L injection of 3.5 mg/mL ampicillin penilloic acid 3, and (b) 20 $\mu$L injection of solution in (a) after incubation at 65 °C for 30 minutes.

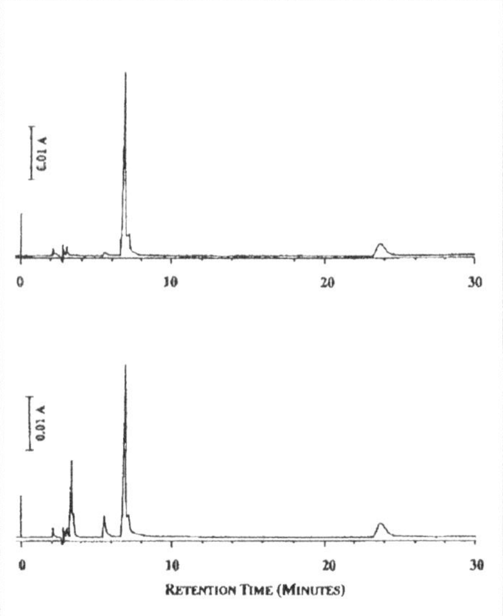


Figure 6: The UV absorbance chromatograms of (a) 20 $\mu$L injection of 2 mg/mL ampicillin penilloic acid 4, and (b) 20 $\mu$L injection of the solution in (a) after incubation at 50 °C for 15 minutes.

Penilloic acids are thermally labile compounds. The incubation of penilloic acids at low temperature (60 °C for 15 min.) yields many undetermined degradation products. Siegel et al. have used thermospray mass spectrometry as a flow reactor for studying *in situ* condensed-phase thermal degradation and mechanisms for decomposition of β-lactam antibiotics [55]. Although thermospray is a soft ionization process, a considerable amount of heat is applied to evaporate the solvent and form ions. The thermospray mass spectra show little or no abundance of the protonated ampicillin penilloic acids. Therefore, thermospray LC/MS/MS might not provide the information for the structural elucidation of penilloic acids because thermospray ionization provides an insufficient abundance of the protonated penilloic acid, making analysis of the fragment ions necessary. In contrast, the FAB mass spectra of ampicillin penilloic acids show a significant abundance of the protonated molecules, which can be used for obtaining CAD spectra.

Quantitative analysis of ampicillin in urine by using thermospray LC/MS was also examined [56]. Selected-ion monitoring of the characteristic fragment ions was used to acquire the data. Cephalexin was used as a non-isotopically labeled internal standard. The coefficient of variation for the standard urine samples (from 50 to 500 ng injections) ranges from 5 to 25 % with or without the internal standard. A 5 to 25 % coefficient of variation is larger than other HPLC methods that have been reported in the literature [57, 58, 59, 60, 61, 62]. This large coefficient of variation is partially due to the fluctuation of solvent pressure during the pumping cycle. The precision of the method may be improved by using a suitable isotopically labeled internal standard and a syringe-type solvent delivery system, which has virtually no fluctuation of solvent pressure. A syringe type delivery system should lower the detection limits in the thermospray LC/MS analysis by reducing the background noise. The detection limit for the determination of ampicillin in urine is approximately 10 ng (with a signal-to-noise ratio of 4:1). The sensitivity of the thermospray LC/MS method is better than that which was reported by using HPLC methods with UV detection. Limits of 0.1 to 2 $\mu g$ were reported when using fluorometric detection [46, 61, 62]. In addition, a linear response for the amount of ampicillin injected (correlation coefficient = 0.9998) was demonstrated in the 0.1 to 6.0 $\mu g$ range of ampicillin by using thermospray LC/MS [56]. This compares with results obtained by using a UV detector (correlation coefficient = 0.9949).

## 3. Determination of Peptides

Many biological samples are polar compounds that are not readily amenable to analysis by mass spectrometry. Peptide molecules have several structural features that contribute to their polarity. Carbonyl and amide groups in the same or different peptide chains can hyrdogen-bond to one another. There are also zwitterionic charges in a peptide chain owing to the presence of the protonated N-terminus cation and C-terminus carboxylate anion. Polar side chains may be present as well.

Various LC/MS interfacing techniques were used to analyze peptide and protein samples. A laser desorption mass spectrometer was coupled to a liquid chromatograph using a moving belt interface to detect peptide samples [63]. Under partial vacuum, a thermospray vaporizer was used to spray the effluent onto the moving belt to identify gramicidin S and Trp-Met-Asp-Phe-amide. Oligopeptide derivatives were sequenced by coupling HPLC and CI mass spectrometry using a moving belt interface [64].

The continuous-flow fast-atom bombardment (CFFAB) interface, which is discussed elsewhere in this volume, was used to analyze peptide samples following enzymatic digestion and HPLC separation. The technique was effective in analyzing enzymatic reactions [65]. Microbore HPLC was coupled with the CFFAB interface to separate and analyze a mixture of peptides produced by proteolytic hydrolysis of proteins [66, 67]. A particle-beam LC/MS system using a CFFAB interface was used to detect somastatin and the β–chain of bovine insulin [68].

The development of ionization techniques that yield multiply charged ions provides mass spectra of large peptides on instruments whose upper mass range is less than 2000 u. Electrospray was used to analyze gramicidin-S and other peptides [15, 69]. Electrospray and tandem mass spectrometry were coupled to provide structural information for the β-chain of insulin [70]. Angiotensin I and a few other small peptides were detected by using an ion spray interface with atmospheric pressure ionization (API) mass spectrometry [71].

A significant amount of research was conducted by using thermospray LC/MS for peptide analysis. Some of the first work included the analysis of amino acids and small peptides; for these, the vaporizer was heated with oxy-hydrogen flames [19, 72]. Thermospray mass spectra of peptide samples were obtained for full mass scans and subnanogram quantities with selected ion monitoring (SIM) [73]. Rapid on-line analysis of protein sequences was developed by using a combination of immobilized enzymes on a liquid chromatography (LC) column and thermospray LC/MS [74]. Polypeptides were sequenced by using this on-line enzymatic hydrolysis and HPLC separation that was coupled to the thermospray interface. [75].

The direct electrical heating of the vaporizer probe, using cartridge heaters, improves the performance and provides better stability than previous thermospray interfaces [11]. This vaporizer was used to sequence peptides by employing on-line enzymatic hydrolysis, HPLC, thermospray ionization and quadrupole mass spectrometry [76, 77]. Carboxypeptidase Y was immobilized on glass beads and packed into an HPLC column. A small volume of peptide solution was injected and the free amino acids from the C-terminus were directly transported into the thermospray mass spectrometer. This method is similar to that used for amino acid sequence determination of peptides from tooth pulp extracts using ultraviolet (UV) detection [78]. Columns containing other immobilized enzymes can be prepared and peptide mapping studies can be conducted by using endopeptidases [79]. Sequence information from proteins was determined by injecting a small amount of the peptide on an immobilized endopeptidase and the fractions separated by an HPLC column. The peptide fractions were then sequenced by using an immoblized exopeptidase column connected to the thermospray interface [74, 80].

Peptides with molecular weights up to 4000 u can be analyzed directly by using quadrupole instruments, by taking advantage of recent advances in quadrupole technology and the introduction of instruments with extended mass ranges [81]. Glucagon (MW 3480) and mellitin (MW 2847) were analyzed directly by thermospray LC/MS by using a quadrupole with an upper mass range of 4000 u. The use of tandem mass spectrometry with thermospray ionization can yield sequence information for peptide samples as well.

A quadrupole mass spectrometer (Hewlett-Packard Model 5988A) and a gradient LC system (Scientific Systems, Inc. Model 230) were interfaced with a standard Vestec Thermospray™ ion source and an ion source that was modified. Modifications to our thermospray vaporizer probe and ion source were made that enhance ion evaporation resulting in increased sample ion current [82, 83]. The modified ion source (Model 721), diagrammed in Figure 7, has an auxiliary tip heater and a desolvation chamber smaller than the standard ion source [84]. Additionally, the modified ion source was altered to contain either a blunt-tip or a needle-tip repeller electrode. Sulfonated azo dyes were used as model compounds to determine the effect of these modifications on the ion evaporation process [83]. Mass spectra of disulfonated dyes show a number of molecular ions and adduct ions that provide unequivocal molecular weight information. The interaction and effect of probe-tip size, repeller voltage and flow rate on sulfonated azo dye detection were studied [83]. An increase in signal response for sulfonated azo dyes was observed. Restriction of the thermospray vaporizer decreases the droplet size of the spray entering the mass spectrometer ion source. The decreased droplet size increases the charge-per-unit area on the thermospray droplets and a subsequent increase in ion evaporation results. A needle repeller in the thermospray ion source enhances sulfonated azo dye detection. Results indicate that the repeller has a direct effect on ion-extraction efficiency [83]. Both restriction of the vaporizer exit aperture and the needle-tip repeller are needed to obtain mass spectra of sulfonated azo dyes. An efficient chromatographic separation procedure for sulfonated azo dyes, which is compatible with thermospray mass spectrometric detection, was also presented [83]. Five disulfonated azo dyes were separated and detected by using selected ion monitoring [83].

### 3.1. Detection and Characterization of Peptides

For the methods of thermospray mass spectrometry described below, peptides are detected in the negative ion mode as their deprotonated molecules ($(M - H)^-$). The products arising from molecular fragmentation in the thermospray ion source have proven very difficult to assign, therefore the instrument is operated under conditions that minimize fragmentation. The mass spectrum arising from a peptide under such conditions, although rarely a single peak, is usually simple. Indeed, if only a single peak is detected for a peptide, it may be difficult to determine whether that peak corresponds to $(M - H)^-$, or to a differently ionized form of the peptide. For example, in those cases where the mass of the peptide exceeds the mass range of the quadrupole for singly charged ions (typically $m/z$ = 2000), multiply charged ions are the only species that may be observed.

Peptide ions other than $(M - H)^-$ that are observed in negative ion thermospray mass spectra are commonly of two types [74]: multiply deprotonated molecules, $(M - 2H)^{2-}$, $(M - 3H)^{3-}$, etc.; and those species in which the molecule is both multiply deprotonated and also associated with an alkali ion, $(M - 2H + Na)^{1-}$, $(M - 3H + Na)^{2-}$, etc. The set of peaks arising from such

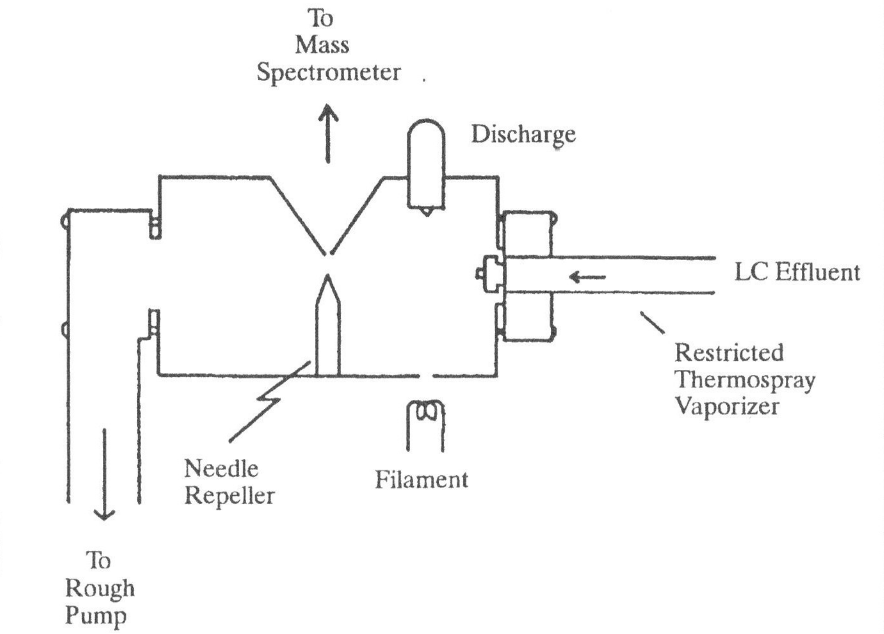


Figure 7: Schematic of the modified thermospray ion source used for peptide studies.

different ionic forms of a single molecule is often the best evidence for the identity of the molecule's true mass. For example, the members of the series: $(M - 2H)^{2-}$, $(M - 3H)^{3-}$, $(M - 4H)^{4-}$, etc. will appear as a set of peaks whose mass-to-charge ratios (*m/z*) are in approximate simple fractional ratios. It is relatively easy to determine the value of *m* from such a series. Peptide ions associated with an alkali ion (usually $Na^+$, but $K^+$ is also common) appear as satellites of the simple deprotonated species bearing the same charge. The distance between the pair of peaks is characteristic of the charge state. That is, $(M - H)^-$ and $(M - 2H + Na)^-$ will be separated by 22 mass units; $(M - 2H)^{2-}$ and $(M - 3H + Na)^{2-}$ will be separated by 11 mass units; and $(M - 3H)^{3-}$ and $(M - 4H + Na)^{3-}$ will be separated by about 7 mass units. Sets of three associated peaks corresponding to $(M - H)^{-1}$, $(M - 2H + Na)^{-1}$, and $(M - 2H + K)^-$ are also observed. In such cases, assignment of the charges of the ions and of the mass of the peptide is usually straightforward.

### 3.2. PEPTIDE MAPPING

The analysis of mixtures of peptides, for example, the products from the digestion or chemical fragmentation of a protein, was simplified by taking advantage of the LC/MS capabilities of the thermospray mass spectrometer [74, 79, 85, 86]. Both reverse-phase HPLC [85] and high performance ion-exchange chromatography [86] were applied to resolve such mixtures. The effluent from the chromatographic column is introduced directly into the thermospray ion source, and the trace of the total ion current (TIC) of the mass spectrometer serves as the simple chromatogram. The TIC trace is actually a connected set of points derived from summing the signals of individual mass spectra, collected at intervals of a few seconds. Each spectrum is stored separately in computer memory, so it is possible to display the individual spectrum corresponding to a selected portion of any peak, or to sum and display the set of spectra comprising any portion of the chromatogram. From such manipulations, it is easy to determine the homogeneity of any peak and to characterize its components. The overall result is a peptide map that has an extra dimension of information: molecular masses accompany the elution times of each component.

Rapid mapping of peptides and proteins can be achieved by putting the entire process on-line [74, 79, 86]. Endopeptidase columns, in which an enzyme is bound covalently to a solid matrix, can be prepared. For example, trypsin was bound to glycophase-coated controlled pore glass (CPG) beads by means of carbonyl diimidazole, and the derivatized beads were packed in a 2.1 mm bore stainless steel column [86]. The endopeptidase column is placed between the injection port and the HPLC column of the previously described mapping assembly to create a rapid, on-line mapping system. If a peptide or denatured protein is injected into this system, it is digested during passage through the endopeptidase column, the resulting fragments are resolved on the HPLC column, and the chromatogram is analyzed by the mass spectrometer, all in a continuous process.

The endopeptidase columns may be reused many times. Digestion of the peptide or protein by the bound enzyme is often rapid [74]. It may go nearly to completion simply upon passage through the column at flow rates suitable for chromatography (e.g., 1 mL/min.) Occasionally, however, the rate of enzymatic cleavage at a critical site may be slow, or an extremely hydrophobic peptide core may precipitate during digestion. In these cases a longer period of exposure to the enzyme may be needed [79]. To achieve this, the endopeptidase column may be inserted in the position of the sample loop of the sample injector. The substrate can then be injected and left in the column for as long as may be desired, before mapping.

Figure 8 shows tryptic maps carried out on the β-chain of two variants of hemoglobin, hemoglobin A (Hb A) and hemoglobin S (Hb S), by using the on-line LC/MS method [79]. The two chromatograms show one outstanding difference: In the map of Hb A there is a peak at 16.7 min. There is no peak at 16.5 min in the map of Hb S, but the latter does have a major peak at 24 min, which is not found in the Hb A map. These peaks represent peptides of differing sequences arising from the same portion (residues 1-8) of the respective proteins.

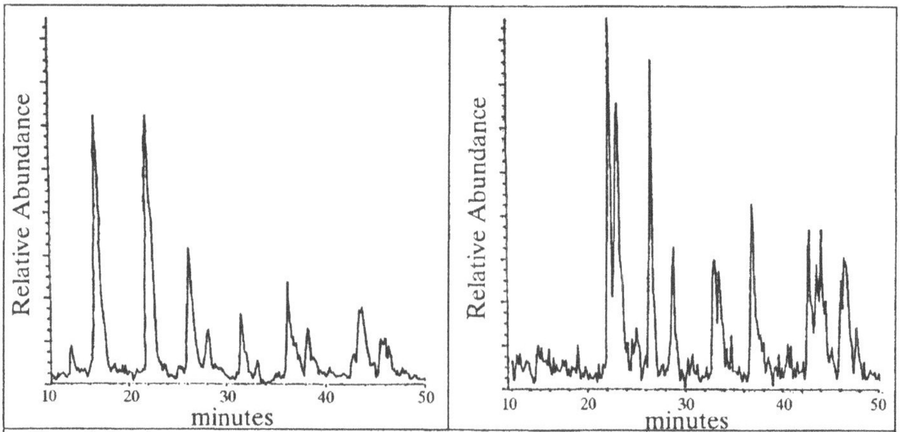


**Figure 8:** Tryptic maps of the β-chains of hemoglobin A (left) and hemoglobin S [right). Each protein chain was digested for 20 min on a column of immobilized trypsin, then chromatographed on a C-4 reverse phase column in 0.1 $NH_4OH$ at pH 6.9. Peptides were eluted with a linear gradient of 0 to 30% propanol.

Figure 9 (left) shows the mass spectrum obtained by summing the individual spectra near 16.7 min in the map of Hb A. The peptide identified by this mass spectrum has an $(M - H)^-$ ion at $m/z = 951$. The mass spectrum shown in Figure 9 (right) was obtained for the 24 min peak of Hb S and it indicates a peptide that is 30 u lighter. The mass difference corresponds exactly to the accepted difference between Hb A and Hb S: the 1-8 sequence of Hb S possesses a Val residue in a position where Hb A possesses a Glu residue. Thus, in a single mapping experiment the difference between the two proteins can be identified [79]. Although there appear to be minor differences in other parts of the two maps, comparison of the mass spectra corresponding to the peaks involved in each case makes it possible to confirm that none of these differences are significant. Such comparisons are not time consuming (they require only 1-2 minutes each), and serve as a good check on experimental results.

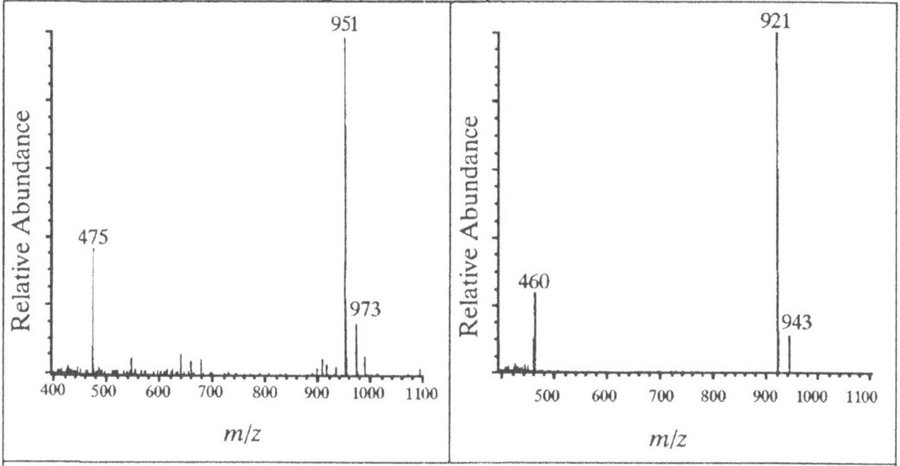


**Figure 9:** (left) The averaged, baseline subtracted mass spectrum taken over the interval corresponding to the peak at 16.7 min in the tryptic map of the β-chain of hemoglobin A (see Figure 8). The peak at $m/z = 951$ corresponds to the $(M - H)^-$ ion of the 1-8 tryptic peptide. The peak at $m/z = 475$ is the $(M - 2H)^{2-}$ ion, and the peak at $m/z = 973$ is the $(M - 2H + Na)^-$ ion. (right) The averaged, baseline subtracted mass spectrum taken over the interval corresponding to the peak at 24 min in the tryptic map of the β-chain of hemoglobin S (see Figure 8). The peaks at $m/z = 921$, 460 and 943 correspond to the ions $(M - H)^-$, $(M - 2H)^{2-}$, and $(M - 2H + Na)^-$, respectively, of the 1-8 tryptic peptide, as above.

### 3.3. PEPTIDE AND PROTEIN SEQUENCING

Partial sequencing of individual peptides may be achieved by using an on-line assembly similar to that for mapping [74, 86]. In this procedure a sequencing column is prepared. The sequencing column contains an exopeptidase bound to a solid support. Carboxypeptidase Y (CPY; for C-terminal sequencing) [86] and aminopeptidase M (APM; for N-terminal sequencing) [74] were the exopeptidases of choice, because each is relatively non-specific. Carboxypeptidase B has also proven useful [74]. A peptide passed through such a column will undergo partial digestion, giving rise to a set of molecules missing one, two, three, or more amino acid residues from the appropriate terminus. The set of peptides emerge as a single peak in the TIC trace if no chromatography is used, but the mass spectrum summed over this peak clearly indicates each component. A peptide sequence can be deduced from the mass spectrum by reading from the highest to the lowest mass detected. Each successive mass detected represents a peptide that is separated from the next highest in the series by the mass of the amino acid residue that was removed to generate it from its predecessor. In principle, only the residues leucine and isoleucine cannot be distinguished from such data.

The mapping and sequencing operations can be combined in a single on-line system, consisting of an endopeptidase column, an HPLC column, an exopeptidase column and the mass spectrometer, all in series [74]. When a protein is injected in this system, it is fragmented and mapped, and then each peptide is individually subjected to the sequencing protocol. The TIC trace looks the same as it does in the standard mapping procedure, but the mass spectra of individual peaks will now consist of the set of sequence ions of the peptide that maps at that position. The experiment can be run again by using different endopeptidase or exopeptidase columns, and both new and overlapping sequences can be generated. Although no single experiment is likely to discover large segments of the sequence, the large number of column combinations possible, and the speed with which successive analyses can be carried out, means that large amounts of sequence information can be generated rapidly [74]. The success of the experiments is not dependent upon whether the starting protein has an unblocked terminus, and by pairing experiments, partial sequences of each peptide can be determined from each terminus. Because the experiments are carried out totally on-line, extensive sample preparation is not required for each peptide analyzed.

An example of sequencing is demonstrated by utilizing the Hb S 1-8 peptide. As stated above, this peptide possesses a Val residue, and therefore differs from the 1-8 peptide of Hb A (which possesses a Glu residue in the corresponding position.) The Hb A 1-8 peptide, however, possesses two Glu residues, which are contiguous (positions 6 and 7). Without sequence data, it is not possible to state which of these residues is different between Hb S and Hb A. In the sequencing experiment [79], the Hb S 1-8 peptide was diverted into a column containing immobilized CPY as it emerged from the HPLC column. (Digestion of this peptide was rather slow, and extensive sequence data was desired, so in this case the exopeptidase column was placed in the sample loop position of an six-way valve located just posterior to the HPLC column. Once the peptide had entered the exopeptidase column, it was switched off line for a 20 min digestion period. The column was subsequently returned on line, and the contents flowed directly into the thermospray source.) The mass spectrum of the digested 1-8 peptide of Hb S is shown in Figure 10. The mass differences between successive sequence ions make it clear that Val is at position 6 in this peptide, whereas position 7 is occupied by Glu.

### 3.4. Disulfide Bridge Location

One of the most common forms of post-translational modification of proteins is the formation of disulfide bridges. Determination of the correct pairings of Cys residues is normally considered the final step of protein sequencing. Such pairings cannot be determined from gene sequencing or standard peptide sequencing techniques, including the method described here. The peptide mapping capacity of the thermospray LC/MS, however, can simplify and speed the task of disulfide assignment.

The basic strategy requires employing the same endopeptidase to prepare two peptide maps. The first is prepared in the usual manner from the reduced protein. A second map is prepared starting from protein that still has its disulfide bonds intact. When the second map is compared to the first, there should be one new peak (the bridged fragment) and two 'missing' peaks (the two halves of the bridged fragment, present in the first map) for each disulfide bond of the protein. The mass of each bridged fragment should correspond to the sum of two 'missing' fragments (minus two hydrogens) and thereby identify the pairing.

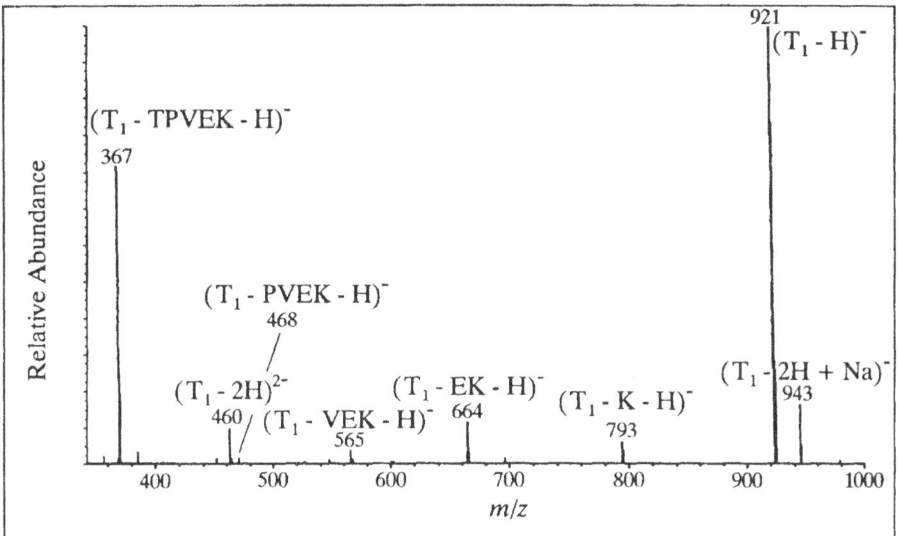


**Figure 10:** The mass spectrum of the carboxypeptidase Y digest of the 1-8 tryptic peptide of the β-chain of hemoglobin S. The sequence peptides derived from the parent ($T_1$) are identified by the residues lost as a result of CPY digestion. (E = Glu, K = Lys, P = Pro, T = Thr, V = Val.) The C-terminal sequence may be read from right to left, by determining which new residue was removed in each step of digestion in order to create each successively smaller ion.

Although the description of the experiment is simple, the experiment itself may not be. Digestion of proteins possessing intact disulfide bridges is often difficult. Endopeptidases that function under denaturing conditions, such as thermolysin, are often required. Incomplete digestion, or multiple alternative cleavage sites, may complicate the analysis. In general, such an analysis cannot be carried out on-line. On the other hand, the use of an LC/MS system permits the monitoring of prolonged enzyme digestions on an essentially real-time basis. Maps of the protein digest may even be developed rapidly enough to follow the course of a digestion and to trace the appearance and disappearance of specific components [87]. The identification of specific peptides and their relationships is simplified, and optimum digestion times are more readily determined.

### 3.5. CONSTRAINTS

Analysis of peptides and proteins by thermospray LC/MS is subject to certain constraints. As with any LC/MS system, the buffers used must be volatile. The range of the ionic strength of the buffer is also limited. In the case of ammonium acetate, buffer concentrations below approximately 0.05 M give inadequate peptide ionization, whereas concentrations of 0.5 M and above cause excessive fragmentation of the peptides. This limits the use of ionic gradients, for example, in the development of ion-exchange chromatograms.

The development of reverse-phase chromatograms is limited by the solvent compatibility of the endo- and exopeptidases. Even relatively low concentrations of acetonitrile can irreversibly denature most of the enzymes commonly used for mapping and sequencing. Propanol has proven to be somewhat more compatible with these enzymes, but above 15% concentration, it too can have deleterious effects [88]. As a result, chromatography on C-4 rather than C-18 reverse-phase columns was employed, so that hydrophobic peptides can be eluted with lower co-solvent concentrations.

The limits of sensitivity of the thermospray technique, and methods for improving sensitivity were discussed above. Although much progress was made, practical peptide analyses of the type described above are still best carried out on samples of 1 nanomole or more per injection.

Thermospray analysis of peptides is still only practical on quadrupole mass spectrometers. The mass range has an upper limit for most available instruments of about $m/z$ = 2000. Large peptides and proteins can be detected only as their multiply charged ions, which may be relatively minor species. This further reduces sensitivity. In such cases, electrospray techniques may prove to be more successful. On occasion peptide dimers may also be detected [88, 89]. Dimers in thermospray mass spectra were found for peptides containing glutamic acid or aspartic acid residues, which may form hydrogen bonded pairs. Such dimers may be mistaken for larger peptides.

## 4. Acknowledgements

The authors wish to acknowledge support for this work from the National Institutes of Health (Grant GM - 32772), the Environmental Protection Agency, the National Aeronautics and Space Administration, the Robert A. Welch Foundation and the Texas Advanced Research and Technology Program. The thermospray ion sources used in these studies were provided by the Vestec Corporation. In addition, we gratefully acknowledge many helpful discussions with Dr. Marvin Vestal.

## 5. References

1. (a) Present address: Sandoz, Inc., E. Hanover, NJ 07936. (b) Vestec Corp., Houston, TX 77054. (c) The Clorox Company, Pleasanton, CA 94566. (d) Pfizer Chemicals, Groton, CT. (e) University of Gdansk, Institute of Chemistry, Gdansk 6, Sobieskiego, Poland. (f) Department of Chemistry, University of Colorado at Denver, Denver, CO 80217-3364.
2. Halpern, B. (1981) 'Biomedical Application of Gas Chromatography/Mass Spectrometry', *CRC Crit. Rev. Anal. Chem.*, **11**, 49-78.
3. Vestal, M. L. (1984) 'High-Performance Liquid Chromatography/Mass Spectrometry', *Science*, **226**, 275-281.
4. Arpino, P. J.; Guiochon, G. (1979) 'LC/MS Coupling', *Anal. Chem.*, **51**, 682A-698A.
5. Edmonds, C. G.; McClosky, J. A.; Edmonds, V. A. (1983) 'A Bibliography of Combined Liquid Chromatography Mass Spectrometry', *Biomed. Mass Spectrom.*, **10**, 237-249.
6. Crowther, J. B.; Covey, T. R.; Henion, J. D. (1986) 'Mass Spectrometry as an On-Line Detector for HPLC', *Chem. Anal.*, **89**, 292-331.
7. Scott, R. P. W.; Scott, C. G.; Munroe, M.; Hess, J., Jr. (1974) 'Interface for On-Line Liquid Chromatography/Mass Spectroscopy Analysis', *J. Chromatogr.*, **99**, 395-405.
8. Yang, L.; Fergusson, G. J.; Vestal, M. L. (1984) 'A New Transport Detector for High-Performance Liquid Chromatography Based On Thermospray Vaporization', *Anal. Chem.*, **56**, 2632-2636.

9. McLafferty, F. W.; Knutti, R.; Venkataraghavan, R.; Arpino, P. J.; Dawkins, B. G. (1975) 'Continuous Mass Spectrometric Monitoring of a Liquid Chromatograph with Subnanogram Sensitivity using an On-Line Computer', *Anal. Chem.*, **47**, 1503-1505.
10. Lee, E. D.; Henion, J. D. (1985) 'Micro-Liquid Chromatography/Mass Spectrometry with Direct Liquid Introduction', *J. Chromatogr. Sci.*, **23**, 253-264.
11. Vestal M. L.; Fergusson, G. J. (1985) 'Thermospray Liquid Chromatograph/Mass Spectrometer Interface with Direct Electric Heating of the Capillary', *Anal. Chem.*, **57**, 2373-2378.
12. Blakley, C. R.; Carmody, J. J.; Vestal, M. L. (1980) 'A New Soft Ionization Technique for Mass Spectrometry of Complex Molecules', *J. Am. Chem. Soc.*, **102**, 5931-5933.
13. Carroll, D. I.; Dzidic, I.; Stillwell, R. N.; Horning, M. G.; Horning, E. C. (1974) 'Subpicogram Detection System for Gas-Phase Analysis Based upon Atmospheric Pressure Ionization (API) Mass Spectrometry', *Anal. Chem.*, **46**, 706-710.
14. Yamashita, M.; Fenn, J. B. (1984) 'Negative Ion Production with the Electrospray Ion Source', *J. Phys. Chem.*, **88**, 4671-4675.
15. Whitehouse, C. M., Dreyer, R. N.; Yamashita, M.; Fenn, J. B. (1985) 'Electrospray Interface for Liquid Chromatographs and Mass Spectrometers', *Anal. Chem.*, **57**, 675-679.
16. Willoughby, R. C.; Browner, R. F. (1984) 'Monodisperse Aerosol Generation Interface for Combining Liquid Chromatography with Mass Spectrometry', *Anal. Chem.*, **56**, 2626-2631.
17. Aschroft, A. E. (1987) 'Micro-and Microbore Liquid Chromatography-Fast Atom Bombardment/Mass Spectrometry (LC/FAB/MS) Analysis of Peptides and Antibiotics', *Org. Mass Spectrom.*, **22**, 754-757.
18. Games, D. E.; Pleasance, S.; Ramsey, E. D.; McDowall, M. A. (1988) 'Continuous Flow Fast Atom Bombardment Liquid Chromatography/Mass Spectrometry: Studies Involving Conventional Bore Liquid Chromatography with Simultaneous Ultraviolet Detection', *Biomed. Environ. Mass Spectrom.*, **15**, 179-182.
19. Blakley, C. R.; Carmody, J. J.; Vestal, M. L. (1980) 'Liquid Chromatograph/Mass Spectrometer for Analysis of Nonvolatile Samples', *Anal. Chem.*, **52**, 1636-1641.
20. Blakley, C. R.; Carmody, J. J.; Vestal, M. L. (1980) 'A New Soft Ionization Technique for Mass Sepctrometry of Complex Molecules', *J. Am. Chem. Soc.*, **102**, 5931-5933.
21. Gartiez, D. A.; Vestal, M. L. (1985) 'Thermospray LC/MS Interface: Principles and Applications', *LC*, **3**, 334-346.
22. Bellar, T. A.; Budde, W. L. (1988) 'Determination of Nonvalatile Organic Compounds in Aqueous Environmental Samples using Liquid Chromatography/Mass Spectrometry', *Anal. Chem.*, **60**, 2076-2083.
23. Taylor, G. W.; Watson, D. (1987) 'Biomedical Applications of Thermospray Liquid Chromatography/Mass Spectrometry', *J. Chromatogr.*, **394**, 135-146.
24. Thomson, B. A.; Iribarne, J. V. (1979) 'Field-Induced Ion Evaporation from Liquid Surfaces at Atmospheric Pressure', *J. Chem. Phys.*, **71**, 4451-4463.
25. Butfering, L.; Schmelzeisen-Redeker, G.; Rollgen, F. W. (1987) 'Studies with a Dual-Beam Thermospray Interface in High Performance Liquid Chromatography/Mass Spectrometry', *J. Chromatogr.*, **394**, 109-116.
26. Smith, R. W.; Parker, C. E.; Johnson, D. M.; Bursey, M. M. (1987) 'Eluent pH and Thermospray Mass Spectra: Does the Charge on the Ion in Solution Influence the Mass Spectrum', *J. Chromatogr.*, **394**, 261-270.

27. Alexander, A. J.; Kebarle, P. (1986) 'Thermospray Mass Spectrometry. Use of Gas-Phase Ion/Molecule Reactions to Explain Features of Thermospray Mass Spectra', *Anal. Chem.*, **58**, 471-478.
28. Bursey, M. M.; Parker, C. E.; Smith, R. W.; Gaskell, S. J. (1985) 'Gas-Phase Ionization of Selected Neutral Analytes during Thermospray liquid Chromatography/Mass Spectrometry', *Anal. Chem.*, **57**, 2597-2599.
29. Ballard, J. M.; Betowski, L. D. (1986) 'Thermospray Ionization and Tandem Mass Spectrometry of Dyes', *Org. Mass Spectrom.*, **21**, 575-588.
30. Betowski, L. D.; Pyle, S. M.; Ballard, J. M.; Shaul, G. M. (1987) 'Thermospray LC/MS/MS Analysis of wastewater for Disperse Azo Dyes', *Biomed. Environ. Mass Spectrom.*, **14**, 343-354.
31. Voyksner, R. D.; McFadden, W. H.; Lammert, S. A. (1987) 'Application of Thermospray HPLC/MS/MS for Determination of Triazine Herbicides', *Chem. Anal.*, **91**, 247-258.
32. Voyksner, R. D. (1987) 'Thermospray HPLC/MS for Monitoring the Environment', *Chem. Anal.*, **91**, 146-160.
33. Voyksner, R. D.; Bursey, J. T.; Hines, J. W. (1985) 'Analysis of Candidate Anticancer Drugs By Thermospray High-Performance Liquid Chromatography/Mass Spectrometry', *J. Chromatogr.*, **323**, 383-394.
34. Watson, D.; MacDermot, J.; Wilson, R.; Cole, P.; Taylor, G. W. (1986) 'Purification and Structural Analysis of Pyocyanin and 1-Hydroxyphenazine', *Eur. J. Biochem.*, **159**, 309-313.
35. Chen, T.; Coutant, J. E.; Sill, A. D.; Fike, R. R. (1987) 'Thermospray High-Perfromance Liquid Chromatogrphic/Mass Spectrometric Analysis of the Degradation Products of Piroximone', *J. Chromatogr.*, **396**, 382-388.
36. Lee, E. D.; Henion, J. D.; Brunner, C. A.; Wainer, I. W.; Doyle, T. D.; Gal, J. (1986) 'High-Performance Liquid Chromatography Chiral Stationary Phase Separation with Filament-On Thermospray Mass Spectrometric Identification of the Enantiomer Contaminant (*S*)-(+)-methamphetamine', *Anal. Chem.*, **58**, 1349-1352.
37. Edholm, L. E.; Lindberg, C.; Paulson, J.; Walhagen, A. (1988) 'Determination of Drug Enantiomers in Biological Samples by Coupled Column LC and LC/MS', *J. Chromatgr.*, **424**, 61-72.
38. Miners, J. O. (1985) 'The Analysis of Penicillins in Biological Fluids and Pharmaceutical Preparations By High-Performance Liquid Chromatography: a Review', *J. Liq. Chromatogr.*, **8**, 2827-2843.
39. Rouan, M. C. (1985) 'Antibiotic Monitoring in Body Fluids', *J. Chromatogr.*, **340**, 361-400.
40. De Silva, J. A. F. (1985) 'Analytical Strategies for Therapeutic Monitoring of Drugs in Biological Fluids', *J. Chromatogr.*, **340**, 3-30.
41. Masada, M.; Nakagawa, T.; Uno, T. (1979) 'A New Metabolite of Ampicillin in Man', *Chem. Pharm. Bull.*, **27**, 2877-2878.
42. Masada, M.; Kuroda, Y.; Nakagawa, T.; Uno, T. (1980) 'Structural Investigation of New Metabolites of Amino-Penicillins Excreted in Human Urine', *Chem. Pharm. Bull.*, **28**, 3527- 3536.
43. Uno,T.; Masada, M.; Yamaoka, K.; Nakawa, T. (1981) 'High Performance Liquid Chromatographic Determination and Pharmacokinetic Investigation of Amino-Penicillins and Their Metabolites in Man', *Chem. Pharm. Bull.*, **29**, 1957-1968.
44. Bird, A. E.; Cutmore, E. A.; Jennings, K. R.; Marshall, A. C. (1983) 'Structure Re-Assignment of a Metabolite of Ampicillin and Amoxycillin of Their Penicilloic Acids', *J. Pharm. Pharmacol.*, **35**, 138-143.

45. Everett, J. R; Jennings, K. R.; Woodnutt, G.; Buckingham, M. J. (1984) 'Spin-Echo $^{1}$H NMR Spectroscopy: a New Method for Studying Penicillin Metabolism', *J. Chem. Soc. Chem. Comm.*, 894-895.
46. Haginaka, J; Wakai, J. (1986) 'Sensitive High Perfromance Liqid Chromatographic Assay of a Metabolite of Ampiciillin in Man', *J. Pharm. Pharmacol.*, **38**, 225-226.
47. Borders, D. B.; Carter, G. T.; Hargreaves, R. T.; Siegel, M. M. (1985) 'Recent Applications of Mass Spectrometry To Antibiotic Research', *Mass Spectrom. Rev.*, **4**, 295-367.
48. Barber, M.; Bordoli, R. S.; Sedgwick, R. D.; Tyler, N.; Green, B. N.; Parr, V. C.; Gower, J. L. (1982) 'Fast Atom Bombardment Mass Spectrometry of Some Penicillins', *Biomed. Mass Spectrom.*, **9**, 11-17.
49. Ohashi, M.; Barron, R. P; Benson W. R. (1983) 'In-Beam Electron Ionization Mass Spectra of Penicillins', *J. Pharm. Sci.*, **72**, 508-513.
50. Barbalas, M. P.; McLafferty, F. W.; Occolowitz, J. L. (1983) 'Targeted Class Analysis of ß-Lactam Antibiotics By Tandem Mass Spectrometry', *Biomed. Mass Spectrom.*, **10**, 258-261.
51. Roy, T. A.; Szinai, S. S. (1976) 'Pyrolysis GLC Identification of Food and Drug Ingredients. II. Qualitative and Quantitative Analysis of Penicillins and Cephalosporins', *J. Chromatogr. Sci.*, **14**, 580-584.
52. Müller, M. D.; Seibl, J.; Simon, W. (1978) 'Characterization of Some Penicillins and Cephalosporins By Pyrolysis Mass Spectrometry', *Anal. Chim. Acta*, **100**, 263-269.
53. McDowall, M. A.; Games, D. A.; Gower, J. L. (1983) 'Mass Spectra and LC/MS Studies of Beta Lactam Antibiotics and Pseudomonic Acids', *Int. J. Mass Spectrom. Ion Phy.*, **48**, 157-160.
54. Suwanrumpha, S.; Freas, R. B. (1989) 'Identification of Metabolites of Ampicillin using Liquid Chromatography/Thermospray Mass Spectrometry and Fast Atom Bombardment Mass Spectrometry', *Biomed. Env. Mass Spectrom.*, **18**, 983-994.
55. Siegel, M. M.; Isensee, R. K.; Beck, D. J. (1987) 'Thermospray Mass Spectrometer Interface Used as a Flow Reactor for *In Situ* Thermal Degradation Studies: Application to β-Lactam Antibiotics', *Anal. Chem.*, **59**, 989-995.
56. Suwanrumpha, S.S. (1989) Ph.D. Dissertation, University of Houston.
57. Haginaka, J.; Wakai, J.; Yasuda, H.; Uno, T.; Takahashi, K.; Katagi, T. (1987) 'HPLC Determination of Ampicillin and Its Metabolited in Rat Plasma, Bile and Urine By Post-Column Degradation with Sodium Hypochlorite', *J. Chromatogr.*, **400**, 101-111.
58. Hikal, A. H.; Jones, A. B. (1985) 'Determination of Ampicillin in Plasma By Paired Ion HPLC', *J. Liq. Chromatogr.*, **8**, 1455-1464.
59. Vree, T. B.; Hekster, Y. A.; Baars, A. M.; Van Der Kleijn, E. (1978) 'Rapid Determination of Amoxycillin and Ampicillin in Body Fluids of Many By Means of HPLC', *J. Chromatogr.*, **145**, 496-501.
60. Ibrahim, E. A.; Abdel-Hamid, M. E.; Abuirjeie, M. A.; Hurani, A. M. (1988) 'Rapid HPLC Determination of Ampicillin in Human Urine', *Anal. Lett.*, **21**, 423-434.
61. Miyazaki, K.; Ogino, O.; Arita, T. (1974) 'Fluorometric Determination of Ampicillin', *Chem. Pharm. Bull.*, 22, 1910-1916.
62. Haginaka, J.; Wakai, J. (1985) 'HPLC Assay of Ampicillin, Amoxicillin and Ciclacillin in Serum and Urine using a Pre-Column Reaction with 1,2,4-Triazole and Mercury(II) Chloride', *Analyst*, **110**, 1277-1281.
63. Hardin, E. D.; Fan, T. P.; Vestal, M. L. (1984) 'Laser Desorption Mass Spectrometry with Thermospray Sample Deposition for Determining Nonvolatile Biomolecules', *Anal. Chem.*, **56**, 2-7.

64. Yu., T. J.; Schwartz, H. A.; Cohen, S. A.; Vouros, P.; Karger, B. L. (1984) 'Sequence Analysis of Derivatized Peptides By High-Performance Liquid Chromatography Mass Spectrometry', *J. Chromatogr.*, **301**, 425-440.
65. Caprioli, R. M. (1988) 'Analysis of Biochemical Reactions with Molecular Specificity using Fast Atom Bombardment Mass Spectrometry', *Biochemistry*, **27**, 513-521.
66. Caprioli, R. M.; DaGue, B.; Fan, T.; Moore, W. T. (1987) 'Microbore HPLC/Mass Spectrometry for the Analysis of Peptide Mixtures Using a Continuous Flow Interface' *Biochem. Biophys. Res. Commun.*, **146**, 291-299.
67. Caprioli, R. M.; Fan, T. (1986) 'High Mass Sensitivity Mass Spectrometric Determination of Peptides: Direct Analysis of Aqueous Solutions', *Biochem. Biophys. Res. Commun.*, **141**, 1058-1065.
68. Sanders, P. E.; Willoughby, R. C.; Mitrovich, S. (1989) 'Particle Beam LC/MS by FAB', presented at the 37th Annual Conference on Mass Spectrometry and Allied Topics, Miami Beach, FL, 110-111.
69. Meng, C. K.; Mann, M. ; Fenn, J. B. (1988) 'Applications of electrospray liquid chromatography/mass spectrometry', *Z. Phys. D.-Atoms, Molecules, and Clusters*, **10**, 361-368.
70. Loo, J. A.; Barinaga, C. J.; Edmonds, C. G.; Udseth, H. R.; Smith, R. D. (1989) 'Electrospray ionization-mass spectrometry and tandem mass spectrometry', presented at the 37th Annual Conference on Mass Spectrometry and Allied Topics, Miami Beach, FL; 586-587.
71. Bruins, A. P.; Covey, T. R.; Henion, J. D. (1987) 'Ion Spray Interface for Combined Liquid Chromatography/Atmospheric Pressure Ionization Mass Spectrometry. ', *Anal. Chem.*, **59**, 2642-2646.
72. Blakley, C. R.; Vestal, M. L. (1983) 'Thermospray Interface for Liquid Chromatography/Mass Spectrometry', *Anal. Chem.*, **55**, 750-754.
73. Blakely, C. R.; Carmody, J. J.; Vestal, M. L. (1980) 'Combined Liquid Chromatograph/Mass Spectrometer for Involatile Biological Samples', *Clin. Chem.*, **26**, 1467-1473.
74. Stachowiak, K.; Wilder, C.; Vestal, M. L.; Dyckes, D. F. (1988) 'Rapid Protein Sequencing by the Enzyme-Thermospray LC/MS Method', *J. Am. Chem. Soc.*, **110**, 1750-1765.
75. Kim, H. -Y.; Pilosof, D.; Vestal, M. L.; Dyckes, D. F.; Kitchell, J. P.; Dvorin, E. (1983) 'Amino Acid Sequence of Polypeptides By Enzymatic Hydrolysis and Direct Detection using a Thermospray LC/MS', *Peptides: Structure and Function*, **8**, 719-722.
76. Pilosof, D.; Kim, H. -Y.; Vestal, M. L.; Dyckes, D. F. (1984) 'Direct Monitoring of Sequential Enzymatic Hydrolysis of Peptides By Thermospray Mass Spectrometry', *Biomed. Mass Spectrom.*, **11**, 403-407.
77. Kim, H. -Y.; Pilosof, D.; Dyckes, D. F.; Vestal, M. L. (1984) 'On-Line Peptide Sequencing By Enzymatic Hydrolysis, High Performance Liquid Chromatography, and Thermospray Mass Spectrometry', *J. Am. Chem. Soc.*, **106**, 7304-7309.
78. May, H. E.; Tanzer, F. S.; Fridland, G.; Wakelyn, C.; Desiderio, D. M. (1982) 'High Performance Liquid Chromatography and Proteolytic Enzyme Characterization of Peptides in Tooth Pulp Extracts', *J. Liq. Chromatogr.*, 5, 2135-2154.
79. Stachowiak, K; Dyckes, D. F. (1989) 'Peptide Mapping using Thermospray LC/MS Detection: Rapid Identification of Hemoglobin Variants', *Peptide Research*, **2**, 267-274.
80. Voyksner, R. D.; Chen, D. C.; Swaisgood, H. E. (1989) 'Optimization of On-Line Enzymatic Hydrolysis HPLC/MS for Peptide Mapping and Disulfide Bond Location',

presented at the 37th Annual Conference on Mass Spectrometry and Allied Topics, Miami Beach, FL, 943-944.

81. Rudewicz, P. J. (1988) 'Analysis of Middle Mass Peptides By Thermospray LC/MS', *Biomed. Environ. Mass Spectrom.*, **15**, 461-463.
82. Fink, S. W.; Freas, R. B. (1989) 'Enhanced Analysis of Poly(ethylene glycols) and Peptides using Thermospray Mass Spectrometry', *Anal. Chem.*, **61**, 2050-2054.
83. McLean, M. A.; Freas, R. B. (1989) 'Enhanced Analysis of Sulfonated Azo Dyes using Liquid Chromatography/Thermospray Mass Spectrometry', *Anal. Chem.*, **61**, 2054-2058.
84. Coutant, J. E.; Ackerman, B. L.; Vestal, M. L. (1987) 'Effects of Source Design On the Nature of Thermospray Mass Spectra', presented at the 35th Annual Conference on Mass Spectrometry and Allied Topics, Denver, CO, 417-418.
85. Pilosof, D.; Kim, H. -Y.; Dyckes, D. F.; Vestal, M. L. (1984) 'Determination of Nonderivatized Peptides by Thermospray Liquid Chromatography/Mass Spectrometry', *Anal. Chem.*, **56**, 1236-1240.
86. Kim, H. -Y.; Pilosof, D.; Dyckes, D. F.; Vestal, M. L. (1984) 'On-Line Peptide Sequencing by Enzymatic Hydrolysis, High-Performance Liquid Chromatography', *J. Am. Chem. Soc.*, **106**, 7304-7309.
87. Stachowiak, K.; Otlewski, J.; Polanowski, A.; Dyckes, D. F. (1990) 'Monitoring Protein Cleavage and Concurrent Disulfide Bond Assignment using Thermospray LC/MS', *Peptide Research*, **3**, 148-154.
88. Stachowiak, K; Dyckes, D. F., unpublished results.
89. Fink, S. W.; Kaczmarek, J.; Burlet, O.; Dyckes, D. F.; Freas, R. B., unpublished results.

# NEUTRALIZATION-REIONIZATION MASS SPECTROMETRY: FUNDAMENTAL STUDIES AND ANALYTICAL APPLICATIONS

M. J. POLCE AND C. WESDEMIOTIS
*Department of Chemistry*
*The University of Akron*
*Akron, Ohio, 44325-3601*
*U.S.A.*

ABSTRACT. Neutralization-reionization mass spectrometry (NRMS) involves the production and analysis of neutral species inside the mass spectrometer. The neutrals are formed by charge exchange or dissociation of mass-selected ions and subsequently reionized for mass-analysis and detection. This method can be particularly useful for the characterization of ion structures, as shown here for $C_2H_7O^+$ isomers. NRMS enables the study of highly reactive or marginally stable molecules in the gas phase, by forming them via neutralization of the corresponding stable cations or anions; for example, unusual biradical intermediates, such as $\cdot CH_2CH_2OCH_2\cdot$, can be generated from readily available distonic ions. By allowing the detection and identification of neutral products from ion fragmentations, NRMS also provides unique mechanistic and molecular structure information.

## 1. Basic Concepts

Neutralization-reionization mass spectrometry (NRMS) is a tandem mass spectrometry technique in which beams of gaseous neutrals are produced by charge exchange (equation 1) [1] or dissociation (equation 2) of mass-selected parent ions. After removal of any un-neutralized parent ions, the neutral beams are reionized into daughter ions that are

$$\cdot CH_2CH_2CH_2CH^+OH \rightarrow \cdot CH_2CH_2CH_2CH\cdot OH \quad (1)$$
$$CH_3CO\text{-}OCH{=}CH_2^{+\cdot} \rightarrow CH_3CO^+ + \cdot OCH{=}CH_2 \quad (2)$$

mass-analyzed and recorded in the neutralization-reionization (NR) spectrum [2-4]. Such NR spectra can be used to study the structures and reactivities of both the parent ion as well as the intermediate neutral species; in addition, they can provide valuable information on ion fragmentation mechanisms and unknown molecular structures [5].

### 1.1. ION STRUCTURES

In conventional MS/MS studies, parent ions are characterized on the basis of the products of their unimolecular dissociations [6]. With NRMS the same ions are converted into neutrals, which usually react by entirely different, more specific pathways, leading often to structurally more informative products than fragmentation of the original ion itself. This diagnostic capability can be essential for the elucidation of ionic compositions [7].



*M. L. Gross (ed.), Mass Spectrometry in the Biological Sciences: A Tutorial*, 303–310.

1.2. ELUSIVE NEUTRAL SPECIES

Many ions with unconventional structures have no stable neutral counterparts. This is true for distonic and ylide ions [8], carbenic radicals ions, and various proton-bound or ion-molecule complexes [9]. Such ions have been identified as common fragments in the mass spectra of several simple organic compounds [10]; their neutralization represents a convenient route to access directly and study in the gas phase the elusive counterneutrals. For example, equation 1 shows that a reactive 1,4-biradical can be prepared from a readily available $\gamma$-distonic radical ion (present in the spectrum of pentanedial); such biradicals are the proposed intermediates of thermal and photochemical reactions of carbonyl compounds and cycloalkanes.

1.3. ANALYTICAL INFORMATION

In MS and MS/MS, molecular structure information is gained by identifying the ionic fragments formed from the molecular or other parent ion [6,11]. The simultaneously produced neutral fragments ($\cdot OCH{=}CH_2$ in equation 2) are not detected. This deficiency is eliminated by NRMS, which allows one to reionize and characterize such neutral fragments, thus leading to new analytical information.

## 2. Experimental Procedure

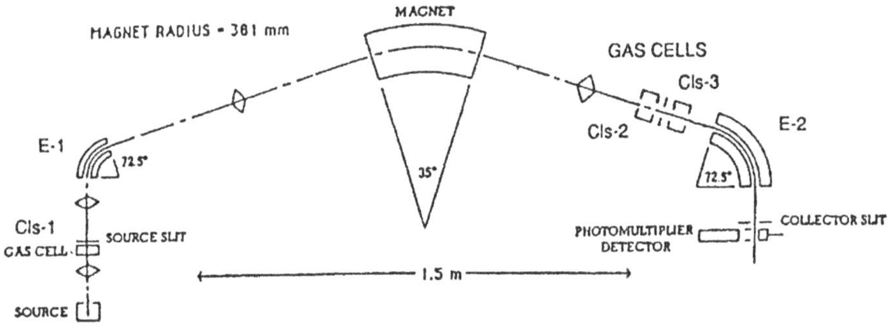


Figure 1. Trisector EBE tandem mass spectrometer.

NR experiments have been performed in the interface region of sector tandem mass spectrometers [2-4]. The data shown in this article were obtained with a modified VG AutoSpec of EBE geometry (Figure 1) [12]. In this instrument the first two sectors serve as MS-1 and the third sector as MS-2; one collision cell (Cls-1) is located in the first field-free region (FFR-1) and two cells (Cls-2 and Cls-3) in FFR-3. Either collision cell can be used for MS/MS experiments via collisionally activated dissociation (CAD) [6]. Cls-2 and Cls-3 house the neutralization and reionization events, respectively, and are 20 mm apart from each other. This distance is traveled by an 8-keV ion (or neutral) of 45 u in 0.11 μs.

Parent ions formed in the ion source or by CAD in Cls-1 are mass-selected through MS-1 and enter Cls-2 where they undergo charge exchanging or dissociating collisions with gaseous targets. Any remaining ions are removed by a deflector electrode located midway between Cls-2 and Cls-3. The resulting neutral beam is collisionally reionized in Cls-3, and the product ions are recorded in the NR spectrum after mass-analysis through MS-2.

The nomenclature used to describe a neutralization-reionization spectrum indicates the charges of the parent and product ions and the collision targets used. Thus, a $^{-}NR^{+}$, $O_2$/He, spectrum of $CH_3CH_2O^-$ is obtained by selecting the $CH_3CH_2O^-$ anion through MS-1, neutralizing it with $O_2$ in Cls-2, deflecting any remaining ions, reionizing the neutrals into cations with He in Cls-3, and mass-analyzing the products in MS-2 [5]. The spectra shown in this study employed transmittances of 80% (20% beam suppression) for each target (single-collision conditions [13]). Electron removal (neutralization of anions, reionization into cations) was effected by $O_2$ or He and electron addition (neutralization of cations, reionization into anions) by Xe. All samples used were commercially available or synthesized following standard procedures.

## 3. Fundamental Studies

### 3.1. ELUCIDATION OF ISOMERIC ION STRUCTURES

CAD of protonated ethanol, $CH_3CH_2OH_2^+$, yields major $C_2H_5^+$ and $OH_3^+$ fragments via losses of water and ethylene, respectively (Figure 2a) [14-17]. Detailed studies of deuterated isomers have shown that these eliminations are accompanied by substantial H scrambling and must proceed, therefore, through an intermediate in which the H atoms can exchange their positional identity. An ideal candidate for this intermediate is $C_2H_4 \cdots H^+ \cdots OH_2$, the proton-bound complex of ethylene and water, to which $CH_3CH_2OH_2^+$ can isomerize after its formation in the ion source.

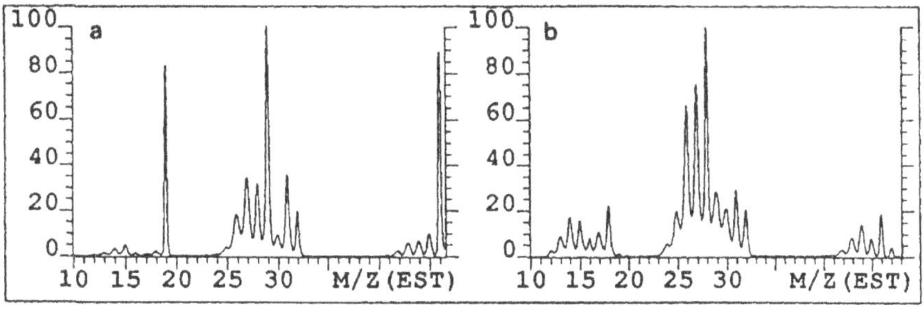


Figure 2. (a) CAD, $O_2$ and (b) $^{+}NR^{+}$, Xe/$O_2$, spectra of $C_2H_7O^+$ ions produced by self-protonation of ethanol.

It is not obvious from CAD whether $C_2H_4 \cdots H^+ \cdots OH_2$ represents a transition state for the H scrambling in $CH_3CH_2OH_2^+$ or a bound and, thus, stable structure that can coexist with $CH_3CH_2OH_2^+$. Here, as shown recently [17], NR spectra can give a more definitive answer because of the different reactivities of the counterneutrals of these two isomers: $CH_3CH_2OH_2^+$ neutralization leads to the hypervalent radical $CH_3CH_2OH_2\cdot$ which is only weakly bound (vide infra); extensive dissociation at the hypervalent center should generate mainly $CH_3CH_2\cdot$ + $OH_2$ and $CH_3CH_2OH$ + $\cdot H$. On the other hand, neutralization of the positive charge in $C_2H_4 \cdots H^+ \cdots OH_2$ should remove any binding force, so that decay into a mixture of $C_2H_4$, ·H, and $OH_2$ follows. Although $OH_2$ is formed from both isomers, $C_2H_4$ can solely arise from the neutralized complex.

The +NR+, Xe/$O_2$, spectrum of protonated ethanol (Figure 2b) shows significant peaks for $C_2H_4^{+\cdot}$ (of *m/z* 28) and $OH_2^{+\cdot}$ (of *m/z* 18), thus proving that the $C_2H_7O^+$ (of *m/z* 47) ion beam entering the neutralization cell contains $C_2H_4\cdots H^+\cdots OH_2$ [17]. This result confirms recent ab initio calculations predicting this proton-bound complex to exist as a stable structure [18]. The NR spectrum of Figure 2b also bears $C_2H_5^+$ (of *m/z* 29) and $CH_3CH_2OH^{+\cdot}$ (of *m/z* 46) ions that can originate from the dissociation products of neutralized $CH_3CH_2OH_2^+$ (vide supra) or during reionization of any surviving $CH_3CH_2OH_2\cdot$ (see next section).

## 3.2. CHARACTERIZATION OF ELUSIVE NEUTRAL SPECIES

NR spectra contain valuable information on the structure and stability of the neutral species accessed upon neutralization [1-5]. A stable neutral species survives unaltered, at least in part, so that upon reionization a recovered parent ion and its diagnostic fragments are observed. The fact that the NR spectrum of Figure 2b does contain a significant recovered parent ion (of *m/z* 47) yielding a diagnostic $CH_3\cdot$ loss (of *m/z* 32) shows that some $CH_3CH_2OH_2\cdot$ has survived intact; consequently, this hypervalent radical must represent a potential energy minimum with a finite barrier towards dissociation. The negligible abundance of $OH_3^+$ (of *m/z* 19) in the NR spectrum further proves that the van der Waals complex $C_2H_4\cdots H\cdot\cdots OH_2$ completely decomposes before reaching the reionization region (lifetime <<0.11 μs) [17].

Many more unstable or unusual neutrals have stable ionic counterparts from which they can be prepared in the mass spectrometer for experimental characterization. For example, $\cdot CH_2CH_2OCH_2\cdot$, a sought biradical intermediate in the thermal and photochemical reactions of oxetane (trimethylene oxide), is not available through conventional experiments [19]. With NRMS it can be formed from distonic $\cdot CH_2CH_2OCH_2^+$, hich is an abundant fragment in the EI mass spectrum of *p*-dioxane [20].

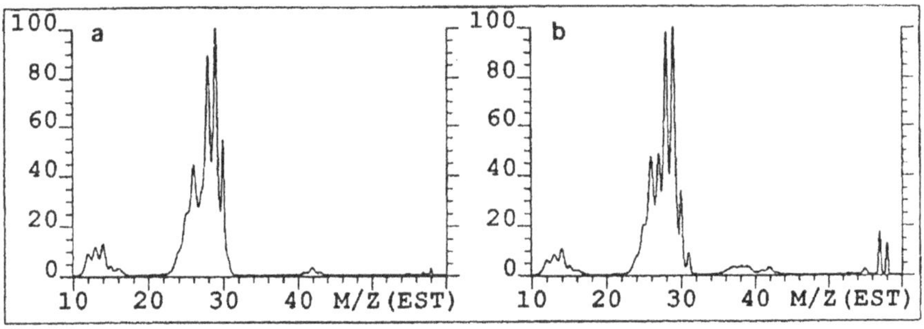


Figure 3. +NR+, Xe/$O_2$, spectra of (a) $\cdot CH_2CH_2OCH_2^+$ and (b) oxetane molecular cations.

Figure 3 compares the NR spectrum of $\cdot CH_2CH_2OCH_2^+$ with that of the isomeric oxetane ion. A weak recovered parent cation (of *m/z* 58) is observed upon reionization of $\cdot CH_2CH_2OCH_2\cdot$ (Figure 3a), indicating that this biradical survives partly undissociated. The recovered molecular ion of oxetane (Figure 3b) is significantly larger, paralleling the higher stability of this closed-shell molecule as compared to the stability of the biradical. Contrary to oxetane, $\cdot CH_2CH_2OCH_2\cdot$ yields no $C_3H_n^+$ fragments (of *m/z* 36-39) and, therefore, has not undergone ring-closure to more stable oxetane (<<5%) in the microsecond time window between neutralization and reionization. Consequently, $\cdot CH_2CH_2OCH_2\cdot$ must represent a stable intermediate with finite barriers towards both isomerization and dissociation [21]. The abundant *m/z* 24-30 ions seen in both spectra can arise from molecules produced above their fragmentation thresholds upon neutralization, so that they decompose into $CH_2{=}CH_2$ + $O{=}CH_2$, which after reionization contribute to these ions; the same ions can also be generated during reionization as they are present in the CAD spectra of the corresponding parent ions.

## 4. Analytical Applications

### 4.1. MOLECULAR STRUCTURES

Unimolecular dissociation coproduces an ionic and a neutral fragment (equation 2). The part that can form the most stable ion carries the charge whereas the complementary part is eliminated as a neutral [11]. In conventional mass spectrometry, only the ionic fragment can be detected; this problem can lead to loss of structural information as demonstrated by the similar mass and tandem mass spectra (Figure 4) of vinyl acetate ($CH_3CO\text{-}OCH{=}CH_2$, equation 2) and biacetyl ($CH_3CO\text{-}COCH_3$). In this case both parent ions form the same ionic products (dominated by $CH_3CO^+$, of *m/z* 43); isomer differentiation is impossible because the structurally characteristic features remain in the eliminated neutrals. With NRMS, however, these neutrals can be reionized (e.g., into anions, Figure 4) and detected, thus allowing for isomeric identification.

This methodology should be generally applicable to samples bearing substituents that cannot stabilize a charge and, therefore, yield no ionic fragments revealing these substituents [5].

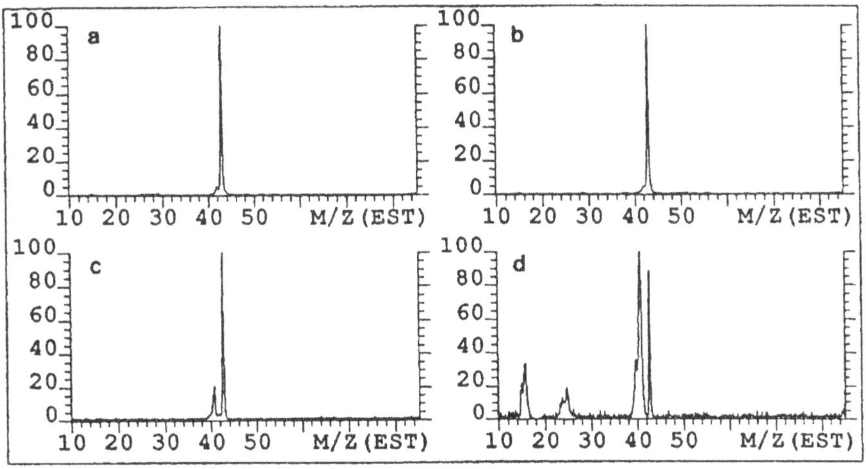


Figure 4. (a,b) CAD, He (ionic fragments) and (c,d) $^+NR^-$, He/Xe, spectra (neutral fragments after reionization into anions) of the molecular ions of (a,c) vinyl acetate and (b,d) biacetyl.

## 4.2. FRAGMENTATION MECHANISMS

Knowledge about the ionic as well as the neutral products from a particular parent ion can be very useful for the postulation or clarification of complex fragmentation mechanisms. This is illustrated for decompositions of the dipeptide alanyl aspartic acid: CAD of $(M + H)^+$ (of *m/z* 205) with He yields the ionic fragments shown in Figure 5a via elimination of the neutral fragments recorded in Figure 5b (after $O_2$ reionization into cations).

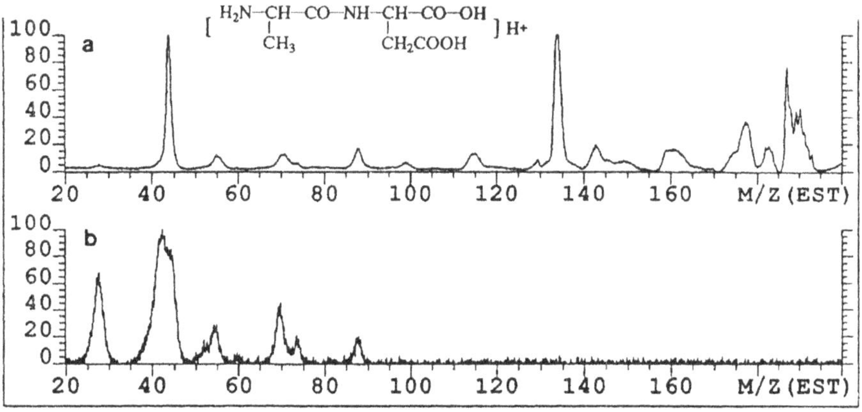


Figure 5. (a) CAD, He, spectrum of $(M + H)^+$ from alanyl aspartic acid (ionic fragments). (b) $^+NR^+$, $He/O_2$, spectrum of the same ion (neutral fragments reionized into cations). $(M + H)^+$ was formed by FAB.

The absence of large ions in the $^+NR^+$ spectrum indicates that many of the smaller fragment ions in the CAD spectrum are not formed directly from $(M + H)^+$ but by consecutive cleavages from larger primary fragments [12]. Based on Table 1, the complementary neutrals of many significant product ions are observed confirming the mechanisms proposed by Costello and Biemann [22]. For peptides with basic amino acids at either terminus, which mainly produce either C- or N-terminal fragments but not both series, such NR spectra could also supply essential structural information by disclosing the complementary pieces eliminated as neutrals.

## 5. Conclusions

Traditionally mass spectrometry has dealt with the generation and detection of *ion* beams. NRMS allows the *neutral* beams formed inside a mass spectrometer to also be "seen." With this technique, valuable analytical information is gained through the capability to identify the neutral products of unimolecular ion decompositions. Complex ion structures can be easier to characterize on the basis of the more specific reactions of the counterneutrals. Finally, uncommon or unstable neutrals can be studied by generating them from their common, stable counterions.

TABLE 1. Important fragment ions *and* fragment neutrals formed by CAD of protonated alanyl aspartic acid.

| Type[a] (*m/z*) | Structure | Via loss of (mass)<br>major *m/z* after reion.[b] |
|---|---|---|
| $b_2$ (*187*) | $H_2N$-CH($CH_3$)-CO-NH-CH($CH_2$COOH)-C≡$O^+$ | $H_2O$ (18)<br>*18* (not shown) |
| $a_2$+1 (*160*) | [$H_2N$-CH($CH_3$)-CO-NH-CH·($CH_2$COOH)]$H^+$ | ·CO-OH (45)<br>*45, 44, 28* |
| $a_2$ (*159*) | $H_2N$-CH($CH_3$)-CO-$NH^+$=CH($CH_2$COOH)<br>(consecutive cleavage from $b_2$) | CO (28)<br>*28* |
| $y_1$ (*134*) | [$H_2N$-CH($CH_2$COOH)-CO-OH]$H^+$ | HN-CH($CH_3$)-C=O (71) (N and C=O joined in a ring)<br>*70, 55, 43, 42* |
| $d_2$ (*115*) | [$H_2N$-CH($CH_3$)-CO-NH-CH=$CH_2$]$H^+$<br>(consecutive cleavage from $a_2$+1) | · CO-OH (45)<br>*45, 44, 28* |
| Immonium (*88*) | $H_2N^+$=CH($CH_2$COOH)<br>(consecutive cleavage from $y_1$) | HCOOH (46)<br>*46, 45, 29* |
| $v_1$ (74) | $H_2N^+$=CH-COOH<br>(consecutive cleavage from $y_1$) | $CH_3$COOH (60)<br>*60, 45, 43* |
| $b_1$ (72) | $H_2N$-CH($CH_3$)-C≡$O^+$ | $H_2N$-CH($CH_2$COOH)-COOH (133)<br>*88, 74, 70, 43, 28* |
| $a_1$ (*44*) | $^+H_2N$=CH($CH_3$)<br>(consecutive cleavage from $b_1$) | CO (28)<br>*28* |

[a] Nomenclature from ref. [22].
[b] Ions expected based on reported NR [2-4] or EI [23] mass spectra.

**Acknowledgements**

We thank Marcela M. Cordero and Jinnan Cai for experimental assistance and Aberra Fura and Fred W. McLafferty for stimulating discussions. This work was generously supported by The University of Akron and the Petroleum Research Fund, administered by the American Chemical Society.

## References


1. Gellene, G. I. and Porter, R. F. (1983) *Acc. Chem. Res.*, **16**, 200.
2. Wesdemiotis, C. and McLafferty, F. W. (1987) *Chem. Rev.*, **87**, 485.
3. Terlouw, J. K. and Schwarz, H. (1987) *Angew. Chem. Int. Ed. Engl.*, **26**, 805.
4. Holmes, J.L. (1988) *Mass Spectrom. Rev.*, **8**, 513.
5. Wesdemiotis, C. and McLafferty, F.W. (1989) *Org. Mass Spectrom.*, **24**, 663.
6. Busch, K. L., Glish, G. L., and McLuckey, S.A. (1988) Mass Spectrometry / Mass Spectrometry, Verlag Chemie, Weinheim.
7. (a) Wesdemiotis, C., Danis, P. O., Feng, R., Tso, J., and McLafferty, F. W. (1985) *J. Am. Chem. Soc.*, **107**, 8059. (b) Zhang, M. -Y., Wesdemiotis, C., Marchetti, M., Ray, J. C., Jr., Carpenter, B. K., and McLafferty, F. W. (1989) *J. Am. Chem. Soc.*, **111**, 8341.
8. Yates, B. F., Bouma, W. J., and Radom, L. (1986) *Tetrahedron*, **42**, 6225.
9. McAdoo, D. J. (1988) *Mass Spectrom. Rev.*, **7**, 363.
10. Hammerum, S. (1988) *Mass Spectrom. Rev.*, **7**, 123.
11. McLafferty, F. W. (1980) Interpretation of Mass Spectra, 3rd ed., University Science Books, Mill Valley.
12. Polce, M. J., Cordero, M. M., Wesdemiotis, C., and Bott, P. A. *Int. J. Mass Spectrom. Ion Processes*, submitted.
13. Todd, P. J. and McLafferty, F. W. (1981) *Int. J. Mass Spectrom. Ion Processes*, **38**, 371.
14. Harrison, A. G. (1987) *Org. Mass Spectrom.*, **22**, 637.
15. Mautner, M. and Sieck, L. W. (1989) *Int. J. Mass Spectrom. Ion Processes*, **92**, 123.
16. Jarrold, M. F., Illies, A. I., Kirchner, N. J., and Bowers, M. T. (1988) *Org. Mass Spectrom.*, **18**, 388.
17. Wesdemiotis, C., Fura, A., and McLafferty, F. W. (1991) *J. Am. Soc. Mass Spectrom.*, **2**, in press.
18. Swanton, D. J., Marsden, B. C. J., and Radom, L. (1991) *Org. Mass Spectrom.*, **26**, 227.
19. Zalotai, I., Hunyadi-Zoltán, Z., Bérces, T., and Márta, F. (1983) *Int. J. Chem. Kin.*, **15**, 505.
20. Wittneben, D. and Grützmacher, H. -F. (1990) *Int. J. Mass Spectrom. Ion Processes*, **100**, 545.
21. Polce, M. J. and Wesdemiotis, C. (1991) Proceedings of the 39th ASMS Conference on Mass Spectrometry and Allied Topics, Nashville, Tennessee.
22. Martin, S. A., Johnson, R. S., Costello, C. E., and Biemann, K. (1988) in C. J. McNeal (ed.), The Analysis of Peptides and Proteins by Mass Spectrometry, John Wiley & Sons, Sons, Chichester, p. 135.
23. McLafferty, F. W. and Stauffer, D. B. (1989) *Wiley/NBS Registry of Mass Spectral Data*, Wiley, New York.

# SOME APPLICATIONS OF THE QUANTUM-MECHANICAL SEMIEMPIRICAL METHODS TO THE GAS-PHASE CHEMISTRY OF BIO-ORGANIC IONS.

N.RUSSO and M. TOSCANO
*Dipartimento di Chimica, Università della Calabria,*
*I-87030 Arcavacata di Rende (CS), Italy.*

## ABSTRACT

Model Hamiltonians of several quantum-mechanical methods together with potential energy hypersurfaces theory are briefly reviewed. The gas-phase chemistry of various bio-organic ions treated with different NDO parametrizations is reported and discussed. In particular, the protonation of deoxynucleosides and nucleobases present in DNAs, and the fragmentation of the fatty acid esters are given as examples of application of modern computational chemistry.

## 1. INTRODUCTION

Recent developments in mass spectrometry techniques and in theoretical methods and computational tools have allowed accurate studies of biological macromolecules both from experimental and theoretical points-of-view [1-5]. The reliability of techniques such as tandem mass spectrometry (MS/MS) in connection with desorption ionization methods has been widely demonstrated (see other chapters of this book). The processes that occur in a mass spectrometer have the advantage to work in absence of intermolecular interactions and solvent effects. In this situation, the results of mass spectrometry studies represent a good test for the reliability of theoretical methods. On the other hand, because it is difficult to have structural information for biomacromolecules, theoretical results can be used as a guideline for the gas-phase chemistry of these systems.

At the theoretical level, treatment of the chemical reactions requires firstly detailed knoweledge of the potential energy of the molecular systems. The Born-Oppenheimer approximation provides a convenient starting point to solve the Scroedinger equation for a system with m electrons and for a fixed set of 3M - 6 parameters (where M is the nuclei number). For systems with a very large number of electrons and nuclei, such as the biomacromolecules, the solution of Hartree Fock (HF) ab-initio equation with reliable basis sets implies the computation of a very high number of multicenter integrals. This means that the calculations of the potential energy hypersurfaces need prohibitive computational times. Nevertheless, in the last decades, a series of semiempirical methods, in which the HF equation has been simplified by the introduction of different levels of approximation and parametrization, has been successfully developed. The most common approximations are the EHT (Extended Huckel Theory) [6], the ZDO (Zero



*M. L. Gross (ed.), Mass Spectrometry in the Biological Sciences: A Tutorial,* 311–322.

Differential Overlap) [7] and the NDO (Neglect of Differential Overlap) [7]. In the former, the electrons are considered to be completely indipendent. In the ZDO approximation, that includes CNDO/2 [7] , INDO [7] and MINDO [8] methods, the bielectronic integrals containing products of non-identical atomic orbitals are neglected. Finally in the NDO approximations (NDDO [7], MNDO [9], and AM1 [10] methods) only the integrals at three and four centers are neglected. Many of these methods (expecially MINDO/3, MNDO and AM1) have been widely used for the determination of potential energy surfaces of several reactions, and reliable results as well as relatively low computational effort for macromolecular systems have been achieved [11-12].

## 2. BASIC THEORY AND MODEL HAMILTONIANS

By introducing the LCAO expansion of one-electron states in terms of atomic orbitals and considering the Born- Oppenheimer approximation, the total energy E for a many-electron system can be written:

$$E = Ro + \sum_{\mu\nu} P_{\mu\nu} H_{\mu\nu} + 1/2 \sum_{\mu\nu\rho\tau} P_{\mu\nu} P_{\rho\tau} [(\mu\nu/\rho\tau)-1/2(\mu\tau/\rho\nu)] + Ecorr$$

where Ro and Ecorr are the nuclear repulsion term and the correlation energy, respectively, and

$$P_{\mu\nu} = \sum_i q_i C^*_{\mu i} C_{\nu i}$$

where $q_i$ is number of electrons in the one-electron states |i>. The electronic energy terms can be obtained by solving a secular equation formally equivalent to the HF one:

$$FC = SC\Lambda$$

with

$$F_{\mu\nu} = H_{\mu\nu} + \sum_{\rho\tau} P_{\rho\tau} [(\mu\nu/\rho\tau) -1/2(\mu\tau/\rho\nu)]$$

where S is the overlap matrix of the atomic orbitals and $\Lambda$ is the diagonal matrix of orbital energies. For the ab-initio methods [13], all the integral elements are explicitly calculated, whereas for the semiempirical ones, approximations are introduced as follows. We can rewrite the Fock operator as:

$$F_{\mu\nu} = F^o_{\mu\nu} + F^q_{\mu\nu}$$

where

$$F^o_{\mu\nu} = H_{\mu\nu} + \sum_{\lambda} n_\lambda [(\mu\nu/\lambda\lambda) -1/2(\mu\lambda/\nu\lambda)]$$

is the Fock matrix for an hypotetical overlap of neutral atoms, and

$$F^{q}_{\mu\nu} = \sum_{\lambda\sigma} (P_{\lambda\sigma} - \delta_{\lambda\sigma}\, n_{\lambda})\, [(\mu\nu/\lambda\sigma) - 1/2(\mu\lambda/\nu\sigma)]$$

is a molecular correction term depending from the electron displacements with n as an average occupation number for the atomic orbital $|\lambda\rangle$ in an appropriate valence state. By using pure atomic orbitals, the $F^{o}_{\mu\nu}$ term can be given in the form:

$$F^{o}_{\mu\nu} = \delta_{\mu\nu}\, \chi_{\mu} + (1 - \delta_{AB})\, \beta_{\mu\nu} + \Sigma\, U^{C}_{\mu\nu}$$

where $\chi_{\mu}$ is the electronegativity of the $|\mu\rangle$ orbital, $\beta_{\mu\nu}$ is a term of interaction between $|\mu\rangle$ and $|\nu\rangle$ orbitals of the A and B atoms, respectively, and $U^{C}_{\mu\nu}$ is the potential due to the neutral atom C (the so-called penetration) with the summation extended to the atoms different from A and B. In the ZDO approximation, taking into account that the orthogonalization aims to compensate the penetration terms, $F^{o}_{\mu\nu}$ can be rewritten as:

$$^{\lambda}F^{o}_{\mu A \nu B} = \delta_{\mu\nu}\, \chi_{\mu} + (1 - \delta_{AB})$$

The $F^{q}_{\mu\nu}$ in this approximation (CNDO) is written as follows:

$$^{\lambda}F^{q}_{\mu A \nu B} = \sum_{\lambda} (P_{\lambda\lambda} - n_{\lambda})[\, \delta_{\mu\nu}\, (\mu\mu/\lambda\lambda)] - 1/2 + P_{\mu\nu}\, (\mu\mu/\nu\nu)$$

whereas in the NDO (i.e., MNDO and AM1) methods, it is:

$$^{\lambda}F^{q}_{\mu A \nu B} = \sum_{\lambda C \sigma D} (P_{\lambda\sigma} - \delta_{\lambda\sigma}\, n_{\lambda})[\sigma_{AB}\, \delta_{CD}(\mu\nu/\lambda\sigma) - 1/2(\mu\lambda/\nu\sigma)]$$

and the integrals at three and four centers are neglected. Furthermore, in the recent AM1 method, the core repulsion function (CRF) is modified as follows:

$$CRF(AB) = Z_A Z_B\, \gamma_{ss}[1 + F(A) + F(B)]$$

where

$$F(A) = \exp(-\,\alpha_A R_{AB}) + \sum_{i} K_{Ai}\, \exp[L_{Ai}(R_{AB} - M_{Ai})^2]$$

$$F(B) = \exp(-\,\alpha_B R_{AB}) + \sum_{j} K_{Bj}\, \exp[L_{Bj}(R_{AB} - M_{Bj})^2]$$

With the respect to the MNDO parametrization, the AM1 one reproduces better the hydrogen bonds and gives more reliable results for biological macromolecules.

## 3.THE POTENTIAL ENERGY HYPERSURFACES

The description of a transformation process from reactants to products needs the knoweledge of the potential energy surface. The potential energy of a molecular system is a function of the nuclear positions. The coordinate set that contains a number of variables large enough to express completely the potential energy is called the internal coordinate system: $q' = q_1....q_{3N-6}$ . For simple molecules, the potential energy surface depends only on a few variables, but for larger cases, the number of such variables increases and the representation becomes more complicated. In these cases it is more appropriate to talk of potential energy hypersurfaces [14, 18]. For one-step chemical reactions, the potential energy profile is simple to obtain. A two-parameter energy function can be visualized by a two dimension contour diagram (see Figure 1) while a multiparametric energy function can be represented as an energy hypersurface in multidimensional space, and it is not easy to visualize. To simplify the problem, a transformation involving many parameters can be described as a simple pathway in which the molecular system goes from the reactants minimum to the products minimum through a transition state, or saddle point, which corresponds to a maximum along the reaction pathway.

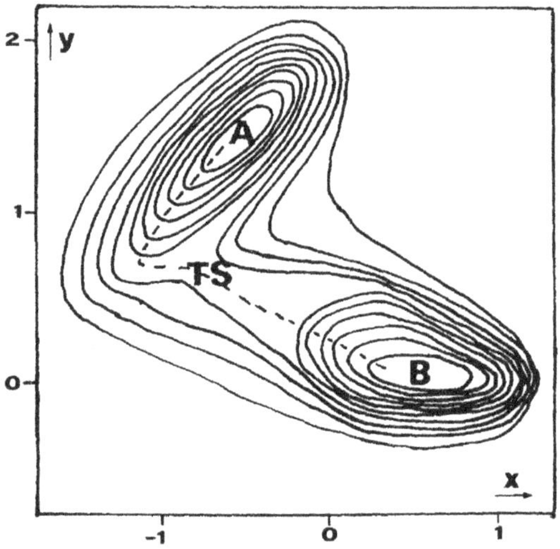


Figure 1. Two-parametric model potential energy surface.

In the two parametric model potential of Figure 1, it is possible to locate two energy minima and one saddle point. The orthogonal trajectory that connects via a common saddle point the two minima is called minimum-energy- reaction-path. This means that the minimum-energy-reaction- path is composed of two steepest descent paths that start from a saddle point in both directions along the principal axis of negative curvature and enter the two minima each along the principal axis of least curvature (Figure 1). Because the energy function is generally not analytically available but computed point by point, locating of the minimum-energy-reaction-path is difficult. In these last years, different methods have been proposed [22-26]. The most common one consists of choosing a suitable structural parameter along which the reaction is driven as the reaction coordinate [26]. For some significant values of this parameter, the total energy of the system is minimized with the respect to all other structural parameters. Although this procedure seems to be very simple, a reliable choise of the reaction coordinate is often quite difficult, and care is advised for cases of doubt on the choice of the possible different reaction paths. The location of the transition state is another problematic point. The saddle points are stationary points on the energy hypersurfaces where the slope vanishes in all directions, but with curvatures that are not all positive.

For this reason, transition states cannot be located by minimization or maximization of the molecular energy with respect to the molecular parameters. Several methods were proposed [17, 18, 27] and for one of them [27] it is suggested that, the energy gradient norm be minimized:

$$\sum_i (\delta E/\delta q_i)^2 = \text{minimum}$$

For this method, it is necessary to know the environment of the transition state, and this is not easy. In any case, the location of a probable transition state will be confirmed by the presence of only a negative eigenvalue in the force costant matrix.

## 4. APPLICATIONS

There has been considerable recent interest in the chemistry of aliphatic epoxide radical cations [28, 29] because they are present in some important systems such as fatty acid esters. As a first example, we consider the study at MNDO and AM1 level of these systems starting from the ethylene oxide radical cation for which ab-initio [30] and experimental ]31] data are available for a comparison. Assuming as a reaction coordinate the r parameter, we calculate the potential energy profile for the reaction depicted in scheme I.

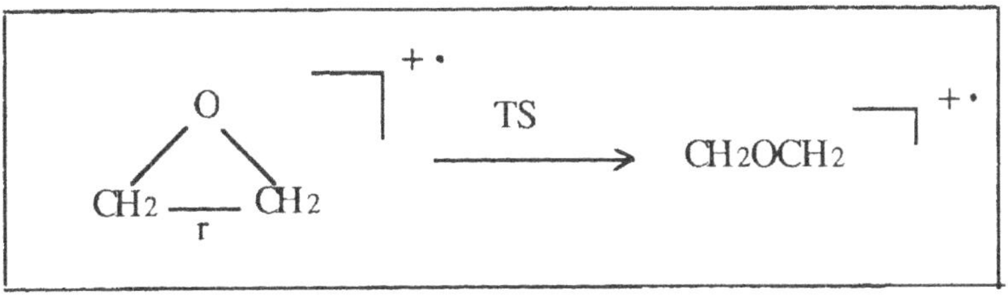


Scheme I

Results are reported in Figure 2.

It is clear from this figure, that the MNDO method strongly overestimates the energy difference between the two minima, but this difference is minor at the AM1 level. The ab-initio computation performed at CAS-Moller-Plesset (MP3) level agrees better with the experimental results. Notwithstanding these differences between semiempirical and ab-initio results, the AM1 and MNDO trends of the reaction are reproduced, and the study of larger systems with these methods, that at the ab-initio MP3 level require very large computational time, can be regarded with confidence, especially from a qualitative point of view.

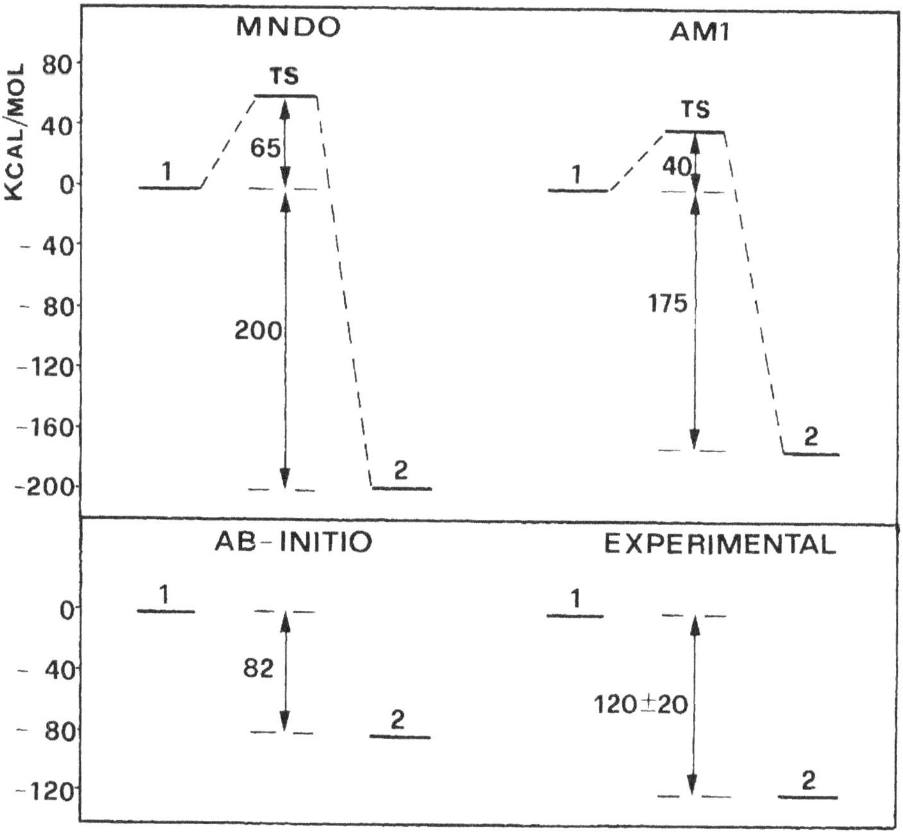


Figure 2. Energy profile for the reaction of scheme I.

Mass spectrometric studies of some aliphatic epoxides and fatty acid esters were reported in the literature [28, 29]. For the latter compounds, different mechanisms of ring-breaking and fragmentation were proposed (see scheme II).

We have investigated the more important processes (1 and 2 in scheme II) employing two model compounds with $R_1$ = $-C_2H_5$ and $R_2$ = $-(CH_2)_2COOCH_3$ and $(CH_2)_4COOCH_3$. A schematic drawing of the reaction pathways is reported in Figure 3. For mechanism 1, a comparison of results shows that transition barriers height does not change significantly with the length of the $R_2$ chain whereas the energy difference between reactants and products increases (31.8 vs 45.6 Kcal/mol).

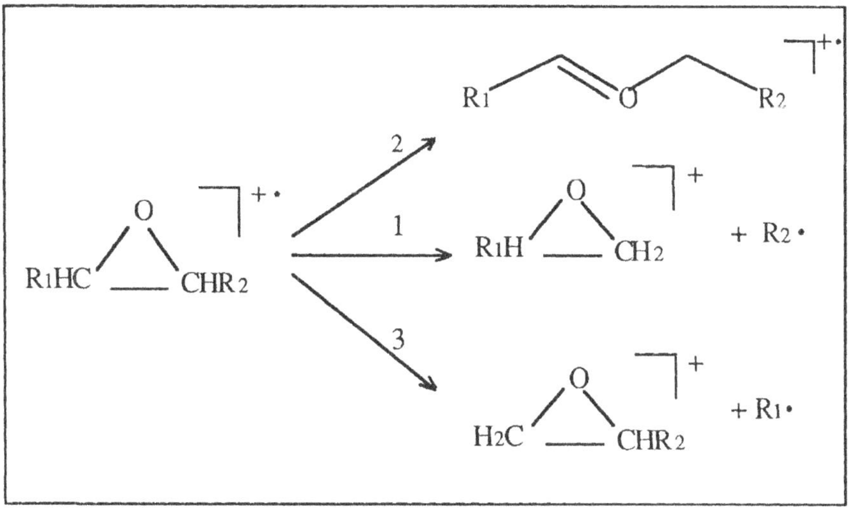


Scheme II

With respect to the case of ethylene oxide radical cation, the height of the transition barrier, for the ring-breaking, decreases more drastically (65.1 vs 40.7 or 41.8 Kcal/mol).

These results show that the barrier for homolytic dissociation of the C-C bond of monoenoic fatty acid esters is definitely lower than that of ethylene oxide radical cation.

In Figure 3, it is evident that the loss of the chain (mechanism 2) is not favoured both from kinetic and thermodynamic points-of-view and is in agreement with the experimental information [28, 29].

As a further example of application of semiempirical methods to the gas-phase bio-organic ion chemistry, we report the proton affinities of nucleobases and deoxyribonuclesides calculated by means AM1 method. It is well-known that protonation of nucleic acid bases at various electronegative centers plays an important role in some biochemical and biophysical processes and has been the subject of extensive experimental and theoretical investigation [32- 39]. Notwithstanding these studies, several fundamental questions remain open for discussion. From a theoretical point of view, only protonation of RNA and DNA bases was investigated, but no study of deoxynucleosides exists until now.

Because the NDO methods generally overestimate the stabilities of the "hydroxy" tautomers with the respect to the "keto" ones [40- 41] we have considered only the structures reported in Figures 4 and 5.

Of course, a more detailed study considering all the possible tautomers is necessary, and it is in progress in our laboratory. Nevertheless, our results can be regarded as a first indication of the validity of the AM1 method to predict reliable proton affinities [42].

Different protonation sites were considered, and in all the calculations, a full geometry optimization was performed because its importance for obtaining realistic results was already demonstrated [39, 42].

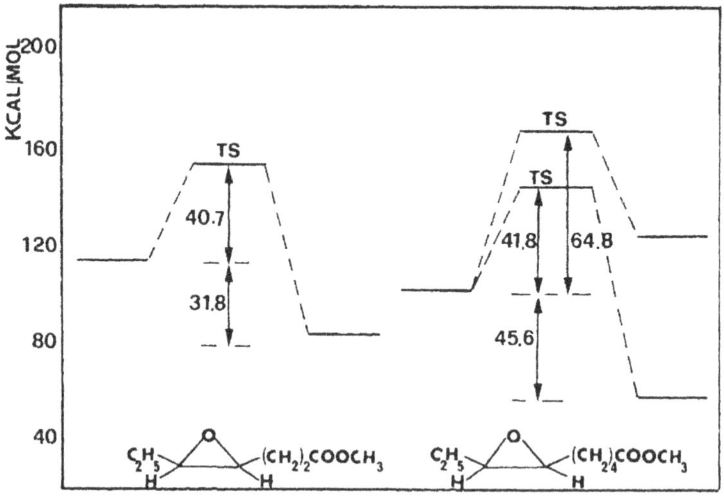


Figure 3. MNDO energy profile for the reactions of scheme II.

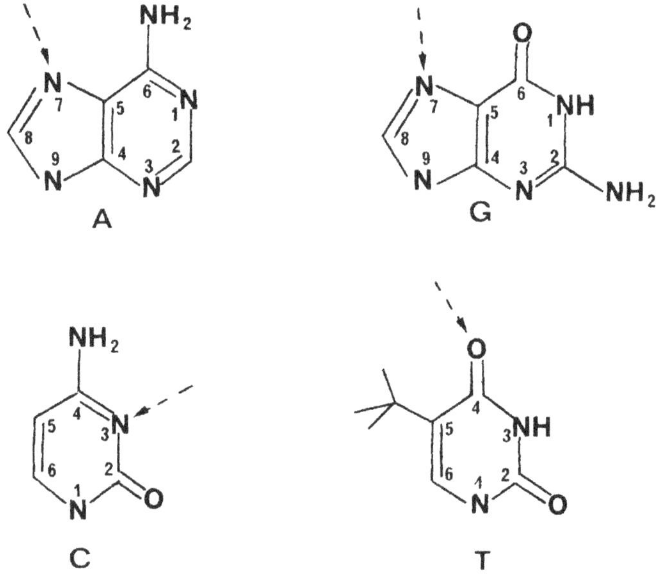


Figure 4. Cytosine, thymine, guanine and adenine nucleobases. The arrows indicate the most stable protonation sites determined at the AM1 level.

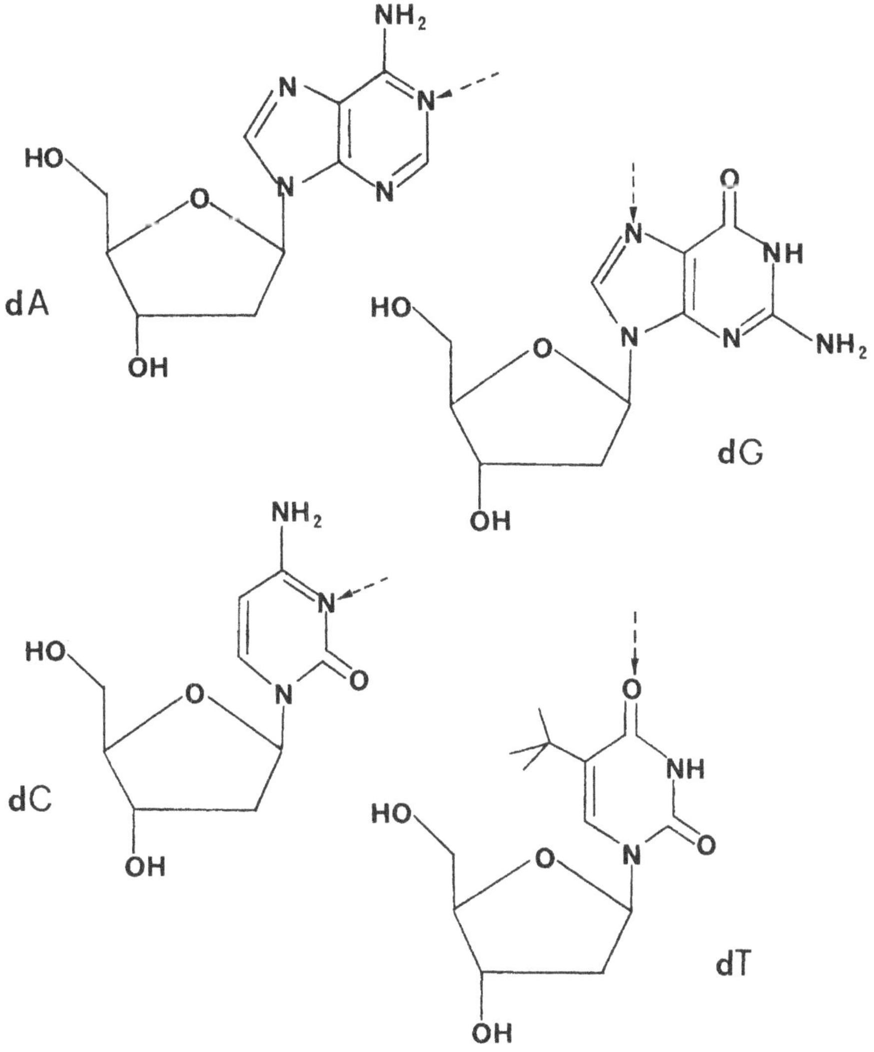


Figure 5. Schematic drawing of deoxyribonucleosides. The arrows indicate the most stable protonation sites computed at the AM1 level.

The most stable protonation sites are shown in Figures 4 and 5 and are in agreement with those indicated in an MS/MS study [32] and with those from previous theoretical ab- initio studies [39]. In table I, our proton affinities (PA) for nucleobases and corresponding deoxynucleosides are collected together with the most recent theoretical and gas-phase experimental values.

Table I. Proton affinities (Kcal/mol) for nucleobases and related deoxyribonucleosides.

| Molecule | AM1 | AB-INITIO[a] | AB-INITIO[b] | EXP[c] | EXP[d] |
|---|---|---|---|---|---|
| Adenine | 214.3 | 241.1 | 286.6 | 224.2 | 223.5 |
| Guanine | 225.7 | 244.7 | 287.2 | 227.4 | 223.0 |
| Cytosine | 227.2 | 249.2 | / | 225.9 | 223.8 |
| Thymine | 204.9 | 211.8 | / | 209.0 | 208.8 |
| dAdenine | 230.7 | / | / | 233.6 | / |
| dGuanine | 231.8 | / | / | 234.4 | / |
| dCytosine | 234.8 | / | / | 233.2 | / |
| dThymine | 212.4 | / | / | 223.5 | / |

a) from ref. 39; b) from ref. 38; c) from ref. 32; d) from ref. 33.

As it is shown in this table, the AM1-values are closer to experiment than those obtained at different ab-initio levels.

The relative proton affinities of nucleobases derived from AM1 calculations follow the order:

Cytosine > Guanine > Adenine >> Thymine

in agreement with the results of an ab-initio study. The experimental order is:

Guanine > Cytosine > Adenine >> Thymine

The discrepance concerns the inversion between guanine and cytosine proton affinities and is probably due to the choice of the tautomers for these molecules. The calculations of proton affinities for other possible tautomers are necessary in order to have more careful results. In the case of deoxyribonucleosides we have found the following order of PA:

dC > dG = dA >> dT

in agreement with the order proposed on the basis of pKa' s values [35]. The gas-phase experimental order is:

dG > dA = dC >> dT

Also in this case we think that an extension of our study to the other tautomers of dA, dC, and dG can be resolutive.

## 5. CONCLUSIONS

In this contribution we have briefly reported the quantum-mechanical semiempirical methods theory and we have given some applications of NDO approximations to the gas-phase chemistry of bio-organic ions. The studies on fragmentation mechanism of the fatty acid esters and the calculations on the proton affinities of nucleobases and deoxynuclesides show that the MNDO and AM1 results agree with experiment and can be used to give better insight in the knoweledge of the potential energy surfaces of large bio-organic systems.

## 6. ACKNOWELEDGEMENT

This work was carried out with financial support of MURST.

## 7. REFERENCES

1. Clementi, E. and Chin, S. (1988.), Biological and Artificial Intelligence Systems, ESCOM Science Publishers B.V., Leiden.
2. Naray-Szabo, G. (1966), Theoretical Chemistry of Biological Systems, Elsevier, Amsterdam.
3. Sindona, G. (1989) in Theofanides, T. (Ed.), Spectroscopy of Inorganic Bioactivators,
NATO ASI Series Vol. 280, Kluwer, Dordrecht, pg. 321.
4. Grese, R.P., Cerny, R.L., and Grass, M.L. (1989) J. Am. Chem. Soc., **111**, 2835.
5. Gaskel, S. (1986), Mass Spectrometry in Biomedical Reserch, J. Wiley & Sons, Chichester, .
6. Hoffmann, R. (1963) J. Chem. Phys., **30**, 1397.
7. Pople, J.A. and Beveridge, D.L. (1970), Approximate Molecular Orbital Theory, McGraw-Hill, New York.
8. Bingham, R.L., Dewar, M.J.S., and Lo, D.H. (1975) J. Am. Chem. Soc., **97**, 1285.
9. Dewar, M.J.S, and Thiel, W. (1977) J. Am. Chem. Soc., **99**, 4899.
10. Dewar, M.J.S., Zoebisch, E.G., Healy, E.F., and Steward, J.P. (1985) J. Am. Chem. Soc., **107**, 3902.
11. Russo, N., Toscano, M. (1986) J. Mol. Struct (Theochem), **148**, 175.
12. Welsh, W.J. (1990) J. Comput. Chem., **11**, 644.
13. Szabo, A., and Ostlund, N.S. (1982), Modern Quantum Chemistry, MacMillan, New York.
14. Muller, K. (1980) Angew. Chem. Int. Ed. Engl., **19**, 1.
15. Larquet, J.C. (1981) Org. Mass Spectrom., **16**, 469.
16. Desouter-Lecomte, M., Dehareny, D., Leyh-Nihant, B., Praet, M. Th., Larquet, A.J., and Larquet, J.C. (1985) J. Phys. Chem., **89**, 214.
17. Fukui, K., Kato, S., and Fujimoto, H. (1975) J. Am. Chem. Soc., **97**, 1.
18. Epiotis, N.D. (1978), Theory of Organic Rections. Reactivity and Structure, Concepts in Organic Chemistry, Vol. 5, Springer, Berlin.
19. Tachibana, A., and Fukui, K. (1978) Theor. Chim. Acta, **49**, 321.
20. Ishida, K., Marokuma, K., and Komarnicki, A. (1977) J. Chem. Phys., **66**, 2153.

21. Basilevsky, M.V. (1977) Chem. Phys, **24**, 81.
22. Halgren, T.A., and Lipscomb, W. (1977) Chem. Phys. Lett., **49**, 225.
23. Pancir, J. (1975) Collect. Czech. Chem. Commun., **40**, 1112.
24. Muller, K., and Brown, L.D. (1979) Theor. Chim. Acta, **53**, 75.
25. Liotard, D., and Punot, J.P. (1980), Colloque sur les Fenomenes Critiques, Springer, Berlin.
26. Dewar, M.J.S., and Kirschner, S. (1971) J. Am. Chem. Soc., **93**, 4290.
27. McIver, J.W., and Koma, A. (1971) Chem. Phys. Lett., **10**, 303.
28. Liguori, A., Romeo, G., Sindona, G., and Uccella, N. (1972) Chim. Ind., **69**, 2.
29. Bouchoux, G., Hoppilliard, Y., and Jaudou, P. (1987) Org. Mass Spectrom., **22**, 98.
30. Bouma, W.J., Poppinger, D., Saebo, S., MacLeod, J.K., and Radom, L. (1984) Chem. Phys. Lett., **104**, 198.
31. Carderman, R.R., Le Breton, P.R., Buttril, S.E., Williamson, A.D., and Beauchamp, J.L. (1976) J. Chem. Phys., **65**, 4929.
32. Liguori, A., Greco, F., Sindona, G., and Uccella, N. (1990) J. Am. Chem. Soc., **112**, 9092.
33. Lias, S.G., Liebmann, J.F., and Levin, R.D. (1984) J. Phys. Chem. Ref. Data, **13**, 695.
34. Izatt, R.M., Christensen, J.J., and Rytting, J.H. (1971) Chem. Rev., **71**, 439.
35. McGhee, J.D., and von Hippel, P.H. (1975) Biochem, **14**, 1281.
36. Lavery, R., Pullman, A., and Pullman, B. (1978) Theor. Chim. Acta **50**, 67.
37. Bonaccorsi, R., Scrocco, E., Tomasi, J., and Pullman, A. (1975) Theor. Chim. Acta, **36**, 339.
38. Mezey, P.G., Ladik, J.J., and Barry, M. (1980) Theor. Chim. Acta, **54**, 251.
39. Del Bene, J.E. (1983) J. Phys. Chem., **87**, 367.
40. Mirek, J., and Sygula, A. (1981) J. Mol. Struct. (Theochem), **86**, 85.
41. Buda, A., and Sygula, A. (1983) J. Mol. Struct. (Theochem), **92**, 255; ibidem 267.
42. Russo, N., and Toscano, M. in preparation.
43. Czerminski, R., Lessyng, B., and Pohorilla, A. (1979) Int. J. Quantum Chem., **16**, 605.

# Part III
# Applications to Biomolecules

# CHARACTERIZATION OF HUMAN HEMOGLOBIN VARIANTS BY MASS SPECTROMETRY

A. Malorni[1], P. Pucci[1,2], P. Ferranti[1] and G. Marino[1,3]
[1]*ICMIB-CNR and Servizio di Spettrometria di Massa, Via Pansini, 5, 80131 Napoli, Italy*
[2]*Dipartimento di Chimica, Università della Basilicata, Potenza, Italy*
[3]*Dipartimento di Chimica Organica e Biologica, Università di Napoli, Italy*

ABSTRACT. A procedure for structural analysis of normal and variant human hemoglobins by using fast atom bombardment mass spectrometry is described. The methodology consists of direct mass spectrometric analysis of proteolytic digests of the variant globin; peptide recognition along the globin chain sequence is accomplished on the basis of their molecular weight. The abnormal peptide(s) will exhibit a mass value different from those of the normal hemoglobin, allowing easy definition of the structural variations. The application of this procedure to the determination of normal globin chains as well as to known and unknown abnormal human hemoglobins is reviewed.

## 1. Introduction

Since the definition of the molecular basis of sickle cell anemia by Pauling and co-workers [1] and the identification of the amino acid substitution by Ingram [2], over 450 human hemoglobin variants have been described. More than one third have been found to have alterations in function and/or stability (for a complete list of human hemoglobin variants see [3]).

Modern methodologies for detection of genetic variants of human hemoglobin rely on electrophoresis and/or isoelectric focusing procedures; further structural characterization of the variant globin is based on HPLC separation of the peptides generated by tryptic digestion, followed by identification of the abnormal peptide, amino acid analysis and peptide sequence determination. This procedure is extremely dependent upon the actual separation of tryptic peptides and, even when successful, it is very time-consuming. Moreover, this structural analysis is limited to peptide fractions exhibiting abnormal chromatographic properties; by using such an approach it might be impossible to detect those modifications not affecting peptide retention times.

In the past few years, several methods of analysing variant hemoglobins by using mass spectrometric techniques have been proposed [4, 5].

In 1983 a new strategy for the analysis of protein structures based on Fast Atom Bombardment Mass Spectrometry (FAB/MS) was developed [6]. The strategy, christened "FAB-mapping", involves the FAB/MS analysis of peptide mixtures generated by enzymatic and/or chemical digestion of proteins. The signals recorded in the mass spectrum are assigned to the corresponding peptides along the protein sequence on the basis of their molecular weights.

The only current limitation so far reported for the use of FAB/MS for peptide mixture analysis is that not all the peptide fragments are seen owing to suppression phenomena occurring in the matrix [7]. It has been shown, however, that low molecular weight proteins, such as human alpha, beta and gamma globins, can be mapped up to over 80% of the entire sequence (see below in the text).

When the FAB-mapping procedure is applied to the analysis of hemoglobinopathies, the



*M. L. Gross (ed.), Mass Spectrometry in the Biological Sciences: A Tutorial*, 325–332.

general strategy consists in the construction of the peptide maps of alpha, beta and gamma normal chains by FAB/MS analysis of the respective tryptic digests. Then the FAB map of the variant globin is compared to that of the normal chain. Those peptides that are modified exhibit an unusual molecular weight, immediately allowing the identification of the globin fragment where the variation has taken place.

Sometimes the mass difference itself indicates where the modification occurred; in other cases, simple protein chemistry procedures (Edman degradation steps, Carboxypeptidase B hydrolysis, subdigestion with other proteolytic enzymes, etc.) carried out on the peptide mixture, followed by FAB/MS analysis, is needed to identify unambiguously the site and the nature of the modification.

## 2. Analysis of normal globin chains

FAB/MS analysis of tryptic digests from human normal alpha, beta and gamma globins are shown in Tables I-III. Signals recorded in the mass spectra are assigned to peptide fragments along the globin sequences.

The FAB map of alpha globin showed 13 intense signals with satellite peaks 15 u lower than the individual parent peak; they correspond to the protonated molecules of the peptides that can be desorbed from the mixture. Out of the 13 $(M + H)^+$ ions, six showed a mass corresponding to the molecular weight of expected tryptic peptides and seven were produced by a further chymotryptic-like cleavage of tryptic peptides 62-90 and 100-141. This is due to the use of a trypsin preparation not treated with any chymotrypsin inhibitor in order to take advantage of aspecific cleavages to produce suitably sized peptides.

The assignments were confirmed by a single step of Edman degradation carried out on the tryptic peptide mixture followed by FAB/MS analysis of the truncated peptides. Tryptic peptides 1-16, 57-61, 91-92 and chymotryptic peptides 100-109, 137-141 were not detected in the FAB map of the normal human alpha-globin chain because of the well-known desorption problems occurring when analyzing peptide mixtures, leading to the mapping of 81% of the entire sequence.

Thirteen signals were identified in the FAB map of normal beta chains, 12 corresponding to predicted tryptic peptides, and the last one at *m/z* 1797 referring to peptide 67-81 plus an extra lysine residue at the N-terminus. The occurrence of such a peptide is due to the incomplete cleavage between lysine 66 and valine 67 because there is a further lysine residue present at position 65, which is the preferential site for tryptic hydrolysis.

The normal beta globin was completely mapped by means of this experiment; the only undetected fragments were the two dipeptides Val-Lys and Tyr-His at positions 60-61 and 145-146, respectively.

FAB/MS analysis of the peptide mixture derived from the tryptic digest of gamma globin allowed the mapping of 86% of the entire sequence. The signal at *m/z* 848, corresponding to peptide 133-141, demonstrates the presence of a glycine residue at position 132, indicating that the globin under examination is a gamma-G chain; no evidence for an alanine residue at position 136 was obtained (gamma-A chain, peptide putative *m/z* value = 862).

TABLE I. Observed $(M + H)^+$ and position in the sequence of tryptic peptides from normal human alpha chain.

| Position in sequence | $(M + H)^+$ | Position in sequence | $(M + H)^+$ |
|---|---|---|---|
| 1- 7 | 729 | 73-80 | 874 |
| 8-11 | 461 | 81-90 | 1078 |
| 12-16 | - | 91-92 | - |
| 17-31 | 1529 | 93-99 | 818 |
| 32-40 | 1071 | 100-109 | - |
| 41-56 | 1833 | 110-122 | 1360 |
| 57-61 | - | 123-127 | 533 |
| 62-68 | 703 | 128-136 | 936 |
| 69-72 | 397 | 137-141 | - |

TABLE II. Observed $(M + H)^+$ and position in the sequence of tryptic peptides from normal human beta globin

| Position in sequence | $(M + H)^+$ | Position in sequence | $(M + H)^+$ |
|---|---|---|---|
| 1- 8 | 952 | 67-82 | 1669 |
| 9-17 | 932 | 83-95 | 1421 |
| 18-30 | 1314 | 96-104 | 1126 |
| 31-40 | 1274 | 105-120 | 1719 |
| 41-59 | 2058 | 121-132 | 1378 |
| 60-61 | - | 133-144 | 1149 |
| 62-65 | 412 | 145-146 | - |
| 66-82 | 1797 | | |

TABLE III. Observed $(M + H)^+$ and position in the sequence of tryptic peptides from normal human gamma chain

| Position in sequence | $(M + H)^+$ | Position in sequence | $(M + H)^+$ |
|---|---|---|---|
| 1- 8 | - | 77-82 | 740 |
| 9-17 | 976 | 83-95 | 1448 |
| 18-30 | 1316 | 96-104 | 1098 |
| 31-40 | 1274 | 105-120 | 1693 |
| 41-59 | 1989 | 121-132 | - |
| 60-61 | - | 133-141 | 848 |
| 62-65 | - | 142-144 | - |
| 66-76 | 1144 | 145-146 | - |
| 67-76 | 1016 | | |

## 3. Analysis of known hemoglobin variants

The ability to detect amino acid substitutions in abnormal globin chains by using the FAB mapping procedure was first tested by investigating two commmon hemoglogin variants, Hb C and Hb D.

The FAB map of tryptic peptides from beta chain of Hb C is shown in Figure 1; the peak at *m/z* 952, corresponding to peptide 1-8, is missing in the spectrum, indicating the occurrence of a modification within this fragment. A new signal occurs at *m/z* 694 indicating that a new cleavage site for trypsin was created, resulting in a shorter peptide. Simple calculations account for the presence of a lysine in position 6, replacing the normal glutamic acid residue; this finding was confirmed by results obtained after a single step of Edman degradation and a Carboxypeptidase B hydrolysis. The mass shift to *m/z* 566, due to the loss of the C-terminal lysine following exopeptidase hydrolysis, gives unambiguous characterization of the variant.

The second example, is the structural analysis of the beta chain from Hb D. Figure 2 shows the partial FAB spectrum of the tryptic peptides derived from a normal beta chain (A) and the variant globin (B). The signal at *m/z* 1377 in the spectrum of the variant chain occurs one mass unit lower than the expected signal at *m/z* 1378 (panel A) corresponding to peptide 121-132. This mass difference accounts only for the substitution of glutamic acid at position 121 with a glutamine residue.

Further confirmation came from the analysis of the truncated peptides following a single step of Edman degradation.

## 4. Analysis of unknown hemoglobin variants

The application of the FAB-mapping strategy to the structural characterization of unknown variants of human hemoglobin is illustrated by two examples among many other investigations carried out in the authors' laboratory. Figure 3 shows the partial FAB spectrum of the tryptic digest of a beta globin variant (panel B) in comparison with the corresponding FAB map of the normal beta chain (panel A). The spectrum of the variant globin shows that the peak at *m/z* 1378, corresponding to peptide 121-132 is absent, and there are two new signals at *m/z* 1249 and 1377. The absence of peptide 121-132 indicates the occurrence of the genetic mutation within this sequence. The new signal at *m/z* 1249 corresponds to peptide 121-132 minus a mass of 129 u, which can be attributed to the loss of the N-terminal glutamic acid. Moreover, the signal at *m/z* 1377 corresponds to the first new signal with an additional mass of 128 u, corresponding to the mass of a lysine or glutamine residue. These two observations suggest that the glutamic acid at the N-terminus of peptide 121-132 was substituted by a lysine residue, thus creating a new site for tryptic cleavage at position 121. This interpretation is supported by results obtained after a single Edman degradation step. The signal at *m/z* 1249 was shifted by 147 u to *m/z* 1102 by the loss of phenylalanine 122, and the signal at *m/z* 1377 changed to 1249, owing to the loss of the N-terminal lysine, confirming the assignments reported above. The variant hemoglobin has a lysine residue at position 121 of its beta chain replacing a glutamic acid: such a variant has been identified as Hemoglobin O-Arab [8].

The ability of the FAB-mapping procedure to give definitive structural information on proteins has been demonstrated by the investigation of Hemoglobin Indianapolis, a very rare beta chain variant [9]. This abnormal hemoglobin was first reported by Adams *et al.* [10] as an extremely labile variant producing the phenotype of a severe beta thalassemia. Recently, the same variant was reported in a clinically normal individual from a Spanish family; in this case the variant did not seem to be biologically unstable [11].

Hb Indianapolis was characterized in an individual from Naples suffering from minor clinical abnormalities, similar to those observed by Baiget *et al.* [11].

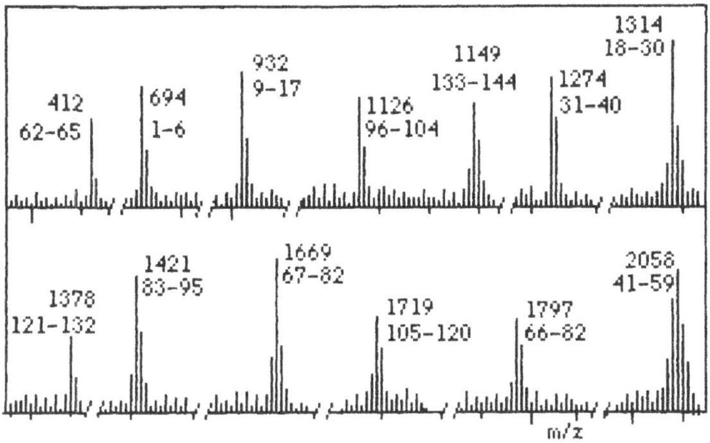


Figure 1. FAB map of the tryptic peptides from the beta chain of Hb C.

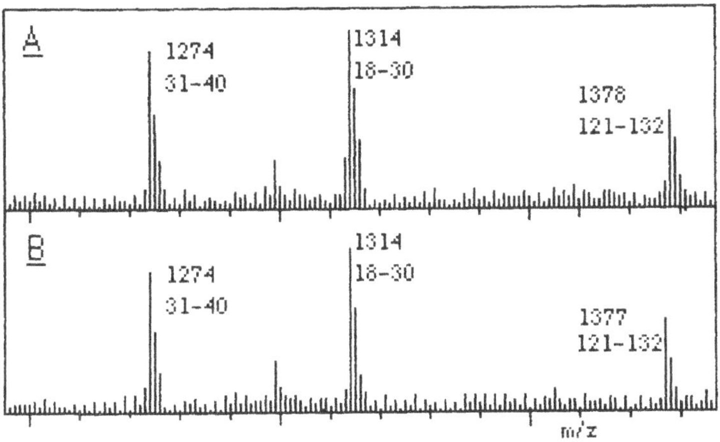


Figure 2. Partial FAB mass spectra of the tryptic peptides from a normal hemoglobin beta chain (A) and from the variant (B) of Hb D.

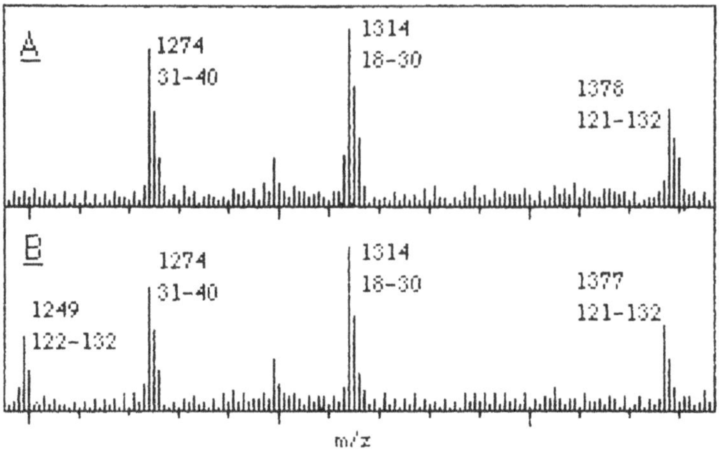


Figure 3. Partial FAB mass spectra of the tryptic peptides from a normal hemoglobin beta chain (A) and from the variant (B).

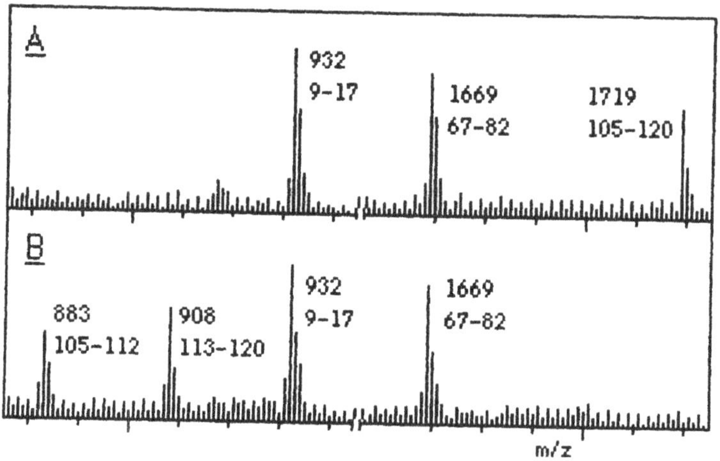


Figure 4. Partial FAB mass spectra of the tryptic digest of a normal hemoglobin beta chain (A) and of the variant (B) from Hb Indianapolis.

Figure 4 shows a partial FAB mass spectrum of the tryptic digest of the beta variant globin (B) and the corresponding FAB map of a normal beta chain (A).

The signals at *m/z* 1719, always observed in the tryptic map of the normal beta chain, were missing; furthermore, two new signals were present in the variant tryptic map at *m/z* 908 and *m/z* 883, respectively. The missing signal at *m/z* 1719 corresponds to sequence 105-120, thus immediately indicating that the substitution occurs within this peptide segment. The peak at *m/z* 908 can be assigned to peptide 113-120, resulting from a new tryptic cleavage site occurring in position 112. Moreover, the peak at *m/z* 883 accounts for peptide 105-112 if the cysteine residue in position 112 had been substituted by an arginine. To confirm the above assignments, the tryptic digest was submitted to a single Edman degradation step and to a Carboxypeptidase B hydrolysis followed by direct analysis of the mixtures by FAB/MS.

Mass spectral data after Carboxypeptidase B hydrolysis allowed us to assign unambiguously arginine to position 112 because of the down-shift of 156 u to *m/z* 727 of the peak at *m/z* 883, whereas the analysis after the Edman degradation step ruled out any possibility of further modifications.

FAB/MS characterization demonstrates that the beta chain variant has an arginine residue in position 112 (i.e., it is a Hemoglobin Indianapolis). This variant, however, did not show any biological instability as was claimed in the first report [10]. The FAB-mapping procedure allowed the mapping of 98% of the entire sequence, leaving little doubt about the occurrence of further modifications. Such a possiblity could not be excluded in the case reported by Adams *et al.* [10] because the extreme lability of the variant did not allow comprehensive structural analysis. This eventuality might explain the discrepancy with the cases reported by Baiget *et al.* [11] and by these authors [9].

## 5. Conclusion

The results presented in this paper clearly show that the FAB mapping strategy can be successfully applied to the characterization of genetic variants of human hemoglobins.

The direct FAB analysis of peptide mixtures generated by proteolytic digests of the variant chain avoids the need of chromatographic separation of peptides. The assignment of the mass values to the corresponding peptides generates a FAB map of the variant protein that does not depend on the chromatographic behaviour of the peptides. With this map, it is possible to identify immediately the site of the substitution, owing to the disappearance of a characteristic peak present in the map of the normal chain, and to the appearance of anomalous signals.

Furthermore, this methodology leads to the mapping of almost the entire sequence of the variant chain, leaving little doubt on whether further modifications have occurred.

## 6. Acknowlegments

This work has been supported partly by grants given by the Italian National Research Council (CNR-Rome) to GM and AM and the Ministero della Pubblica Istruzione (Rome) to GM.

## 7. References

1. Pauling,L., Itano, M. A., Singer, S. J. and Well, I. C. (1949) 'Sickle cell anemia, a molecular disease', *Science* **110**, 543

2. Ingram, V. M. (1956) 'A specific chemical difference between the globins of normal human and sickle-cell anemia haemoglobin', *Nature* **178**, 792.
3. Wrighstone, R. N. (1987) 'A comprehensive list of hemoglobin variants', *Hemoglobin* **11**, 242.
4. Wada, Y.,Hayashi, A., Fujita, J., Matsuo, T., Katakuse, I. and Matsuda, H. (1981) 'Structural analysis of human hemoglobin variants with Field Desorption Mass Spectrometry', *Biochim. Biophys. Acta* **667**, 233.
5. Katakuse, I., Ichihara, T., Nakabushi, H., Matsuo, T., Matsuda, H., Wada, Y. and Hayashi, A. (1984) 'Secondary Ion Mass spectra of tryptic peptides of human haemoglobin chains', *Biomed. Mass Spectrom.* **11**, 386.
6. Morris, H. R., Panico, M. and Taylor, G. W. (1983) 'FAB-mapping of recombinant DNA protein products', *Biochem. Biophys. Res. Commun.* **117**, 299.
7. Naylor, S., Findeis, S., Gibson, B. W. and Williams, D. H. J. (1986) 'An approach towards the complete FAB analysis of enzymatic digests of peptides and proteins', *Am. Chem. Soc.* **108**, 6459.
8. Pucci, P., Carestia, C., Fioretti, G., Mastrobuoni, A. M. and Pagano, L. (1985) 'Protein fingerprint by FAB/MS: characterization of normal and variant human hemoglobins', *Biochem. Biophys. Res. Commun.* **130**, 84.
9. De Biasi, R., Spiteri, D., Caldora, M., Iodice, R., Pucci, P., Malorni, A., Ferranti, P. and Marino, G.(1988) 'Identification by Fast Atom Bombardment Mass Spectrometry of Hb Indianapolis [ß112(G14)Cys->Arg] in a family from Naples, Italy', *Hemoglobin* **12**, 323.
10. Adams, J. G. III, Steinberg, M. H., Boxer, L. A., Baehner, R. L., Forget, B. G. and Tsistrakis, G. A. (1979) 'The structure of Hb Indianapolis [ß112(G14)Arginine]. An unstable variant detectable only by isotopic labeling', *J. Biol. Chem.* **254**, 3479.
11. Baiget, M., Gomez Pereira, C., Jue, D. L., Johnson, M. H., McGuffey, J. E. and Moo-Penn, W. F. (1986) 'A case of Hb Indianapolis [ß112(G14)Cys->Arg] in an individual from Cordoba, Spain', *Hemoglobin* **10**, 483.

# DETECTION OF POST-TRANSLATIONAL MODIFICATION OF PROTEINS BY MASS SPECTROMETRY

G. MARINO[1,2], R. SICILIANO[1], P. PUCCI[1,3], P. FERRANTI[1] and A. MALORNI[1]

[1]*Servizio di Spettrometria di Massa, CNR-Università di Napoli, Via Pansini, 5, 80131 Napoli, Italy*

[2]*Dipartimento di Chimica Organica e Biologica, Università di Napoli, 80134 Napoli, Italy.*

[3]*Dipartimento di Chimica, Università della Basilicata, Potenza, Italy.*

ABSTRACT. The ascertainment of post-translational modifications of proteins constitutes a structural problem often not amenable to conventional protein chemistry methodologies. A procedure that is alternative and complementary to conventional methods is based on the use of mass spectrometry and fast atom bombardment ionization. The molecular weight of the proteolytic fragments of the protein is mapped onto the anticipated protein sequence; post-translational modifications and/or protein processing can be recognized by mass shifts either to lower mass for the loss of signal peptides or to higher mass owing to the addition of various modifications. Application of this strategy to the analysis of several post-translational events is reviewed.

## 1. Introduction

The modern approach to protein sequencing consists of direct stepwise sequencing using the gas-phase Edman technique [1] or in base sequencing the gene coding for the protein [2, 3] followed by translation into the amino acid sequence. The final biological product, however, is commonly the result of a number of post-translational steps, and the assignment of post-translational modified residues is often a difficult task to accomplish without a major commitment in terms of time and sample.

The ascertainment of post-translational modifications, which are usually important in understanding protein function, is becoming increasingly important, and recent research has led to the discovery of several enzyme activities actually involved in modification processes. The importance of post-translational modifications is now well-recognized in such areas of research as blood group antigens (carbohydrates), blood coagulation mechanism (gamma-carboxylation), protein synthesis regulation (ADP-ribosylation) and oncogene research (tyrosine phosphorylation). For a review, see [4]. Clearly, neither DNA sequencing nor the Edman method provide unambiguous data on such modifications.

A complementary procedure to these methods of protein structure characterization consists in the use of mass spectrometry. Traditionally mass spectrometric analysis using electron ionization is limited to relatively low molecular weight compounds having enough volatility and thermal stability to permit vaporization. Application of this technique to peptides requires "ad hoc" derivatization procedures. Despite these limitations, electron ionization mass spectrometry has made a significant contribution in the field of post-translational modifications such as in the characterization of some previously undescribed modifications: acid residues [5], beta-hydroxylation of aspartic acid residues [6], etc.; for a comprehensive review, see [7].

More recently two combined advances have permitted an extension of the application of mass spectrometric methods to biopolymer problems. These advances are related to the development of high field instruments [8] and to the introduction of fast atom bombardment (FAB) as a new ionization method [9]. In FAB high mass spectrometry, sample ions are



*M. L. Gross (ed.), Mass Spectrometry in the Biological Sciences: A Tutorial*, 333–342.

directly generated from a liquid solution by bombardment with a high energy beam of particles (xenon atoms or cesium ions), and mass analysis can be carried out by using magnet sectors suitably configured to measure masses up to 12-15 KDa at full sensitivity. Double focusing sector mass spectrometers offer, at present, a wide range of capabilities making them ideally suited for structural analysis of peptides and proteins.

In 1983, a strategy, christened "FAB mapping", for structural analysis of recombinant proteins was developed by Morris and co-workers [10]. It consists of the direct FAB analysis of a proteolytic digest of the protein. Because there is a lack of fragmentation when nanomolar quantities of multi-component samples are examined, relatively simple FAB mass spectra of the mixture, containing only protonated molecules, are obtained. Although some components may not be desorbed as a protonated molecule on the first run, they can be induced to ionize by altering running conditions or by submitting the mixture to an HPLC separation. The experiment provides the molecular weight of all the peptides produced by enzymatic or chemical digestion of the protein, and these can be mapped onto the anticipated sequence of the protein. Mass shifts deduced from the FAB mass spectrum of the mixture of truncated peptides, obtained after a single step of Edman degradation, can confirm the tentative assignment. Similarly, mass shifts of 128 or 156 u produced by carboxypeptidase B (CpB) digestion of tryptic digest can be used to confirm the assignment already made and can possibly lead to the identification of the carboxy-terminal peptide (the C-terminus will not have been removed by CpB hydrolysis unless it is Lys or Arg). Post-translational modifications or protein processing can be recognized in the FAB map by a shift in the mass to either lower mass than that calculated for the loss of signal peptides or to higher mass due to the addition of various modifying groups, such as carbohydrates, phosphates, etc.

This chapter reviews some of the applications of this strategy to the identification and assignment of post-translational modifications of proteins, including phophorylation, transglutaminase (TGase)-mediated reaction products, as well as deamidation sites leading to the formation of isopeptidic bonds.

## 2. Phosphorylation

Phosphorylation is one of the major post-translational events that occurs at specific sites along the protein sequence involving a recognition mechanism by a protein kinase. The identification of phosphorylated residues by the classical Edman degradation method is a difficult task because loss of the phosphate moiety occurs during the cleavage reaction. By using both the FAB mapping strategy and mass fragmentation, we have assigned the phosphorylation sites in buffalo beta-casein [11].

According to the phosphorylation code, buffalo beta-casein contains five possible modification sites located in two tryptic peptides, four in the icosaheptapeptide 2-28, and one in the hexadecapeptide 33-48. FAB analysis of the thermolytic digest of peptide 2-28 gave, in addition to the expected signals mapping the remainder of the molecule, four signals whose *m/z*-values did not correspond to simple combinations of amino acids in the sequence. Two of these could, however, be assigned to peptides 6-16 and 2-16 with the presence of one phosphate ester (80 u) and the remaining two to peptides 6-22 and 12-28 if four phosphate esters were assumed to be present in each peptide. Because peptides 6-16 and 2-16 contain a single hydroxylated residue, Ser-15, an unambiguous assignment was possible. Peptides 6-22 and 12-28, however, contain five and six hydroxylated residues respectively, including the phosphoserine just assigned. Results are summarized in table I.

To assign the remaining phosphorylated residues, peptide 2-28 was digested by *Staphylococcus aureus* V-8 protease. Figure 1 shows the FAB mass spectrum of the resulting peptide mixture; extending the logic used in interpreting the thermolytic digest data, three

TABLE I. Observed $(M + H)^+$, sequence and position in the sequence of peptides generated by thermolytic digest of buffalo beta-casein peptide 2-28.

| $(M + H)^+$ | Sequence | Position in the sequence |
|---|---|---|
| 360 | I-V-E | 12-14 |
| 370 | I-T-H | 23-25 |
| 374 | I-N-K | 26-28 |
| 519 | E-L-E-E | 2-5 |
| 628 | L-N-V-P-G-E | 6-11 |
| 969 | L-N-V-P-G-E-I-V-E | 6-14 |
| 1249 | L-N-V-P-G-E-I-V-E-S-L + 1 P | 6-16 |
| 1469 | E-L-E-E-L-N-V-P-G-E-I-V-E | 2-14 |
| 1749 | E-L-E-E-L-N-V-P-G-E-I-V-E-S-L- + 1 P | 2-16 |
| 2095 | L-N-V-P-G-E-I-V-E-S-L-S-S-S-E-E-S + 4 P | 6-22 |
| 2192 | I-V-E-S-L-S-S-S- E-E-S- I-T-H-I-N-K + 4 P | 12-28 |

TABLE II. Observed mass values and position in the sequence of V-8 protease digest of native calmodulin.

| $(M + H)^+$ | Position in the sequence |
|---|---|
| 1288 | 1-11 |
| 423 | 12-14 |
| 1844 | 15-31 |
| 1702 | 32-47 |
| 862 | 48-54 |
| 1349 | 55-67 Asn 60 |
| 1350 | 55-67 Asp/isoAsp 60 |
| 2061 | 68-84 |
| 417 | 85-87 |
| 1859 | 88-104 Asn 97 |
| 2269 | 105-123 |
| 1926 | 105-120 |
| 548 | 124-127 |
| 995 | 141-148 |

signals can only be assigned to the 2-28 sequence by assuming the presence of phosphate groups. The signal at *m/z* 1249 confirms the assignment of a phosphoserine residue at position 15. The related signals at *m/z* 1879 and 2008 can be assigned to peptides 6-20 and 6-21, each containing four phosphate groups. The 6-21 sequence contains only four hydroxylated amino acids, the serine residues in positions 15, 17, 18 and 19. Thus, each of these must have been modified by a phosphate group to fit the molecular weight observed.

The results were confirmed by incubating the unfractionated peptide mixtures with alkaline phosphatase; the shifts in mass of the respective signals account for the loss of the released phosphate moieties in each of the phosphopeptides assigned.

The peptide 33-48 was analyzed before and after digestion with alkaline phosphatase: an 80 mass unit shift was observed. Because the peptide has a single Ser in position 35 and because there is specificity of the enzyme, the modification site was easily assigned.

### 3. TGAse-mediated reaction products

TGases are $Ca^{+2}$-dependent enzymes linking glutamine residues of peptides to primary amines such as polyamines, or the epsilon-amino group of lysine residues [12]. The biological significance of TGase-mediated reactions remains an open question; it is well-known that cross link formation catalyzed by TGase is essential for fibrin clot formation, but the significance of covalent binding of polyamines to proteins is still obscure.

The identification of both acyl donor and acceptor substrates, as well as the detection of the epsilon (gamma-glutamyl) lysine crosslinks and (gamma-glutamyl) amine derivatives, are some of the most important aspects in the field of TGase-mediated reactions. FAB mass spectrometric detection of derivatives produced by TGase in vitro was carried out using two biologically active peptides: Substance P and beta-Endorphin [13, 14].

Figure 2 shows the FAB mass spectra of Substance P after treatment with TGase. Figure 2A demonstrates that Substance P was not modified following six hours of incubation in the presence of the enzyme and $Ca^{+2}$. In fact, only a single $(M + H)^+$ at *m/z* 1347, corresponding to the unmodified peptide, was observed in this spectrum. The absence of any significant signal at *m/z* 1330 and 2676 indicates that neither intra- nor inter-molecular (dimeric) crosslinks were formed between lysine and glutamine residues. In contrast, when Substance P was incubated with TGase, in the absence of $Ca^{+2}$, and in the presence of spermine, the FAB mass spectrum (Figure 2B) showed a further signal at *m/z* 1532, corresponding to the covalent adduct between Substance P and spermine. In fact, polyamine adds an extra 185 u to the molecular weight of Substance P. Because other peaks are absent in the FAB mass spectrum, the occurrence of further adducts, such as the covalent binding of two spermine molecules to the peptide or the formation of cross bridges between spermine and two molecules of Substance P, is ruled out. The FAB mass spectrometric fragmentation pattern of Substance P fully modified by spermine indicates that Gln-5 is the site of TGase-induced modification of the peptide.

The partial FAB mass spectrum of the tryptic digest of beta-Endorphin following incubation with TGase in the absence (panel A) or the presence (panel B) of $Ca^{+2}$ is shown in Figure 3. The two signals at *m/z* 1133 and 1005 correspond to beta-Endorphin fragments 10-19 and 10-18, respectively. The data shown in panel A indicate that the neuropeptide was unmodified when it was incubated with TGase in the absence of calcium. When the enzyme activator was present in the reaction mixture, however, two new signals at *m/z* 1320 and 1448 were evident in the spectrum (panel B). The new mass values correspond to the fragments 10-18 and 10-19, respectively, crosslinked to the fragment 29-31.

These findings clearly demonstrate that TGase catalyzes the formation of specific isopeptide bonds between Gln-11 and Lys-29.

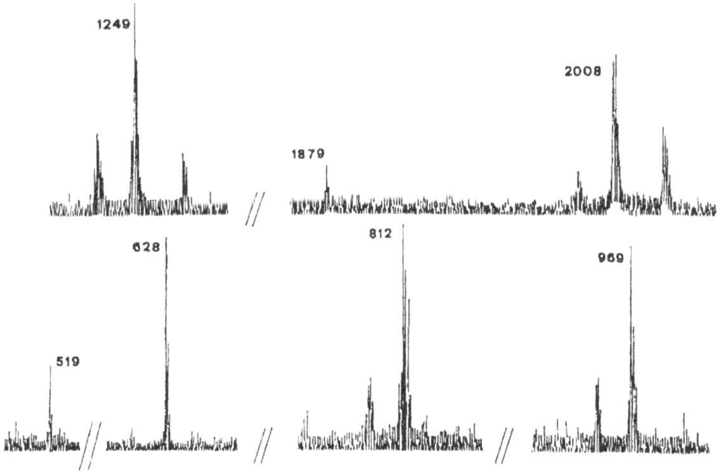


Figure 1. FAB mass spectrum of the unfractionated mixture of peptides produced by digestion of Tp1 with V-8 protease.

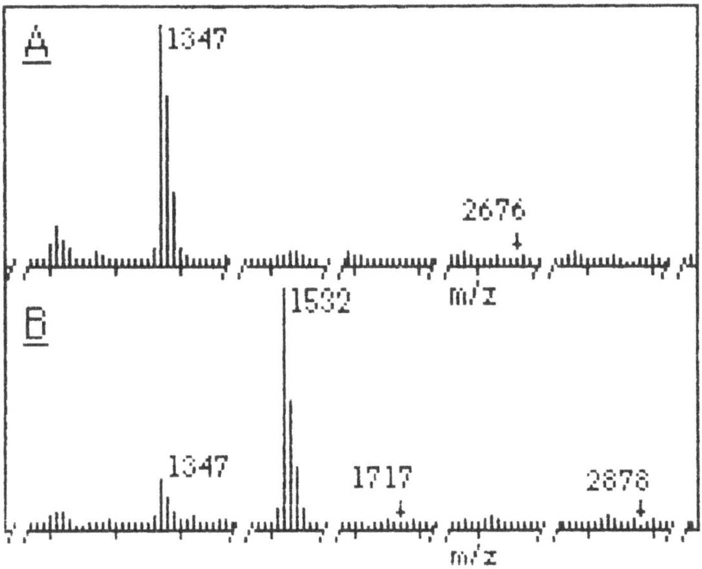


Figure 2. FAB mass spectra of Substance P after treatment with TGase. (A) is the spectrum after six hours of incubation in the presence of $Ca^{2+}$ and (B) is that obtained when spermine was used instead of $Ca^{2+}$.

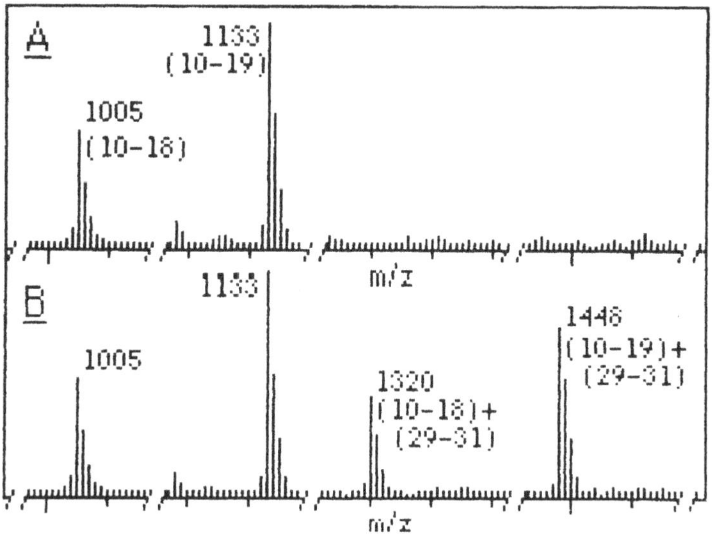


Figure 3. FAB mass spectra of the tryptic digest of beta-Endorphin following incubation in the absence (A) or presence (B) of $Ca^{2+}$.

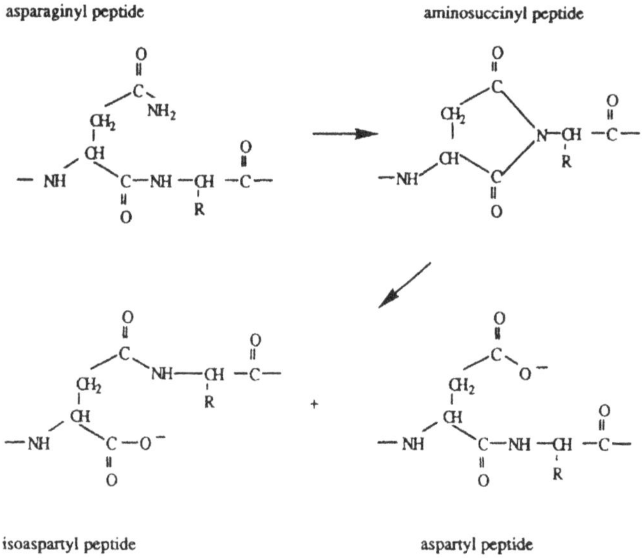


Figure 4. Deamidation of asparagine residues in proteins when the residue is followed by glycine.

## 4. Deamidation

Deamidation is the simplest modification occurring within a polypeptide chain; it involves the formation of aspartic and/or glutamic acid from the corresponding amide residues (i.e., asparagine and glutamine).

It was reported that the deamidation of asparagine residues in proteins takes place most frequently when these residues are followed by glycine in the protein sequence [15]. In this particular case, the deamidation reaction occurs via a cyclic succinimidic intermediate, which in turn opens to give rise to two different products, as shown in Figure 4. Following the opening of the succinimide ring, a portion of the polypeptide chain again forms a normal $\alpha$-peptide bond between the newly formed aspartic acid and the following glycine, whereas in the portion of the protein that contains an abnormal iso-peptide bond, it is the $\beta$-carboxyl group of the aspartic acid that links the amino group of the following glycine residue.

When using FAB/MS analysis of proteolytic digests of deamidated asparagine-containing proteins, it is possible to discriminate between the native and the modified protein; the deamidated peptide(s), in fact, will exhibit a mass shift of 1 u in the FAB map (i.e., the mass difference between asparagine and aspartic acid) so that the deamidation site(s) can be detected and located. The two isobaric products formed by the opening of the succinimide ring cannot be distinguished in such a way.

It is well-known, however, that the enzyme Protein Carboxy Methyl Transferase (PCMT) catalyzes the methylesterification of *L*-isoaspartyl residues in proteins [16]. Following incubation of the deamidated peptides with this enzyme, the isoaspartyl-containing fragment(s) will be shifted by 14 u owing to the methyl ester formation, and the aspartate and iso-aspartate residues can be distinguished. In this way, it has been possible to locate the deamidation and methylation sites in a small polypeptide, the porcine Adrenocorticothropic Hormone (ACTH), and in a medium-sized protein, the Calmodulin from bovine brain.

Porcine ACTH was digested with elastase, and the resulting peptide mixture was analyzed by FAB/MS. The Asn-Gly sequence present at position 25-26 in the ACTH primary structure was mapped by the signal at *m/z* 1007 corresponding to peptide 23-31. The peptide mixture was then deamidated by incubation in ammonium hydroxide for 3 h and re-analyzed by FAB/MS. The resulting FAB map was identical to the previous one except for the signal at *m/z* 1007 which moved to *m/z* 1008, indicating that Asn-25 was deamidated to a mixture of Asp/isoAsp. The deamidated peptides were enzymatically methylated by incubation with PCMT in the presence of *S*-Adenosyl-Methionine as methyl donor; FAB/MS analysis of the resulting peptide mixture is shown in Figure 5. All the recorded signals are identical to those of the deamidated ACTH with the exception of the peak at *m/z* 1022. This signal was assigned to peptide 22-31 for which the isoAsp residue at position 25 had been modified by the enzyme. Figure 5 shows that the signal at *m/z* 1008 is still present in the FAB spectrum; this signal refers to the portion of peptide 22-31 containing the normal $\alpha$-aspartyl residue, which is not recognized by PCMT.

Native calmodulin from bovine brain was submitted to proteolytic digestion with V-8 protease, and the resulting peptide mixture was analyzed by FAB/MS. The FAB map is shown in Table II, and accounts for over 90% of the protein sequence. The two Asn-Gly sequences, present at positions 60-61 and 97-98 were mapped into peptides 55-67 and 88-104 with mass signals at *m/z* 1349 and 1859, respectively. It should be noted that a further signal is present at m/z 1350 indicating that the commercial preparation of Calmodulin is partially deamidated at Asn-60.

Two samples of Calmodulin, A and B, were deamidated in the absence and in the presence of 0.5 mM $CaCl_2$, respectively, to observe any effect on the deamidation reaction, due to the $Ca^{2+}$ binding. The two Asn-Gly sequences, in fact, are both involved in two different $Ca^{2+}$

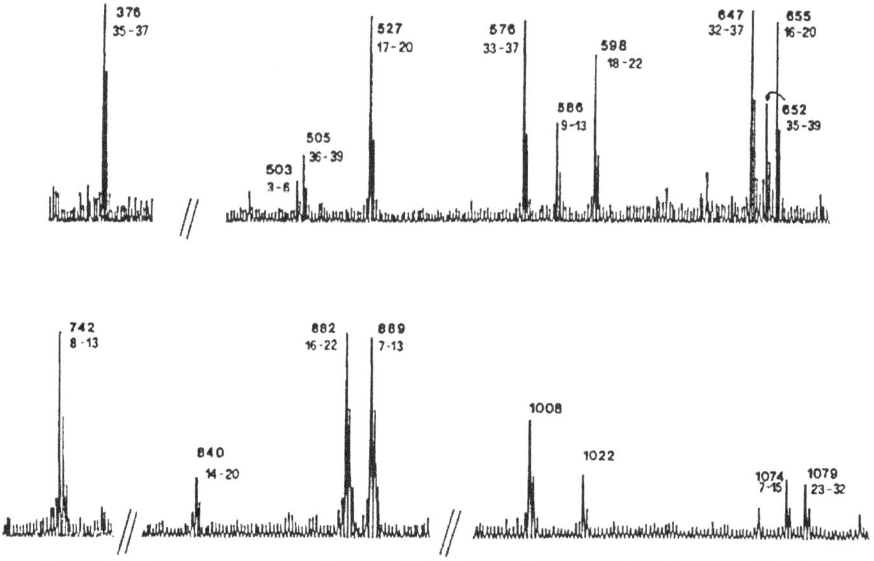


Figure 5. FAB mass spectrum of a peptide mixture from porcine ACTH that had been first deamidated and then enzmatically methylated (see text for details).

TABLE III. Observed mass values of V-8 protease digested calmodulin following deamidation and methylation in the absence (Sample A) and in the presence (Sample B) of calcium ions.

| Peptide | $(M + H)^+$ Native Sample | $(M + H)^+$ Deam. Sample A | $(M + H)^+$ Deam. Sample B | $(M + H)^+$ Deam. met. Sample A | $(M + H)^+$ Deam. met. Sample B |
|---|---|---|---|---|---|
| 1-11 | 1228 | 1228 | 1228 | 1228 | 1228 |
| 12-14 | 423 | 423 | 423 | 423 | 423 |
| 15-31 | 1844 | 1844 | 1844 | 1844 | 1844 |
| 32-47 | 1702 | 1702 | 1702 | 1702 | 1702 |
| 48-54 | 862 | 862 | 862 | 862 | 862 |
| 55-67 | 1349/50 | 1350 | 1349/50 | 1350 | 1349/50 |
| 55-67 | | | | 1364 | |
| 68-84 | 2061 | 2061 | 2061 | 2061 | 2061 |
| 85-87 | 417 | 417 | 417 | 417 | 417 |
| 88-104 | 1859 | 1860 | 1859 | 1860 | 1859 |
| 88-104 | | | | 1874 | |
| 105-120 | 1926 | 1926 | 1926 | 1926 | 1926 |
| 105-123 | 2269 | 2269 | 2269 | 2269 | 2269 |
| 124-127 | 548 | 548 | 548 | 548 | 548 |
| 141-148 | 955 | 955 | 955 | 955 | 955 |

binding domains in the Calmodulin structure.

Table III shows the results obtained by FAB/MS analysis of the V-8 protease digests of samples A and B. The Asn residues at position 60 and 97 were both deamidated in sample A as indicated by one u shift of the signals corresponding to peptides 55-67 and 88-104. On the other hand, analysis of sample B showed that, due to the presence of $Ca^{2+}$, the Asn residues did not deamidate. Therefore, the resulting FAB map is identical to that of native Calmodulin.

The two samples were methylated by PCMT according to the procedure described previously.

The peptide map of sample A showed the presence of two new signals at *m/z* 1874 and 1364, 14 u higher than those correponding to the deamidated peptides 88-104 and 55-67. These two signals demonstrate that the isoAsp residues formed during deamidation of Asn-60 and Asn-97 had been methylated by the enzyme. A portion of the peptides, however, still contains the normal $\alpha$-Asp residues as indicated by the presence of the signals at *m/z* 1860 and 1350.

Analysis of sample B did not show the presence of any methyl-derivatives, again demonstrating the protection effect of $Ca^{2+}$ ions on the deamidation of the two Asn residues.

## 5. Conclusions

A FAB mass spectrometric procedure has been developed to provide an additional tool for protein structure characterization. The FAB mapping strategy is complementary to classical Edman chemistry and/or DNA sequencing in confirming protein sequence and assigning post-translational modifications. The unique capability of FAB mass spectrometry in analyzing peptide mixtures is an obvious advantage because purification procedures are often the rate determining step in structural analysis of proteins.

The identification of the site, the nature, and the extent of post-translational modifications is one of the most important requirements of modern protein research both in academic and industrial laboratories. The mass spectrometric approach described here, based on the double focusing sector mass spectrometry, constitutes a fast, thorough, and cost-effective procedure for solving a variety of structural problems not amenable to conventional methods.

Although the cases discussed deal with only a few modifications, it is evident that the same approach can be used for many others, as long as they induce changes in peptide mass.

## 6. Acknowledgements

This work has been supported partly by grants given by the Italian National Research Council (CNR-Rome) to GM and AM and the Ministero della Pubblica Istruzione (Rome) to GM.

## 7. References

1. Hewick, R. M., Hunkapiller, M. W., Hood, L. E. and Dryer, W. J. J. (1981) 'A gas-liquid phase peptide and protein sequenator', *Biol. Chem.* **256**, 7990.
2. Sanger, F., Nicklen, S. and Coulson, A. R. (1977) 'DNA sequencing with chain-terminating inhibitors', *Proc. Natl. Acad. Sci. USA* **74**, 5463.
3. Maxam, A. M. and Gilbert, W. (1977) 'A new method for sequencing DNA', *Proc. Natl. Acad. Sci. USA* **74**, 560.

4. Zappia, V., Galletti, P., Porta, R. and Wold, F. (1988) Advances in Post-Translational Modifications of Proteins and Ageing, Plenum Press, London.
5. Morris, H. R., Dell, A., Petersen, T. E., Sottrup-Jensen, L. and Magnusson, S. (1976) 'Mass spectrometric identification and sequence location of the ten residues of the new amino acid (t-carboxy-glutamic acid) on the N-terminal region of prothrombin', *Biochem. J.* **153**, 663.
6. Drakenberg, T., Fernlund, P., Roepstoff, P. and Stenflo, (1983) 'ß-hydroxyaspartic acid in vitamin k-dependent protein C', *Proc. Natl. Acad. Sci. USA* **80**, 1802.
7. Carr, S. and Biemann, K. (1974) 'Identification of post-translational modified amino acids in proteins by mass spectrometry', *Methods Enzymol.* **106**, 29.
8. Morris, H. R., Dell, A. and McDowell, R. A. (1981) 'Extended performance using a high field magnet mass spectrometer', *Biomed. Mass Spectrom.* **8**, 463.
9. Barber, M., Bordoli, R. S., Sedgwick, R. D. and Tyler, A. N. (1981) 'Fast atom bombardment (FAB): a new ion source for mass spectrometry', *J. Chem. Soc. Chem. Commun.* **325**.
10. Morris, H. R., Panico, M. and Taylor, G. W. (1983) 'FAB-mapping of recombinant DNA protein products', *Biochem. Biophys. Res. Commun.* **117**, 299.
11. Petrilli, P., Pucci, P., Morris, H. R. and Addeo, F. (1986) 'Assignment of phosphorylation sites in buffalo ß-casein by fast atom bombardment mass spectrometry', *Biochem. Biophys. Res. Commun.* **140**, 28.
12. Lorand, L., and Conrad, S. M. (1984) 'Transglutaminases', *Mol. Cell Biol.* **58**, 9.
13. Porta, R., Esposito, C., Metafora, S., Pucci, P., Malorni, A. and Marino, G. (1988) 'Substance P as a transglutaminase substrate: identification of the reaction products by fast atom bombardment mass spectrometry', *Anal. Biochem.* **172**, 499.
14. Pucci, P., Malorni, A., Marino, G., Metafora, S., Esposito, C. and Porta, R. (1988) 'ß-endorphin modification by transglutaminase *in vitro*: identification by FAB/MS of glutamine-11 and lysine-29 as acyl donor and acceptor sites', *Biochem. Biophys. Res. Commun.* **154**, 735.
15. Clarke, S. (1988) 'Mass spectrometric identification and sequence location of the ten residues of the new amino acid ( -carboxylglutamic acid) in the N-terminal region of prothrombin', *Int. J. Peptide Protein Res.* **30**, 808-821.
16. Aswad, D. W. (1984) 'Stoichiometry methylation of porcine adrenocorticotropin by protein carboxyl methyltransferase requires deamidation of asparagines', *J. Biol. Chem.* **259**, 10714.

# MANIPULATION OF EXPERIMENTAL VARIABLES FOR FAB MS AND MS/MS TO FACILITATE IDENTIFICATION OF A POST-TRANSLATIONAL MODIFICATION IN RABBIT ELONGATION FACTOR 1α

C. E. COSTELLO
*Mass Spectrometry Facility, Department of Chemistry*
*Massachusetts Institute of Technology*
*Cambridge, MA 02139*
*USA*

ABSTRACT. Seven post-translationally modified residues have been identified in rabbit elongation factor 1α (EF-1α), a cytosolic protein which catalyzes the GTP-dependent binding of aminoacyl-tRNA to ribosomes. Five of these modifications had been shown to involve methylation of lysines. Rigorous identification of the remaining two modifications required FAB MS and MS/MS, both in the positive and negative ion modes, the use of varied solvent/matrix combinations and periodate oxidation.

## 1. Introduction

Elongation factor lα (EF-lα) plays a vital role in eukaryotic protein synthesis and is an abundant cellular component. Although cDNA and/or genomic clones of EF-1α have been obtained from a variety of sources, the utility of the deduced amino acid sequences for structure/function studies is limited unless the effects of post-translational modifications on the activity of the protein are also considered. Post-translational modifictions to seven of the 452 amino acid residues in rabbit EF-1α have recently been identified: five involved lysine methylations and the remaining two were shown to result from the attachment of glycerylphosphorylethanolamine to glutamic acid [1]. The latter modification was defined primarily through a set of FAB MS and MS/MS experiments, the course and rationale for which are described here.

## 2. Experimental Methods

### 2.1. SAMPLE PREPARATION.

Explicit details are provided in reference [1]. Rabbit reticulocyte lysate was centrifuged and the supernatant was used for the preparation of EF-1α. Isolation of the protein was carried out using DEAE-cellulose, phosphocellulose, dialysis, and CM-cellulose separation steps with a series of buffers. Separation was followed by UV monitoring and gel electrophoresis. The purified protein was reduced and carboxyamidated before tryptic digestion. Tryptic fragments were separated by HPLC by using a Waters μBondapak C18 reverse phase column, with a nonlinear gradient of acetonitrile into 0.05% TFA. Adjacent fractions having similar absorbance at 220 and 255 nm were pooled, and the amino acid sequences of the peptides were determined using



*M. L. Gross (ed.), Mass Spectrometry in the Biological Sciences: A Tutorial*, 343–350.

an Applied Biosystems Model 477A protein microsequencer with an on-line PTH analyzer. The fraction containing residue 301 was further digested with *Staphylococcus aureus* V8 protease to truncate the (23-residue plus modification) peptide to a size (16-residue plus modification) more convenient for FAB MS/MS analysis. The products of this digestion were separated by HPLC by using a Beckman Ultrasphere ODS C18 column and a nonlinear acetonitrile/0.05% TFA gradient. A second protein aliquot was carboxyamidated and digested with CNBr and the resulting peptides were separated and analyzed in the same manner as the tryptic digestion products. The purified peptides that included the residues 301 (the hexadecapeptide from trypsin/V8 digestion) and residue 374 (a tryptic pentapeptide) were furnished as lyophilized samples (5 and 8 nmol, respectively) for FAB MS analysis, because these peptides included the sites believed to be post-translationally modified.

### 2.2. OXIDATION OF THE MODIFIED PENTAPEPTIDE.

A solution of $NaIO_4$ (100 μL, 8 mM) was added to the solution of the peptide remaining after FAB MS analysis (*ca.* 4 μg in 2 μL MeOH), and the mixture was allowed to stand 26 hr at room temperature. After the reaction was quenched with 5 μL ethylene glycol, the mixture was again allowed to stand overnight at room temperature. Reaction products were separated by HPLC by using a Waters μBondapak column, a water(0.05% TFA)/acetonitrile(0.035% TFA) gradient and detection at 214 nm.

### 2.3. FAB MS AND FAB MS/MS.

FAB MS was carried out using the first two sectors (MS-1) of the JEOL HX110/HX110 tandem mass spectrometer, operated at 1:2000 resolution (scanning). The 6-keV $Xe^0$ primary beam was generated with a JEOL FAB gun. Accelerating voltage was ± 10 kV, and ± 18 kV post-acceleration was used at the detector. Electron multiplier voltage was 1.4 kV for measurement of the spectrum of the larger peptide, and ± 1.0 kV for the smaller pepetide. FAB MS/MS was performed using all four sectors ($E_1B_1$-$E_2B_2$), and collisions took place in a cell located between MS-1 and MS-2. The collision cell was floated at ± 3 kV with respect to ground, so collisions took place at 7 keV. Post-acceleration voltage was ± 18 kV and electron multiplier voltage was + 1.5 kV. MS-1 resolution was 1:2000 or 1:1000 (static); MS-2 resolution was 1:1000 (scanning). Helium was used as the collision gas at a pressure sufficient to reduce precursor ion signal to 25% of its initial value. Collisionally activated decomposition (CAD) spectra were recorded during linked scans of the MS-2 electric and magnetic sectors at constant B/E. MS-1 was calibrated with CsI clusters; MS-2 was calibrated with LiI/NaI/CsI for the positive ion mode and CsI/glycerol for the negative ion mode [2].

For the initial FAB MS analysis, 1-2 μL of 5% acetic acid in water was added to the lyophilized samples and an equal volume of glycerol was added to the tubes containing these solutions. About 0.3 μL of the mixture was used for each MS or MS/MS analysis. For the later FAB MS analysis, the new sample of the modified pentapeptide was dissolved in 4 μL methanol and 1 μL of this solution was added to 0.5 μL of *m*-nitrobenzyl alcohol or triethanolamine on the FAB probe tip.

## 3. Results and Discussion

Two initial goals were formulated when this mass spectral analysis was undertaken: (1) to determine whether post-translational modifications of the two peptides could be detected, and whether both had the same modification and (2) to identify the nature of the modification. Because the amino acid sequences of the modified peptides submitted for FAB MS analysis were already known, the measurement of the FAB mass spectrum was expected to answer the first question, because a shift in the molecular ion from that calculated for the unmodified peptides would indicate that a change had occurred. Fragment ions, if any could be observed above the background, might indicate the location of the modification.

The FAB mass spectrum of the longer peptide, that which contained residue 301, is shown in Figure 1. The $(M+H)^+$ was observed at *m/z* 1827.9, $(M+Na)^+$ at *m/z* 1849.8, and $(M+K)^+$ at *m/z* 1865.8. The $(M+H)^+$ calculated for the sequence Ala-Leu-Ser-Glu-Ala-Leu-Pro-Gly-Asp-Asn-Val-Gly-Phe-Asn-Val-Lys is *m/z* 1630.8. A shift of 197.1 u was evident. Fragment ions present at low abundance could be assigned to $y_5$-$y_{12}$ (using the Biemann modification of the Roepstorff nomenclature [3]), and indicated, therefore, that the modification had occurred to one of the first four amino acids at the N-terminus. This observation was consistent with results obtained with the Edman gas-phase sequenator, which gave low recovery for Glu, suggesting that the modification had occurred at that residue [1]. Only a few other sequence-related ions were present ($z_6$ *m/z* 647, $w_7$ *m/z* 717, $x_9$ *m/z* 975), and the spectrum was dominated by clusters of the glycerol matrix at *m/z* (92n+ 1) and matrix plus a major contaminant at *m/z* (92n + 337).

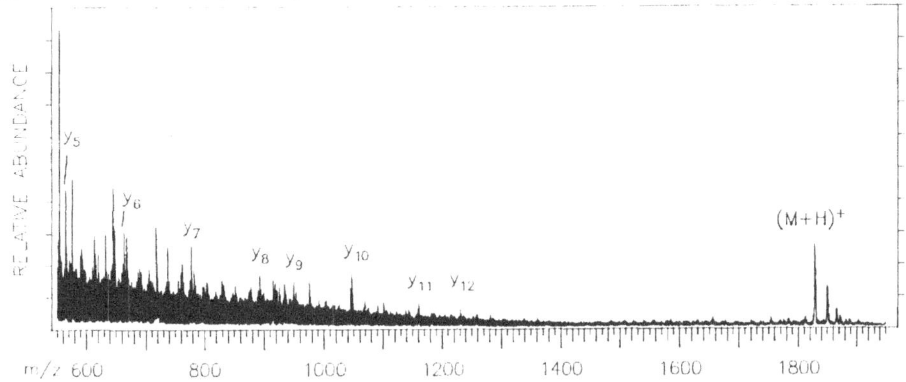


Figure 1. FAB mass spectrum (*m/z* 550-1950) of the peptide containing residues 298-313 of rabbit EF-1α, obtained by digestion with trypsin and *Staph. V8* protease.

The FAB mass spectra of the pentapeptide that contained residue 374 were recorded in both negative and positive ion modes. In these spectra, whose molecular ion regions are shown in Figure 2, $(M-H)^-$ was observed at *m/z* 802.4 and $(M+H)^+$ at *m/z* 804.4. Sodium adducts appeared in both spectra, although 5% acetic acid had been used as the solvent in an attempt to minimize such peaks. The calculated value for $(M+H)^+$ for the amino acid sequence Phe-Ala-Glu-Leu-Lys is *m/z* 607.3, a difference of 197.1 u from the observed value. The presence of a high background from matrix, salts and buffers precluded the observation of positive ion fragment ions, and obscured most of the negative ion spectrum as well. Ions at *m/z* 560 and 673 in the negative ion spectrum were tentatively assigned to $c_3$ and $c_4$, suggesting that neither residue 4 nor 5 was modified, again allowing for modification of glutamic acid. This result indicated that the two peptides had undergone the same modification.

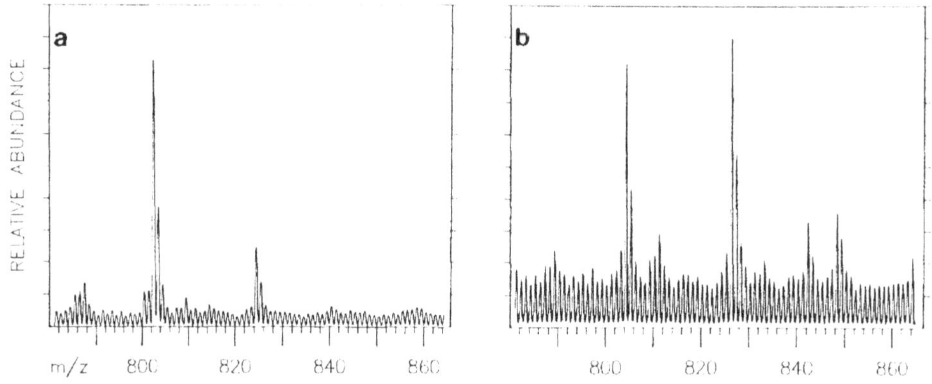


Figure 2. Molecular ion regions of the (a) negative and (b) positive ion FAB mass spectra of the tryptic peptide containing residues 372-376 of rabbit EF-1α

Despite the presence of contaminants, the molecular ions of the sample had sufficient abundance for CAD MS/MS analysis, and this approach offered to provide definitive information about the modification. The CAD mass spectra obtained for the pentapeptide are shown in Figure 3. In the CAD mass spectrum of the $(M - H)^-$ *m/z* 802.4 (Figure 3a), abundant product ions were observed which could be assigned to $PO_3^-$ (*m/z* 78.8), and $H_2PO_4^-$ (*m/z* 96.9). The presence of ions at *m/z* 170.9 and 728.6 $(M - H - 74)^-$ suggests that the phosphate is bound to a terminal glycerol moiety. Other product ions at *m/z* 196.9 and 214.0 indicate that the phosphate is also esterified to ethanolamine. Earlier experiments that had used radiolabeled phosphate and ethanolamine had suggested that these moieties were incorporated into the modified protein [4]. The product ions at *m/z* 268.0 and 297.0 are consistent with the attachment of ethanolamine in an amide linkage to the glutamic acid side chain, because these ions would correspond to the complete modified side chain and to a species that included NH=CH from the backbone of the residue, respectively. Assignments of the product ions formed by cleavages within the modified Glu residue are shown below.

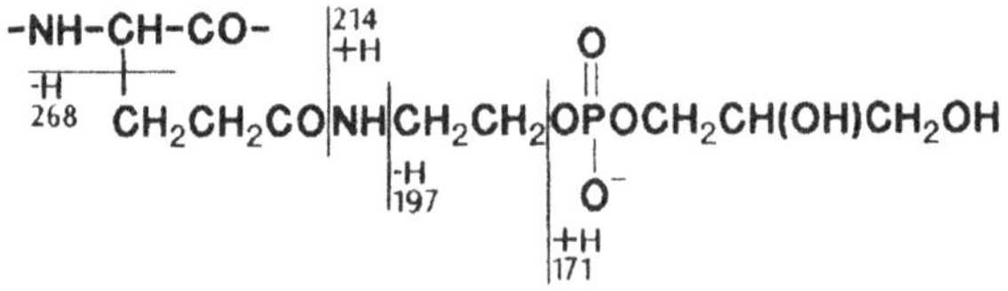

Some sequence ions were observed, as noted on Figure 3a. Because the negative charge would be most likely located on the phosphate group, it was not surprising that only those sequence ions that included the modified Glu were present.

The CAD mass spectrum of $(M+H)^+$ *m/z* 804.5 (Figure 3b) is consistent with the structure assigned above. Losses of 74 u and 172 u can be attributed to cleavages on either side of the phosphate group. Loss of the modified side chain led to the ion at *mz* 534.3, and its cleavage yielded the $\mathbf{w}_3$ ion at *m/z* 314.2. It is noteworthy that sequence ions that include the phosphate group have a greater than usual abundance for the species containing an additional hydrogen.

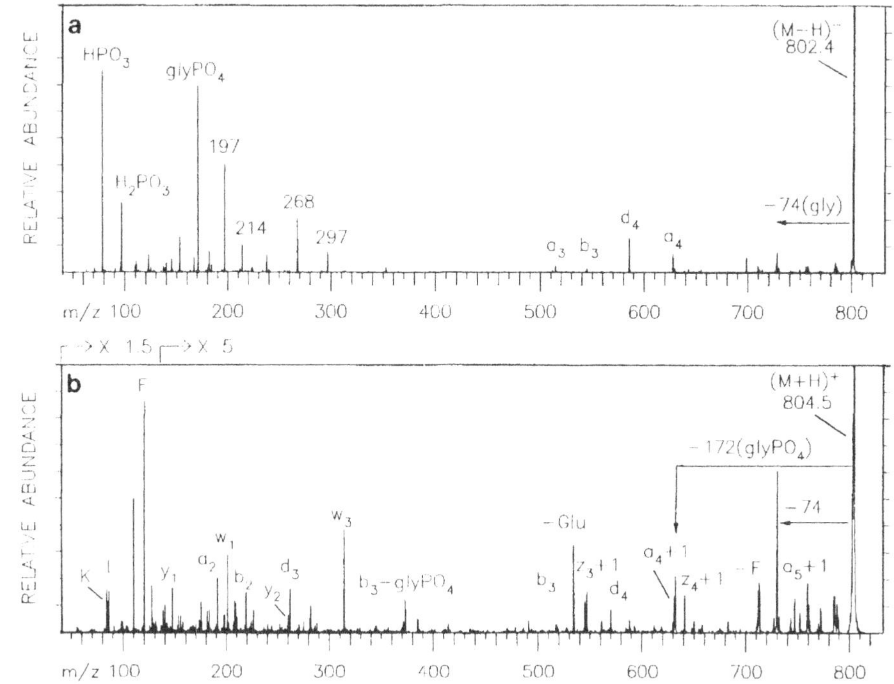


Figure 3. CAD mass spectra of the tryptic peptide corresponding to rabbit EF-1α residues 372-376. (a) $(M-H)^-$ *m/z* 802.4. (b) $(M+H)^+$ *m/z* 804.5.

The CAD mass spectrum of the modified hexadecapeptide $(M+H)^+$ *m/z* 1827.9, shown in Figure 4, has features similar to those of the pentapeptide. Losses of 74 and 172 u from the molecular ion were also observed, to produce abundant ions at *m/z* 1753.7 and 1655.7, respectively. C-terminus sequence ions are the most abundant, particularly the **w**-series, and are consistent both with the known amino acid sequence and the assignment of the modification at Glu.

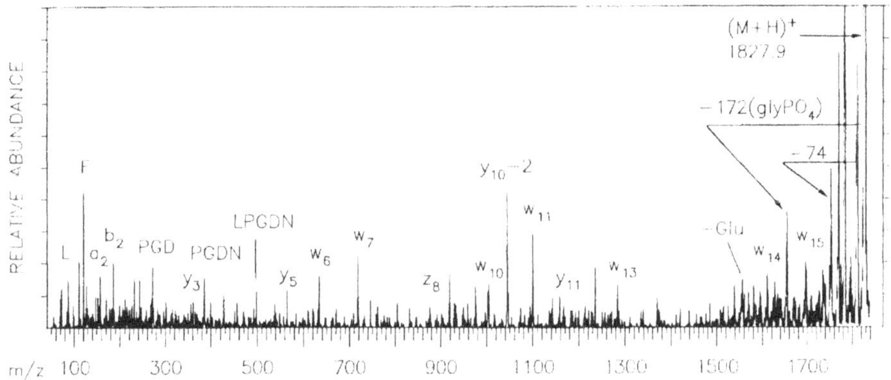


Figure 4. CAD mass spectrum of the $(M+H)^+$ *m/z* 1827.9 of the peptide corresponding to rabbit EF-1α residues 298-313.

Although glycerylphosphate and glycerylphosphorylethanolamine have been recognized as important structural components in biological systems, concern arose as to whether the glycerol moiety identified in this instance could be an artifact introduced during the mass spectral analysis by esterification of a native free phosphate group with the glycerol matrix. Because the peptide samples initially supplied had been entirely mixed with the matrix, it was necessary to obtain a fresh sample for an independent determination. A new 5 nmol sample of the pentapeptide was prepared from a fresh digest of the protein, and a portion of it was analyzed using methanol as the solvent and either triethanolamine or *m*-nitrobenzyl alcohol (NBA) as the matrix. The molecular ion region of the negative ion mass spectrum in methanol/NBA is shown in Figure 5a. In both matrices, the $(M-H)^-$ was still observed at *m/z* 802.4, and no signal appropriate for a species that lacked glycerol was present.

The remainder of this sample was used to determine the attachment site of the phosphate group on the glycerol moiety. For this purpose, periodate oxidation was carried out. This reaction cleaves the C-C bond in vicinal diols, producing aldehydes. This mild reaction has proved very useful for linkage determination in the FAB MS analysis of small quantities of carbohydrates [5]. In this case, it would not occur if the phosphate were bound at the central hydroxyl of glycerol. The molecular ion region of the mass spectrum obtained for the HPLC fraction containing the major product from oxidation of the modified pentapeptide is shown in Figure 5b. The $(M-H)^-$ shifted to *m/z* 770.4, a loss of 32 u, $HCH_2OH$. The $(M-H+G)^-$ *m/z* 862.5 has an abundance equal to $(M-H)^-$, as is frequently observed for aldehydes. This result indicates that diol cleavage had occurred, and that the phosphate is, therefore, bound to a terminal hydroxyl group.

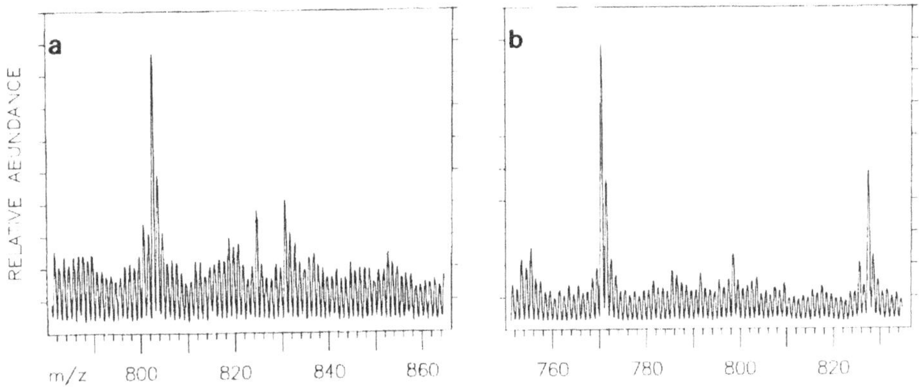


Figure 5. (a) Molecular ion region of the negative ion FAB mass spectrum of the tryptic peptide corresponding to EF-1α residues 372-376 in methanol/NBA. (b) Molecular ion region of the negative ion FAB mass spectrum of the major product obtained by periodate oxidation of the sample shown in the left panel, dissolved in methanol/glycerol. The ion at *m/z* 827 is the $(glycerol_9 - H)^-$ cluster ion.

## 4. Conclusions

The use of several types of mass spectral analysis - positive and negative ion mode measurement of both FAB MS and FAB MS/MS spectra, and use of a variety of solvents and matrices, combined with microscale chemistry, can result in the structural determination of unknown or partially unknown samples of biological origin, even when one has available only a very small quantity of material that is still somewhat impure, and when quite unexpected features are present. Preliminary clues about possible substructures provided by radiolabeling and amino acid sequencing may help to suggest appropriate mass spectral experiments or to facilitate interpretation of the data. In the project described here, such an experimental approach led to the identification of the post-translational modification of two sites in rabbit EF-1α as glycerylphosphorylethanolamine amidation of glutamic acid side chains.

## 5. Acknowledgements

Protein isolation, chemical and enzymatic cleavages and separation of the cleavage products were carried out by W. C. Merrick and his colleagues at Case Western Reserve University, Cleveland, OH. S. Y. Chan of the Harvard School of Public Health, Boston, MA, performed the oxidation of the modified pentapeptide. The MIT Mass Spectrometry Facility is supported by Grant No. RR00317 (to K. Biemann) from the NIH National Center for Research Resources.

## 6. References

1. Dever, T. E., Costello, C. E., Owens, C. L., Rosenberry, T. L. and Merrick, W. C. (1989) 'Location of Seven Post-translational Modifications in Rabbit Elongation Factor 1α Including Dimethyllysine, Trimethyllysine, and Glycerophosphorylethanolamine', *J. Biol. Chem.*, **264**, 20518-20525.
2. Sato, K., Asada, T., Kunihiro, F., Kammei, Y., Kubota, E., Costello, C. E., Martin, S. A., Scoble, H. A. and Biemann, K. (1987) 'High Performance Tandem Mass Spectrometry: Calibration and Performance of Linked Scans of a Four Sector Instrument', *Anal. Chem.*, **59**, 1652-1659.
3. Johnson, R. S. and Biemann, K. (1987) 'The Primary Structure of Thioredoxin from *Chromatium vinosum* Determined by High Performance Tandem Mass Spectrometry', *Biochemistry.*, **26**, 1209-1214.
4. Rosenberry, T. L., Krall J. A., Dever, T. E., Haas, R., Louvard, D. and Merrick, W. C. (1989) 'Biosynthetic Incorporation of [$^3$H] Ethanolamine into Protein Synthesis Elongation Factor 1α Reveals a New Post-translational Protein Modification', *J. Biol. Chem.*, **264,** 7096-7099.
5. Angel, A.-S., Lindh, F. and Nilsson, B. (1990) 'Determination of Binding Positions in Oligosaccharides and Glycosphingolipids by Fast-Atom-Bombardment Mass Spectrometry', *Carbohydr. Res.*, **168**, 15-31.

# USE OF FAB MS AND MS/MS IN THE STRUCTURAL DETERMINATIONS OF COCKROACH NEUROPEPTIDES

C. E. COSTELLO
*Mass Spectrometry Facility, Department of Chemistry*
*Massachusetts Institute of Technology*
*Cambridge, MA 02139*
*USA*

ABSTRACT. FAB MS and MS/MS were employed to elucidate structural features which could not be determined from Edman sequencing of three neuropeptides isolated at the 80-100 pmol level from the brains of the virgin female cockroach *Diploptera punctata.* Mass spectral results showed that all three were amidated at the C-terminus and provided the complete amino acid sequence in the case where the Edman results were ambiguous.

## 1. Introduction

Mass spectrometry is most effectively used for the solution of biological problems when it is part of an integrated approach that utilizes complementary techniques as well. The isolation and purification of peptides with high biological activity can be carried out at a sample level that places great demands on the analytical methods used for structural characterization. In the project for which the mass spectral measurements are described here, 80-100 pmol each of four allatostatic peptides were isolated from 1000-1200 brains of the virgin female cockroach *Diploptera punctata.* The peptides were characterized with an *in vitro* assay on the basis of their ability to inhibit juvenile hormone synthesis by the corpora allata and were subjected to automated Edman sequencing. Mass spectrometry was required to determine whether the C-terminus was amidated and to establish the amino acid sequence of one of the peptides for which incomplete amino acid sequence information had been obtained and which appeared to have a modification to the Tyr residue. An account of the isolation, structural determination and biological activity of these peptides has been published [1]. Mass spectral results that were summarized in that paper are presented here at greater length as an illustrative example.



*M. L. Gross (ed.), Mass Spectrometry in the Biological Sciences: A Tutorial,* 351–356.

## 2. Experimental Methods

### 2.1. SAMPLE PREPARATION.

Explicit details are provided in reference [1]. Brains were dissected, homogenized in 0.9% NaCl and heat-treated at 100 °C for 5-10 min. The supernatant obtained after centrifugation of multiple batches of 20-30 brains each was stored at -80 °C and was pooled until adequate material for analytical studies (from 1000-1200 brains) was accumulated. The extracts were processed first with $C_{18}$ cartridges (Waters), then with three HPLC separations using $C_{18}$ columns and increasingly more shallow gradients of acetonitrile in water: 0-50%, 10-35%, and 11-44%. Fractions were collected manually and assayed for allatostatic activity.

### 2.2. AMINO ACAD SEQUENCE ANALYSIS.

Four active peaks (100 pmol each) were sequenced by Edman degradation using either gas-phase or liquid-phase sequenators.

### 2.3. FAB MS AND MS/MS.

FAB MS was carried out using the first two sectors (MS-1) of a JEOL HX110/HX110 tandem mass spectrometer with EBEB configuration. The primary $Xe^0$ beam was produced with 6 keV energy by a JEOL FAB gun. FAB gun filament current was 2 A and emission current 10 mA. MS-1 was operated with 10 kV accelerating voltage. MS/MS experiments used all four sectors. Collision-induced decomposition took place in a collision cell located between MS-1 and MS-2. Helium was used as the collision gas at a pressure sufficient to reduce the precursor ion signal by 75%. The collision cell was floated at 3 kV with respect to ground potential, so collisions took place at 7 keV. Before the detectors, -18 kV post-acceleration was applied. Instrument control and data acquisition and processing were handled by the JEOL DA5000 data system. For FAB MS, MS-1 was operated at 1:1000 resolution (scanning)and a single spectrum for each sample was acquired during a magnetic scan at the scan rate which would cover the range 0-6000 u in 2'30". The electron multiplier voltage was 1.0 kV and the preamp was x1. For FAB MS/MS, MS-1 was operated at 1:000 resolution (static) and MS-2 was operated at 1:500 resolution (scanning). The electron multiplier voltage was 1.8 kV and the preamp was set at x10. Data were acquired during 25-second linked scans at constant B/E ratio. The spectrum shown is the sum of two scans.

The samples were supplied as three HPLC fractions of 500-700 μL 3:1 acetonitrile/water with 1.1% TFA, each containing 70-100 pmol of peptide and were shipped frozen. The amount of peptide in each fraction was estimated on the basis of HPLC UV detector response. Each solution was transferred to a small conical tipped tube into which 2 μL glycerol had been placed. The solutions were frozen by immersing the conical end of the tube in liquid nitrogen. The tubes were then placed in a Speed-Vac Concentrator® and the acetonitrile and water were removed. About 0.5 μL of the resulting glycerol solution (20-25 pmol) was applied to the stainless steel FAB probe for MS or MS/MS analysis.

## 3. Results and Discussion

The amino acid sequences determined for the four active peptides by Edman degradation using either gas or liquid-phase sequenators were: (**1**) Ala-Pro-Ser-Gly-Ala-Gln-Arg-Leu-Tyr-Gly-Phe-Gly-Leu, (**2**) Gly-Asp-Gly-Arg-Leu-Tyr-Ala-Phe-Gly-Leu, (**3**) Gly-Gly-Ser-Leu-Tyr-Ser-Phe-Gly-Leu and (**4**) ?-Arg-Leu-?Tyr-Ser-Phe-Gly-Leu-?. The uncertainties indicated by "?" in the sequence of **4** resulted from a malfunction during the analysis, which prevented recording of the data for the first residue, low response for Tyr which suggested that this residue might have been post-translationally modified, and lack of information about whether the second Leu was at the C-terminus. Two of the peptides for which complete amino acid sequences had been obtained (**1** and **3**) were furnished as test samples to be used prior to the MS analysis of the partially unknown peptide **4**, in order to verify that the amounts were adequate and that the samples were free of contaminants that might interfere with or suppress the peptide signals. A secondary goal of the MS analysis was the determination whether carboxyl or amide groups were present at the C-terminus of all three peptides.

The molecular ion regions of the positive ion FAB mass spectra of **1, 3** and **4** are shown in Figure 1a, 1b and 1c, respectively. The calculated value for the $(M+H)^+$ of sample **1** with a C-terminal carboxyl group is *m/z* 1336.7. Observation of the $(M+H)^+$ at *m/z* 1335.7 and a low abundance $(M+Na)^+$ at *m/z* 1357.7, (Figure 1a), indicates that the C-terminus had been post-translationally modified to the amide. Similarly, the $(M+H)^+$ of **3** is at *m/z* 899.5 (Figure 1b), whereas the peptide with C-terminal carboxyl group would have $(M+H)^+$ at *m/z* 900.5.

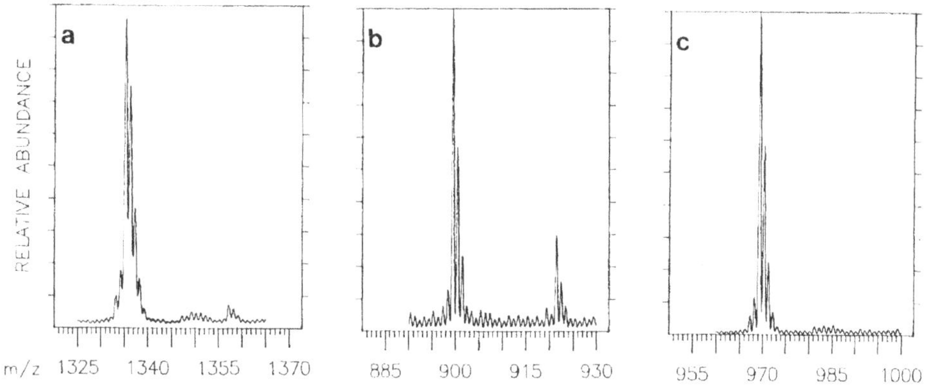


Figure 1. Molecular ion regions of the FAB mass spectra of three allatostatic cockroach neuropeptides (a) **1**, $(M+H)^+$ *m/z* 1336.7, (b) **3**, $(M+H)^+$ *m/z* 899.5, (c) **4**, $(M+H)^+$ *m/z* 969.5.

Because excellent signal-to-noise ratios were obtained for both these samples and no major contamination appeared in either spectrum, it was appropriate to proceed with analysis of the unknown peptide. Figure 1c shows the molecular ion region of the spectrum of this compound, with the $(M+H)^+$ at *m/z* 969.5. This $^{12}C$ ion was then selected for MS/MS analysis by manually adjusting the magnet of MS-1 so that the *m/z* 969.5 ion beam was transmitted into the collision cell located between MS-1 and MS-2. Helium was admitted to the collision cell, and the electric and magnetic fields of MS-2 were scanned at constant B/E. A CAD spectrum was recorded using the sample still remaining on the probe from the first analysis. The quality of this spectrum was acceptable, but since unused sample remained, a fresh 0.5 μL of the glycerol solution was applied to the probe tip and a second CAD spectrum, that shown in Figure 2, was obtained with greater signal intensity. Very high quality MS and MS/MS spectra were thus both recorded with a total consumption of only half of the estimated 78 pmol sample furnished by the investigator.

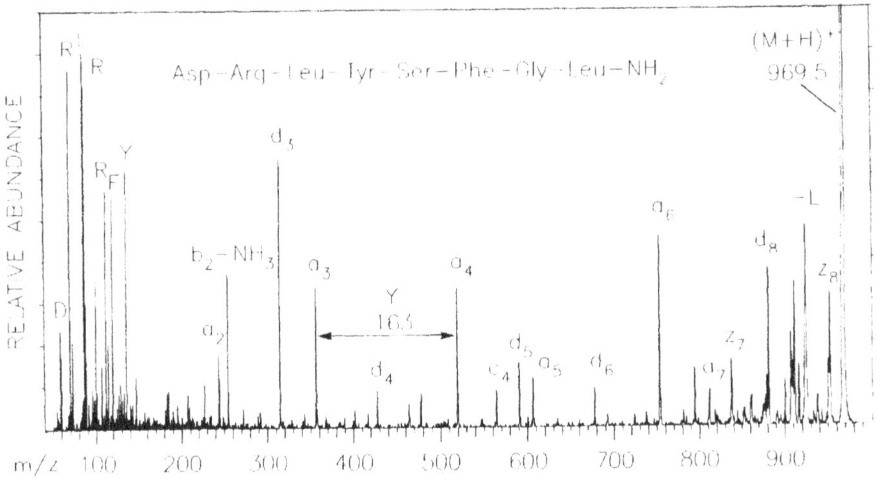


Figure 2. CAD mass spectrum of the $(M+H)^+$ *m/z* 969.5 of peptide **4**.

From the CAD mass spectrum, it was possible to resolve all the questions about the structural details of the peptide **4**. The structure was determined to be Asp-Arg-Leu-Tyr-Ser-Phe-Gly-Leu-$NH_2$. Assignments of product ions are indicated in Figure 2. (Nomenclature for product ions follows the Biemann modification of the Roepstorff symbols [2].) At the low mass end of the spectrum, which was recorded from *m/z* 50, immonium ions were observed at *m/z* 60.0 (Ser), 86.1 (Leu), 120.1 (Phe), 129.1 (Arg), 136.1 (Tyr), which could be assigned to the amino acids known to be present. (The immonium ion for Gly would occur at *m/z* 30.0, below the mass range scanned.) The presence of an Asp residue is suggested by the immonium ion at *m/z* 88.0. A complete **a** series of ions was present. The $\mathbf{a_2}$ ion at *m/z* 244.0 and the $\mathbf{z_7}$ ion at *m/z* 837.5 confirm that Asp is indeed the previously undefined N-terminal amino acid. An abundant ion at *m/z* 255.1 can be assigned to $\mathbf{b_2}$-$NH_3$. This fragment is routinely observed in the CAD mass spectrum of peptides that have Arg in the penultimate position with respect to the N-terminus [3]. The interval of 163 u between $\mathbf{a}_3$ (*m/z* 357.2) and $\mathbf{a}_4$ (*m/z* 520.2) is appropriate for an unmodified Tyr residue, and, therefore, it could be concluded that no post-translational modification had occurred at this site. The remainder of the CAD sequence information is consistent with the Edman sequenator data. The CAD data indicated that it was unnecessary to consider any extension of the sequence beyond Leu-8. The presence of the $\mathbf{d}_3$ ion at *m/z* 315.2 and the $\mathbf{d}_8$ ion at *m/z* 882.3 each 42 u below the corresponding $\mathbf{a}_3$ and $\mathbf{a}_8$ ions at *m/z* 357.2 and 924.4, respectively, affirmed that the intervals of 113 u both indicated Leu (and not Ile) [4]. With the amino acid sequence established, the only point not yet specifically addressed was the assignment of the group at the carboxyl terminus. A peptide with the sequence determined would have an $(M+H)^+$ at *m/z* 970.5 if the C-terminus were carboxyl. Observation of the $(M+H)^+$ at *m/z* 969.5 thus indicates that the C-terminus of this peptide, like the others, was amidated.

The four peptides share as common features five amino acid residues, three at the C-terminus and the fifth and sixth residues on that side, and the C-terminus amidation. Three of the peptides have the seventh residue on the C-terminus in common as well. The shorter peptides are not truncated versions of the longer ones, and Woodhead *et al.* therefore regard them as a "family" of neuropeptides, such as have been found in other invertebrates [1 and references therein]. Subsequent synthesis and biological testing of these three peptides, and of the fourth, which was not analyzed by mass spectrometry, confirmed that the allatostatic activity was restricted to the C-terminal amides, and was not observed for the C-terminal carboxyl analogs [1].

## 4. Conclusions

Careful sample manipulation and the prior examination of appropriate model or related compounds isolated from biological sources facilitate the successful mass spectral analysis of minute quantities of valuable samples. Information provided by ancillary techniques helps in the selection of experimental conditions for the MS analysis and in the interpretation of the spectra. Even with electron multiplier detection, the molecular weights and sequences of peptides having about ten residues may be determined by using FAB MS and MS/MS at the level of tens of pmol. In the example given here, three allatostatic cockroach neuropeptides were shown to have C-terminal amides and the amino acid sequence was determined for the one whose sequence had been unknown.

## 5. Acknowledgements

The samples of cockroach neuropeptides were isolated and purified by B. Stay and her colleagues at the University of Iowa, Iowa City, IA. Mass spectra were acquired with the assistance of S. A. Martin. The MIT Mass Spectrometry Facility is supported by Grant No. RR00317 (to K. Biemann) from the NIH Center for Research Resources.

## 6. References

1. Woodhead, A. P., Stay, B., Seidel, S. L., Khan, M. A. and Tobe, S. S. (1989) 'Primary Structure of Four Allostatins: Neuropeptide Inhibitors of Juvenile Hormone Synthesis', *Proc. Natl. Acad. Sci. USA*, **86**, 5997-6001.
2. Johnson, R. S. and Biemann, K. (1987) 'The Primary Structure of Thioredoxin from *Chromatium vinosum* Determined by High Performance Tandem Mass Spectrometry', *Biochemistry,* **26**, 1209-1214.
3. Papayannopoulos, I. A., personal communication.
4. Johnson, R. S., Martin, S. A., Biemann, K., Stults, J. T. and Watson, J. T. (1987) 'Novel Fragmentation Process of Peptides by Collision-Induced Decomposition in a Tandem Mass Spectrometer: Differentiation of Leucine and Isoleucine', *Anal. Chem.*, **59**, 2621-2625.

# THE MASS SPECTROMETRIC DETERMINATION OF RETRO-INVERSO LINEAR PEPTIDES

F. DE ANGELIS
*University of l'Aquila, Department of Chemistry*
*Via Assergi, 4*
*I-67100 l'Aquila*
*Italy*

ABSTRACT. The title compounds are attractive, biologically stable analogs of bioactive peptides. A reliable methodology for their analysis is required because there are difficulties associated with their synthesis. Furthermore, there is a need for the analysis of their metabolites present in the biological fluids when the peptides are used as drugs. In this paper, the mass spectrometric behaviour of this class of compounds is reviewed, taking into consideration three kinds of approaches: i) a thorough investigation of the direct fast atom bombardment mass spectrum; ii) the direct analysis, without separation, of the peptide mixture generated through selective, partial acid hydrolysis of the peptide analog; iii) sequence analysis by tandem mass spectrometry. The study is performed by using a few compounds as tests of the methodology. The implications arising from the mass spectrometric analysis of the modified peptides are also tentatively discussed in terms of the stability of such compounds.

## 1. Introduction

Mass spectrometry (MS) and tandem mass spectrometry (MS/MS) [1], in combination with fast atom bombardment (FAB) ionization [2], are important tools for peptide and protein sequence analysis [3]. These techniques show potential as the method of choice especially for those proteins that are blocked or modified [4], for those that have undergone post-translational modifications, or for those that are genetically engineered proteins and there is a need to confirm the expected primary structure [5].

Generally, there is also a particular need for application of MS to peptides arising from chemical synthesis. The major drawback of the synthesis of a protein, whether it is performed in solution or in the solid phase, is the yield of each synthetic step, which should always approximate one hundred per cent. Lower yields, especially for those compounds that consist of many amino acid residues, have a dramatic impact on the final overall yield. Thus, extensive purification procedures are required for the protein of interest, and also a full structural characterization of all intermediates and by-products is necessary, to check which step of the synthesis is to be improved.

Structural determinations of intermediates and artifacts of the synthesis are of fundamental importance when the normal protein backbone is modified to impart more favorable biological properties to the protein. This is the case for the synthesis of the partially modified retro-inverso (PMRI) peptide isomers.



*M. L. Gross (ed.), Mass Spectrometry in the Biological Sciences: A Tutorial*, 357–369.

## 2. Retro-Inverso Peptides

The retro-inverso modification of the peptide backbone consists in the reversal of the direction of one or more of the amide bonds (defined as those from the carbonyl carbon atoms to the nitrogen atoms) [6]. As a consequence of this manipulation, the chirality of the amino acid residues in the reversed portion is inverted in order to maintain the correct side-chain topology (Figure 1).

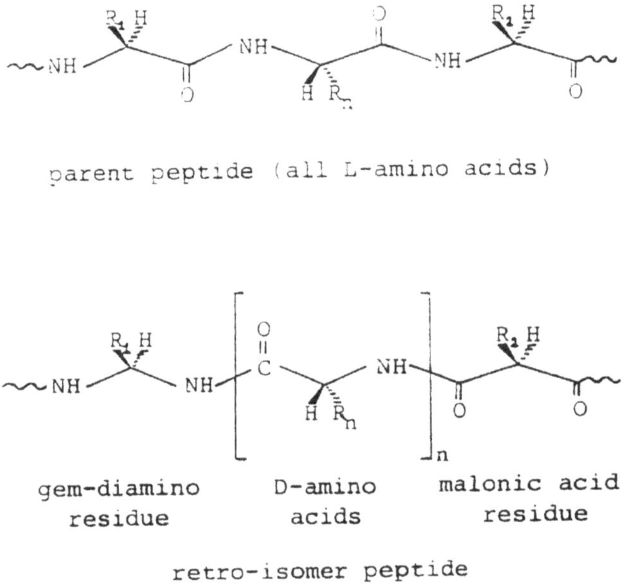


Figure 1. Structural description of the retro-inverso peptide approach (n = 0,1,...).

This approach appears very effective for obtaining long- lasting analogs of bioactive peptides [7]. This is because PMRI peptides exhibit enhanced stability towards proteolytic enzymes while maintaining substantial biological activity.

The synthesis of retro-inverso peptides requires, in the inverted sub-structure, the assembly of two new "amino acid" entities (besides the D-amino acid residues): a *gem*- diaminoalkyl residue (*gem*-a.a.) and a malonic acid residue (*m*-a.a.), as shown in Figure 1.

Considerable efforts have been made to achieve a reliable, high yield, and general synthesis of PMRI peptide isomers. Solution syntheses [8] and solid phase schemes [9] were proposed. Both synthesis methods require reliable and rapid analytical tools for accurate control of the intermediates as well as of the products and by-products of the synthesis. It is obvious that the classical Edman degradation technology and proteolytic digestion are not the methods of choice.

## 3. Sequence Analysis Using FAB-MS

### 3.1 . GENERAL METHOD

The experimental methodology we review here is a standard one. The peptide of interest is purified by high-performance liquid chromatography (HPLC) and directly analyzed by a Kratos MS 80 instrument equipped with a FAB xenon source, using glycerol or bis-(2-hydroxyethyl)sulphide [10] as the matrix. The mass spectrum reported in Figure 2 as an example, refers to a bradykinin-potentiating peptide 5a analog [11].

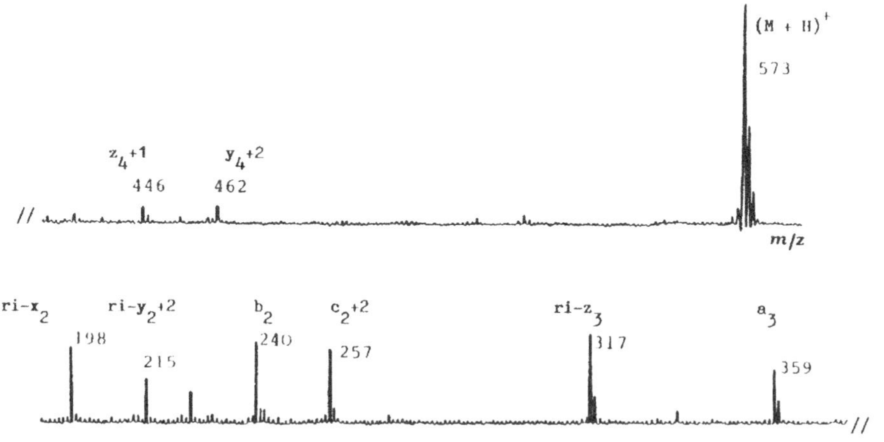


Figure 2. FAB Mass spectrum of a bradikynin-potentiating peptide retro-inverso analog (Pro-Lys-gPhe-mAla-Pro-OH).

In the FAB mass spectrum, there are several peaks due to ions carrying sequence information. The interpretation procedure is, for convenience, divided into two parts: i) analysis of those ions that refer to the normal part of the backbone; ii) analysis of those ions characteristic of the modified part in the molecule.

As to the former part of the analysis, product ions can be interpreted according to the classical fragmentation pattern [12], as shown in Scheme 3. The corresponding ions are designated following the common notations [13, 3]. It is worth noting that ions formed by cleavage in the proximity (i.e., at the left-hand $\alpha$-amino group) of g-Phe are particularly abundant. This situation appears general for those product ions that arise by fragmentation at the modified part of the molecule (vide infra).

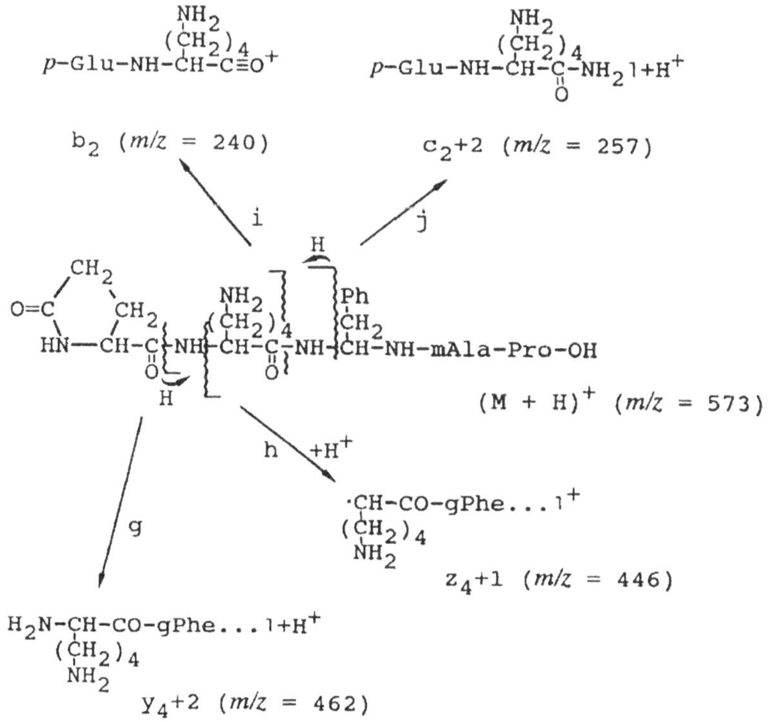


Scheme 1. Fragmentation scheme for the normal portion of the peptide.

The most informative fragmentations, giving information on the backbone structure, are those that directly involve the modified amino acids (as in $c_2 + 2$). They give rise to ions carrying the reversed portion of the peptide. Accordingly, the fragmentation scheme (reported in Scheme 2) contrasts with that of normal peptides.

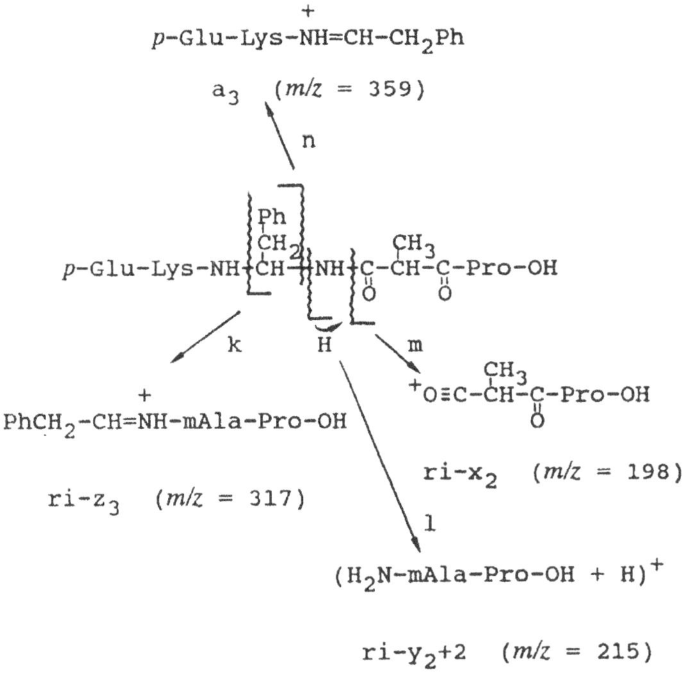


Scheme 2. Fragmentation scheme pertaining to the reversed portion of the peptide.

Fragmentation of the $(M + H)^+$ ions occurs at one of the bonds in the polyamide backbone to produce several ion types. According to convention [3], they can be designated a, b and c (pathways n, i and j, respectively) when the charge is retained by the N-terminal fragment, and y, z (pathways g and h, respectively) and x when the charge is retained by the C-terminus, even if these ion families carry the modified moieties. Numbering, and +1 or +2 notations, should also follow the conventional system.

Some difficulties arise when the modified bonds are directly involved in the fragmentation process. To avoid confusion, we would like to keep, with minimum modifications, the common nomenclature (ri is used as a prefix to indicate retro-inversion). That is, ion types ri-x (pathway m) are C-terminal acylium cations (x ions are best depicted by $^+$O≡C-NH~); ri-y+2 ions (pathway l), where the charge is on the C-terminus, are protonated amides (y+2 ions are protonated amines); ri-z ion types (acyl immonium ions, pathway k), which also contain the C-terminus of the peptide, terminate with the $\alpha$-carbon that carries the g-amino acid side chain (z ions also terminate with the $\alpha$-carbon).

It follows that in-chain fragment ions, which are absent in the direct FAB spectrum of the PMRI peptide under study, could in principle be named by using the conventional two-letter coding [3], taking into account the proposed modifications.

As a general consideration, we wish to emphasize the fact that ions formed by fragmentation at, or in the close proximity of, the gem-amino acid are particularly abundant with respect to fragment ions of normal peptides, which are usually scarcely abundant or even absent. This observation is tentatively interpreted in terms of a lower bond strength in the PMRI compounds.

The conclusions given here are general for all PMRI peptides [14], and appear particularly useful for the determination of such compounds by tandem mass spectrometry (vide infra).

3.2. PARTIAL ACID HYDROLYSIS

Despite the fact that PMRI peptides are said to be remarkably stable to hydrolysis [15], the gem-diamino bonds appear more susceptible to acid hydrolysis than the normal amide bonds. The compounds hydrolyze to amides, aldehydes and ammonia (ammonia is generated only when the gem-amino acid is a terminal one) at a rate higher than for normal peptide bonds (depending on the reaction conditions). This particular feature appears useful for the structural characterization of such compounds, especially when studying unknowns arising as by-products in the synthesis, or in the analysis of metabolites, when investigating these compounds as drugs. The partial, selective hydrolysis [16] of the PMRI peptide generates new, ideally complementary peptides. This allows one to add up the fragments and to reconstruct the complete sequence of amino acids in the parent peptide, thus also locating the retro-inversion sites. This approach is obviously based on the capability of FAB mass spectrometry to analyze peptide mixtures.

For example, a partial acid hydrolysis can be performed with aqueous 5% HCl, heating the solution to 90 $^{o}$C for 0.25-1.0 h in a closed vial. The hydrolysate is then placed on the FAB target and ionized, resulting in a mass spectrum characterized by the presence of the $(M + H)^+$ ions of the hydrolysis products. Generally, the hydrolysis fragments deriving from cleavage at both sites of the gem-amino acid are more abundant than the others, thus directly allowing the retro-inversion site to be located. When the hydrolysis is performed in aqueous 5% $HCl/CH_3OH$ (5:2 by vol.) and the solution heated in a closed vial at 90 $^{o}$C for 15 min, the FAB mass spectrum shows groups of peaks, each of them separated by 14 mass units from the others. These peaks, which are easily recognized in the spectrum, are due to the $(M + H)^+$ of the hydrolysis (free carboxy termini) and methanolysis (carboxy methyl esters) products of the gem- amino acid moiety.

The analysis becomes more difficult when other acid labile functionalities, besides the gem-diamino group, are present in the PMRI peptide. Examples are amidated peptides or those containing a pyroglutamic residue. In this event, results are complicated by the appearance in the spectrum of multiple peaks due to products arising by partial solvolysis of the primary amide or of the pyroglutamic moiety.

To simplify the analysis procedure, milder and more selective hydrolysis conditions can be used [14]. By treating the PMRI peptide with 80% aqueous trifluoroacetic acid at 60 $^{o}$C for 3 h, only partial hydrolysis of the gem-amino acid occurs. Data for a PMRI analogue of a Substance P fragment are reported in Scheme 3.

$$p\text{-Glu-Phe-gPhe-mGly-Leu-Met-NH}_2 \xrightarrow{\text{TFA}}$$

$(M + H)^+$ ($m/z$ = 724)

$$\longrightarrow p\text{-Glu-Phe-NH}_2 \quad + \quad \text{H}_2\text{N-mGly-Leu-Met-NH}_2$$

$(M + H)^+$ ($m/z$ = 276) $\qquad$ $(M + H)^+$ ($m/z$ = 347)

Scheme 3. Hydrolysis products generated by aqueous 80% TFA treatment at 60 °C for 3 h of a PMRI analogue of a Substance P fragment.

## 4. Analysis on a Triple Quadrupole Mass Spectrometer

### 4.1. GENERAL ASPECTS

MS/MS experiments are generally used to sequence unequivocally peptides. Magnetic deflection, four sector instruments [17] or triple stage quadrupole mass spectrometers [18] can be employed (both categories of instruments are reviewed in Section I of this book). The major advantages of the tandem technique reside with two aspects: i) the possibility of studying peptide mixtures: MS-1 can be focused on the $(M + H)^+$ ion of the peptide of interest which is then fragmented by collision activated dissociation (CAD), and finally the product ions analyzed in MS-2; ii) the mass spectrum shows selected peaks, which are fairly intense and diagnostic as to the amino acid sequence of the peptide. The interpretation of this type of mass spectral data can make use of the notations outlined in the previous sections, also with reference to the PMRI peptide analogs.

As to this kind of compounds, we wish to show how the tandem approach can also be used in combination with the direct FAB analysis to make predictions of the strength of the reversed peptide bond in relation to its position in the molecule, and hence on the stability of the latter.

### 4.2. INTERPRETATION OF THE CAD SPECTRA

Linear peptides contain end groups. Reversal of the direction of the peptide bond at the terminal amino acid should confer to the molecule enhanced resistance against exopeptidases. This is the case of a gem-diaminoalkyl residue (*g*-a.a.) at the N-terminus. Retro-inverso peptides of this type (isolated through preparative HPLC), when analyzed by direct FAB mass spectrometry, generally show a very intense peak 17 u lower than the expected value (Figure 3), its intensity being several times that of the putative protonated molecule.

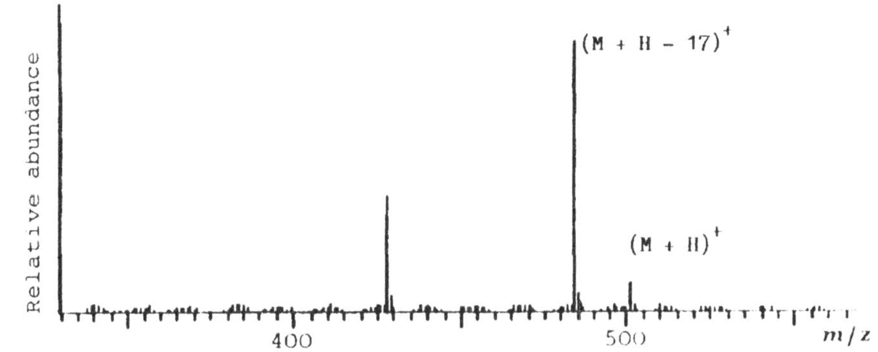


Figure 3. FAB Mass spectrum of the N-terminal, gem-diaminoalkyl peptide H-gThr-mLys-Pro-Arg-OH.

There could be three reasons for this anomalous behaviour: i) the peptide markedly decomposes during the freeze-drying process or during storage according to the following equation (this is highly unlikely) [15]:

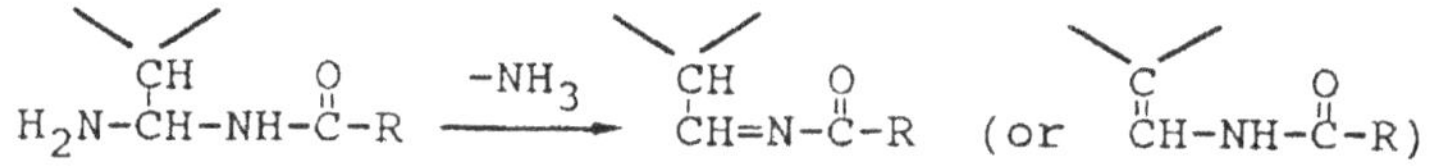


(1)

ii) the molecule decomposes in the matrix under the high energy atom beam; iii) the molecule spontaneously decomposes in the gas phase.

To evaluate the foregoing possibilities, we have carried out a few experiments using N-(1-aminoalkyl)amides of type **1** as model compounds of retro-inverso peptides partially modified at the amino terminus [9].

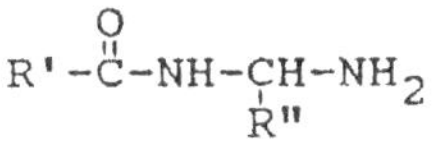

1

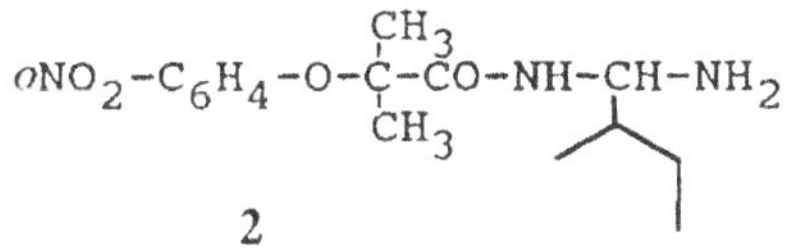

2

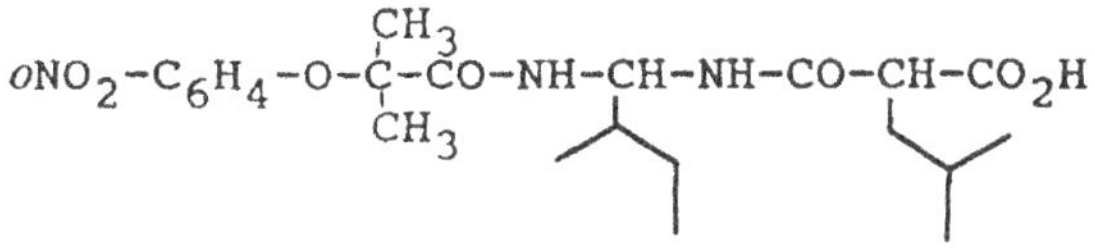

3

The amide **2** (protected *g*-Ile) was dissolved in thioglycerol and deposited on the FAB target of a triple quadrupole mass spectrometer (Finnigan-MAT instrument), and the resulting mixture was subjected to bombardment by a beam of 8-Kev $Cs^+$ ions, generated from a $Cs^+$ ion gun. In the MS/MS experiment, the first quadrupole mass filter, Q1, was focused to transmit the $(M + H)^+$ ions to the quadrupole collision cell Q2, filled with $Ar^o$ to a pressure of $3x10^{-4}$ torr. The product ion spectrum was recorded by scanning the third quadrupole mass filter, Q3.

The mass spectrum at the top of Figure 4, taken by operating in the MS mode, shows that most of the total ion current is carried by the ion at *m/z* 293 (differing by 17 u from the $(M + H)^+$ ion), which appears to be too abundant to be interpreted with confidence as a fragment ion. Nevertheless, the product ion spectrum of the $(M + H)^+$ ion, *m/z* 310, shown at the bottom of Figure 4, reveals that the loss of ammonia from the protonated molecule is a facile process for this type of compound. Even the spectrum taken without collisional activation shows the same fragmentation with loss of ammonia.

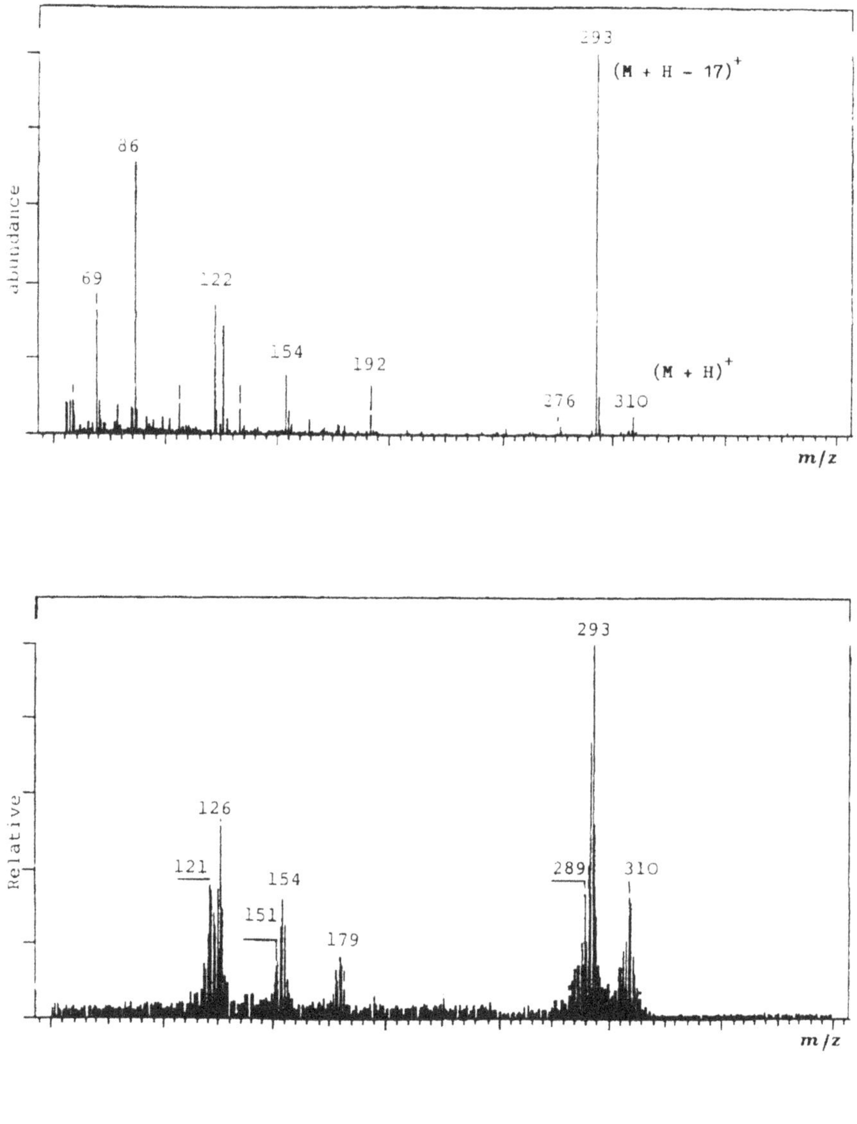


Figure 4. FAB mass spectrum (top) and CAD spectrum of the *m/z* 310 ion of compound **2**.

Thus, the ion at *m/z* 293 could well represent a fragment ion of type a (see Scheme 2), rather than the protonated molecular ion of the N-acylimine (or of the N- acyleneamine) likely to be present as a by-product in the mixture from the synthesis of **2** (see eq. 1).

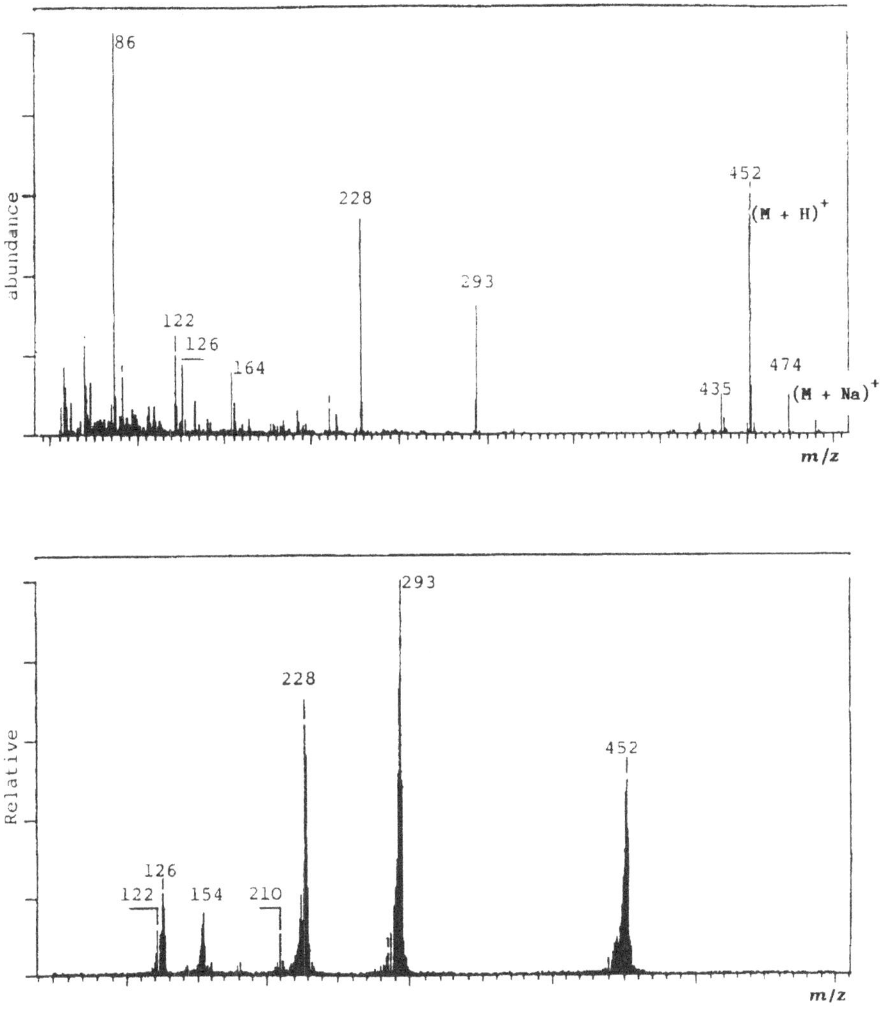


Figure 5. FAB mass spectrum (top) and CAD spectrum of the *m/z* 452 ion of compound **3**.

On the other hand, when the g-amino acid is an in-chain residue, the abundances of fragment ions are markedly different. This is the case, for instance, for compound **3** (N-1 malonyl derivative of **2**), whose PMRI peptidic structure mimics that of the protected dipeptide H-Ile-Leu-OH. The corresponding mass spectrum and low energy CAD mass spectrum are shown in Figure 5.

Two fragment ions diagnostic of the g-Ile presence in the peptide appear in both spectra, and they could be interpreted respectively as an $a_1$ ion ($m/z$ = 293) and a ri-$z_2$ ion ($m/z$ = 228). The fragmentation patterns for both compounds **2** and **3** are summarized in Scheme 4.

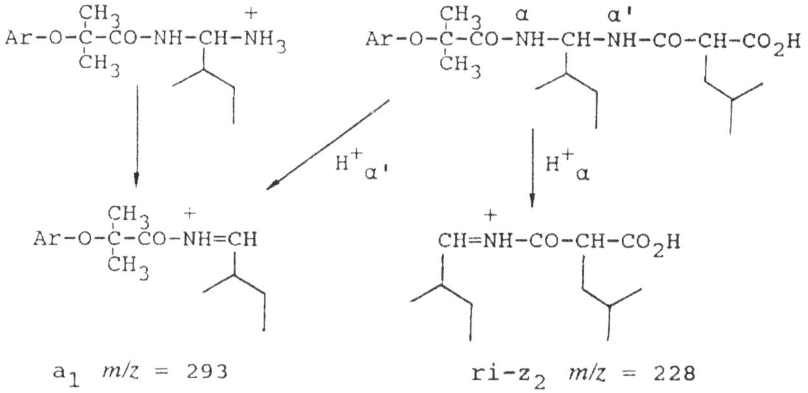


Scheme 4. Fragmentation upon collisional activation of compounds **2** (left) and **3** (right).

Absence of an ion of type ri-z in the CAD mass spectrum of **2** ($m/z$ = 84) means that the driving force for decomposition of its $(M + H)^+$ ion to form the $a_1$ ion resides in the case of protonation of the terminal free amino group, as compared to the other nitrogen atom present in the molecule. This protonation makes also the C-N$\alpha$ bond particularly labile, thus explaining the high abundance of the product ion in the normal FAB spectrum.

It should also be emphasized that the two $a_1$ and ri-$z_2$ fragments in the CAD spectrum of **3**, besides the fact that their corresponding peaks clearly emerge from the other peaks, are ideally complementary; as a consequence, fragmentations of this kind, when present, can be used to locate rapidly the retro-inversion site in the PMRI peptide.

## 4. Conclusions

Analytical techniques like those described in this chapter can be used as tools for obtaining structural information as to the nature and the position of the inverted sites in PMRI peptide isomers. Once again [20], we would like to point out that for complex structural problems, a single analytical method is not sufficient to solve the problem. Possibly, a combination of several instrumental and chemical techniques like those described above, in conjunction with others such as enzymic cleavages [21], must be cleverly used.

Moreover, a few conclusions can be put forward with reference to the stability of these modified peptide analogs, taking into account that FAB-MS can be considered as a sort of chemical ionization technique, $H^+$ being the ionizing species: i) in our case, abundant fragment ions are formed by cleavage in the close proximity of the retro-inverso site, this being an indication of the site of protonation in the molecule; ii) in general, more facile fragmentation is observed with respect to normal peptides. It follows that the PMRI peptide analogs, despite the fact that are more stable to proteases, are less stable under acidic conditions, and the fragmentation depends on the position of the retro-inversion in the peptide backbone.

## 5. Acknowledgements

Direct FAB mass spectra were taken with a Kratos MS 80 instrument at the Department of Chemistry of the University of Rome "La Sapienza". The author is greatly indebted to Prof. C. Heizmann, of the University of Zürich, for providing him the opportunity to perform the MS/MS experiments, and to Mr. W. Staudenmann for assistance.

The author also thanks Prof. A.S. Verdini (Italfarmaco, Milano) and his staff for the PMRI peptides and for helpful discussions.

## 6. References

1. Busch, K. L., Glish, G. L., and McLuckey, S. A. (1988) Mass Spectrometry/Mass Spectrometry: Techniques and Applications of Tandem Mass Spectrometry, VCH, New York.
2. Barber, M., Bordoli, R. S., Sedgewick, R. D., and Tyler, A. N. (1981) *J. Chem. Soc., Chem. Commun.*, 325.
3. Biemann, K. and Martin S. A. (1987) *Mass Spectrom. Rev.*, **6**, 1.
4. Williams, D. H., Santikarn, S., De Angelis, F., Smith, R. J., Reid, D. G., Oelrichs P. B., and MacLeod, J. K. (1983) *J. Chem. Soc., Perkin Trans. 1*, 1869.
5. Morris, H. R. and Greer, F. M. (1988) *Trends in Biotech.*, **6**, 140.
6. Goodman, M. and Chorev, M. (1979) *Acc. Chem. Res.*, **12**, 1.
7. Groggia, A., Pessi, A., Sisto, A., Verdini, A. S., Virdia, A., and Viscomi, G. C. (1985) in C. M. Deber, U. J. Hruby, and K. D. Kpoole (eds.), Proceedings of the Ninth American Peptide Symposium, Pierce Chemical Co., Rockford, Ill., USA, p. 659.
8. Chorev, M., Willson, C. G., and Goodmann, M. (1977) *J. Am. Chem. Soc.*, **99**, 8075.
9. Pessi, A., Pinori, M., Verdini, A. S., and Viscomi, G. C. (1983) *J. Chem. Soc., Chem. Commun.*, 195.
10. De Angelis, F., Nicoletti, R., and Santi, A. (1988) *Org. Mass Spectrom.*, **23**, 800.
11. Verdini, A. S., and Viscomi, G. C. (1985) *J. Chem. Soc., Perkin Trans. 1*, 697.
12. Williams, D. H. (1983) *Gazz. Chim. Ital.*, **113**, 27.
13. Roepstorff, P. and Fohlman, J. (1984) *Biomed. Mass Spectrom.*, **11**, 601.
14. De Angelis, F., Ceccarelli, S., Viscomi, G. C., Pinori, M., and Verdini, A. S. (1989) *Biomed. Environ. Mass Spectrom.*, **18**, 867.
15. Loudon, G. M., Almond, M. R., and Jacob, J. N. (1981) *J. Am. Chem. Soc.*, **103**, 4508.
16. De Angelis, F., Botta, M., Ceccarelli, S., and Nicoletti, R. (1986) *Biochem. J.*, **236**, 609.
17. Scoble, H. A., Martin, S. A., and Biemann, K. (1987) *Biochem. J.*, **245**, 621.
18. Hunt, D. F., Yates III, J. R., Shabanowitz, J., Winston, S., and Hauer, C. R. (1986) *Proc. Natl. Acad. Sci. USA*, **83**, 6233.
19. De Angelis, F. and Verdini, A.S. (1991), a more comprehensive study is being published.
20. De Angelis, F., Botta, M., and Nicoletti, R. (1987) *Biochem. J.*, **245**, 523.
21. Morris, H.R., Panico, M., and Taylor, G.W. (1983) *Biochem. Biophys. Res. Commun.*, **117**, 299.

# MASS SPECTROMETRY IN STUDIES OF *O*-LINKED GLYCOPEPTIDES ISOLATED FROM NATURAL AND RECOMBINANT GLYCOPROTEINS

C. A. SETTINERI AND A. L. BURLINGAME
*Department of Pharmaceutical Chemistry*
*Mass Spectrometry Facility*
*University of California, San Francisco, CA 94143-0446*

ABSTRACT. Characterization of the carbohydrates attached to a glycoprotein requires determination of the primary structure of the oligosaccharides as well as their sites of attachment. Since there is no consensus sequence to identify attachment sites of *O*-linked carbohydrates and there is no generic enzyme which cleaves all *O*-linked sugar structures, the identification of the structures at each given site for *O*-linked oligosaccharides is a challenging analytical task. Recently, however, progress has been made in this area with the analysis of intact glycopeptides by high-performance tandem mass spectrometry (MS/MS). In this paper we review the use of high-performance MS/MS to determine the oligosaccharide sequence, the peptide sequence, and in certain cases the site(s) of attachment of *O*-linked sugars to intact glycopeptides from recombinant human platelet-derived growth factor B-chain (rhPDGF-B) and bovine fetuin.

## 1. Introduction

The complete characterization of a natural or recombinant glycoprotein involves determination or verification of the amino acid sequence as well as the structural elucidation of the attached carbohydrates. Glycosylation is thought to confir many functional roles upon proteins including cell adhesion and motility, cell-surface recognition, receptor binding, clearance rate and immunogenicity [1-4]. In addition there are clear effects on protein solubility and stability [4]. Carbohydrates can be attached to proteins via the amino acid side chains of asparagine (*N*-linked), and serine or threonine (*O*-linked). In addition, glycans can be attached to the carboxy-terminus of glycoproteins as a glycoinositol phospholipid moiety, thus providing protein anchorage in the cell surface membrane [5].

Potential attachment sites for *N*-linked carbohydrates can be identified from the consensus sequence Asn-Xaa-Ser/Thr [6,7] in the protein; however, no consensus sequence is known for *O*-linked carbohydrates, so potential attachment sites cannot be identified from inspection of the protein sequence. This is further complicated by the fact that *O*-glycosylation has generally been found to occur in serine and threonine rich sequences, making it more difficult to pinpoint which serines or threonines may be modified. In addition, structural analysis of *O*-linked glycopeptides is thwarted by the lack of a generic enzyme which cleaves all *O*-linked sugar structures, as is the case for *N*-linked glycopeptides using peptide:N-glycosidase F (PNGase F) [8]. Therefore, the identification of both the sites, and structures at each given site, for *O*-linked glycoproteins is a challenging analytical task.

The classical methodology available for characterization of *O*-linked carbohydrates of a glycoprotein involves chemical release of the oligosaccharides from the glycoprotein or intact



*M. L. Gross (ed.), Mass Spectrometry in the Biological Sciences: A Tutorial*, 371–381.

glycopeptide by β-elimination [9] using base and mild reducing conditions. The reduced oligosaccharides can then be separated by gel filtration [10] or high-pH anion exchange chromatography [11] and the individual components either permethylated [12] or acetylated [13] for analysis by liquid secondary ion (LSIMS) or fast atom bombardment mass spectrometry (FAB-MS). Additionally, methylated [14] released oligosaccharides can be analyzed for composition and linkage of the individual sugar residues by GC-MS [15]. However, all information regarding the attachment site of the oligosaccharides is lost in these analyses, and the amino acid chains often suffer degradation [16].

Alternatively, the glycoprotein can be digested with a specific protease to yield glycopeptides, or with a non-specific protease such as pronase to yield glycosylated amino acids and small glycopeptides. The glycopeptides can be analyzed intact by Edman degradation [17], where a blank cycle infers the site of carbohydrate attachment, but a single pure glycopeptide is required, and other methods are needed to determine the structure of the carbohydrate at any particular site. FAB-MS has been employed for molecular weight determination of intact unmodified *O*-linked glycopeptides [18-21], as well as the permethylated reduced oligosaccharides [14,18]. However, chemical sequencing is required to establish the site of glycosylation when more than one serine or threonine is present in the glycopeptides.

Recently, however, glycopeptides have also been analyzed by high-performance tandem mass spectrometry (MS/MS) [22, 23] using LSIMS and collision-induced dissociation (CID). We have reported previously the use of MS/MS for structural characterization of intact *O*-linked glycopeptides [22, 23]. This method has been used to determine the location of seven attachment sites of mannose monomers and dimers attached to recombinant human platelet-derived growth factor B-chain (rhPDGF-B) expressed in yeast [22].

In addition, we have found that CID spectra of glycopeptides can be enhanced by derivatization with a hydrophobic group such as *t*-Boc-tyrosine, which decreases the hydrophilicity of glycopeptides [23]. *O*-linked glycopeptides from bovine fetuin were derivatized at their *N*-terminus with *t*-Boc-tyrosine. This resulted in an increased sensitivity in positive ion LSIMS of a factor of ten over the N-terminally free glycopeptides, and enabled us to sequence the carbohydrate and peptide portions of these glycopeptides. In this paper we review the use of high-performance MS/MS to determine the oligosaccharide sequence, the peptide sequence, and in certain cases the site(s) of attachment of *O*-linked sugars to intact glycopeptides from rhPDGF-B and bovine fetuin.

## 2. Materials and Methods

### 2.1. PURIFICATION AND DIGESTION

Recombinant human PDGF-B dimer was produced via the α-factor secretion pathway in *Saccharomyces cerevisiae* [24], and reduced and carboxymethylated as described elsewhere [22]. Bovine fetuin was obtained from Calbiochem (San Deigo, CA), digested under sterile conditions with pronase, and a pooled carbohydrate containing fraction of glycopeptides was obtained as described earlier [25].

Trypsin digestions of rhPDGF-B were performed using 5% TPCK-trypsin (Serva, Heidelberg, West Germany) in 75 mM ammonium bicarbonate buffer (pH 7.8) with 0.1 mM calcium chloride. Digestions were performed at 37°C for 16 hours. Endoproteinase glu-C (*Staphylococcus aureus* V8) (Boehringer Mannheim, Indianapolis, IN) digestion was performed in 50 mM ammonium bicarbonate (pH 7.8) at a weight ratio of 4% for 18 hours at 37°C. A fresh aliquot of glu-C (4%) was added to the reaction mixture and the incubation was continued for 8 hours (total 26 hours). Extreme conditions were necessary to obtain complete digestion as assessed by use of sodium dodecylsulfate-polyacrylamide gel electrophoresis (SDS-PAGE) of samples taken at various time points during the digestion. Pronase digestion of the bovine fetuin carbohydrate-containing fraction was performed using Calbiochem-Behring pronase at a weight ratio of 1% in 50 mM 4-(2-hydroxyethyl)-1-piperazineethanesulfonic acid buffer containing 70 mM sodium chloride, 50 mM calcium chloride, and 40 mM sodium citrate, pH 7.9. Digestions were performed at 37°C for five days with daily additions of 0.5% pronase.

### 2.2. CHROMATOGRAPHY

Reverse phase chromatography of rhPDGF-B was performed on a Waters (Milford, MA) HPLC system using Vydac (The Separations Group, Hesperia, CA) $C_{18}$ reverse phase columns (4.6 x 250 mm). Gradients were generated using the following solvents: eluent A) 0.1% trifluoroacetic acid (TFA) (w/v) in water and eluent B) 80% acetonitrile (v/v) in solvent A. Enzymatic digests were titrated to pH 2 with 25% (w/v) TFA before injection. Peak detection was achieved at 230 nm. The pronase digest of the fetuin glycopeptides was first chromatographed on a Sephadex G-50 column (2.5 x 200 cm) equilibrated in pyridine-acetate buffer, pH 4.7. The neutral sugar containing fractions were pooled and desialylated as described earlier [25]. The *t*-BOC-Tyr-glycopeptides (*t*-BOC-Tyr derivatization described in [25]) were separated by reverse phase HPLC on a chromatograph consisting of two Gilson Model 302 pumps controlled by Gilson Gradient Manager, using a Supelcosil (Supelco) 5 µm column (4.6 x 250 mm).

### 2.3. MASS SPECTROMETRY

LSIMS analysis was carried out using a Kratos MS50S double focusing mass spectrometer (Kratos Inc., Manchester, UK) equipped with a high field magnet and a cesium ion source [26] and gun at 10 keV [27]. Concentrated HPLC fractions (1-2 µl) were added to 1 µl of liquid matrix (1:1 thioglycerol:glycerol with 1% TFA) for analysis. MS/MS experiments were carried out using a Kratos Concept IIHH four sector EBEB tandem mass spectrometer [28], equipped with a cesium ion source and a multichannel array detector that allows the simultaneous detection of ions over a 4% mass window by sequential stepping of the magnetic field at MS-2 under computer control [28, 29]. Tandem spectra were acquired while operating at a resolution of approximately 2000. The collision cell was floated at 2 keV and filled with helium so that the parent ion intensity was decreased to one third of its initial value [28]. Again, 1-2 µl of sample was used with the same matrix as above or with meta-nitrobenzyl alcohol. CID spectra were interpreted without the aid of computer analysis.

### 2.4. EDMAN DEGRADATION

Edman degradations were performed using an Applied Biosystems 470A protein sequencer (Applied Biosystems Inc., Foster City, CA) with an on-line 120A phenylthiohydantoin (PTH) amino acid analyzer. Standard programs and reagents were used during these analyses [30].

### 2.5. CARBOHYDRATE COMPOSTION ANALYSIS

Carbohydrate analysis was performed using a Dionex (Sunnyvale, CA) AI-450 Model II high performance anion exchange chromatography system with pulsed amperometric detection [31]. Eluent A was 200 mM NaOH and eluent B was 100% water. Identity of the type of sugars was determined by co-injection of a standard sugar(s) with the hydrolysis sample.

## 3. Results and Discussion

Tryptic and endoproteinase glu-C (*Staphylococcus aureus* V8) peptide mapping using mass spectrometry were performed to determine the structure of rhPDGF-B derived from expression in yeast cells [22]. From this mapping, several fractions were identified which contained molecular ions corresponding to peptides with masses 162 or 324 Da greater than the masses of predicted proteolytic peptides, indicating that they may have been modified with one or two hexose residues, respectively. Therefore, tandem mass spectrometry was used to determine if these were indeed glycopeptides, and if so, to identify the site(s) of modification and the nature of the glycosylation attached. By looking for diagnostic sequence ions with a mass shift indicating the attachment of one or two hexose residues, it could be determined from the CID spectra where the hexose residues were attached, and how many hexoses were attached at a particular site. Using this method in combination with Edman sequence analysis, seven different sites of glycosylation, including monomannosyl and dimannosyl modifications were identified in the rhPDGF-B sequence. Edman sequence analysis of the rhPDGF-B glycopeptides was used to confirm specific sites of glycosylation by assuming that a blank cycle indicates a modified amino acid (i.e. the glycosylated amino acid). Six of the seven glycosylation sites were unambiguously determined by MS/MS, however, Edman sequence analysis was required for confirmation of one of the glycosylation sites. In addition, MS/MS was used to determine the structure of a derivatized *O*-linked glycopeptide from bovine fetuin which contained two gal-galNAc disaccharides attached to two different sites in the glycopeptide.

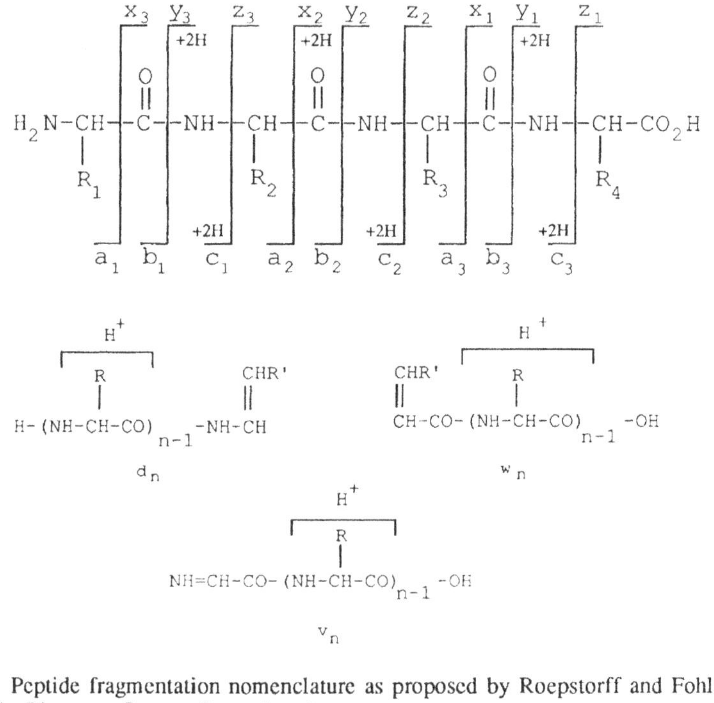


Figure 1. Peptide fragmentation nomenclature as proposed by Roepstorff and Fohlman, and modified by Biemann. See text for explanation.

The nomenclature for putative peptide fragments, as proposed by Roepstorff and Fohlman [32], and modified by Biemann [33], is shown in Figure 1. As shown, **a**, **b**, **c** and **d** ions are formed with charge retention on the amino-terminal portion of the peptide, and **x**, **y**, **z**, **w** and **v** ions are formed with charge retention on the carboxy-terminal portion. The subscript numbers identify the residue number where the fragment is formed. The nomenclature for putative carbohydrate fragments, as proposed by Domon and Costello [34], is shown in Figure 2. The fragments with charge retention on the non-reducing terminus (drawn left) are designated **A**, **B** and **C**, while the reducing terminal fragments (drawn right) are similarly designated **X**, **Y** and **Z**. Fragments **A** and **X** contain superscript numbers designating the two bonds involved in the ring cleavage which would account for the observed masses of these fragment series. Again, subscript numbers identify the residue number.

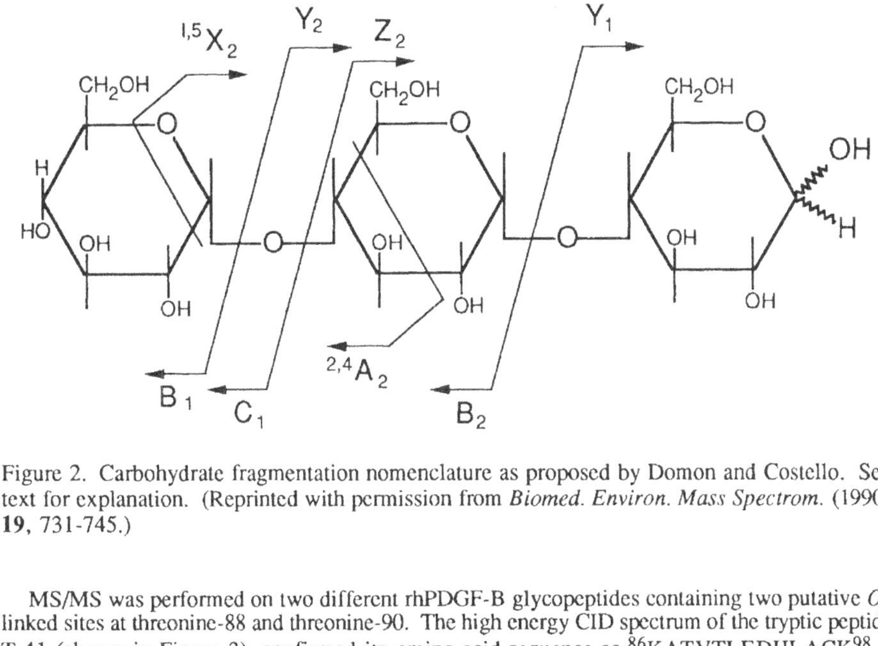


Figure 2. Carbohydrate fragmentation nomenclature as proposed by Domon and Costello. See text for explanation. (Reprinted with permission from *Biomed. Environ. Mass Spectrom.* (1990) **19**, 731-745.)

MS/MS was performed on two different rhPDGF-B glycopeptides containing two putative *O*-linked sites at threonine-88 and threonine-90. The high energy CID spectrum of the tryptic peptide T-41 (shown in Figure 3), confirmed its amino acid sequence as $^{86}$KATVTLEDHLACK$^{98}$ + mannose$_2$ ($MH^+$ = m/z 1810.8), and revealed that both threonine-88 and 90 were modified with one mannose residue each. The diagnostic sequence ions $y_9$, $z_9$, $x_{10}$, $y_{10}$, $z_{10}$, $v_{10}$, $w_{10}$, $v_{11}$ and $w_{11}$ were shifted by 162 Da and sequence ions $x_{12}$ and $y_{12}$ were shifted by 324 Da, indicating that each threonine was modified with one mannose residue.MS/MS of another peptide, V-42, shown in Figure 4, with the sequence $^{77}$IVRKKPIFKKATVTLE$^{92}$ + mannose$_2$ ($MH^+$ = m/z 2195.3), which covers the same region of protein sequence as T-41, also indicated one mannose residue on each threonine. Amino-terminal sequence ions which contained only threonine-88 were shifted by 162 Da, while those which contained threonine-88 as well threonine-90 were shifted by 324 Da. MS/MS of tryptic peptide T-53 from rhPDGF-B, confirmed its amino acid sequence as $^{57}$NVQCRPTQVQLR$^{68}$ + mannose$_1$ ($MH^+$ = m/z 1661.8), and showed that threonine-63 was modified with a single mannose residue (data not shown). MS/MS of rhPDGF-B peptide T-56,

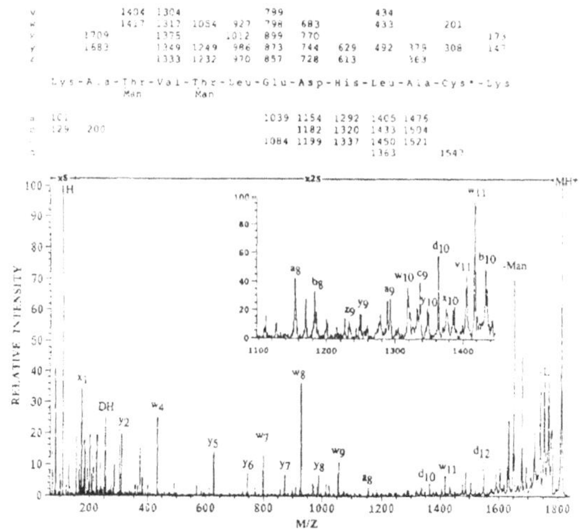


Figure 3. CID spectrum of tryptic peptide T-41, KA(Man)TV(Man)TLEDHLACK, MH+ = 1810.8, from rhPDGF-B. The insert shows an expanded region of the spectrum illustrating the diagnostic sequence ions. All ions identified are in the ion table above the spectrum where the m/z values for ions belonging to the same ion series are listed by rows. For clarity, only the most abundant ions are labeled in the spectrum. (Spectrum reproduced with permission from [22].)

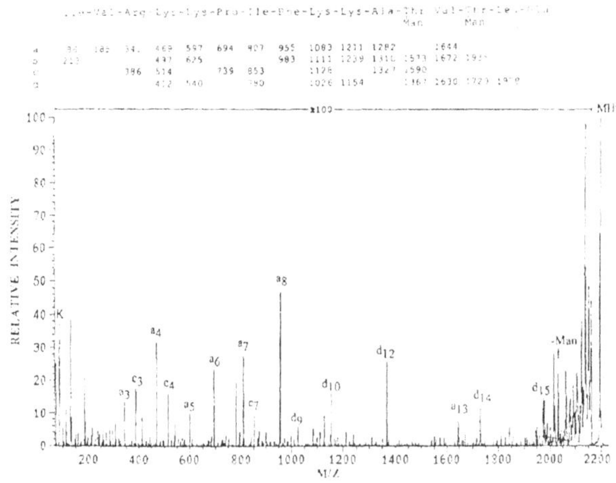


Figure 4. CID spectrum of glu-C peptide V-42, IVRKKPIFKKA(Man)TV(Man)TLE, MH+ = 2195.3, from rhPDGF-B. All ions identified are in the ion table above the spectrum where the m/z values for ions belonging to the same ion series are listed by rows. Only the most abundant ions are labeled in the spectrum. (Spectrum reproduced with permission from [22].)

with the sequence $^{20}$TEVFEISR$^{27}$ + mannose$_2$ (MH$^+$ = m/z 1304.6), showed that both threonine-20 and serine-26 were glycosylated with one mannose residue. The *C*-terminal sequence ions identified, listed in Table 1, inicated that serine-26 is modified with only one mannose residue, and therefore threonine-20 is modified with one mannose as well. MS/MS of peptide T-33, with the sequence $^{99}$CETVAAAR$^{106}$ + mannose, MH$^+$ = m/z 1040.4, shown in Table 1 (spectrum published elsewhere [35]), confirmed that this peptide was glycosylated on threonine-101 with one mannose residue. The final glycosylation site in rhPDGF-B, threonine-6, was determined using Edman sequence analysis. Therefore, MS/MS was used to determine 6 of the 7 glycosylation sites in rhPDGF-B.

TABLE 1. CID Fragmentation of rhPDGF-B Tryptic Peptides T-56 and T-33.

**T-56** MH$^+$= m/z 1304.6

| | | | | | | | | |
|---|---|---|---|---|---|---|---|---|
| v | | | 868 | | 592 | 479 | | |
| w | | | 881 | | 591 | 492 | 229 | |
| x | | 1067 | | 839 | 692 | 563 | | 201 |
| y | | | 912 | 813 | 666 | 537 | | 175 |
| z | | | | | 650 | 521 | | 159 |
| | Thr - | Glu - | Val - | Phe - | Glu - | Ile - | Ser - | Arg |
| | Man | | | | | | Man | |

**T-33** MH$^+$= m/z 1040.4

| | | | | | | | | |
|---|---|---|---|---|---|---|---|---|
| v | | 542 | 443 | | | | | |
| w | | 804 | 555 | 456 | | | | |
| x | | | | 513 | 414 | 343 | 272 | 201 |
| y | | 879 | 750 | 487 | | 317 | 246 | 175 |
| z | | | 734 | 471 | | 301 | 230 | 159 |
| | Cys* - | Glu - | Thr - | Val - | Ala - | Ala - | Ala - | Arg |
| | | | Man | | | | | |
| a | 134 | 263 | | | | | | |
| b | | 291 | 554 | | | | | |

Cys* indicates carboxymethyl-cysteine.
The m/z values for ions belonging to the same ion series are listed by rows.

In order to explore the possible extension of this method to glycopeptides from natural sources, which contain different carbohydrate structures from a protein expressed in yeast, we examined the CID spectrum of a glycopeptide isolated from bovine fetuin. In addition, in an attempt to decrease its hydrophilicity, this glycopeptide was derivatized at its *N*-terminus with *t*-BOC-tyrosine [23]. This derivatization resulted in an increase in the sensitivity in positive ion LSIMS by a factor of ten over the *N*-terminally free glycopeptide [36]. MS/MS of this glycopeptide, with the sequence (Y)-GPTPSA + Hex$_2$HexNAc$_2$, (MH$^+$ = m/z 1522.8) which contains two possible sites of glycosylation, yielded a CID spectrum which was quite different from the CID spectra discussed above. The spectrum was dominated by the fragmentation of the carbohydrate side chains rather than the peptide backbone. As shown in Figure 5, "ion pairs" were observed 100 Da apart at m/z 1522 and 1422; 1360 and 1260; 1157 and 1057, due to rearrangement losses of the *t*-BOC moiety from the glycopeptide. The ions at m/z 1422, 1260, 1057, 895 and 692 indicate the loss of the *t*-BOC-group, followed by the sequential losses of a hexose, N-acetyl-hexosamine, hexose, and N-acetyl-hexosamine from the intact glycopeptide.

The ion at m/z 366 is a non-reducing terminal carbohydrate oxonium ion, $B_2$, which represents a hexosyl-N-acetyl-hexosamine disaccharide. No larger non-reducing terminal ions, such as an ion representing a tetrasaccharide like the Gal[Gal-GlcNAc]-GalNAc which has been isolated from bovine fetuin [37], were detected. Glycopeptides containing this tetrasaccharide exhibit a carbohydrate oxonium ion at m/z 731 [36], and this ion was not present in the spectrum. Therefore, the only possible interpretation is that two identical hexosyl-N-acetyl-hexosamine glycans were attached to this peptide. Since there are two possible sites of glycosylation in this glycopeptide, each one must be modified with a single HexHexNAc moiety. In addition, carbohydrate composition analysis revealed the presence of only galactose and N-acetyl-galactoamine attached to this glycopeptide (data not shown). Therefore, the hexosyl-N-acetyl-hexosamine glycans must be galactosyl-N-acetyl galactosamine glycans.

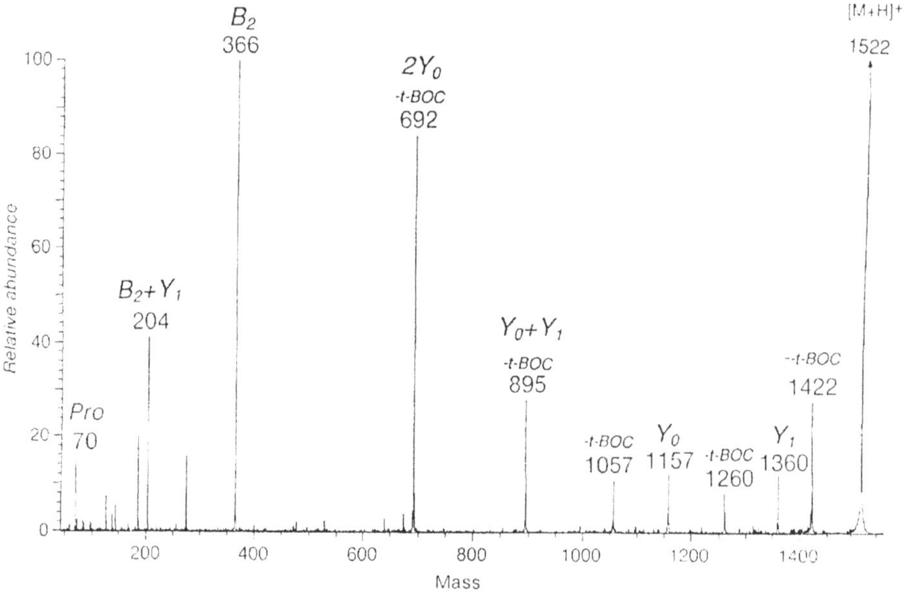


Figure 5. CID spectrum of the derivatized fetuin *O*-linked glycopeptide, t-BOC-YGP(Gal-GalNAc)TP(Gal-GalNAc)SA, $MH^+$=1522.8. (Spectrum reproduced with permission from [23].)

The ion at m/z 692 in Figure 5 represents the gas-phase deglycosylated and *N*-deprotected peptide. MS/MS was performed using this ion as the selected precursor, yielding the CID spectrum shown in Figure 6; which confirmed the sequence (Y)-GPTPSA for this peptide. Therefore, the structure of the intact glycopeptide was determined from the mass spectra and the carbohydrate composition analysis as:

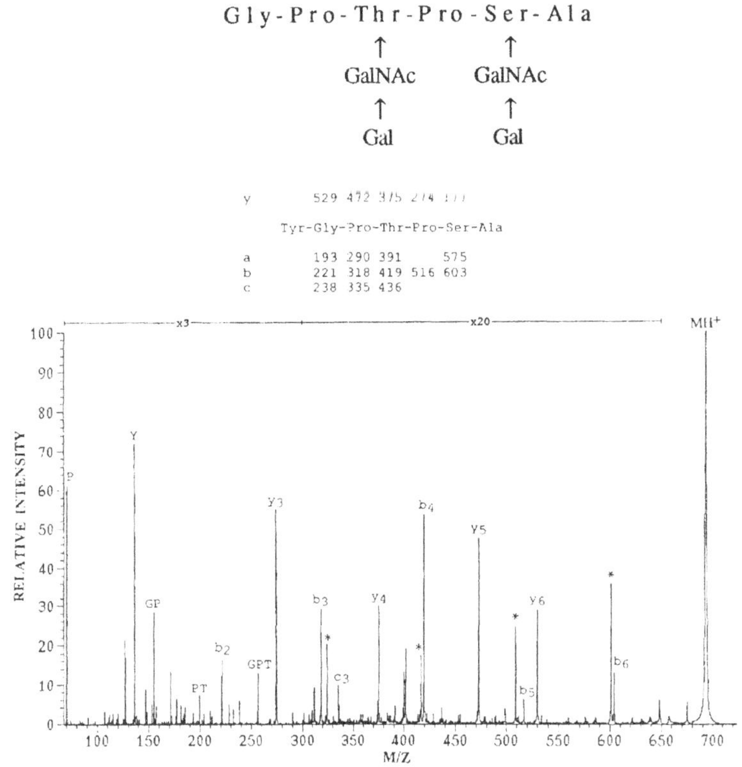


Figure 6. CID spectrum of the peptide fragment at m/z 692.4 of the fetuin glycopeptide. The fragment ions labelled with an asterisk are present due to the glycerol loss from a background peak with the same nominal mass as the parent ion [35]. (Reproduced with permission from [23].)

## 4. Conclusion

It is clear that complete characterization of *O*-linked glycosylation of natural or recombinant proteins has been a particularly difficult task, which has required a variety of analytical methods used in concert; including various chromatographic and electrophoretic techniques, Edman degradation, mass spectrometry and others. We have shown that high performance tandem mass spectrometry can be used to facilitate the characterization of intact *O*-linked glycopeptides. In addition, we have shown that hydrophobic derivatization of glycopeptides enhances the sputtering efficiency and therefore the sensitivity of such glycopeptides in positive ion LSIMS. While this method cannot determine the specific type of sugar moieties (e. g. hexose vs. mannose, galactose, etc.) attached to the glycopeptide or the linkage positions, it can rapidly reveal the sequence of the oligosaccharide(s) attached to intact glycopeptides without intervention of chemical degradation.

The data presented indicate, however, that the CID fragmentation can differ depending on the size of the oligosaccharide attached to the glycopeptide. It appears that glycopeptides containing only one hexose moiety attached at an *O*-linked site can be fully characterized from a single high energy CID spectrum. However, these data indicate that with increasing numbers of sugar units attached to the same residue in a peptide and with increasing numbers of glycosylated amino acids in the same peptide, the CID spectrum may become dominated by carbohydrate fragmentation rather than by peptide fragmentation. Thus, the possibility of obtaining both the sequence of the peptide and the location of glycosylation sites directly from the same CID spectrum may be thwarted. However, fragmentation (in LSIMS) of glycopeptides may yield abundant deglycosylated peptide fragments, providing the opportunity to obtain the peptide sequence from the same fraction which contains the glycopeptide by selecting the deglycosylated peptide fragment ion for CID analysis. Futher developments in derivatization strategies may make MS/MS the method of choice for simultaneously determining the carbohydrate and the peptide sequence of larger *O*-linked glycopeptides in the near future.

## 5. Acknowledgements

This work was supported by grants RR01614 (A. L. B.) from the National Center for Research Resources, and DIR8700766 (A. L. B.) from the U.S. National Science Foundation. Travel to the NATO Advanced Study Institute was supported by a travel grant (to C. A. S.) from the National Science Foundation.

## 6. References

1. Kobata, A., (1984) *in* Biology of Carbohydrates, vol. 2, V. Ginsburg and P. W. Robbins, Eds., pp. 87-162, Wiley, New York.
2. Fukada, M. (1985) *Biochim. Biophys. Acta*, **780**, 119-150.
3. Smedsrod, B., Einarsson, M., and Pertoft, H. (1988) *Thromb. Haemostasis*, **59**, 480-484.
4. Rademacher, T. W., Parekh, R. B., and Dwek, R. A. (1988) *Ann. Rev. Biochem.*, **57**, 785-838.
5. Cross, G. A. M. (1990) *Ann. Rev. Cell Biol.*, **6**, 1-40.
6. Lehle, L., and Bause, E. (1984) *Biochim. Biophys. Acta*, **799**, 246-251.
7. Bause, E., and Lehle, L. (1979) *Eur. J. Biochem.*, **101**, 531-540.
8. Tarentino, A. L., Gomez, C. M. and Plummer, T. H. (1985) *Biochem.*, **24**, 4665-4671.
9. Carlson, D. M. (1968) *J. Biol. Chem.*, **243**, 616-626.
10. Angel, A.-S. and Nilsson, B. (1990) *Biomed. Environ. Mass Spectrom.*, **19**, 721-730.
11. Townsend, R. R., Hardy, M. R. and Lee, Y. C. (1989) *in* Methods in Enzymology, **179**, pp. 65-76, Academic Press, San Diego.
12. Sasaki, H., Bothner, B., Dell, A., and Fukuda, M. (1987) *J. Biol. Chem.*, **262**, 12059-12076.
13. Carr, S. A., Roberts, G. D., Jurewicz, A. and Frederick, B. (1988) *Biochimie*, **70**, 1445-1454.
14. Hakomori, J. (1964) *J. Biochem. (Tokyo)*, **55**, 205-208.
15. Hellerqvist, C. G. and Sweetman, B. J. (1990) *in* Biomedical Applications of Mass Spectrometry, vol. 34, C. H. Suelter and J. T. Watson, Eds., pp.91-143, Wiley, New York.
16. Harbon, S., Herman, G. and Clauser, H. (1968) *Eur. J. Biochem.*, **4**, 265-272.
17. Conradt, H. S., Geyer, R., Hoppe, J., Grotjahn, L., Plessing, H. and Mohr, H. (1985) *Eur. J. Biochem.*, **153**, 255-261.
18. Sasaki, H., Ochi, N., Dell, A. and Fukuda, M. (1988) *Biochem.*, **27**, 8618-8626.
19. Carr, S. A., Anderegg, R. J. and Hemling, M. E. (1988) *in* The Analysis of Peptides and Proteins by Mass Spectrometry, C. J. McNeal, Ed., pp. 95-114, John Wiley and Sons.

20. Linden, H.-U., Klein, R. A., Egge, H., Peter-Katalinic, J., Dabrowski, J. and Schindler, D. (1989) *Biol. Chem. Hoppe-Seyler*, **370**, 661-672.
21. Matsuura, H., Takio, K., Titani, K., Green, T., Levery, S. B., Salyan, M. E. K. and Hakamori, S. (1988) *J. Biol. Chem.*, **263**, 3314-3321.
22. Settineri, C. A., Medzihradszky, K. F., Chu, C., George-Nascimento, C, Masiarz, F. R.and Burlingame, A. L. (1990) *Biomed. Environ. Mass Spectrom.*, **19**, 665-676.
23. Medzihradszky, K. F., Gillece-Castro, B. L., Settineri, C. A., Townsend, R. R., Masiarz, F. R.and Burlingame, A. L. (1990) *Biomed. Environ. Mass Spectrom.*, **19**, 777-781.
24. Brake, A. J., Merryweather, J. P., Coit, D. G., Heberlein, U. A., Masiarz, F. R., Mullenbach, G. T., Urdea, M. S., Valenzuela, P. and Barr, P. J. (1984) *Proc. Natl. Acad. Sci. USA*, **81**, 4642-4646.
25. Townsend, R. R., Hilliker, E., Li, Y.-T., Laine, R. A., Bell, W. R. and Lee, Y. C. (1982) *J. Biol. Chem.*, **257**, 9704-9710.
26. Aberth, W., Straub, K. M., Walls, F. C. and Burlingame, A. L. (1982) *Anal. Chem.*, **54**, 2029-2034.
27. Falick, A. M., Wang, G. H. and Walls, F. C. (1986) *Anal. Chem.*, **58**, 1308-1311.
28. Walls, F. C., Baldwin, M. A., Falick, A. M., Gibson, B. W., Kaur, S., Maltby, D. A., Gillece-Castro, B. L., Medzihradszky, K. F., Evans, S. and Burlingame, A. L. (1990) *in* Biological Mass Spectrometry, A. L. Burlingame and J. A. McCloskey, Eds., pp. 197-216. Elsevier, Amsterdam.
29. Cottrell, J. S. and Evans, S. (1987) *Anal. Chem.*, **59**, 1990-1995.
30. Hunkapiller, M. W. (1985) *Applied Biosystems User Bulletin Number 14*, Applied Biosystems, Foster City, CA.
31. Hardy, M. R., Townsend, R. R. and Lee, Y. C. (1988) *Anal. Biochem.*, **170**, 54-62.
32. Roepstorff, P. and Fohlman, J. (1984) *Biomed. Mass Spectrom.*, **11**, 601.
33. Biemann, K. (1988) *Biomed. Environ. Mass Spectrom.*, **16**, 99-111.
34. Domon, B. and Costello, C. E. (1988) *Glycoconjugate J.*, **5**, 397-409.
35. Falick, A. M., Medzihradszky, K. F. and Walls, F. C., (1990) *Rapid Commun. Mass Spectrom.*, **4**, 318-322.
36. Gillece-Castro, B. L., Townsend, R. R., Hardy, M. R., and Burlingame, A. L. unpublished results from this laboratory.
37. Edge, A. S. B. and Spiro, R. G. (1988) *J. Biol. Chem.* **262**, 16135-16141.

# NUCLEOSIDES AND OLIGONUCLEOTIDES: STRUCTURES AND PRINCIPLES OF REACTIVITY FROM MASS SPECTROMETRY

GIOVANNI SINDONA
*Dipartimento di Chimica*
*Università della Calabria*
*I-87030 Arcavacata di Rende(CS)*
*Italy*

ABSTRACT. Mass spectrometric methodologies play a fundamental role in the structure determination of nucleic acids components. Ribo and deoxyribo nucleosides present in natural RNAs and DNAs have been characterized by the combined use of appropriate derivatization procedures and of well-established ionization methods such as electron and chemical ionization. The sequence and the composition of natural and modified oligonucleotides can be routinely evaluated by means of desorption ionization techniques, whereas spray methods enable the molecular weight of intact RNA molecules to be determined. The evaluation of the gas-phase chemistry of nucleosides, nucleotides and oligonucleotides by means of MS/MS techniques has provided rules for the interpretation of the mechanisms that drive the unimolecular dissociations of the examined molecules, thus allowing their structure determination.

## 1. Introduction

Nucleic acid components, such as nucleobases, nucleosides, nucleotides and oligonucleotides, are molecules amenable to analysis by mass spectrometry (MS).

**$R^1$ = Nucleobase: purines and pyrimidines**
**$R^2$= $R^3$= $R^4$ = H: deoxyribonucleoside**
**$R^2$= OH; $R^3$= $R^4$= H: Ribonucleoside**
**$R^3$= $HOPO_2^-$: 3'-Mononucleotide**
**$R^4$= $HOPO_2^-$: 5'-Mononucleotide**

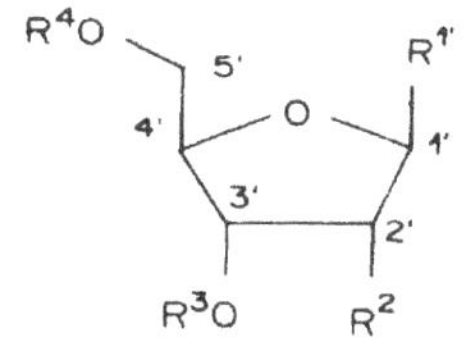


**Chart**

Ionization methods requiring the evaporation of the analyte can be used for smaller constituents of DNAs and RNAs, often in conjuction with a derivatization procedure (1, 2) whereas desorption (DI) (3, 4) and electrospray ionization (ESI) methods (5) are well- suited for larger molecules. DI and ESI methods are, however, of more general applicability because they allow DNA or RNA molecules to be analyzed regardless their molecular complexity. At the present stage of MS capabilites, oligomers approaching 25 kDa (6) have been detected by DIMS,



*M. L. Gross (ed.), Mass Spectrometry in the Biological Sciences: A Tutorial,* 383–405.

whereas structure information has been obtained for species in the range 500-3000 Da (7). Molecular weight determination of RNAs in the range of 100 kDa can be almost routinely performed by ESI (5).

The present review does not give a full coverage of the literature in this field, because many specialized accounts were recently published (1, 2, 8-12) but aims to provide an overview of the capabilities of MS in nucleic acid chemistry.

## 2. Nucleosides and Nucleotides

Natural nucleosides afford after electron ionization (EI) radical cations whose preferred fragmentation path corresponds to the dissociation of the glycosil bond. This behaviour was first recognized in the early application of the methodology (13) and represents useful information to look for when nucleosides are submitted to MS analysis by any of the available ionization methods (Scheme 1). The low vapor pressure of un-derivatized nucleosides prevented EI

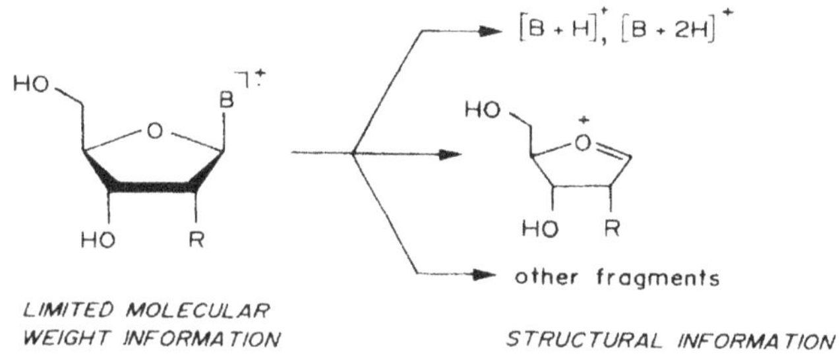


Scheme 1

to become a method of analysis of general applicability. Nevertheless, some general rules for discriminating among ribo and deoxyribo nucleosides have been suggested.

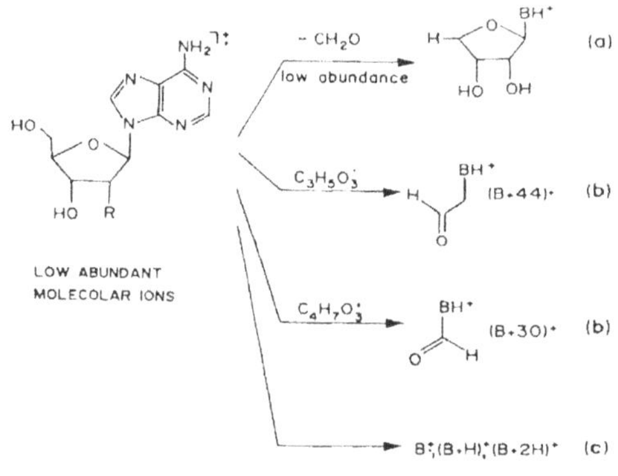


Scheme 2

In particular it was recognized that the loss of formaldehyde (path a) was not found when the 5'-OH of the examined derivative of adenine was not present or when it was in the trans position with respect to the base; the formation of $(B + 44)^+$ and $(B + 30)^+$ was an indication of the presence of the 2'-OH function, whereas the already known formation of various types of protonated nucleobases could be used for the identification of unknowns (Scheme 2) (14). It must be pointed out, however, that a direct analysis of nucleoside by EIMS often fails in providing straightforward information on its molecular composition. The molecular ion peak is, in fact, often absent or accounts for a minor perecentage of the total ion current. The preparation of volatile derivatives, by proper masking the polar functions, facilitates the use of EIMS. Trimethylsilyl, *t*-butyldimethylsilyl, *O*-isopropilidene, acetyl, trifluoroacetyl, *O*, *N*-permethyl functionalized nucleosides have been analyzed by mass spectrometry with different degree of success. The choice of the derivatization procedure is often connected with the structure of the species being examined and with the type of experiment to be performed. A powerful application of EIMS in connection with a selected derivatization procedure has been represented by the structural determination of modified nucleosides present in natural RNAs (1).

Molecular weight information on simple nucleosides can be obtained by field desorption (FD). The FD mass spectrum of guanosine (Figure 1) shows that the molecular radical cation $(M^+)$ at *m/z* 283 and a very abundant $[M+H]^+$ species at *m/z* 284 are formed.

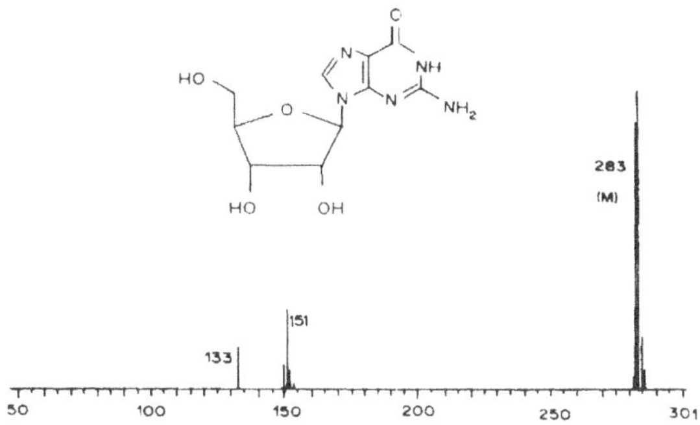


**Fig. 1. Field desorption spectrum of adenosine**

The formation of protonated sugar residue (*m/z* 133) and $(B + H)^+$ ions, account for the expected reactivity of the $M^+$ species; however no $(B + 2H)^+$ fragments at *m/z* 152 are formed as it would be expected from the gas-phase chemistry of the $(M + H)^+$ ions (15). A major drawback of FDMS is represented by the formation of unreproducible fragment ions on the emitter surface (16, 17) preventing, in the case of unknowns, a straightforward structural assignment. This behaviour can be exemplified by the FD spectra of 3'-adenilic acid (3'-AMP) taken at different anode temperature (16) and reported in figure 2. At the best-anode temperature (BAT) the conjugated acid of 3'-AMP is the only species present in the spectrum. Fragmentation of the molecule can be induced by raising the emitter current; however, the number and the relative abundances of the fragment ions change drastically with the emitter temperature.

Several normal and modified nucleosides have been investigated by positive chemical ionization (CI) (18). The combined use of reagent gases with different proton affinity (PA) values allows the establishing of gas-phase reactivity data useful in the structure determination. By analogy with the chemistry undergone by the radical cations produced by EI, the protonated nucleosides exhibit the main fragments $(B + 2H)^+$, $S_1$ $(B + 44)^+$ and $S_2$ $(B + 30)^+$ (Scheme 3).

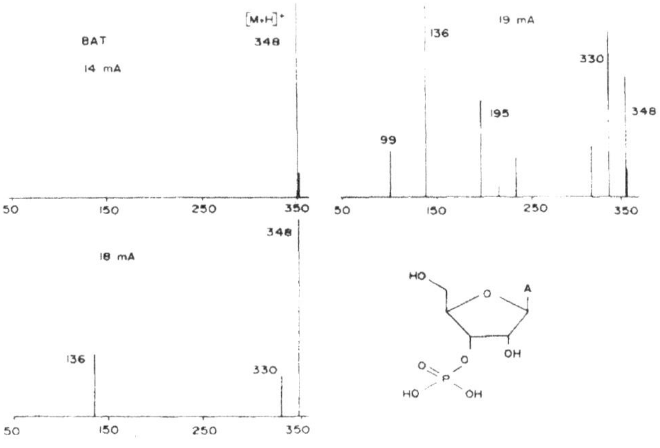


**Fig.2. Field Desorption spectra of 3'-adenilic acid at different anode temperatures**

Similar information were obtained by desorption chemical ionization (DCI), depositing the analyte on a tungsten wire that was inserted into the plasma of the ammonia reagent gas and heated by passing a current of 70 to 400 mA (19).

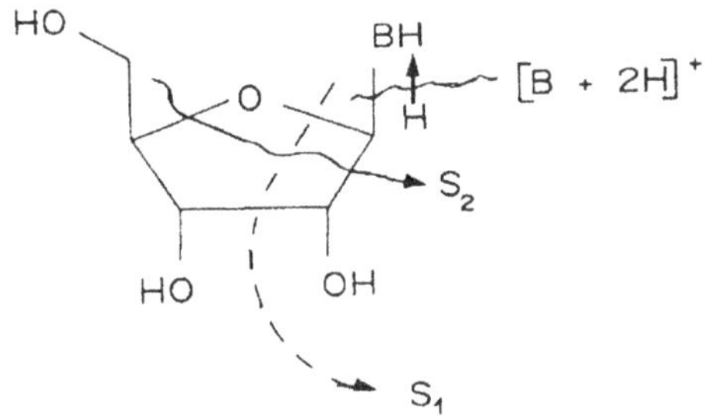


**Scheme 3**

CIMS provides spectra containing information on both the composition and the structure of the analytes. Protonated molecular ion, in fact, can be obtained (for guanosine in the desorption ionization mode) together with abundant fragments that are protonated nucleobases. This behaviour is typical of any MS methodology that brings about the formation of the $(M + H)^+$ species from the analyte and is due to the gas-phase chemistry of protonated nucleosides, which undergo preferential "depurination" by releasing the sugar moiety as a neutral species (15).

The analysis of nucleic acid molecules has been carried out also by coupling liquid chromatography with mass spectrometry (LCMS), via a thermospray interface (20). Structural assignment, perfomed according to scheme 4, leads to the identification of a number of modified nucleosides present in tRNA$^{val}$ of rabbit liver (21).

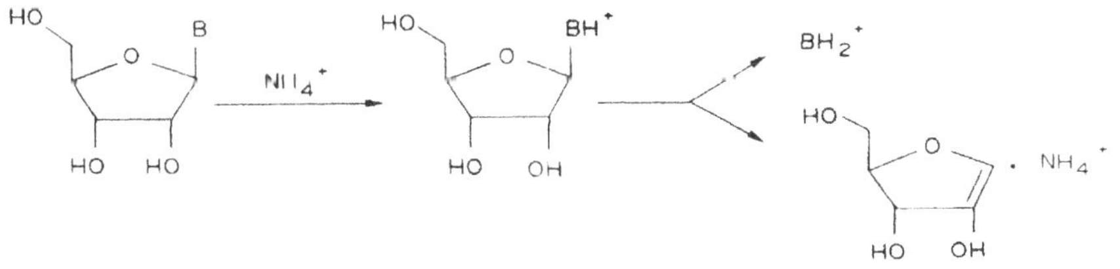


**Scheme 4**

In the early application of this methodology, it was also shown that spectra of nucleotides such as 5'-adenilic acid can be obtained either in the positive or in the negative ionization mode (22).

The ionization methods previously examined have been applied with various degree of success, mainly in the analysis of nucleobases and nucleosides, which are the more volatile and less polar species among nucleic acid molecules. A routine use of these methods, however, often requires the derivatization of the analyte to obtain enough reproducible structural data. The differentiation of 2'-, 3'- and 5'- deoxynucleosides, by either EI or CI, in fact, can be better achieved if the precursors are extensively trimethylsylilated (23). The introduction of desorption ionization methodologies (4) and, among them, the rapid adoption of fast-atom-bombardment (FAB) (24) in many mass spectrometry laboratories has favored a rapid growth of applications in nucleic acid field. A major limitation in the application of FAB to the analysis of low molecular weight nucleosides and nucleotides might be represented by the presence in the spectra of the chemical noise associated with the matrix background. This can be particularly true when unknowns are to be analyzed. Nevertheless, because the contribution of the matrix is known, FABMS can be routinely used in the analysis of nucleosides and nucleotides even with single focussing instruments. The synthesis of the nucleoside phosphorothioate **2** (Figure 3) was monitored by FABMS. Figure 3 shows the mass spectrum of negative ions, taken in the single focussing mode of a BE mass spectrometer, of the crude mixture obtained after hydrolysis and chloroform extraction. The formation of the expected compound is clearly evidenced by the presence of peaks at *m/z* 783, 743, 657 and 481, corresponding to the intact anion and to the ionic species formed: (i) by replacing a chlorine atom with an hydrogen radical, (ii) by releasing of the base and (iii) by elimination of the 5' protecting group, respectively. A metastable ion transition is clearly shown that correspond to the releasing of the base portion from the $(M-H)^-$ species, as unambigoulsy confirmed by the MIKE spectrum of the parent ions (25). The fragmentation pattern of a protonated nucleoside produced by FAB is similar to that observed when any ionization methods that brings about the formation of protonated species is used (26). This observation provides evidences that most of the fragments present in a FAB spectrum are formed by gas-phase processes. A detailed investigation of a gas-phase reaction undergone by a given ion can be carried out by means of the well-established MS/MS techniques (27). From an analytical point of view, this approach allows the evaluation of possible overlapping of analyte and matrix peaks, the detection of the relevant ionic species with a more favorable signal-to-noise ratio and the "finger-printing" of the analyte. Moreover, an accurate investigation of the gas-phase dissociations undergone by the sampled ionic species can be performed as a function of their internal energy, thus enabling a complete structure characterization.

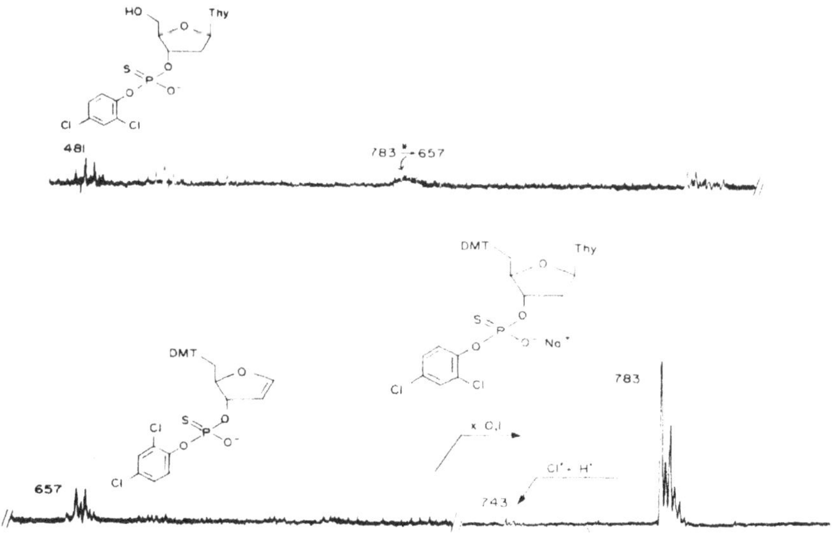


**Fig.3 Negative ions FAB spectrum of nucleoside phosphorothioate 2.**

The MIKE spectrum (15) of the $(M + H)^+$ ions that are FAB desorbed from glycerol/water solution of 2'-deoxyadenosine (**3**, Figure 4) looks very simple.

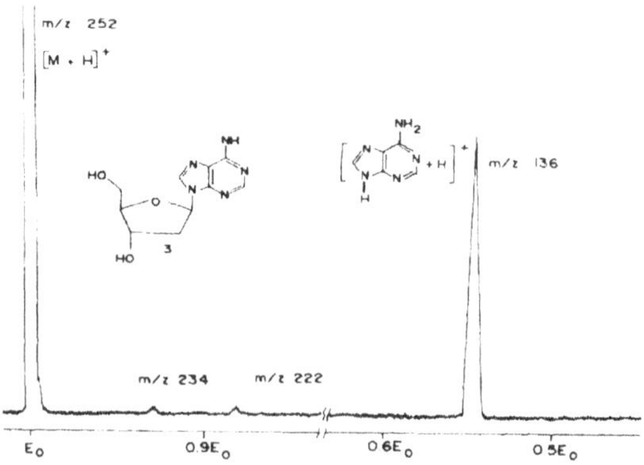


**Fig.4. MIKE spectrum of protonated 2'-deoxyadenosine**

(MIKES , Mass-analized-Ion-Kineticis Energy-Spectrum, is one form of MS/MS that is easily executed on a reverse geometry, BE, sector mass spectrometer). Nearly a single product ion is displayed, which corresponds to the formation of the protonated adenine base. The same precursor, after collision with helium, in a MIKE-CAD (Collisional Activated Dissociation) experiment, afforded additional fragments, two of which, of *m/z* 162 and *m/z* 117 (Figure 5), provide additional evidence on the structure of the analyte.

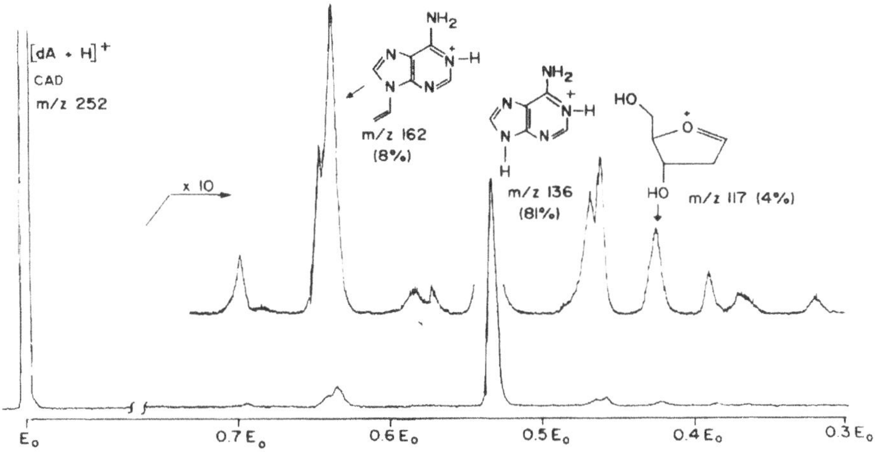


**Fig.5. CAD spectrum of protonated 2'-deoxyadenosine**

Similar chemistry was shown by the other nucleosides that are the constituents of DNA, thus allowing one to conclude that the formation of $(B + 2H)^+$ ions represents the lowest-critical-energy pathway allowed for the unimolacular dissociation of unmodified protonated nucleosides (15).

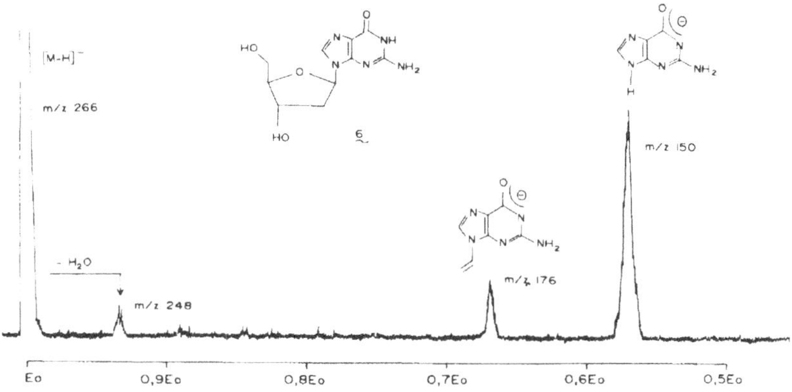


**Fig.6. MIKE spectrum of the (M - H)- ions of 2'-deoxyguanosine**

The unimolecular dissociations of protonated deoxynucleosides, reacting with different internal energy contents, can be used either as a "finger-print" for a reference compound or as a source of reactivity data to be employed in the structure determination of a similar species. 5'-benzoyl (**4**) and 6-benzoyl-2-deoxyadenosine (**5**) are synthetic nucleosides used as building blocks in the chemical synthesis of oligonucleotides. The MIKE spectra of **4** and **5** demonstrates the formation of protonated adenine and 6-benzoyladenine, respectively. The straightforward differentiation of the two isomers by FAB-MS/MS has been used as analytical tool for the synthesis of **4** (28). A similar approach allowed the differentiation of the anomeric precursors of bredinin, a nucleoside having immunosuppresive activity (29). The lability of the glycosil bond of nucleosides is also the driving force of the gas-phase dissociations of their $(M - H)^-$ conjugated bases (30, 31). The main peak of the MIKE spectrum of the conjugated base of deoxyguanosine (**6**) (Figure 6) corresponds to the formation of the deprotonated nucleobase at *m/z* 150, the latter is also very abundant in the spectrum of the primary desorbed ions. The fragments $(B + 2H)^+$ and $B^-$ fragments are, therefore, structurally significant species usually present in the positive and negative FAB spectra of simple nucleosides. The chemical modification of either the nucleobase or the sugar moiety can cause competitive or alternative reaction paths to be taken:a very simple case is represented by 2-isobutyryl-deoxyguanosine ($dG^{ib}$, **7**). The FAB spectrum of the negative ions of **7** displayed two sets of peaks (32), one including the species at *m/z* 336, 246 and 220 and the other the fragments at *m/z* 266, 176 and 150 (Figure 7).

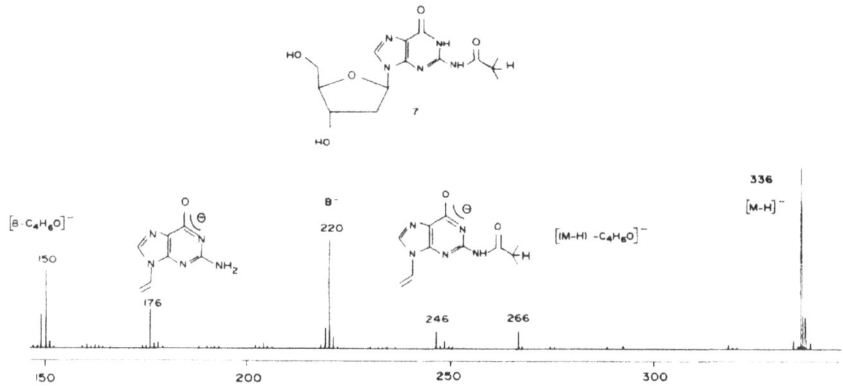


**Fig.7 FAB spectrum of the negative ions of 2-isobutyryl-deoxyguanosine**

These ions are the species $(M-H)^-$, $[(B + 27)-H]^-$ and $B^-$ which can be obtained from the conjugated base of **7** (MW 337) and from unprotected deoxyguanosine **6** (MW 267), respectively. The first set of ions provides information on the structure of the analyte, the second could lead to the conclusion that impurities of deoxyguanosine, which is the starting material in the preparation of 7, might be present. The sampled species, a common building block in oligonucleotide synthesis, purified by silica gel chromatography, however, did not contain traces of deoxyguanosine. As a matter of fact, the MIKE spectrum of the *m/z* 336 ion (Figure 8) showed that the species at *m/z* 266, which corresponds to the $(M - H)^-$ of deoyguanosine, is a fragment formed in the gas-phase rearrangement of the parent ion of dG$^{ib}$, by releasing of formal dimethylketene units.

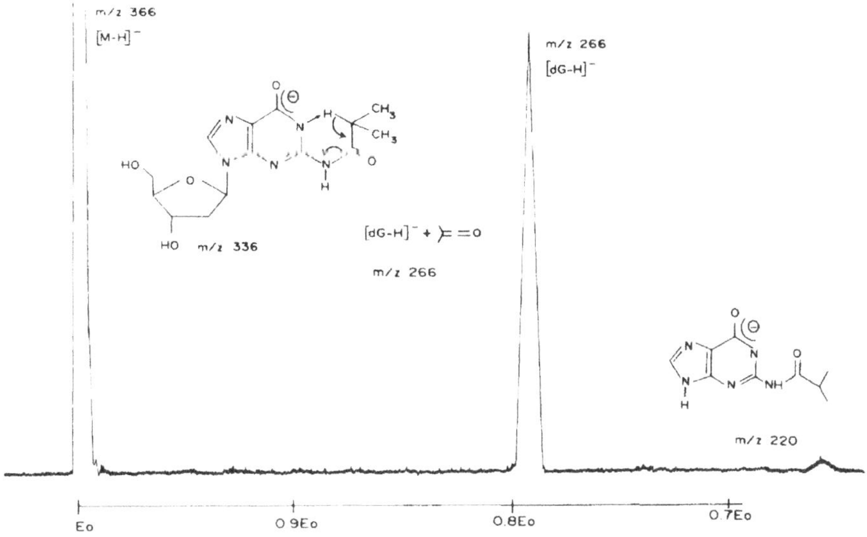


**Fig.8 MIKE spectrum of the (M - H)⁻ ions of 2-isobutyryl-deoxyguanosine**

Therefore, whatever the mechanism of formation of the *m/z* 266 ion in the gas phase (Figure 8), the presence of the side chain in the aglycone moiety of the nucleoside deeply affects the reactivity of the molecule. In this particular case, in fact, the lowest activation energy pathway is not represented by the heterolysis of the glycosil bond. Moreover, the MIKE spectrum alone does not provide complete information on the structure of the analyte.

The occurence of the glycosil bond breakage of an ionized nucleoside is usually evidenced by the presence in the mass spectrum of appropriate ions due to the nucleobase. In the case of protonated deoxyadenosine, the retention of the positive charge on the nucleobase is kinetically and thermodynamically favored over the formation of the oxonium ions (Scheme 5) (15).

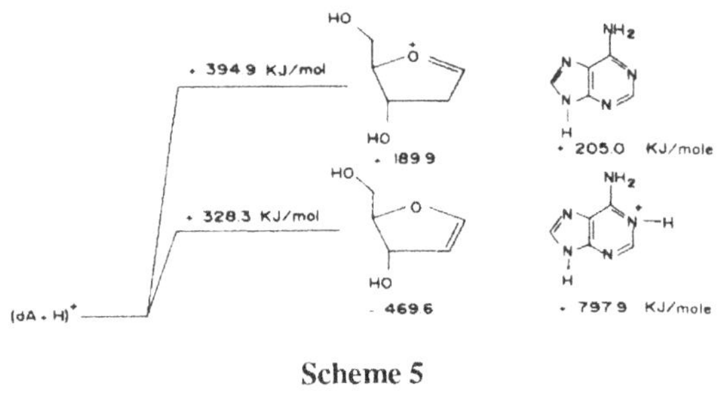


**Scheme 5**

The competition between the formation of protonated nucleobases and its release as a neutral can be deeply affected by the presence of side chains in the sugar residues. In fact, a benzoyl group on the 5' position of deoxyadenosine, as in the case of compound **4** (scheme 6) enhances the rate of formation of the oxonium ions at *m/z* 221. This reaction channel is preferentially taken in the unimolecular dissociations of the $(M + H)^+$ species of 3',5'-dibenzoyl adenosine (**8**), and leads to the formation of the *m/z* 325 ion (Scheme 6) (31).

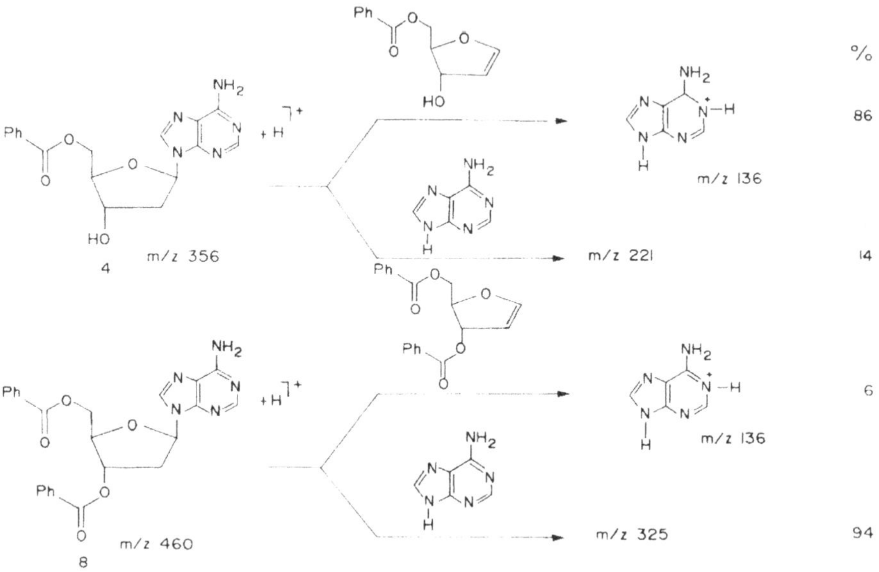


**Scheme 6**

It was already mentioned that when the benzoyl group is present on the nucleobase, as in the case of compound **5**, protonated 6-benzoyladenine is nearly the only ion present in the MIKE spectrum of the parent $(M + H)^+$ species, thus showing a specific charge retention on the purine moiety. The highly competitive retention of the charge on the sugar moiety in the depurination of protonated **8** might seem in contrast with the data previously discussed. The experimentally determined proton affinity (PA) of **8** (33), 235.4 kcal/mole, however, is nearly 2 kcal and 1 kcal higher than those of deoxyadenosine (34) and 5'-benzoyldeoxyadenosine (33), respectively. The enhanced stabilization of the protonated species upon benzoylation of the sugar hydroxyl groups, suggests, in the case of the $(M + H)^+$ of **8**, the establishment of intramolecular hydrogen bonding as reported in scheme 7. The elimination of neutral adenine from **8**, therefore, occurs through the same mechanism involved in the formation of protonated nucleobases from the conjugated acids of nucleosides (Scheme 7). It can be concluded, therefore, that the dissociation of the glycosil bond of nucleosides is a rearrangement process that is not affected by the type and location of the charge and can occur also remote from the charge site (31).

Mononucleotides can be easily determined by FAB in the negative ion mode in connection also with MS/MS methodologies. The unimolecular dissociations of all the set of natural 3'- and

5'- deoxy and ribomononucleotides are characterized by the release of the nucleobase as a neutral species.

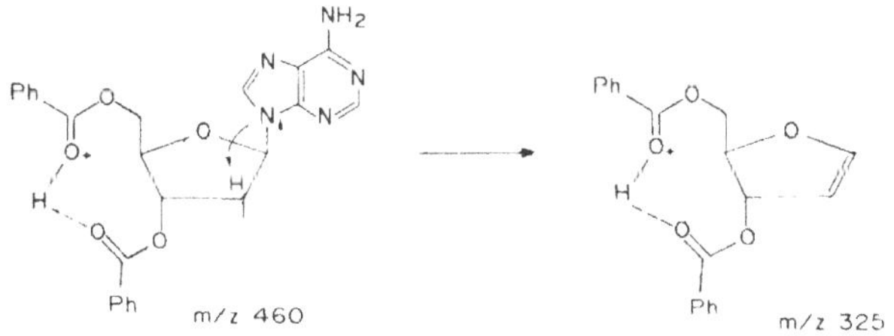


**Scheme 7**

This process is favored when the phosphate group is placed on the 3'-position of the sugar ring as a consequence of the enhanced acidity of the 2'-protons (35). The same approach can be used in the characterization of the protected mononucleotides employed in the chemical synthesis of DNA oligomers. The negative FAB spectrum of the mononucleotide **10** (Figure 9), sampled as triethylammonium salt, displayed the molecular ion, which corresponds to the base peak of the spectrum and a series of fragments, which enables the identification of the analyte.

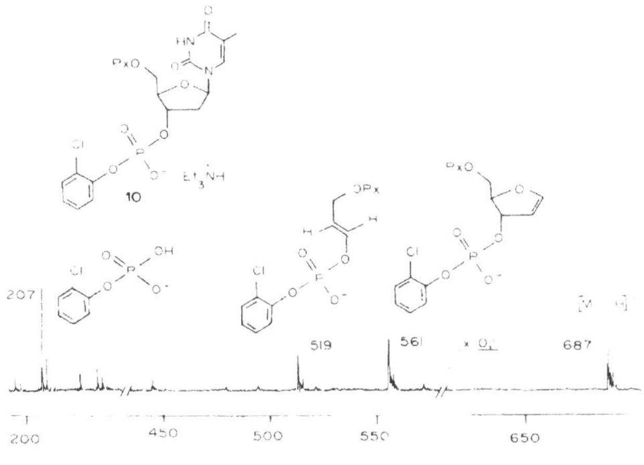


**Fig.9. FAB spectrum of the negative ions of mononucleotide 10**

Compound **10** is perfectly suited to be analyzed by negative ion FAB because here the formation of the anions does not require deprotonation reactions. This class of molecules usually gives rise to spectra that are less contaminated by the matrix background. The situation is different when all the polar functions of the nucleotide are completely masked. Compound **11** (Figure 10) has been used to develop a thiophosphate internucleotidic linkage (36, 37). The negative ion FAB spectra of **11**, obtained from different matrices, does not give information on the molecular weight of the molecule, and the spectra are affected by the nature of the matrix itself (38).

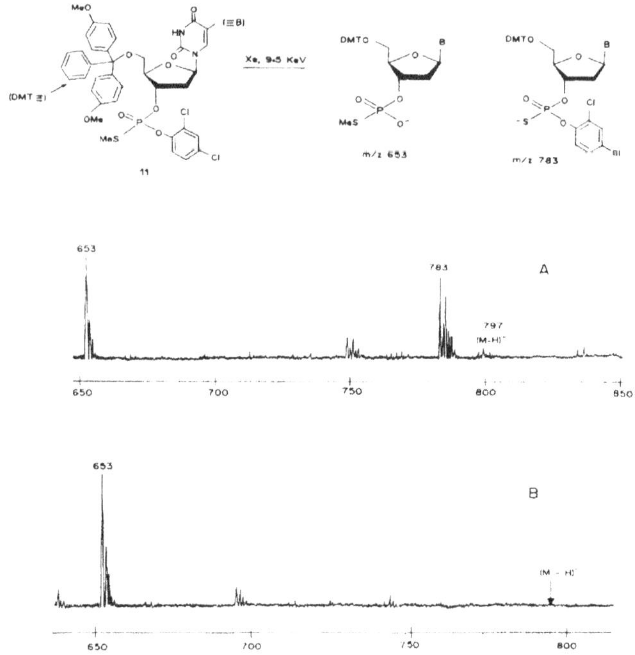


**Fig.10. Negative ion FAB spectra of mononucleotide 11: (A) glycerol-thioglycerol (1:1); (B) glycerol**

The interaction of the analyte with the matrix can lead to a variety of chemical reactions (3). When the critical energy for the abstraction of a proton from the substrate is higher than that of other competing bimolecular processes, the (M - H)$^-$ species are not formed or they account only for a minor percentage of the total ion current (39). In the examined case, the solvolysis of the phosphate moiety of **11**, competes with Bronsted-type equilibria and leads to species at *m/z* 653 and 783 as a function of the nucleophilicity of the matrix (Figure 10).

Another aspect of DI methods is the formation of clusters. The formation of a dimeric aggregate of a fully protected hexanucleotide (40) was recognized in the early application of $^{252}$Cf plasma desorption mass spectrometry (CFPDMS) (41, 42) and various types of clusters have been observed in the FABMS of nucleosides and oligonucleotides. The formation of protonated aggregates of nucleosides and other basic compounds can be deliberately induced by exposing to atom bombardment proper mixtures of both substrates in suitable matrices. The unimolecular dissociations of proton-bound heterodimers formed between nucleobases, deoxyribonucleosides, ribonucleosides and appropriate amines of known PA's provide data that can be used to establish, in the absence of solvent effects, the proton affinities of the nucleic acid components. In figure 11 is the MIKE spectrum of the clusters formed between deoxycitidine and triethylamine. Only ions due to the partitioning of the proton among the two interacting molecules are displayed, and the relative abundances of the two product ions reflect the different proton affinities of the molecules. The unknown PA of dC was determined by applying the equation reported in figure 11, using triethyl and tripropyl amine as reference compounds (34).

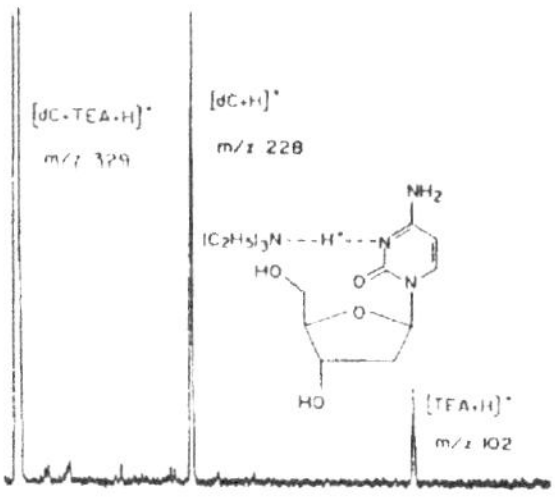


$$PA_N = \frac{\ln(k_N/k_{a1})\,PA_{a2} - \ln(k_N/k_{a2})\,PA_{a1}}{\ln(k_N/k_{a1}) - \ln(k_N/k_{a2})}$$

**Fig.11 MIKE spectrum of the proton-bound dimer formed from deoxycytidine (dC) and triethylamine (TEA)**

All the DI methods available today have been used for the determination of nucleic acid constituents. Actually, mononucleotides are considered as a sort of test compounds in establishing the performance of a given ionization method. The broad field of applications of FABMS, either in the positive or in the negative ionization mode, however, make this technique the method of choice for the routine analysis of nucleosides and nucleotides. An important role is still played by the conventional ionization methods of EI and CI, combined with chromatographic devices, in the qualitative and quantitative analysis of small ammounts of natural nucleosides, which, however, need to be derivatized.

## 3. Oligonucleotides

The detailed analysis of a DNA or RNA strand by mass spectrometry requires, as for any organic molecule, the determination of both the structure and molecular composition. These two goals can not be easily achieved by applying a single ionization technique particularly when high molecular weight compounds are to be investigated. The use of mass spectrometry in the oligonucleotide field has become increansigly important in connection with the chemical synthesis of natural and modified oligomers. The characterization of such complex organic molecules, in fact, hardly can be performed with other analytical methodologies. Furthermore, for the synthesis of oligonucleotides, it is important to have reliable means for establishing the structure of growing oligomers bearing protecting groups, whose charged species often show a more complex gas-phase chemistry than the corresponding unblocked ones. Therefore, the issues that should be addressed by any MS analysis of an oligonucleotide are represented by the determination of the molecular weight and by the sequencing of natural and modified strands both in the protected and in the unprotected form.

## 3.1 STRUCTURE DETERMINATION OF BUILDING BLOCKS FOR OLIGONUCLEOTIDE SYNTHESIS

Fully protected dinucleoside monophosphates afford by $^{252}$Cf plasma desorption MS in the negative ion mode, spectra enabling the identification of the two nucleotide components and the evaluation of the nature of the phosphate protecting group (43). Similar results were obtained by using FAB (44). In the common practice of many laboratories, medium-sized fully-protected oligodeoxiribonucleotides are, indeed, routinely analyzed by this technique. In Figure 12 is reported the negative ion FAB spectrum of an unnatural DNA strand, having the d(TUTU) sequence, where the polar groups are masked by the classic protecting groups employed in the phosphotriester approach to oligonucleotide synthesis (45). The spectrum shows all the relevant sequence peaks, accompanied by species due to the replacement of the chlorine atom of the phosphate protecting group by an hydrogen radical. This side reaction, already encountered in the $^{252}$Cf PDMS experiments, was interpreted (46) as due to an electron capture from the analyte followed by the explusion of $Cl^-$ anions giving rise to a neutral radical which, finally, abstracts an hydrogen radical, presumably, from the matrix. This unusual loss of 34 daltons from chlorine containing molecules has been found also in the FAB mass spectra of some chlorinated nucleosides (47).

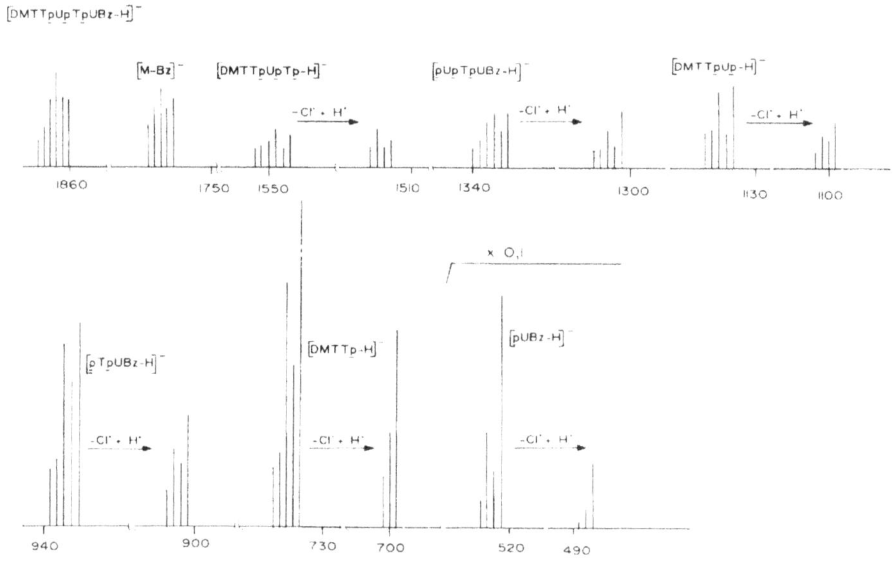


**Fig.12. FAB spectrum of the negative ions of d(TUTU)**

The detection, by FAB, of the $(M - H)^-$ species of a dinucleotide building block is facilitated when a phosphodiester moiety is attached at one of the two ends of the growing oligonucleotide such as in the case of species **12** and **13**.

When all the set of natural deoxyribodinucleotide diphosphate of type structure **12**, protected in the usual way, were analyzed by negative ion FAB, molecular weight and structural information were easily obtained (44).

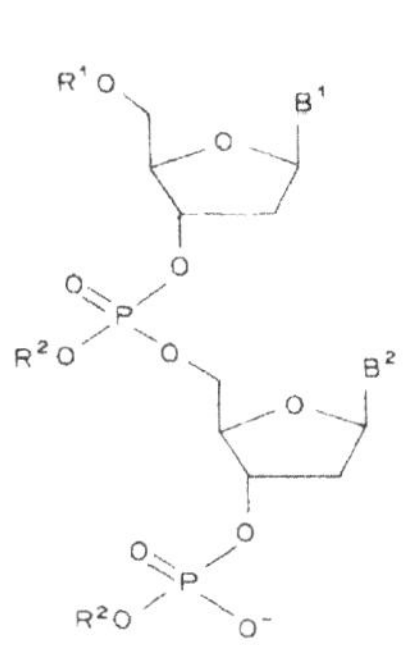

12

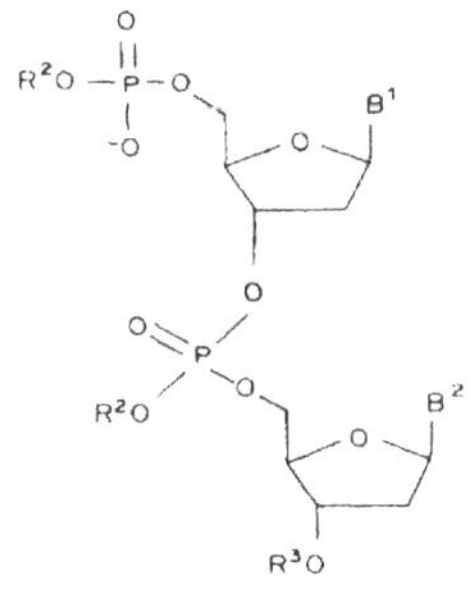

13

The breakdown pattern of species **12** and **13** depends on the sequence of the oligomer and on the position of the terminal phosphate groups, whereas it seems less affected by the nature of the protecting groups at the sugar hydroxyl moieties (48).

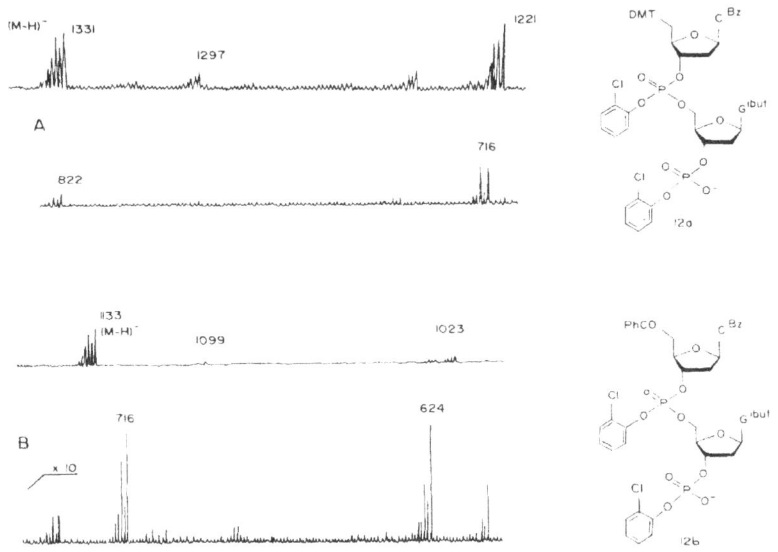


**Fig.13 FAB spectra of the negative ions of 12a and 12b**

The negative FAB spectra of the dinucleotides DMTC$^{Bz}$pG$^{ib}$p (**12a**) and PhCOC$^{Bz}$pG$^{ib}$p

(**12b**), bearing the classic 2-chlorophenyl group at the phosphate moieties (Figure 13), showed similar gas-phase chemistry. The replacement of one chlorine and of one chlorophenyl groups by an hydrogen atom, gives rise to ions of *m/z* 1297, 1221 and *m/z* 1099, 1023, respectively. The sequencing of both dimers was easily achieved by the apperance of peaks at *m/z* 822, 716 and *m/z* 716, 624, respectively. The internucleotidic bond dissociation seems to be highly affected by the sequence of the analyte. In fact, the dimer DMTG$^{ib}$pC$^{Bz}$p (**12c**), an isomer of **12a**, exhibited only one sequence ion at *m/z* 828 accompanied by a species at *m/z* 718 due to the replacement of the 2-chlorophenyl group which results from a selective internucleotidic bond breakage from the 5' end. A more pronounced effect is produced by a phosphodiester moiety at the 5' end of the dinucleotide. The negative ion FAB spectrum of pG$^{ib}$pC$^{Bz}$ (**13a**) gave exhaustive information on the composition of the molecule and on the nature of the protecting groups and no information on the sequence. Fully protected diribonucleotides electrosprayed on aluminized polyester film and exposed to a pulsed beam of $Cs^+$, in a typical SIMS experiment (4), afford positive and negative ion mass spectra, enabling both structure and composition to be determined (49). Under similar experimental conditions, the ribo and arabino isomers of branched trinucleotides provide SIMS spectra that allow the differentiation of the two diasteroisomers (50). Longer, fully-protected oligomers can be analyzed by the above mentioned techniques. PDMS was applied early in the detection of the cationized dimeric cluster of a dodecadeoxyribonucleotide (40); the complementary use of both positive and negative PDMS spectra enables the characterization of medium-sized protected oligomers (51, 52), whereas species of similar size can be routinely analyzed by FAB, especially when a terminal phosphodiester moiety is present at the 3'-phosphate end of the oligomer (44), and by SIMS. Complementary information is obtained from both positive and negative ion spectra (53). The analysis of a fully protected DNA or RNA oligomer can be, therefore, confidently carried out by applying the DI methodologies. This approach requires only small amount of material and provide a quick answer to organic chemists especially when they are using new synthetic strategies. The new frontiers of the chemical synthesis of oligonucleotides are represented by the challenge to obtain "antisense" oligomers for therapeutic purposes (54). The structure of carbamate analogs of type structure **14** and **15** was confirmed by negative FABMS (55).

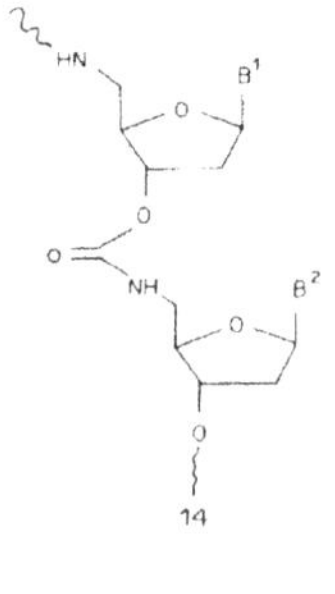

14

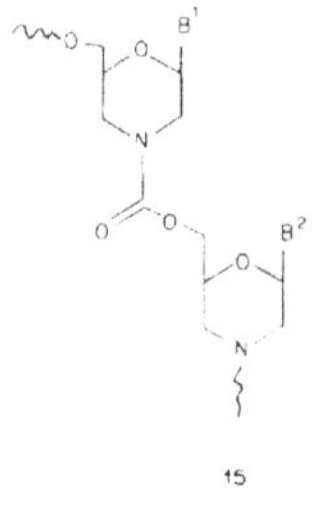

15

The sequencing of short oligomers of structure **14** is possible because appropriate ions appear in the mass spectrum, whose formation is explained according to scheme 8.

The production of 5' sequence ions was assumed to occur from the $M^-$ species formed by capture of low-energy electrons into the LUMOs of the carbonyl groups. Longer oligomers provide molecular weight information but less predictable fragmentation patterns. Under the same experimental conditions *N*-carbamoylmorpholine-linked nucleic acid oligomers of structure **15** give rise to negative FAB spectra, allowing the reading of the sequence from both ends of the modified strand (56).

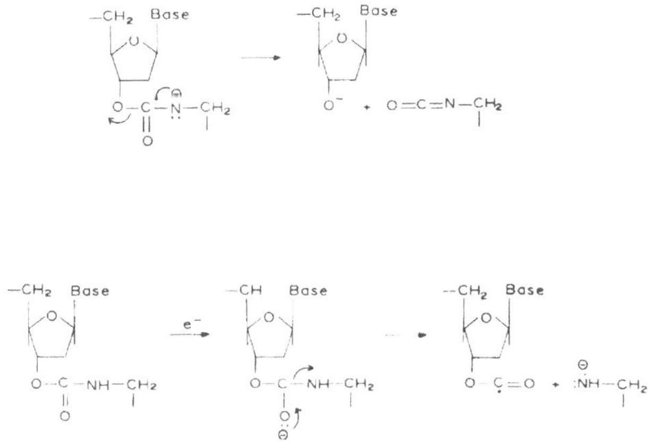

**Scheme 8**

In this case two $M^{-}$ radical anions were considered as the precursors of the major fragment ions. The occurence of charged specie other than $(M - H)^{-}$ in the negative FAB spectra of oligonucleotides was demonstrated by the deconvolution of the corresponding molecular anion peak cluster (57) The matching of the distribution of peaks within the experimental cluster with that calculated by applying an appropriate algorithm revealed that $(M\text{-}2H)^{-}$and $M^{-}$ radical anions and $(M+H)^{-}$ anions contribute to the shape of the peak envelope and that the $(M - H)^{-}$ species provide always the major contribution.

## 3.2 UNPROTECTED OLIGOMERS

Mass spectrometry can provide fast and sensitive methods for the evaluation of either the molecular weight of large DNA and RNA polymers or the molecular weight and the structure of reasonably sized oligonucleotides. Matrix assisted laser desorption (LD) (6) and electrospray ionization (ESI) (5) techniques allow an accurate identification of large oligomers at the picomole level. ESI spectra of intact t-RNAs, with molecular weight in the range of 65,000 Da, show a number of polycharged molecular anions, from which the molecular weight can be evaluated with reasonable accuracy (5). Coupling of ESI with MS/MS methods (5, 58) might allow, in the near future, the acquisition of structural data from polycharged species;however, at the present stage of mass spectrometric development a complete determination, in terms of structure and composition, of a DNA or RNA segment can be carried out on oligomers with molecular weight in the range of few kDa by means of DI methods.
Early applications of FD, in the positive ion mode (59) showed that diribonucleotides such as CpA afforded low abundance of $(M + H)^{+}$ species and sequence information from the appearance of appropriate cyclophosphate fragments (60). In the early application of FAB to nucleic acid

components (61), the positive ion mass spectrum of (3'-5')-dithymidilic acid (**16**), sampled as triethylammonium salts (Figure 14 A), provide exhaustive data on the composition of the analyte and too many fragments, some of which are not easily rationalized.

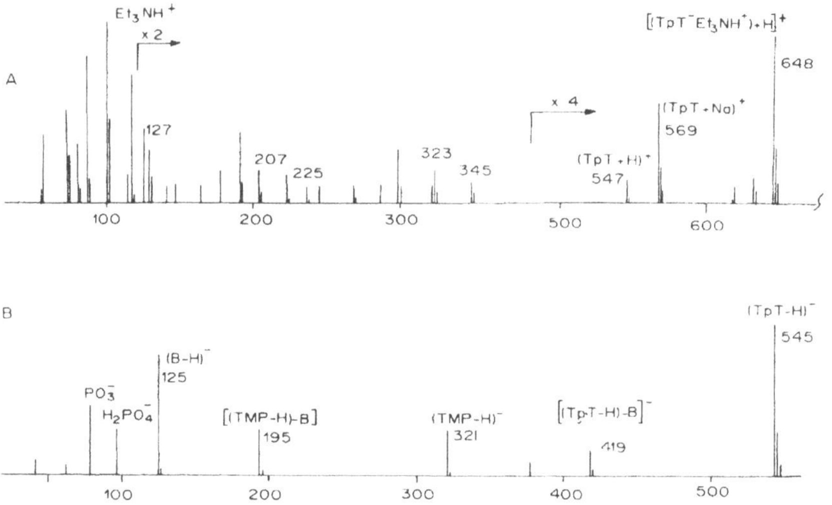


**Fig.14 FAB spectra of (3'-5')dithymidilic acid: (A) positive ions, (B) negative ions.**

In the negative ion mode (fig. 14 B) the FAB spectrum provide data concerning the composition and the structure of the dinucleotide, by means of signifcant peaks that can help in the structure determination (61).

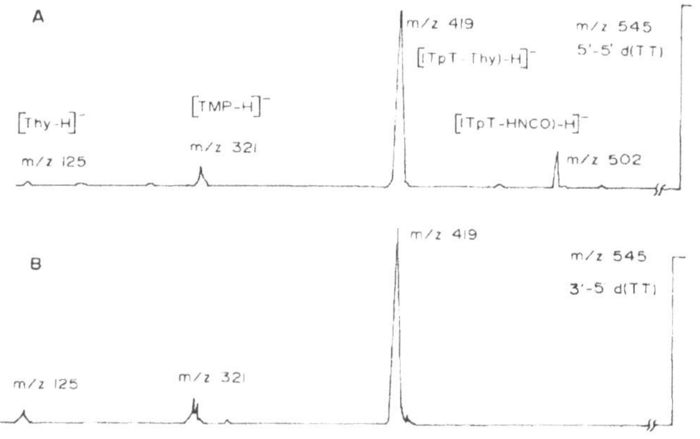


**Fig.15. MIKE spectra of the isomeric dithymidilic acids 16 and 17**

Negative ion FAB was therefore, proposed as a valuable MS method in the oligonucleotide field.

Moreover, the combined use of MS/MS methods allow the finger-printing of the dimer, thus provinding a tool for the discrimination of regioisomeric (5'-5') (**17**, Figure 15 A) and (3'-5') (**16**, Figure 15 B) deoxythymidilic acids (62). An important observation deriving from the MIKE spectra of **16** and **17** is that the lowest-critical-energy pathway for the unimolecular dissociations of a dinucleotide is associated with the elimination of the base unit as a neutral species. The relative yields of products from base elimination and internucleotidic bond brekage in the mass and in the MIKE spectrum of **16** clearly show that the loss of the aglycone occurs by a rearrangment of the ionized molecule characterized by an higher activation enthropy than the formation of the nucleoside monophosphate, which still originates from a proton transfer to the departing species (Scheme 9).

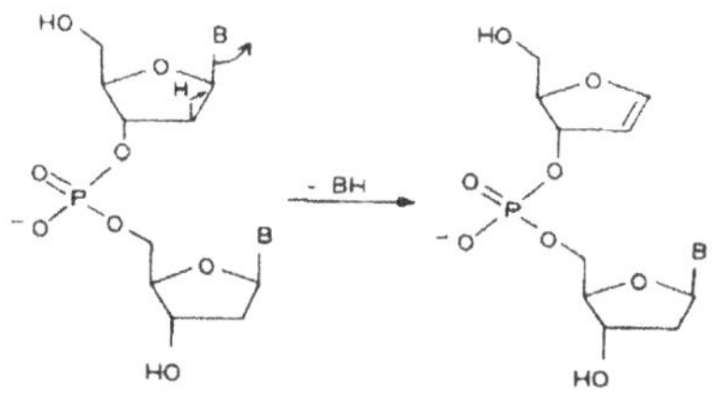


**Scheme 9**

In the case of the homodinucleotide **16** the loss of neutral thymine can originate either from the 5'-end or from the 3'-end nucleoside portions. Experiments, therefore, were devised to ascertain if some selectivity can be observed in the base elimination reaction from heterodinucleotides. The metastable ion and collisional activated dissociations of full set of ribo and deoxyribo dinucleotides showed that the elimination of the base unit occurs preferentially from the 5' end suga (63). Moreover, the gas-phase chemistry of the (M - H)$^-$ ions FAB desorbed from glycerol/water solutions of the chemically similar dinucleotides d(TU) (**18**) and d(UT) (**19**), show the same trend, thus confirming that the preferential elimination of the 5'-base is not affected by the nature of the aglycones which, in this particular experiment, are very similar indeed (64) (Figure 16).

The application of DI methods to larger strands of DNA and RNA molecules leads always to abundant molecular ion clusters dominated by the presence of (M - H)$^-$species. Polycharged species are usually absent or account for a very low percentage of the total ion current (65). The analytes, at the pH of the experiments and in the condensed phase have the structure of polyanions balanced by appropriate cations. Therefore, during the sputtering event, an oligomer bearing **n** negative charges undergoes specific protonation at **n - 1** sites . A similar observation was made from the crystallization of the poly A fragment, $A_3$, in the acid form. To achieve the neutrality, in the absence of counterions, the trimer crystallizes, at pH 3, as a zwitterionic species bearing two negative phosphates and two protonated adenine rings (66). Later reinterpretation (67) of the data showed that the actual structure of the molecule in the solid state is $AH^+p\_Ap\_AH^+$.

FAB-MS/MS investigations on d(TACC) and d(GGTA) oligomers, analyzed as triethylammonium salts, suggest that the gaseous (M - H)$^-$ species should possess a zwitterionic structure, where the **n** negative phosphates were balanced by **n - 1** protonated bases (68). Therefore, when the production of the molecular anions takes place in the selvedge (4), Bronsted equilibria occurs to bring about the protonation of the basic sites of the oligomer. A parallel may be drawn, therefore, between the phase transitions from solution to solid and solution to gas states. The achievement of neutrality, in the former, and the energetics of the sputtering process, in the latter, both favor the production of oligonucleotides in zwitterionic form. This observation is in agreement with the experimental findings that more sensitivity is achieved in FAB when the oligomers are sampled as ammonium salts (7) or when *p*-toluensulfonic acid is added (69). In this case, in fact, acidic hydrogens are available to bring about the protonation of the nucleobases.

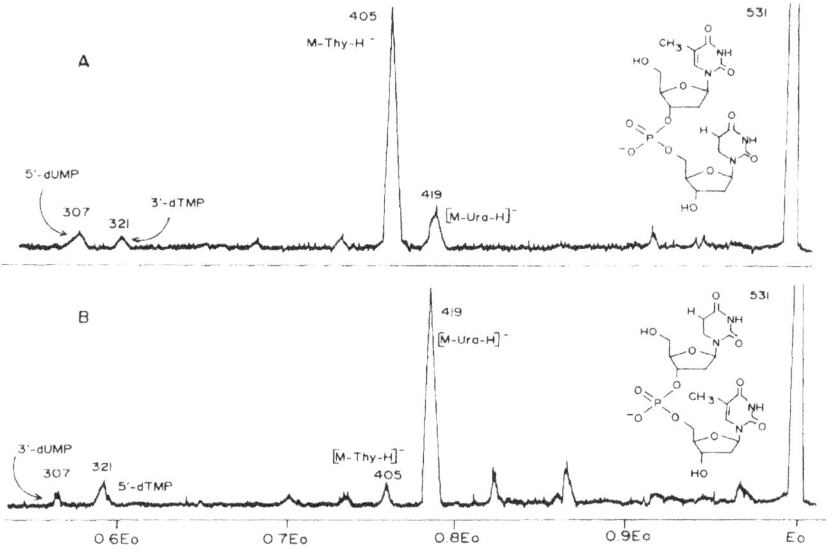


**Fig.16. MIKE spectra of (A) 3'-5'-d(TU) and (B) 3'-5' d(UT)**

A detailed structure of DNA strands up to decamers was provided by the pioneering experiments perfomed by Grotjahn (7). The FAB spectra of the triethylammonium salts of a number of oligonucleotides provided, in the negative ion mode, the composition and the structure of the analytes. Abundant fragments were displayed in the spectrum of the stable ions, and these allow the bidirectional sequence analysis of the oligomer. It was also observed that the abundance of 3' sequence ions was always higher than that of the 5' ones (7). This criterion, however, has not general applicability (68) and another suggestion was made for the unambiguous reading of the sequence of an uknown oligomer (12).
As previously observed for dinucleotides, the combined use of FAB and MS/MS experiments may provide important insights into the sequencing of a given oligomer. The occurence of sequence peaks in the MIKE and CAD spectra (68, 70) allows, indeed, the establishemnt of the structure of the molecule but makes sequence determination less straightforward. The data provided by the negative FAB spectra of the stable ions produced by medium sized oligomers,

combined with the information available from the selective detection of the unimolecular decays of the metastable fragments, however, could give a complete account on the structure of oligonucleotides.

## 4. Conclusions

In the last thirty years, mass spectrometry has become an unique tool for chemists and biologists dealing with nucleic acid molecules. The know-how acquired in the characterization of small components such as nucleosides and nucleotides has enabled the identification of a relevant number of natural species often bearing exstensive structural modification with respect to the five common deoxyribo and ribosides present in DNA and RNA. The advent of DI, and more recently, of spray techniques has allowed the detection of intact natural polymers and the identification of medium sized oligomers. Mass spectrometry is today the analytical method of choice in the characterization of synthetic modified nucleosides and "antisense" oligonucleotides because it offers the possibility of carrying out a full structure determination in "real time" on quantities of material as small as those that can be scraped from a TLC plate. The future seems promising for the development of mass spectrometry as an alternative method for the sequencing of natural nucleic acid molecules.

## 5. References


1) McCloskey, J.A. In:"Mass Spectrometry in Biomedical Research"; Gaskell, S.J., Ed.;John Wiley & Sons: New York, 1986, chp. 6
2) Schram K.H. In: "Biomedical Applications of Mass Spectrometry", Suelter, C.H. and Watson J. T., Eds.; John Wiley & Sons: New York, 1990, pp. 203-287.
3) Fenselau, C.; Cotter, R.J. *Chem.Rev.* **1987**, *87*, 501-512.
4) Pachuta, S.J.; Cooks, R.G. *Chem. Rev.* **1987**, *87*, 647-669.
5) Smith, R.C.; Loo J.A., Edmomds C.G., Barinaga C.J., Udseth H.R. *Anal. Chem.* **1990**, *62*, 882-899.
6) Hillenkamp, F.; Karas, M.; Ingendoh, A.; Stahl, B. In: "Biological Mass Spectrometry", Burlingame A. and McCloskey, J.A., Eds., Elsevier: New York, 1990.
7) Grotjahn, L.; Bloecker, H.; Frank, R. *Biomed. Mass Spectrom.* **1985**, *12*, 514.
8) McCloskey, J.A. In: "Mass Spectrometry in the Health and Life Sciences", Burlingame, A.L. and Castagnoli Jr., N., Eds.; Elsevier: Amsterdam, 1985, pp. 521-544.
9) Jankowski, K; Jocelyn Parè, J.R.; Wightman, R.H. In: "Advances in Heterocyclic Chemistry", Katritzky, A.L.. Ed.; Academic Press: Orlando, 1986, vol. 39, pp. 79-116.
10) Tomer, K.B. *Mass Spectrom. Rev.* **1989**, *8*, 445-482
11) Sindona, G. In: "Spectroscopy of Inorganic Bioactivators. Theory and Appplications", Theophanides, T., Ed.; Kluwer: Dordrecht, 1989, pp. 321-345.
12) Crain, P.F. *Mass Spectrom. Rev.* **1990**, *9*, 505-554.
13) Biemann, K.; McCloskey, J.A. *J. Am. Chem. Soc.* **1962**, *84*, 2005.
14) Shaw, S.J.; Desiderio, D.M.; Tsuboyama, K.; McCloskey, J.A. *J. Am. Chem. Soc.* **1970**, *92*, 2510.
15) Liguori A.; Greco F.; Sindona G.; Uccella, N. *Org. Mass Spectrom.* **1990**, *25*, 459

16) Budzikiewicz, H.; Linscheid, M. *Biomed. Mass Spectrom.* **1977**, *4*, 103.
17) Linscheid, M.; Burlingame, A.L. *Org. Mass Spectrom.* **1983**, *18*, 245.
18) Wilson, M.S.; McCloskey, J.A. *J. Am. Chem. Soc.* **1975**, *97*, 3426.
19) Esmans. E.L.; Luyten, Y.; Alderweireldt, F.C. *Biomed. Mass Spectrom.* **1983**, *10*, 347.
20) Vestal, M.L. *Mass Spectrom. Rev.* **1983**, 2, 447-480.
21) Edmomds, C.G.; Vestal, M.L.; McCloskey, J.A. *Nucleic Acids Res.* **1985**, *13*, 8197.
22) Blakley, C.R.; Carmody, J.J.; Vestal, M.L. *Anal. Chem.* **1980**, *52*, 1636.
23) Reimer, M.L.J.; McClure, T.D.; Schram, K.H. *Biomed. Environ. Mass Spectrom.* **1989**, *18*, 533.
24) Barber M.; Bordoli, R.S.; Sedgwick, R.D.; Tyler, A.N. *J. Chem. Soc. Chem. Commun.* **1981**, 325.
25) Procopio, A.; Sindona, G.; Uccella N. unpublished results.
26) Eicke, A.; Sichtermann, W.; Benninmghoven, A. *Org. Mass Spectrom.* **1980**, *15*, 289.
27) Busch, K.L.; Glish, G.L.; McLuckey, S.A. "Mass Spectrometry/Mass Spectrometry: Techniques and Applications of Tandem Mass Spectrometry"; VCH Publishers: New York, 1988.
28) Liguori, A.; Perri, E.; Sindona, G.; Uccella, N. *Tetrahedron* **1988**, *44*, 229.
29) Banoub, J.H.; Mackenzie, G.; Descotes, G.; Humble, R.W.; Shaw, G.; Becchi, M.; Fraisse, D. *Biomed. Environ. Mass Spectrom.* **1990**, *19*, 97.
30) Crow, F.W., Tomer, K.B.; Gross, M.L.; McCloskey, J.A.; Bergstrom, D.E. *Anal. Biochem.* **1984**, *139*, 243.
31) Liguori, A.; Sindona, G.; Uccella, N. *Nucleosides Nucleotides* **1990**, *9*, 373.
32) Sindona G.; Uccella N. unpublished results.
33) Greco, F.; Liguori, A.; Sindona, G.; Uccella N. "2nd International Symposium on Applied Mass Spectrometry in the Health Sciences", Barcelona (Spain), April 17-20, 1990.
34) Greco, F.; Liguori, A.; Sindona, G.; Uccella, N. *J. Am. Chem. Soc.* **1990**, *112*, 9092.
35) Cerny, R.L.; Gross, M.L.; Grotjahn, L. *Anal. Biochem.* **1986**, *156*, 424.
36) Caruso, P.B.; Liguori, A.; Sindona, G.; Uccella, N. *Synth. Commun.* **1988**, *18*, 841.
37) Caruso, P.B.; Liguori, A.; Sindona, G.; Uccella, N. *Gazz. Chim. Ital.* **1988**, *118*, 253.
38) Liguori, A.; Sindona, G.; Uccella, N. *J. Chem. Soc., Perkin I* **1988**, 1661.
39) Liguori, A.; Sindona, G.; Uccella, N. *J. Am. Chem. Soc.* **1986**, *108*, 7488.
40) McNeal, C.J.; Macfarlane, R.D. *J. Am. Chem. Soc.* **1981**, *103*, 1609.
41) Macfarlane, R.D.; Togerson, D.F. *Science* **1976**, *191*, 920.
42) Sundqvist B.; Macfarlane R.D. *Mass Spec. Review*, **1985**, *4*, 421-460
43) McNeal, C.J.; Ogilvie, K.K.; Theriault N.Y.; Nemer M.J. *J. Am. Chem. Soc. 1982*, *104*, 972.
44) Grotjahn, L.; Frank, R.; Heisterberg-Moutsis G.; Bloecker, H. *Tetrahedron Lett.* **1984**, *25*, 5373.
45 Gait M.J. "Oligonucleotide Synthesis: A Practical Approach" ; IRL Pres: Oxford, 1984.
46) Williams, D.H.; Findeis, A.F.; Naylor, S.; Gibson, B.W. *J. Am. Chem. Soc.* **1987**, *109*, 1980.
47) Sethi, S.K.; Nelson, C.C.; McCloskey, J.A. *Anal. Chem.* **1984**, *56*, 1977.
48) Greco F.; Liguori, A.; Sindona G.; Uccella N. *Adv. Mass Spectrom.* **1985**, *10B*, 1455.
49) Ens, W.; Standing, K.G.; Westmore, J.B.; Ogilvie, K.K.; Nemer, M.J. *Anal. Chem.* **1982**, *54*, 960.
50) Lafortune, F.; Standing,K.G.; Westmore, J.B.; Damha, M.J.; Ogilvie, K.K. *Org. Mass Spectrom.* **1988**, *23*, 228.

51) McNeal, C.J.; Ogilvie, K.K.; Theriault, N.Y.; Nemer, M.J. *J. Am. Chem. Soc.* **1982**, *104*, 976.
52) McNeal, C.J.; Ogilvie, K.K.; Theriault, N.Y.; Nemer, M.J. *J. Am. Chem. Soc.* **1982**, *104*, 981.
53) Beavis, R.; Ens, W.; Nemer, M.J.; Ogilvie, K.K.; Standing, K.G.; Westmore, J.B. *Int. J. Mass Spectrom. Ion Phys.* **1986**, *46*, 475.
54) Uhlmann, E.; Peyman, A. *Chem. Rev.* **1990**, *90*, 543-584.
55) Griffin, D.; Laramèe, J. A.; Deinzer, M.L.; Stirchak, E.P.; Weller, D.D. *Biomed. Environ. Mass Spectrom.* **1988**, *17*, 105.
56) Laramèe, J.A.; Arbogast, B.; Deinzer, M.L.; Stirchak, E.P.; Weller, D.D. *Org. Mass Spectrom.* **1990**, *25*, 33.
57) Laramèe, J.A.; Arbogast, B.; Deinzer, M.L. *Anal. Chem.* **1989**, *61*, 2154.
58) Mann, M. *Org. Mass Spectrom.* **1990**, *25*, 576-587
59) Beckey, H.D.; Levsen, K.; Roellgen, F.W.; Schulten, H.R. *Surf. Sci.* **1978**, *70*, 325.
60) Schulten, H.R.; Schiebel, H.M. *Nucl. Acids Res.* **1976**, *3*, 2027.
61) Sindona, G.; Uccella, N.; Weclawek, K. *J. Chem Res.(S)* **1982**, 184.
62) Neri, N.; Sindona G.; Uccella, N. *Gazz. Chim. Ital.* **1983**, *113*, 197.
63) Cerny, R.L.; Gross, M.L.; Grotjahn, L. *Anal. Biochem.* **1986**, *156*, 424.
64) Liguori, A.; Sindona, G.; Uccella, N. *Biomed. Environ. Mass Spectrom.* **1988**, *16*, 451.
65) Williams, D.H.; Bradley, C.; Bojesen, G.; Santikarn, S.; Taylor L.C.E. *J.Am. Chem. Soc.* **1981**, *103*, 5700.
66) Suck, D.; Manor, P.C.; Saenger, W. *Acta Cryst.* **1976**, *B32*, 1727.
67) Taylor, R.; Kennard, O. *J. Mol. Struct.* **1982**, *78*, 1-28
68) Panico, M.; Sindona, G.; Uccella, N. *J. Am. Chem. Soc.* **1983**, *105*, 5607.
69) Moser, H.; Wood, G.W. *Biomed. Environ. Mass Spectrom.* **1988**, *15*, 547.
70) Cerny, R.L.; Tomer, K.B., Gross M.L.; Grotjahn L. *Anal. Biochem.* **1987**, *165*, 175.

# APPLICATION OF TANDEM MASS SPECTROMETRY TO CARCINOGEN-MODIFIED NUCLEOSIDES

RONALD L. CERNY and MICHAEL L. GROSS
*Midwest Center for Mass Spectrometry*
*Department of Chemistry*
*University of Nebraska-Lincoln*
*Lincoln, Nebraska 68588-0362*

ABSTRACT. Tandem mass spectrometry was used in the structure determinations of adducts formed between nucleosides and the potent carcinogens benzo[*a*]pyrene and 7,12-dimethylbenz[*a*]anthracene. Adduct isomers are differentiated based on the fragmentations observed in their CAD spectra.

## 1. Introduction

The previous chapter provides a detailed review of the use of mass spectrometry in the study of nucleosides. In this chapter, we wish to demonstrate how the knowledge gained from the fundamental studies can be applied to efforts on biological systems. We present an overview of our involvement in the structure determinations of nucleosides modified by some potent carcinogenic polycyclic aromatic hydrocarbons (PAH).

It is now evident that in chemical carcinogenisis, covalent binding of the chemical agent to biomolecules such as DNA is the first step in the complex process that leads to tumor formation and cancer [1, 2]. Most PAH are not carcinogenic in their original form. They require metabolic activation to an intermediate species capable of chemically reacting with a biomolecule to form an adduct. For some time it was believed that the principal activation in PAH carcinogenesis involved monooxygenation of the PAH to produce bay region diol epoxides [3, 4]. In biological systems, this process is catalyzed by cytochrome P-450. Monooxygenation to diol epoxides is probably not the only activation. Some PAH, in which bay-region diol epoxide formation is not possible, still display high carcinogenicity [5].

Researchers at University of Nebraska Medical Center have proposed another activation mechanism. Cavalieri and Rogan [6, 7] suggested that one-electron oxidation of PAH to produce reactive radical cations is an alternative activation process. One electron oxidation can also be catalyzed by Cytochrome P-450.

The structure of the carcinogen adducts can be used to differentiate between proposed mechanisms. In this vein, we began a collaboration with Professors Cavalieri and Rogan to use mass spectrometry, particularly tandem mass spectrometry, to search for evidence of adducts formed by radical cations of PAH.

## 2. Synthetic Adducts of Benzo[*a*]pyrene

Initial efforts were directed at the synthesis of possible adducts of benzo[*a*]pyrene with deoxyguanosine by electrochemical oxidation of the PAH in solution with the nucleoside. This approach was taken to simplify the reaction system and to facilitate identification of any adducts formed. Adducts of mg quantities can be produced and purified, providing ample quantities for NMR and good CAD reference spectra. Furthermore, these adducts are then



*M. L. Gross (ed.), Mass Spectrometry in the Biological Sciences: A Tutorial*, 407–411.

used as markers for retention times in HPLC separations to identify adducts formed in *in vitro* studies. A more detailed discussion of the synthesis, separation, and structure determination of these adducts was published [8].

The reaction of deoxyguanosine with the BP radical cation produced three covalently bound adducts. By NMR analysis the most abundant product was determined to be covalently bound between the C-6 position of BP and the N-7 position of guanine (N7-Gua). The FAB spectrum of this product contained an $(M + H)^+$ ion of *m/z* 402, indicating that the deoxyribose moiety was not included in the structure. The ion of *m/z* 402 was collisionally activated and analyzed by using a Kratos MS-50 triple analyzer, which has an ESA for MS2. The CAD spectrum of the $(M + H)^+$ ion (Figure 1D) shows formation of two major fragment groups, one centered at *m/z* 251 ($C_{20}H_{11}^+$ of BP) and another at *m/z* 277 (BP---$NC^{+\cdot}$). Several other fragment ions are also produced by a series of losses of $C_nH_{n-4}$, $C_nH_{n-5}$, or $C_nH_{n-3}$, in a manner analogous to fragmentations of unsubstituted PAH.

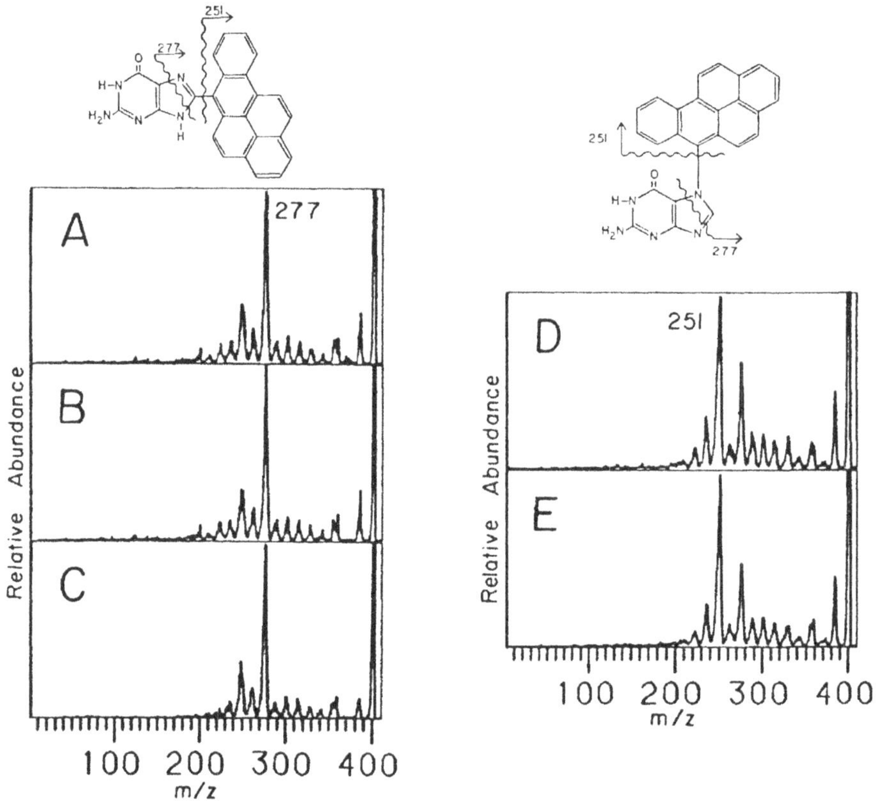


Figure 1. CAD spectra of *m/z* 402 ions from (A) C8-Gua, (B) C8-dG, (C) C8-G, (D) N7-Gua, (E) adduct isolated from HRP catalyzed reaction of BP with calf-thymus DNA. Reprinted from reference [8] with permission.

A second product of the reaction of the BP radical cation with deoxyguanosine is formed by covalent bonding of the C-6 position of BP to the C-8 position of guanine (C8-Gua). FAB produced an $(M + H)^+$ of *m/z* 402, again indicating that the adduct does not contain the deoxyribose moiety. Upon collisional activation (Figure 1A), the $(M + H)^+$ ion produces the same three major fragment ions as the N7 adduct. The relative abundances of these fragments are, however, significantly different than those for the N7 adduct. For the C8 adduct the fragment ion of *m/z* 277 (BP-$CN^{+\cdot}$) is the most abundant. The greater abundance of this fragment ion from C8-Gua as compared with N7-Gua relates to the greater stability of BP-$CN^{+\cdot}$ over BP-$NC^{+\cdot}$. These differences in the relative abundance of the ions of *m/z* 251 and of 277 are reproducible and provide a basis for the differentiation of the two isomeric species.

The third product of $BP^{+\cdot}$ with deoxyguanosine again involves the C8 position of guanine and the C6 position of BP. Upon FAB, this adduct produces an $(M + H)^+$ ion of *m/z* 518, indicating the presence of the deoxyribose unit. An abundant fragment ion of *m/z* 402 arises by H transfer and loss of the sugar unit and is formally the same ion as the $(M + H)^+$ of C8-Gua. The CAD spectrum of this ion of *m/z* 402 (Figure 1B) confirms that substitution occurs at C8 and not N7.

The same synthetic approach was applied to the ribonucleoside guanosine to ascertain if the nature of the sugar moiety has any influence on adduct formation. The reaction of $BP^{+\cdot}$ with guanosine produces three adducts, N7-Gua, C8-Gua, and C8-Guanosine. The spectra of the first two adducts are consistent with those discussed previously. Upon FAB, the third adduct produces an $(M + H)^+$ of *m/z* 534 indicating the presence of the ribose moiety. A fragment ion of *m/z* 402 is also produced. The CAD spectrum of this ion (Fig. 1C) is identical to that of C8-Gua, indicating that the BP moiety is bound at the C8 position of the guanine base. This was confirmed by NMR studies.

### 3. Structure Determination of Adduct formed by *In Vitro* Processes

After the characterization of the synthetic isomers was completed, *in vitro* studies were undertaken to evaluate the proposed mechanism. If the one-electron oxidation mechanism for adduct formation is valid, then the reaction of BP with DNA in a biological system should produce BP-modified adducts. Horseradish peroxidase (HRP) is known to catalyze one-electron oxidation. A sample of calf-thymus DNA was incubated with BP in the presence of HRP for 1 hour at 37 $^{o}$C. HPLC separation sugggested one component as a possible BP adduct. This component was further purified by HPLC and collected. Six individual incubations resulted in 10 ng of adduct. This amount is insufficient for NMR analysis but is sufficient for mass spectral analysis. Upon FAB, the adduct produces an $(M + H)^+$ ion of *m/z* 402. The CAD spectrum of the ion of *m/z* 402 (Figure 1E) is identical to that of the synthetic N7 Gua. The detection limit was approximately 40 ng.

### 4. Synthetic Adducts of 7,12-Dimethylbenz[*a*]anthracene

The electrochemical synthetic approach was also applied to an even more potent carcinogen 7,12-dimethylbenz[*a*]anthracene (DMBA) [9]. Again, NMR was used as a complementary technique to characterize the adducts. When deoxyguanosine was reacted with the radical cation of DMBA, three adducts form. The first adduct produces an $(M + H)^+$ ion of *m/z* 522 upon FAB, indicating the adduct contains an intact deoxyguanosine moiety and DMBA. Upon collisional activation, the $(M + H)^+$ ion fragments to produce ions indicative of the structure; an ion of *m/z* 405 results from elimination of the deoxyribose moiety, an ion of *m/z*

255 that is the (M - H)$^+$ of 7,12-DMBA, ions of *m/z* 239 and 240 resulting from cleavages between the methylene unit and the aromatic moiety, and an ion of *m/z* 152, protonated guanine. From studies of 7,12-DMBA-pyridine adducts, it was shown that the ratio of the ions of *m/z* 239 and 240 in the CAD spectra of source produced ions of *m/z* 255 can be used to determine if the covalent bond is to the C7 or C12 methyl group of 7,12-DMBA. For this deoxyguanosine-7,12 DMBA adduct, the ion of *m/z* 240 is more abundant than the ion of *m/z* 239. This confirms adduct formation occurs at the 12-$CH_3$ position of 7,12-DMBA. We cannot determine the site of attachment on the guanosine moiety based on the CAD spectrum as we could for the BP adducts. Bonding via a methylene unit precludes the production of the type of fragment ions observed for the BP adducts. Cleavages on either side of the methylene unit (to produce the fragment ions of *m/z* 239, 240, and 255) are much more facile processes than extraction of the subunits of NC or CN from the purine ring.

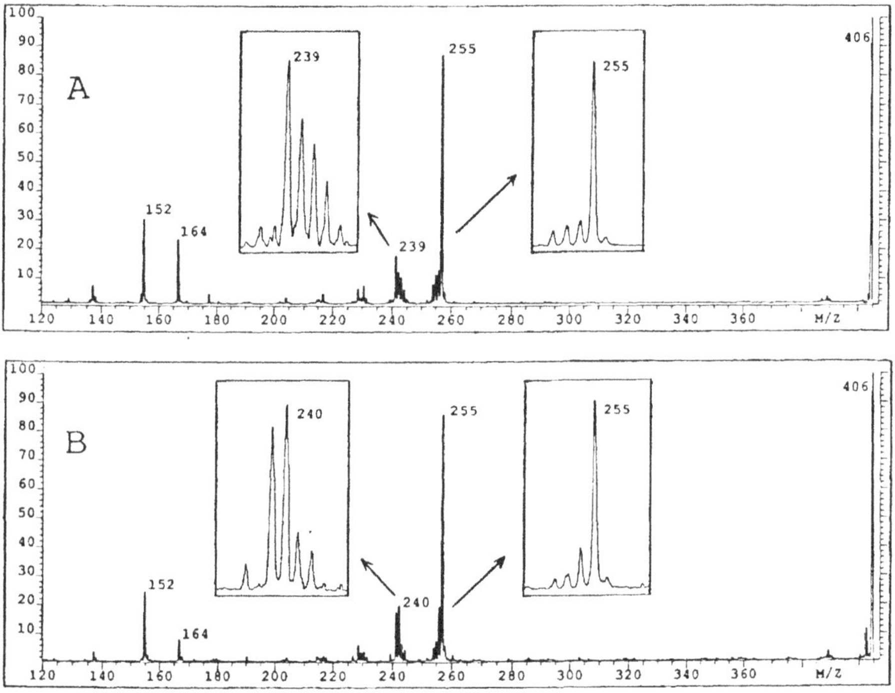


Figure 2. CAD spectra of (M + H)$^+$ ions of *m/z* 406 of (A) 12-methyl-BA-C7-$CH_2$-N7 Gua, and (B) 7-methyl-BA-C12-$CH_2$-N7 Gua.

Two other adducts were isolated, and they produce (M + H)$^+$ ions of *m/z* 406 indicating that deoxyribose is not present in the adduct. The CAD spectra of the two (M + H)$^+$ ions, obtained on a four-sector ZAB-T tandem mass spectrometer, are distinct (Figure 2). The ions of *m/z* 255 and of *m/z* 239 and 240 arise from the DMBA moiety. The ions of *m/z* 152 are protonated guanine, and the ions of *m/z* 164 are (Gua - H + $CH_2)^+$, indicating the attachment of Gua occurs at one of the methyl groups of DMBA. The relative abundances of the ions of *m/z* 239 and 240 are indicative of whether attachment is at the C7 or C12 methyl group. In Figure 2A, the ion of *m/z* 239 is more abundant than the ion of *m/z* 240, indicating the guanine adduct is bound at C7 The spectrum in Figure 2B shows a more abundant ion of *m/z* 240 that is characteristic of an adduct bound at C12.

## 5. Conclusion

A systematic approach to the structure determination of PAH-nucleoside adducts starts with synthetic model adducts. The adducts can be characterized by the complementary techniques of NMR and MS. Fragmentation processes occurring upon CA are identified that are specific for the adduct isomers. Adducts formed in *in vitro* processes utilizing biological systems can then be identified based on their CAD spectra. Improvements in both ionization efficiency and in detection methods should make the structure determination of adducts formed *in vivo* possible.

## 6. References


1. Miller, J. A. (1970) *Cancer Res.* **30**, 559.
2. Miller, E. C., Miller, J. A. (1981) *Cancer (Philadelphia)* **47**, 2327.
3. Sims, P., Grover, P. L., in Polycyclic Hydrocarbons and Cancer, Gelboin, H. V., Tso, P. O. P., Eds., Academic: New York, 1981, Vol. 3, pp. 117-181.
4. Conney, A. H. (1982) *Cancer Res.* **42**, 4875.
5. Hertzberg, R. P., Dervan, P. B. (1982) *J. Am. Chem. Soc.* **104**, 313.
6. Cavalieri, E. L., Rogan, E. G., in Free Radicals in Biology, Pryor, W. A., Ed., Academic: New York, 1984, Vol. VI, pp. 323-369.
7. Cavalieri, E., Rogan, E., in Polycyclic Hydrocarbons and Carcinogenesis, Haney, R. G., Ed., ACS Symposium Series 283, ACS: Washington, DC 1985; pp. 290-305.
8. Rogan, E. G., Cavalieri, E. L., Tibbels, S. R., Cremonesi, P., Warner, C. D., Nagel, D. L., Tomer, K. B., Cerny, R. L. and Gross, M. L. (1988) *J. Am. Chem. Soc.* **110**, 4023.
9. RamaKrishna, N. V. S., Cavalieri, E. L., Rogan, E. G., Dolnikowsky, G., Cerny, R. L., Gross, M. L., Jeong, H., Jankowiak, R., and Small, G. J. (1991) *J. Am. Chem. Soc.*, submitted.

# STRUCTURE DETERMINATION OF LIPIDS: COMPARISON OF CLASSICAL METHODS AND NEW APPROACHES INVOLVING CHARGE-REMOTE FRAGMENTATION

KENNETH A. CALDWELL and MICHAEL L. GROSS
*Midwest Center for Mass Spectrometry*
*Department of Chemistry*
*University of Nebraska-Lincoln*
*Lincoln, Nebraska 68588-0362*

ABSTRACT. Classical methods for determining structure of complex lipids are briefly reviewed. The practical aspects of FAB MS and FAB MS/MS coupled with charge-remote fragmentation for analysis of lipids are discussed. FAB MS is useful for obtaining molecular weight information. Phosphoglycerides can be successfully analyzed by FAB MS in the negative ion mode directly without derivitization. Free fatty acids or those liberated from complex lipids can be analyzed directly without the need for a reference standard by using FAB MS/MS and charge-remote fragmentation.

## 1. Introduction

The advent of fast atom bombardment mass spectrometry (FAB MS) created new frontiers for routine mass spectral analysis [1-3]. The use of a vacuum compatible liquid matrix produced long lived ion currents and introduced the particle bombardment ion source to sector instruments. Reproducible, high quality mass spectra of polar and/or thermally sensitive materials could be obtained in minutes and many biomolecules could be subjected to mass spectral analysis for qualitative assay.

The FAB ion source also spurred the further development of MS/MS techniques, which were being used on sector instruments in ion chemistry studies. In the MS/MS experiment, a specific ion is selected from among all other ions produced in the source, and allowed to undergo collisions with a target gas. A spectrum of the product ions resulting from collisionally activated decomposition (CAD) of the precursor ion is obtained by scanning a second-stage mass analyzer, hence the term MS/MS [4]. The continued merger of particle bombardment ionization with MS/MS also has motivated the development of new instrumentation, specifically four sector tandem mass spectrometers and recently a four sector mass spectrometer with spectrograph capabilities [5].

The use of a collision gas to induce the decomposition of the ion selected for MS/MS analysis allows high energy fragmentation chemistry to be explored. In particular, a new class of fragmentation termed "charge-remote fragmentation" was discovered in this lab for ions with localized charge and containing long alkyl groups, such as the anions of fatty acids [6-11] and of cholesterol sulfate [12]. Initial studies were aimed at identifying the mechanism of the decomposition and investigating applications. Subsequently, the energetics of the process were investigated [13], and an analogy between charge-remote fragmentation and thermal decompositions was established [14]. In what can be considered method development, strategies were investigated to localize better the charge on fatty acids by using picolinyl esters [15, 16] and on fatty alcohols and fatty acids by in situ cationization [13, 17-20].



*M. L. Gross (ed.), Mass Spectrometry in the Biological Sciences: A Tutorial,* 413–425.

In this chapter, we emphasize the *practical* aspects of FAB MS and FAB MS/MS coupled with charge-remote fragmentation for the structural analysis of biomolecules containing long chain hydrocarbon groups, in particular fatty acids and some complex lipids. The speed and ease of obtaining molecular weight and structural information by using FAB MS- and FAB MS/MS-based methodologies are the principal advantages over classic methods of analysis. Additionally, the structure of free fatty acids can be determined directly by using FAB MS/MS, without the need for an authentic reference standard for comparison. This is a considerable advantage for studying novel or modified fatty acids.

## 2. Classification and Biological Importance of Selected Lipids

Lipids have been empirically defined as "water-insoluble organic biomolecules that can be extracted from cells and tissues by nonpolar solvents e.g., chloroform, ether, or benzene" [21, 22]. Lipids share a common structural feature: a portion of the molecule is hydrocarbon in nature. There are two types of lipids. Complex lipids contain fatty acid constituents and yield fatty acid salts upon saponification. These saponifiable lipids are classified according to the backbone structures to which the fatty acids are linked via ester, ether, or amide bonds. (See Table 1). For example, fatty acids are esterified to the two free hydroxyl groups of glycerol 3-phosphate, the backbone structure of phosphoglycerides. Simple lipids do not contain fatty acid constituents, although some simple lipids, such as the prostaglandins, are derived from certain essential fatty acids. The molecular structure of many lipids may be found in reference 21.

Table 1. Classification of lipids.

| Lipid Type | Backbone |
|---|---|
| Complex (saponifiable) | |
| Acylglycerols | Glycerol |
| Phosphoglycerides* | Glycerol 3-phosphate |
| Sphingolipids | Sphingosine |
| Waxes | Nonpolar alcohols of high molecular weight |
| Simple (nonsaponifiable) | |
| Terpenes | |
| Steroids | |
| Prostaglandins | |

*Phosphoglycerides are sometimes loosely referred to as phospholipids. However, the term phospholipid refers to all lipids containing phosphorous and includes phosphorous containing sphingolipids.

Lipids have important biological functions. The hydrocarbon portion of the molecule allows intermolecular hydrophobic interactions to occur between lipids and/or proteins. Also, the extreme hydrophobicity of some lipids allows them to be stored in an unhydratated form essential for efficient energy storage. The hydrocarbon portion of the simple lipids serves as a template that may be functionalized to produce a particular biological activity or function.

Most cell membranes are 50-60 percent protein and 40-50 percent lipid [21]. A large portion of the lipids are phosphoglycerides, complex lipids found almost exclusively in cell membranes. These polar lipids are arranged in a membrane-forming bilayer with the hydrophobic fatty acid tails of the molecules together [23]. Along with providing cell structure and aqueous compartmentalization, the bilayer provides a mechanism to anchor certain enzymes and membrane transport systems to the cell membrane by hydrophobic interactions [23]. Phosphatidylcholine and phosphatidylethanolamine are the most abundant phosphoglycerides in higher plants and animals [21].

Another example of an important complex lipid is the triacylglycerols (fats), which consist of fatty acids esterified to a glycerol backbone and serve as nature's fuel storage molecules, providing depots for energy rich fatty acids [24, 25]. Seventy percent of the energy for a resting heart comes from oxygenation of free fatty acids transported from adipose (fatty) tissue. These free fatty acids are cleaved from the glycerol backbone by the action of lipases in the fat cells of the adipose tissue and are transported through the blood in a lipoprotein complex with serum albumin [25].

The length, degree of unsaturation, and (possible) incorporation of branched methyl groups, etc. of the fatty acids comprising the lipids of cell membranes and fats varies among different organisms [21] and is influenced by the ambient temperature of the organism's environment [24]. Unsaturated fatty acids have lower melting points than saturated ones, and are often found in cold-water fish. Although fatty acids are major components of complex lipids, unesterified free fatty acids do not occur in large concentrations in cells and tissues. The fatty acids may be liberated in vitro, however, by alkaline hydrolysis or by the action of lipases.

Prostaglandins are simple lipids. These 20-carbon molecules are derived from cyclization of $C_{20}$ polyunsaturated fatty acids such as arachidonic acid, which is itself synthesized from the essential fatty acid linoleic acid [24]. Among the functionalities encountered along the hydrocarbon portion of the molecule are double bonds, hydroxy, and ketone groups. Prostaglandins are important because natural and many synthetic prostaglandins are biologically active and have potential as pharmacological agents in humans (see reference 26 and references therein).

## 3. Classic Methods of MS Analysis of Fatty Acids and Phospholipids

By way of comparison to an MS/MS based methodology, classic mass spectrometric methods of fatty acid and phospholipid analysis will be briefly reviewed. A more exhaustive account of fatty acid and phospholipid analysis has been presented elsewhere (see references 27 and 28, respectively).

### 3.1. FATTY ACIDS

Traditional MS methods of structural analysis of fatty acids are not direct. The free fatty acid is converted to a volatile derivative, such as the methyl ester, and analyzed by gas-liquid chromatography. The retention times of the unknown fatty acid are compared to authentic standards. If the detector is a mass spectrometer, then the molecular weight can be determined, especially if chemical ionization is employed. If the GC detector is the flame-ionization detector, the peak areas are proportional to the amount of each fatty acid in the mixture, which is an advantage for semiquantitative analysis.

Structural information from EI or CI mass spectra of unsaturated fatty acid methyl or ethyl esters is limited. The degree of unsaturation can be inferred from the molecular weight, but it is virtually impossible to determine the location of double bonds owing to migration of double bonds along the alkyl chain as a result of ionization [27]. This long-standing problem in mass spectrometry has been addressed in several research labs. Some of the approaches to eliminate migration of the double bond after ionization of the fatty acid methyl esters are: (1) derivatization of the double bond, (2) alternate derivatization of the acid function, (3) combination of one or both of the these two approaches with chemical ionization, and (4) *in situ* mass spectral derivitization utilizing ion-molecule reactions in a chemical ionization source [27]. Ideally, double bond or head group derivatization prevents charge migration and allows the position of the double bond to be inferred by the relative abundance of the ions in the mass spectrum or reveals the double bond position by directing specific fragmentation chemistry to occur at that site. This is true of other functionalities, such as branch points, cyclopropane rings, hydroxy or methoxy groups, etc., which may be encountered in naturally occurring fatty acids.

### 3.2. PHOSPHOGLYCERIDES

Phosphoglycerides are less volatile than fatty acids because these complex lipids have a higher molecular weight which, in combination with the polar head group, makes these biomolecules less amenable to GC analysis. Early mass spectrometric studies of intact phosphoglycerides that were volatilized by heating the lipid to *ca.* 300 °C revealed that low abundance molecular ions are produced by 70-eV electron ionization because extensive fragmentation and/or pyrolysis occurs [29]. These problems were addressed by using lower ionizing energy electrons [29], by introducing the sample on a gold support [30], and by using in-beam ionization of samples introduced on a tungsten support [31].

As in the case of fatty acids, the analysis of phosphoglycerides is not direct. Wet chemistry employing chromatographic purification of the lipid extract is followed by treatment of the sample with phospholipase enzymes that cleave specific constituents of phosphoglycerides; the site of action of these enzymes is indicated in Figure 1 for phosphatidylcholine. Derivatization of phosphoglycerides for GC/MS analysis usually involves cleaving the diglyceride portion from the intact lipid by phospholipase C hydrolysis of the bond between the phosphate group and the oxygen at the 3 carbon. The diglyceride product is subsequently derivatized for GC/MS analysis. These techniques were described in several texts [32-34].

Nonderivatized phospholipids are more successfully analyzed by using soft ionization techniques, such as chemical ionization (especially if the sample is introduced into the mass spectrometer on a metal support), field desorption, laser desorption, FAB, etc. [28]. Separation of mixtures can be accomplished by using various chromatography techniques including TLC, column chromatography, and HPLC as well as LC/MS. The mixture components can be identified by comparing retention times and/or the mass spectra to authentic standards or synthetic models.

## 4. Practical Aspects of FAB MS of Selected Lipids

The integration of a vacuum compatible liquid solvent, the "matrix", with the ionization process distinguishes fast atom bombardment mass spectrometry from other desorption ionization methods and especially from conventional EI mass spectrometry. A tribute to fast atom bombardment ionization is the short time required for it to gain widespread acceptance as an ionization technique. Some of the reasons for the rapid success of FAB ionization are:

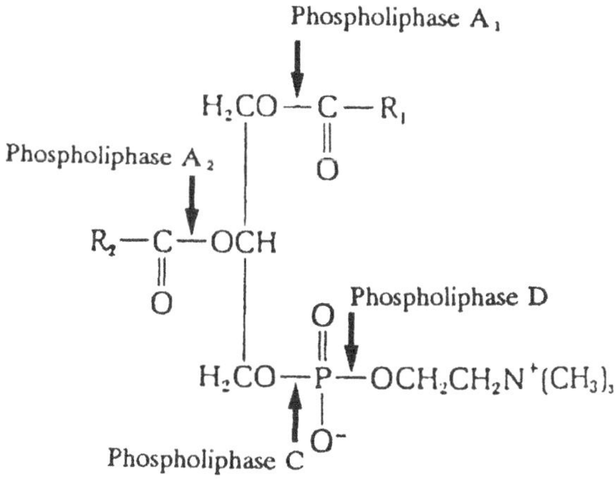


Figure 1. Phosphatidylcholine. The indicated bonds can be cleaved by the action of phospholipase enzymes.

(1) the ability to retrofit existing mass spectrometers with a FAB source, (2) the need for a dependable ionization source not requiring thermal volatilization of the sample to produce gas-phase ions, and (3) the ease of performing the FAB MS experiment. The last reason is a very practical one. In principle, to obtain a FAB mass spectrum of an isolated sample, one needs only to apply a thin layer of liquid matrix to a direct insertion FAB probe, mix the sample with the matrix, and insert the probe into a FAB source of a scanning mass spectrometer. The experiment, however, is complicated by the dependence of the experiment's success on the choice of a suitable liquid matrix.

The requirements for a FAB matrix were discussed elsewhere [35, 36]. The most important requirement is vacuum compatibility. Glycerol, although it was not mentioned explicitly, was the first reported FAB matrix [1,2], owing to its low vapor pressure at room temperature and good solvent properties. In addition to the matrices listed in references 35 and 36, the use of other matrices has been reported including 1-dodecanol [37] and 2-hydroxyethyl disulfide [38].

For a successful FAB mass spectrometric analysis, the analyte should be dissolved, or at least well dispersed, in the liquid matrix. For nonpolar fatty acids and the prostaglandins, triethanolamine works well as a matrix for negative ion spectra, and 3-nitrobenzyl alcohol, usually with an added alkali or alkaline earth salt, has been used in studies of the positive ions. Puzo and co-workers [39] reported the use of monobutyltriethylene glycol as a matrix for the analysis of selected glycolipids. This matrix allows for more hydrophobic interactions with the nonpolar portion of the glycolipids, leading to improved performance with this matrix because the solubility of the analyte in this matrix is better than in glycerol.

The ability to use different liquid matrices can be used to advantage in an analysis. Fast atom bombardment with and without metal salts is a general strategy that is very useful for determining the molecular species present in a sample. For example, the gas-phase protonated molecule ion of fatty acids and prostaglandins dissolved in a 3-nitrobenzyl alcohol matrix are less abundant than the water loss ions. Adding an alkali salt such as lithium iodide to the matrix allows the exchangeable protons to be replaced, at least in part, by lithium cations. $[M + Li]^+$ and $[M - H + 2Li]^+$ ions, which do not readily decompose by water loss, are desorbed. Alkaline earth metals may also be used. A solution of 3-nitrobenzyl alcohol and barium acetate produces upon FAB abundant $[M - H + Ba]^+$ ions for fatty acids and prostaglandins. The iodide and acetate salts are more soluble in FAB matrices than are the more common chloride salts. Because the solubility of the acetate salts in most FAB matrices is high, some care must be taken to ensure that the salt concentration in the matrix is not too high. A saturated solution should be incrementally diluted to determine the optimum concentration for a particular class of compounds. This frees up matrix molecules previously involved with solvation of the salt for solvation of the analyte.

It was established that concentration-dependent discrimination occurs in the analysis of fatty acid mixtures [37] and that this discrimination is much reduced when a 1-dodecanol matrix is used. This was attributed to the bulk solubility of the fatty acids in dodecanol. The liquid 2-hydroxyethyl disulfide is also a good matrix for analysis of fatty acids and other nonpolar, lipid-like materials. This latter matrix provides longer lived ion signals than from either 3-nitrobenzyl alcohol or 1-dodecanol. It was reported by De Angelis et al. [38] that surface activity discrimination in the analysis of peptide mixtures is not as severe in this matrix as in other matrices, presumably owing to the good solvating properties of 2-hydroxyethyl disulfide.

The importance of analyte solubility in the matrix cannot be overstressed. This point is made explicitly clear when free fatty acids are placed in a neat glycerol matrix. Fatty acids are not well-dissolved in glycerol, and the secondary ion beam is erratic. Adding KOH to the glycerol increases the solubility of the fatty acids, and the secondary ion beam is stable. In the negative ion mode, abundant $[M - H]^-$ ions are observed [37] and in the positive ion mode both $[M + K]^+$ and $[M - H + 2K]^+$ species are observed. Likewise, Budzikiewicz and co-workers [40] observed that phosphoglycerides analyzed by using a glycerol matrix had to be first dissolved in a cosolvent system and mixed with the glycerol prior to analysis before useful results could be obtained.

Various matrix combinations have been used to study phosphoglycerides. For analysis of the negative ions, diethanolamine or triethanolamine is used almost universally. For most phosphoglycerides analyzed in the negative ion mode, the $[M - H]^-$ ion is easily detected in the high mass region of the spectrum. Phosphatidylcholines, however, do not yield $[M - H]^-$ but three other characteristic negative ions $[M - CH_3^+]^-$, $[M - HN(CH_3)_3^+]^-$, and $[M - HN(CH_3)_3^+ - C_2H_2]^-$ [40, 41]. Analysis of the positive ions of phosphatidylcholines reveals abundant $[M + H]^+$ ions [40, 42-43]. In these studies, glycerol with a cosolvent to aid dissolution, thioglycerol, or 3-nitrobenzyl alcohol, was used as a matrix. Although it does not appear to have been reported in the literature, 2-hydroxyethyl disulfide is probably a useful matrix for FAB MS of phospholipids, especially if positive ions are to be studied.

## 5. FAB MS/MS and Charge-Remote Fragmentation

Charge-remote fragmentations are gas-phase ion decompositions that occur at sites of the ion that are distant from the location of the charge and are relatively uninfluenced by the charge. Although this class of fragmentation chemistry was first identified and extensively examined for decompositions of fatty acid carboxylates, the processes are found in the fragmentations of steroids, prostaglandins, complex lipids, peptides, carbohydrates, and certain antibiotics. The cumulative literature was recently reviewed by Adams, [44] and the recent literature was reviewed by Gross [45] with comments on the status of this area of research.

5.1. FATTY ACIDS

As an example of the application of FAB MS and FAB MS/MS combined with charge-remote fragmentation for the analysis of selected lipids, consider the analysis of free fatty acids. Fatty acids yield abundant [M - H]⁻ species upon FAB in the negative ion mode (triethanolamine matrix) and mono and dilithiated species in positive ion FAB (3-nitrobenzyl alcohol matrix with added LiI). These species allow the molecular weight to be determined and, therefore, the degree of unsaturation or ring content. To determine possible branch points and/or locations of double bonds, another dimension of information is required.

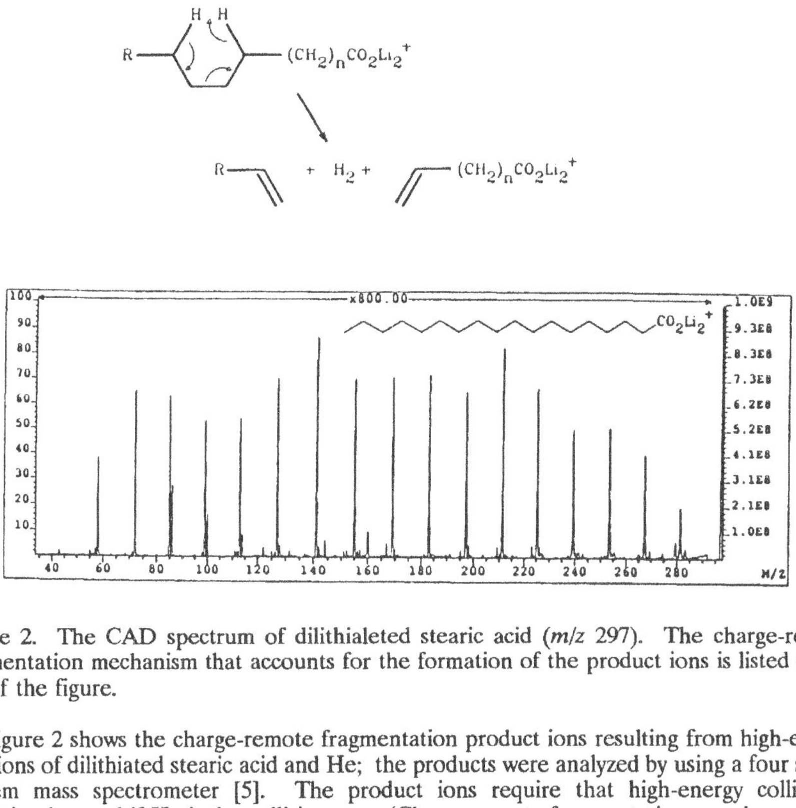


Figure 2. The CAD spectrum of dilithialeted stearic acid (*m/z* 297). The charge-remote fragmentation mechanism that accounts for the formation of the product ions is listed at the top of the figure.

Figure 2 shows the charge-remote fragmentation product ions resulting from high-energy collisions of dilithiated stearic acid and He; the products were analyzed by using a four sector tandem mass spectrometer [5]. The product ions require that high-energy collisional activation be used if He is the collision gas. (Charge-remote fragmentations can be made to occur in triple stage quadrupoles and hybrids if the center-of-mass energy and the number of collisions are made to be high). The continuous pattern of major ions formed by losses of the elements of $CH_4$ (*m/z* 281) through $C_{16}H_{34}$ can be accounted for by considering the 1,4-elimination mechanism given at the top of the figure. By this mechanism, loss of the elements of an alkane occurs by the actual loss of the elements of an alkene along with $H_2$ along the hydrocarbon backbone, distant from the localized charge. For high-energy keV collisions, charge-remote fragmentation is the dominant decomposition pathway of fatty acids provided that the charge is well-localized. The two other ions in the region of $CH_4$ loss arise from the interesting losses of $H_2O$ and $CH_2$ from the dilithiated precursor.

Although these fragmentations occur upon FAB desorption [9, 46], the ion abundances are too small for reliable interpretation, thus collisional activation and MS/MS are required. The other requirments for MS/MS are the need to separate the structure specific ions from the matrix ions, to assign unambiguously the origin of fragment ions, and to deal directly with mixtures. (The MS/MS experiment is ultimately limited by interferences from decompositions from background ions co-desorbed along with the analyte ion).

In accord with the proposed mechanism, the fragmentation pattern observed in Figure 2 is interrupted if a methyl branch occurs along the backbone (see Figure 3A). A double bond in the backbone also perturbs the pattern (Figure 3B). Facile cleavages occur allylic to the double bond, at positions b and c, and the location of the double bond can be determined from the corresponding peaks in the spectrum. The dominant charge-remote fragment is produced by cleavage at position a, which produces a product ion with a terminal double bond in conjugation with the oleic acid double bond.

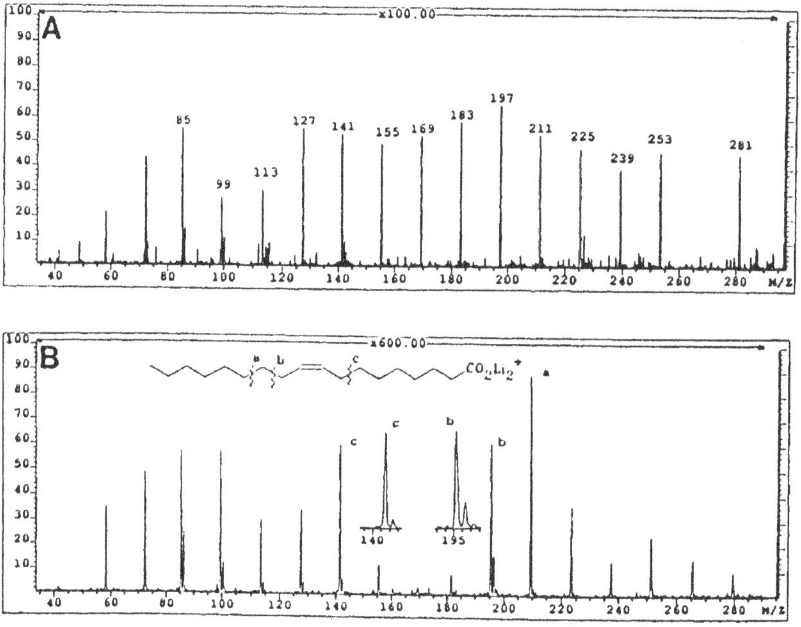


Figure 3. The CAD spectrum of dilithiated (A) 16-methylheptanoic acid (*m/z* 297) and (B) oleic acid (*m/z* 295).

In addition to branched [11,18] and unsaturated [7, 9, 17] fatty acids, the efficacy of collisional activation and charge-remote fragmentation for the determination of the structure of fatty acids with cyclopropane or cyclopropene rings, hydroxy substituents, or epoxide rings was established [11]. Recently, Crockett et al. [47] for the first time investigated a nearly complete series of fatty acids in which the location of the functional group was systematically varied. The members of the series of homoconjugated, 18-carbon fatty acids undergo charge-remote fragmentation that can be used for determination of the positions of the double bonds. In this study, the CAD spectra of $[M - H]^-$, $[M - H + 2Li]^+$, and $[M - H + Ba]^+$ precursor species were compared. While each of these species are sufficient for localizing the charge so that charge-remote fragmentations can be induced, the $[M - H + 2Li]^+$ species yields the most readily interpretable product ion spectra [47].

Fast atom bombardment ionization is not required, of course, to generate fatty acid carboxylate precursor ions for high-energy collisional activation. An example of GC/MS combined with charge-remote fragmentation for structural analysis is that of Promé and co-workers [48-50] who have studied mycobacterial polyunsaturated fatty acids by collisionally activating carboxylate anions produced from negative ion chemical ionization of pentafluorobenzyl fatty acid esters. The carboxylate anion is produced directly by loss of a pentafluorobenzyl radical as a result of electron capture. The product ion spectra of fatty acid carboxylates produced from esters that were separated by gas chromatography are obtained in real time by scanning the electrostatic analyzer of a reverse geometry instrument.

### 5.2. PHOSPHOGLYCERIDES

As an example of an entirely mass spectrometric based approach for the determination of the structure of a complex lipid by using FAB MS and FAB MS/MS, we focus on 1-palmitoyl-2-oleoy-phosphatidylcholine and the isomeric 1-oleoy-2-palmitoyl-phosphatidylcholine. (For some representative examples of previously published studies, see references 41-43,51,52). The spectrum of negative ions (Figure 4A) obtained from cesium ion bombardment [53] of 1-palmitoyl-2-oleoy-phosphatidylcholine displays the three anions characteristic of phosphatidylcholines $[M - CH_3^+]^-$, $[M - HN(CH_3)_3^+]^-$, and $[M - HN(CH_3)_3^+ - C_2H_2]^-$ (or $[M - CH_2{=}CHN(CH_3)_3^+]^-$) [40,41]. Collisional activation of the $[M - CH_3^+]^-$ ion of *m/z* 744 leads almost exclusively to production of the carboxylate anions of the constituent fatty acids, with the fatty acid at the sn2 position being most abundant (Figure 4B). If the analogous ion of *m/z* 744 from the isomeric 1-oleoy-2-palmitoyl-phosphatidylcholine is collisionally activated, the fatty acid at the sn2 position is again the most abundant product ion (Figure 4C). In a limited MS/MS study that included these two phosphatidylcholines and three others, Jensen et al. [41] reported that the β fatty acid carboxylate is the most abundant product ion resulting from collisional activation of either $[M\text{-}CH_3^+]^-$ or $[M\text{-}HN(CH_3)_3^+]^-$. This is not, however, a general rule [52].

If the 1-palmitoyl-2-oleoy-phosphatidylcholine anion of *m/z* 673 is collisionally activated (Figure 4D), in addition to the carboxylate anions, another major ion appears that arises by loss of the oleic acid moiety from the sn2 position to give an ion of *m/z* 391 [41]. In fact, collisional activation of the $[M\text{-}HN(CH_3)_3^+\text{-}C_2H_2]^-$ species provide the most reliable information regarding the relative location of constituent fatty acids in phosphatidylcholines [52]. Gage and co-workers [52] recently completed an MS/MS study of 24 phosphatidylcholines in which it was found that collisional activation of the $[M\text{-}HN(CH_3)_3^+\text{-}C_2H_2]^-$ ions (formally deprotonated phosphatidic acids) always leads to preferential loss of the neutral fatty acid at the sn2 position. Preliminary results of collisional activation of fragment ions corresponding to phosphatidic acid produced from other phosphoglycerides shows that the sn2 fatty acid moiety is consistently preferentially lost over the sn1 fatty acid [52]. This allows

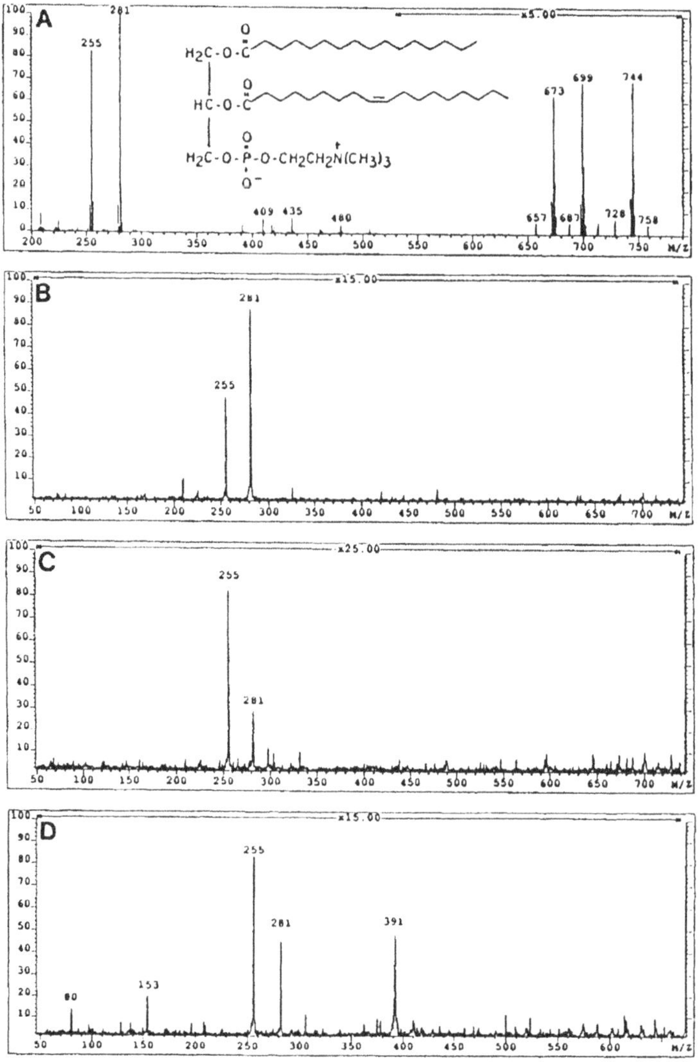


Figure 4. A. Cesium ion bombardment mass spectrum of the negative ions desorbed from 1-palmitoyl-2-oleoyl-phosphatidylecholine (structure shown) dissolved in a triethanolamine matrix. B-D. Product ions resulting from collisional activation of the (B) 1-palmitoyl-2-oleoyl-phosphatidylecholine [M - $CH_3^+$ ]$^-$ ion with *m/z* 744 (C) the 1-oleoyl-2-palmitoyl-phosphatidylecholine [M - $CH_3^+$]$^-$ ion with *m/z* 744 and (D) the 1-palmitoyl-2-oleoyl-phosphatidylecholine [M - $HN(CH_3)_3^+$ - $C_2H_2$]$^-$ ion with *m/z* 673.

the relative position of the constituent fatty acids to be determined. The structure of each fatty acid can be determined by collisionally activating the fatty acid carboxylates both of which are liberated as a result of particle bombardment. Thus, one is able not only to determine the structures of the constituent fatty acids, but also to locate their positions in the complex lipid. This is accomplished by an entirely instrumental method.

The advantage of MS/MS/MS comes to the fore when a mixture of complex lipids is encountered and more than component produces fatty acids of the same mass. In that case, a high mass lipid ion can be activated in the first step of a sequential activation strategy to give the fatty acid anion. That acid anion is in turn activated in a second step (MS/MS/MS) to produce structurally specific charge-remote fragmentations. A recent example of the application of not only FAB MS and FAB MS/MS, but also FAB MS/MS/MS to study complex lipids is the work of Fenselau and co-workers [43] who analyzed the phosphatidylcholines associated with the protein coat of the AIDS virus.

## 6. Conclusion

In this chapter, we have focused on methods for the structure determination of fatty acids and phosphatidylcholines. Charge-remote fragmentations form the basis of an entirely instrumental method for these materials. Some applications are discussed in the next two chapters. The utility of charge-remote fragmentation for other classes of lipids has been also demonstrated in part, and method development continues.

Collisional activation of commercially available phosphoglycerides have been recorded in the literature and future prospects should include more applications of developed methodologies for real world problem solving of complex lipid structures. The field of tandem mass spectrometry is rapidly maturing, and it is expected that mass spectrometrists will be collaborating with structural chemists to use new methods for solving real problems.

## 7. Acknowledgement

This chapter was prepared with support of the U.S. National Science Foundation (Grant CHE - 9017250). KAC also acknowledges the receipt of a travel award from the NSF to attend the NATO ASI.

## 8. References

1. Barber, M., Bordoli, R. S., Sedgwick, R. D., Tyler, A. N (1981) *J. C. S. Chem. Comm.* 325.
2. Barber, M., Bordoli, R. S., Sedgwick, R. D., Tyler, A. N (1981) *Nature*, **293**, 270.
3. Barber, M., Bordoli, R. S., Elliot, G. J., Sedgwick, R. D., Tyler, A. N (1982) *Anal. Chem.* **54**, 645A.
4. Busch, K. L., Glish. G. L., McLuckey, S. A. (1988) Mass Spectrometry/Mass Spectrometry, VCH Publishers, Inc., New York.
5. This instrument was commissioned by the Midwest Center for Mass Spectrometry, a former NSF regional instrumentation facility (Grant No. CHE 8620177). It was designed by Robert Bateman and built by VG Analytical (now Fisons Instruments) for the MCMS. The spectrometer known as a ZAB-T, was briefly described by Gross: Gross, M. L., in *Methods in Enzymology, Mass Spectrometry*; McCloskey, J. A., Ed.; Academic Press: New York, 1990, Vol 193, pp 131-153.
6. Tomer, K. B., Crow, F. W., Gross, M. L. (1983) *J. Am. Chem. Soc.* **105**, 5487.

7. Jensen, N. J., Tomer, K. B., Gross, M. L. (1985) *J. Am. Chem. Soc.* **107**, 1863.
8. Jensen, N. J., Tomer, K. B., Gross, M. L. (1985) *Anal. Chem.* **57**, 2018.
9. Jensen, N. J., Tomer, K. B., Gross, M. L., Lyon, P. A. (1985) in "Desorption Mass Spectrometry Are Sims and FAB the Same", Lyon, P. A., Ed.; American Chemical Society Symposium Series, No. 291: Washington, D. C. pp 194-208.
10. Jensen, N. J., Gross, M. L. (1986) *Lipids* **21**, 362.
11. Tomer, K. B., Jensen, N. J., Gross, M. L. (1986) *Anal. Chem.* **58**, 2429.
12. Tomer, K. B., Gross, M. L. (1988) *Biomed. Environ. Mass Spectrom. 15*, 89.
13. Adams, J. A., Gross, M. L. (1986) *J. Am. Chem. Soc.* **108**, 6915.
14. Adams, J. A., Gross, M. L. (1989) *J. Am. Chem. Soc.* **111**, 435.
15. Deterding, L. J., Gross, M. L. (1987) *Anal. Chim. Acta.* **200**, 431.
16. Deterding, L. J., Gross, M. L. (1988) *Org. Mass Spectrom.* **23**, 169.
17. Adams, J., Gross, M. L. (1987) *Anal. Chem.* **59**, 1576.
18. Adams, J., Gross, M. L. (1988) *Org. Mass Spectrom.* **23**, 307.
19. Adams, J. A., Gross, M. L. (1986) *J. Am. Chem. Soc.* **108**, 6915.
20. Davoli, E., Gross, M. L. (1990) *J. Am. Soc. Mass Spectrom.* **1**, 320.
21. Lehninger, A. L. (1975) *Biochemistry*, 2nd edition, Worth Publishers, Inc., New York, chapter 11.
22. For a discussion of the difficulty of defining lipids, see: Mead, J. F., Alfin-Slater, R. B., Howton, D. R., Popjak, G. (1986) *LIPIDS, Chemistry, Biochemistry, and Nutrition* Plenum Press, New York, chapter 2.
23. Singer, S. J., Nicolson, G. L. (1972) *Science*, **175**, 720.
24. Lehninger, A. L. (1975) *Biochemistry*, 2nd edition, Worth Publishers, Inc., New York, chapter 24.
25. *ibid*, chapter 30.
26. Zirrolli, J. A., Davoli, E., Bettazzoli, L., Gross, M., Murphy, R. C. (1990) *J. Am. Soc. Mass Spectrom.* **1**, 325.
27. Jensen, N. J., Gross, M. L. (1987) *Mass Spectrom. Rev.* **6**, 497, and references therein.
28. Jensen, N. J., Gross, M. L. (1987) *Mass Spectrom. Rev.* **7**, 41, and references therein.
29. Klein, R. A. (1971) *J. Lipid Res.* **12**, 628.
30. Constantin, E., Nakatani, Y., Hueber, R. (1980) *Tetrahedron Lett.* **21**, 4745.
31. Ohashi, M., Iino, H., Tsujimoto, K., Ohashi, Y., Shida, Y. (1985) *Org. Mass Spectrom.* **20**, 642.
32. Christie, W. W. (1982) *Lipid Analysis*, Pergamon Press, New York.
33. Bergelson, L. D., (1980) *Lipid Bio-chemical Preparations*, Elsevier Publishers, Amsterdam.
34. Asselineau, J. (1966) *The Bacterial Lipids*, Holden-Day Inc., San Francisco.
35. Gower, J. L. (1985) *Biomed. Mass Spectrom.* **12**, 191.
36. De Pauw, E. (1986) *Mass Spectrom. Rev.* **5**, 191.
37. Caldwell, K. A., Gross, M. L. (1989) *Anal. Chem.* **61**, 494.
38. De Angelis, F., Nicoletti, R., Santi, A. (1988) *Org. Mass Spectrom* **23**, 800.
39. Rivière, M., Fournié, J. J., Vercellone, A., Puzo, G. (1988) *Org. Mass Spectrom* **16**, 275.
40. Münster, H., Stein, J., Budzikiewics, H. (1986) *Biomed. Mass Spectrom.* **13**, 423.
41. Jensen, N. J., Tomer, K. B., Gross, M. L. (1986) *Lipids*, **21**, 580.
42. Heller, D. N., Murphy, C. M., Cotter, R. J., Fenselau, C., Uy, O. M. (1988) *Anal. Chem.* **60**, 2787.
43. Bryant, D. K., Orlando, R. C., Fenselau, C., Sowder, R. C., Henderson, L. E. (1991) *Anal. Chem.* **63**, 1110.
44. Adams, J. (1990) *Mass Spectrom Rev.* **9**, 141.
45. Gross, M. L. *Int. J. Mass Spec. Ion Proc.*, in press.
46. Sweetman, B. J., Blair, I. A (1988) *Biomed. Environ. Mass Spectrom.* **17**, 337.

47. Crockett, J. S., Gross, M. L., Christie, W. W., Holman, R. T. (1990) *J. Am. Soc. Mass Spectrom.* **1**, 183.
48. Promé, J. C., Aurelle, H., Couderc, F., Savagnac, A. (1987) *Rapid Commun. Mass Spectrom.* **1**, 50.
49. Aurelle, H., Treilhou, M., Promé, D., Savagnac, A., Promé, J. C. (1987) *Rapid Commun. Mass Spectrom.* **1**, 65.
50. Couderc, F., Aurelle, H., Promé, D., Savagnac, A., Promé, J. C. (1988) *Biomed. Envir. Mass Spectrom.* **16**, 317.
51. Ayanoglu, E., Wegmann, A., Pilet, O., Marbury, G. D., Hass, J. R., Djerassi, C. (1984) *J. Am. Chem. Soc.* **106**, 5246.
52. Huang, Z. H., Gage, D. A., Sweeley, C. C. (1991) *J. Am. Soc. Mass Spectrom.* **3**, in press.
53. Bombardment with keV-cesium ions produces the same qualitative result as bombardment with keV atoms. The interested reader may wish to see Aberth, W., Straub, K. M., Burlingame, A. L. (1982) *Anal. Chem.* **54**, 2029, and Falick, A. M., Wang, G. H., Walls, F. C. (1986) *Anal. Chem.* **58**, 1308.

# THE USE OF FAB MS OF CELLULAR LIPIDS FOR THE CHARACTERIZATION OF MEDICALLY IMPORTANT BACTERIA.

R. WAIT
*Division of Pathology,*
*Public Health Laboratory Service Centre for Applied Microbiology and Research,*
*Porton Down,*
*Salisbury, United Kingdom.*

ABSTRACT. The profiling of cellular lipids by GC/MS is a valuable technique for the identification and classification of micro-organisms. Most procedures however entail the hydrolysis and derivatization of complex lipids prior to analysis. Fast atom bombardment mass spectrometry avoids these stages and enables the direct analysis of unmodified polar lipids. By the use of collisional activation and linked scanning, complex mixtures of phospholipids can be characterized without the necessity of a chromatographic separation step, and the disposition of substituents can be established by mass-measurement of the products of phospholipase digestion. This chapter describes the application of these techniques to problems of microbiological identification, classification and physiology.

## 1. Introduction

Conventional methods of microbial identification rely on a range of morphological, nutritional, biochemical and serological approaches. These rather subjective phenotypic characteristics are of use for microbial identification and classification to the extent that they reflect the underlying genetic relationships between species. The ultimate criterion of the relatedness of two organisms is the degree of homology between their respective DNA as measured by hybridization. Because DNA based methods are not at present a practical means of routine identification, there is a need for an approach intermediate between the subjective morphological criteria of classical bacteriology and direct comparison at the level of DNA. Chemotaxonomy, which may be defined as the use of the chemical analysis of cellular components in microbial classification and identification [1, 2], is one such approach. Potential chemotaxonomic markers must fulfil certain criteria; most importantly, a large number of genes should be involved in their synthesis both to ensure biosynthetic stability, and because chemical variation is then more likely to reflect significant underlying genetic differences. Secondly, the chosen cellular materials should be amenable to analysis by rapid and simple techniques so that a reasonably sized data-base of typical organisms can be assembled.

One of the most generally useful chemotaxonomic approaches has been the analysis of cellular fatty acids by gas chromatography and gas chromatography/mass spectrometry. A good example is the identification of isolates of *Legionella pneumophila* and related species, which unusually for Gram-negative organisms, contain high levels of iso and anteiso branched fatty acids, the characteristic pattern of which generally enables identification to genus level [3]. Closely related species however, are often similar in their gross fatty acid composition and cannot be distinguished by this method.

In the intact cell fatty acids occur bound in more complex lipid structures such as lipopolysaccharides and phospholipids, from which they must be liberated by hydrolysis and



*M. L. Gross (ed.), Mass Spectrometry in the Biological Sciences: A Tutorial,* 427–441.

derivatized prior to GC/MS. This hydrolytic step represents a considerable destruction of cellular information, some of which might be relevant to the problem of species discrimination. Hence a method for the routine analysis of intact polar lipids would be of value in chemotaxonomic studies.

Most bacteria contain members of several phospholipid families, varying in their fatty acid composition, which results in mixtures of considerable complexity. Purification (by HPLC or two dimensional TLC) is usually laborious, and further characterization generally requires enzymatic or chemical degradation and identification of the products by GC/MS. Such methods are too cumbersome and time consuming to be of value for routine taxonomic studies of micro-organisms. The direct analysis of mixtures by FAB MS is therefore an attractive alternative method of phospholipid analysis. In particular, as will be shown below, a combination of FAB, linked scanning [4] (or tandem mass-spectrometry) and the use of specific phospholipase enzymes [5] offers a powerful strategy for the rapid analysis of quite complex mixtures.

The structures of some of the more widely distributed phospholipids found in bacterial cells are shown in Figure 1. Most are built according to a common architecture, in which fatty acids are esterified to positions 1 and 2 of glycerol. The fatty acids vary in chain length, degree of unsaturation, and their possession of branches and cyclopropane rings. The phosphate group on position 3 of the glycerol usually bears polar substituents such as ethanolamine (which may be *N*-methylated), choline, glycerol, serine or sugars. In a related group of phosphoglycerides, known as plasmalogens, there is an alk-1-enyl substituent on position 1 of the glycerol. Many more unusual types of lipid are also found in certain bacterial genera [6]; though taxonomically useful their determination will not be discussed here.

It is possible to obtain FAB spectra of polar lipids by applying a suspension of solvent-disrupted bacterial cells directly to the FAB target [7]. In our experience, however, more reproducible data are obtained from chloroform extracts. Suitable extracts can be rapidly prepared by a modification of the procedure of Bligh and Dyer [8], in which about 5 mg (wet weight) of bacteria are shaken at room temperature with a monophasic mixture of chloroform, methanol and water. On the addition of further chloroform and water, phase separation occurs, and the lower (organic) layer, which contains the polar lipids, may be removed and concentrated in a stream of dry nitrogen. The extracts so prepared are generally suitable for direct analysis by FAB MS, because most potentially interfering substances (including alkali metal ions) remain in the aqueous phase.

## 2. Strategies For the Characterization of Polar Lipids by FAB MS

The positive and negative ion mass-spectra of common phospholipids have been extensively described in the literature [9-16] and therefore need only a brief summary here. The negative ion spectra of 3-*sn*-phosphatidylglycerols, 3-*sn*-phosphatidylserines, 3-*sn*-phosphatidylinositols and 3-*sn*-phosphatidylethanolamines are broadly similar and can be considered together. The spectrum of 1,2-dipalmitoyl-*sn*-glycero-3-phosphoglycerol (Figure 2) is typical; the abundant deprotonated molecular ion (of *m/z* 721) fragments, producing the species at *m/z* 647, 483, 465 and 255. The ion at *m/z* 647 is formed by cleavage of the O-$CH_2$ bond of the head group with hydrogen transfer to, and charge retention on the phosphatidic acid moiety. Ions at *m/z* 465 and *m/z* 483 arise by losses of $CH_3(CH_2)_{14}CO_2H$ and $CH_3(CH_2)_{14}CO$ respectively, whereas the *m/z* 255 ion is apparantly a carboxylate anion, $CH_3(CH_2)_{14}CO_2^-$.

3-*sn*-Phosphatidylcholines do not produce $(M - H)^-$ anions; accordingly their relative molecular masses must be determined from three ions $(M - 15)^-$, $(M - 60)^-$ and $(M - 86)^-$ arising from losses of various portions of the choline moiety [11,12]. Carboxylate anions are also

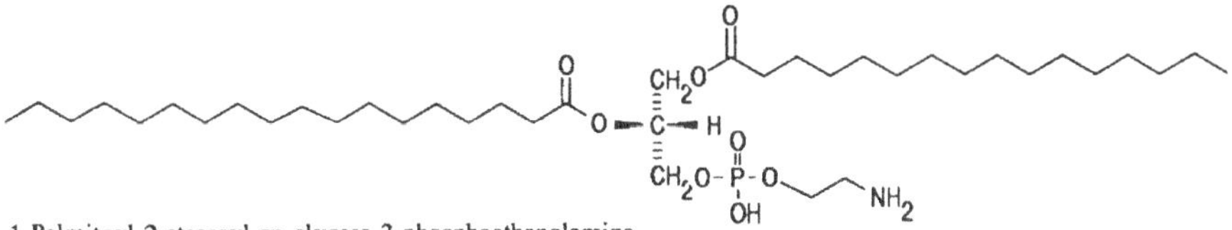

1-Palmitoyl-2-stearoyl-sn-glycero-3-phosphoethanolamine

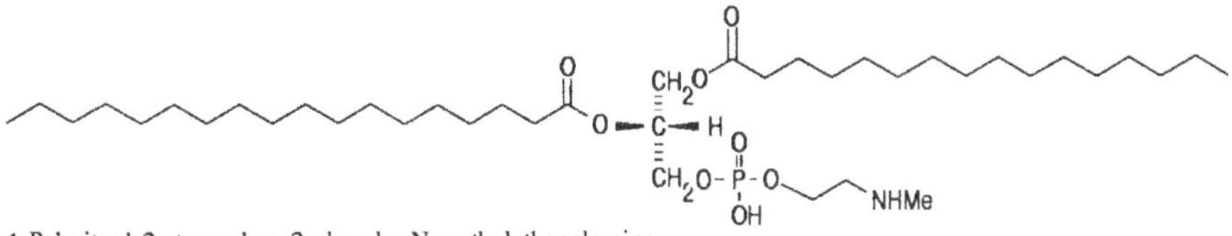

1-Palmitoyl-2-stearoyl-sn-3-phospho-N-methylethanolamine

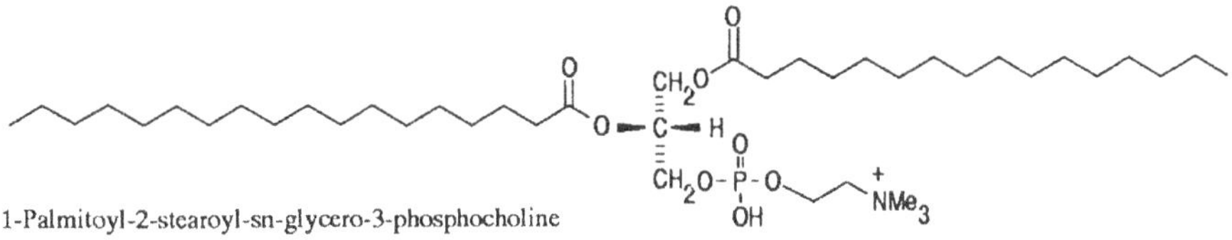

1-Palmitoyl-2-stearoyl-sn-glycero-3-phosphocholine

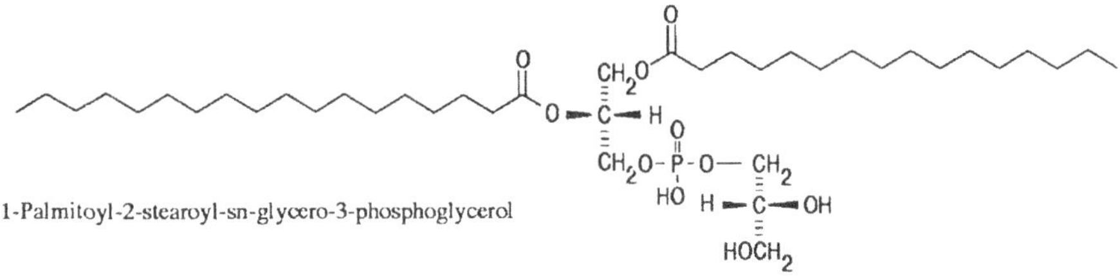

1-Palmitoyl-2-stearoyl-sn-glycero-3-phosphoglycerol

Figure 1. The structures of some of the types of phospholipid most commonly found in bacteria.

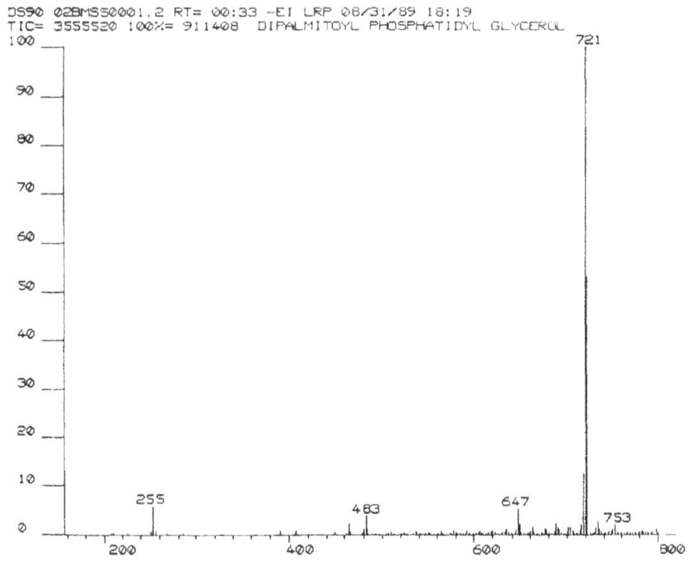


Figure 2. Negative ion FAB spectrum of 1,2-dipalmitoyl-*sn*-glycero-3-phosphoglycerol.

present as in the spectra of other phospholipids. It has been observed that in most cases, the carboxylate anion derived from the substituent on position 2 of the glycerol is more abundant, thus defining the substitution pattern of the intact 3-*sn*-phosphatidylcholine. In mixtures of 3-*sn*-phosphatidylcholines it is possible to mass select and collisionally activate the $(M - 60)^-$ fragments to obtain this information [11]. Collisional activation and MS/MS or linked scanning can also be used to characterize further the carboxylate anions. Charge-remote fragmentations produce cleavages at each $CH_2$ of the alkyl chain, from which features such as methyl branches or double bonds may be located by means of the perturbations they induce in this regular ion series [17,18]. It should be noted, however, that because bacteria frequently contain complex mixtures of isobaric fatty acids, collisional activation of their carboxylate anions can produce mixed spectra which are difficult to interpret.

In the positive ion FAB spectra of phospholipids the dominant fragmentation process is loss of the polar group, usually with charge retention on the glycerol back-bone [9]. The behaviour of dipalmitoyl-3-*sn*-phosphatidylethanolamine (Figure 3) is typical; the prominent $(M+H)^+$ ion fragments by cleavage of the $CH_2$-O bond of the phosphate group, with charge retention on the glycerol backbone ion (of *m/z* 551). The mass of this ion thus reveals the total number of carbon atoms and degree of unsaturation of the two fatty acid substituents. In the case of distearoyl-3-*sn*-phosphatidylethanolamine, for example, a corresponding elimination of 141 produces a backbone fragment at *m/z* 603. The facile loss of the polar head group also occurs on collisional activation, as can be seen in Figure 3b, which shows a B/E scan of protonated dipalmitoyl-3-*sn*-phosphatidylethanolamine. The mass of the eliminated fragment is diagnostic of the polar head group; thus dipalmitoyl-3-*sn*-phosphatidyl-*N*-methylethanolamine loses 155u, and 3-*sn*-phosphatidyl-*N*-*N*-dimethylethanolamines and 3-*sn*-phosphatidylglycerols undergo analogous fragmentations. The process is so facile that in the spectra of 3-*sn*-phosphatidylglycerols $(M+H)^+$ ions are barely visible, although alkali metal cationized species may be more

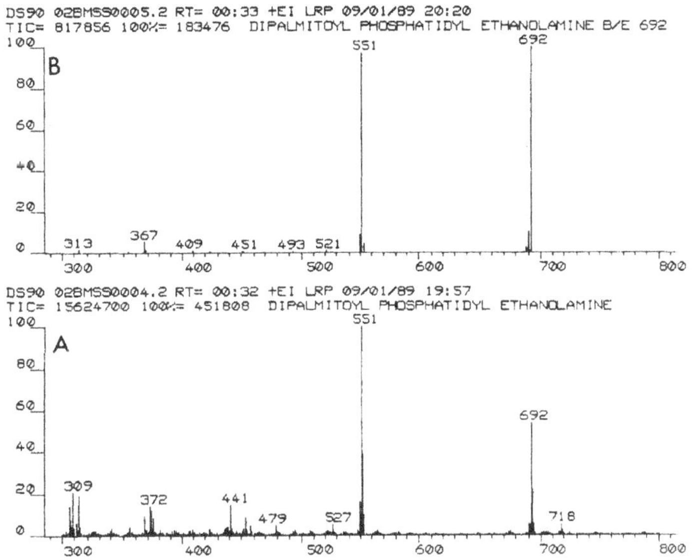


Figure 3. Positive ion FAB spectrum of dipalmitoyl-3-*sn*-phosphatidylethanolamine (A), and B/E linked scan of the $(M + H)^+$ of the same compound (B).

conspicuous. The loss of the polar head group can be exploited for the selective recording of particular classes of phospholipids by means of constant neutral loss scans for the appropriate mass (see also the chapter by van Bremen). This is the case even in instances where the neutral fragment is lost so readily that an $(M+H)^+$ is not observed in the conventional FAB spectrum (for example 3-*sn*-phosphatidylglycerols). Table 1 lists some neutral losses that are specific for common phospholipid families.

TABLE 1. Constant neutral loss scans for different phospholipid classes

| Phospholipid type | Mass of neutral fragment |
|---|---|
| 3-*sn*-Phosphatidylethanolamine | 141 |
| 3-*sn*-phosphatidyl-*N*-methylethanolamine | 155 |
| 3-*sn*-phosphatidyl-*N*-*N*-dimethylethanolamine | 169 |
| 3-*sn*-Phosphatidylglycerol | 172 |

The 3-*sn*-phosphatidylcholines behave somewhat differently in that although cleavage of the $CH_2$-O-P bond occurs, charge is retained on the choline phosphate moiety, giving an ion of *m/z* 184. Hence, they cannot be selectively detected by means of a constant neutral loss scan. Precursor ($B^2/E$) scans of the *m/z* 184 ion might in principle be used instead; however the precursor ion resolution of two sector instruments is so poor that relative molecular masses cannot always be assigned precisely enough to distinguish saturated and mono-unsaturated species. 3-*sn*-phosphatidylcholines in mixtures must therefore be verified by B/E scans, although again poor parent ion resolution may introduce ambiguities.

Summarising and over-simplifying, the negative ion spectra afford more information about the fatty acid substituents, whereas positive ion spectra offer better clues to the nature of the head group. Hence where possible both both should be recorded, because they offer complementary information. When analysing complex mixtures however, such as are obtained from bacteria there are strong arguments for the positive ion mode. For example, with positive ion extraction, discrete $(M + H)^+$ ions are obtained for all components of a mixture of 3-*sn*-phosphatidylcholines and 3-*sn*-phosphatidylethanolamines. In the negative mode however phosphatidyl cholines do not produce $(M - H)^-$ anions; instead they fragment to generate three high mass species, which results in complicated spectra, especially in the case of multi-component mixtures. Furthermore one of these fragments, $(M - 15)^-$, is isomass with the $(M - H)^-$ of the corresponding phosphatidylethanolamine containing two more $CH_2$ units in the alkyl chains of its substituents, thus rendering the interpretation of B/E scans problematic. Another advantage of positive ion spectra is that they allow the use of the constant neutral loss scans described above. This has two advantages; firstly signals for the $(M + H)^+$ ions can be assigned to the appropriate type of phospholipid. Secondly the problem of the poor parent ion resolution of the B/E scan on two-sector instruments can be circumvented, since although the resolution of the neutral fragment in the CNL scan is also poor its mass is generally sufficiently low to enable nominal mass assignments of product or precursor.

Thus by analysing the products of collisional activation by means of CNL scanning we can identify the different types of phospholipid in mixtures and determine the total number of carbon atoms and double bonds in their fatty acids. The experiment will not however define the identities of the individual acids or establish their relative positions on the glycerol backbone. We can obtain this latter information by use of the enzyme phospholipase $A_2$ to remove selectively the substituent from position 2 of the phospholipid, followed by identification of the hydrolysis product by FAB MS. In the case of 1-stearoyl-2-oleoyl-*sn*-glycero-3-phosphocholine for example the $(M + H)^+$ ion of *m/z* 788 shifts to *m/z* 524, which is the *m/z* value of the $(M + H)^+$ of a lysophosphatidylcholine bearing a saturated C18 substituent. This mass shift demonstrates that the unsaturated substituent is on position 2. The application of this strategy to complex mixtures of phospholipids is more problematic, because it may not be possible to associate all digestion products with their precursors. In this case additional experiments are necessary such as treatment of the original sample with phospholipase $A_1$ to remove the substituent from position 1. If ambiguities still remain they may be resolved by HPLC separation of the phospholipid mixture into its individual molecular species prior to enzymatic digestion and FAB MS.

## 3. Microbiological applications

### 3.1. IDENTIFICATION OF BACTERIA

An unidentified clinically isolated organism was shown by classical bacteriological techniques to bear some similarity to *Corynybacterium haemolyticum*. The taxonomic status of this organism is controversial, the extent of its relatedness to *Corynybacterium pyogenes* in particular being disputed [19]. The polar lipid composition of the unknown organism was, therefore, compared to that of the type strains of *C. haemolyticum* and *C. pyogenes*, and the resulting positive ion FAB mass spectra are shown in Figure 4. The data from the unidentified strain and the authentic *C. haemolyticum* are very similar, four major $(M + H)^+$ species being present, at *m/z* 786, 784, 758 and 734. A low mass ion of *m/z* 184 was consistent with the presence of 3-*sn*-phosphatidylcholines, and collisional activation and B/E scanning verified this

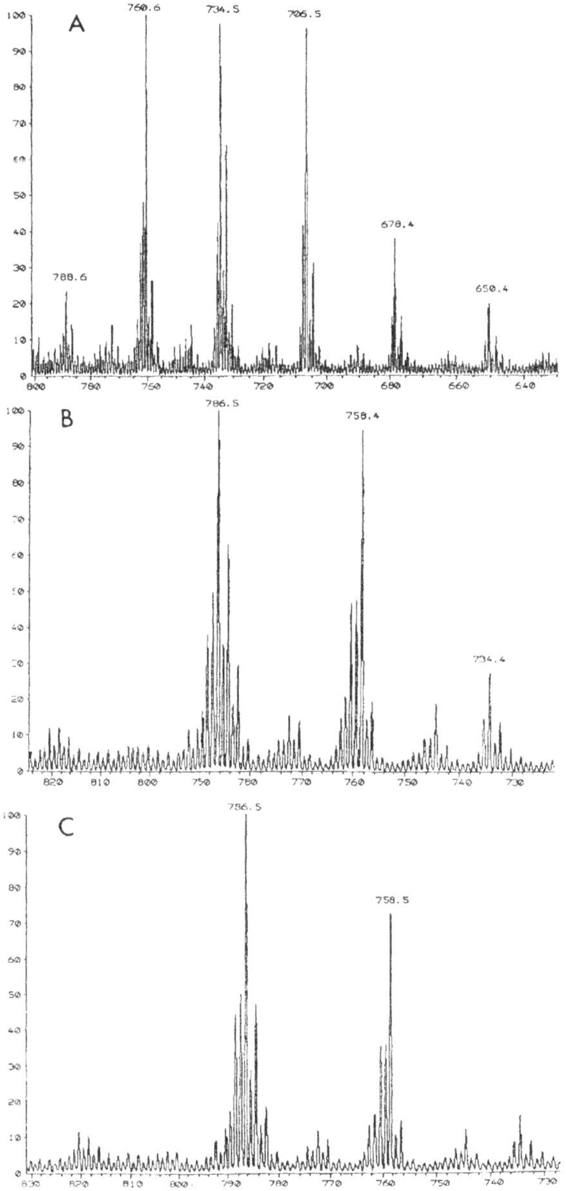


Figure 4. Positive ion FAB spectra of phospholipids extracted from from the type strain of *Corynybacterium pyogenes* (A), from the type strain of *Corynybacterium haemolyticum* (B) and from an unidentified clinically isolated organism (C).

assignment. This was sufficient to define the total number of carbon atoms and double bonds in the substituents, as summarised in table 2 but did not establish their individual identities; indeed many compositions are possible (eg the ion of *m/z* 786 could contain fatty acids 18:1 + 18:1; 18:0 + 18:2; 20:1 + 16:1 etc).

TABLE 2. Phospholipid composition of *C. haemolyticum.*

| Mass of $(M + H)^+$ | Phospholipid type | Substituents |
|---|---|---|
| 786 | 3-*sn*-Phosphatidylcholine | 36:2 |
| 784 | 3-*sn*-Phosphatidylcholine | 36:3 |
| 758 | 3-*sn*-Phosphatidylcholine | 34:2 |
| 734 | 3-*sn*-Phosphatidylcholine | 32:0 |

On treatment with phospholipase $A_2$ the principal $(M + H)^+$ ions shift to *m/z* 524 and *m/z* 496 which are lysophosphatidylcholines substituted in position 1 with octadecanoic and hexadecanoic acids respectively, from which we can deduce that the principle substituent on

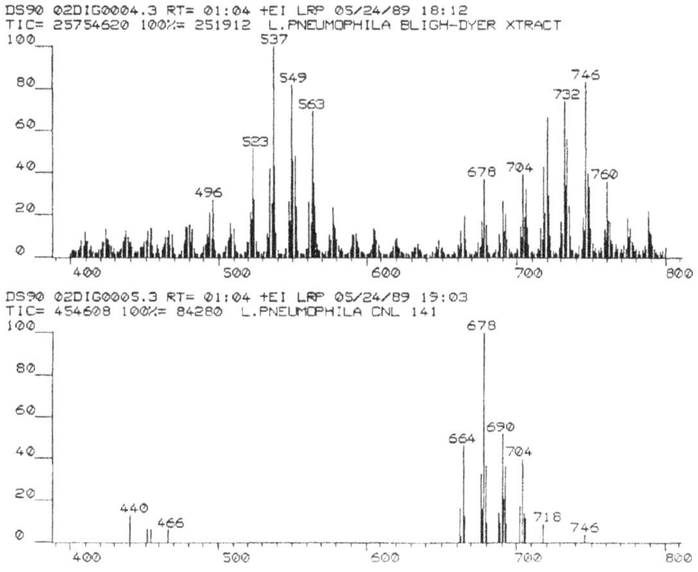


Figure 5. Positive ion FAB spectrum of a mixture of phospholipids extracted from *Legionella. pneumophila* (A), and (B) constant neutral loss scan for 141 u.

position 2 is an octadecadienoic acid (18:2).

Essentially similar approaches can be applied to organisms of much more complex lipid composition. Figure 5a shows the positive ion FAB spectrum of a Bligh-Dyer extract of *L. pneumophila*, (the causative organism of Legionnaires' Disease). At least 15 significant $(M + H)^+$ species are visible in the region *m/z* 664-760. The majority of signals are accompanied by satellites 2 u lower (of significantly greater abundance than expected for $(M - 2H)^+$ fragments),

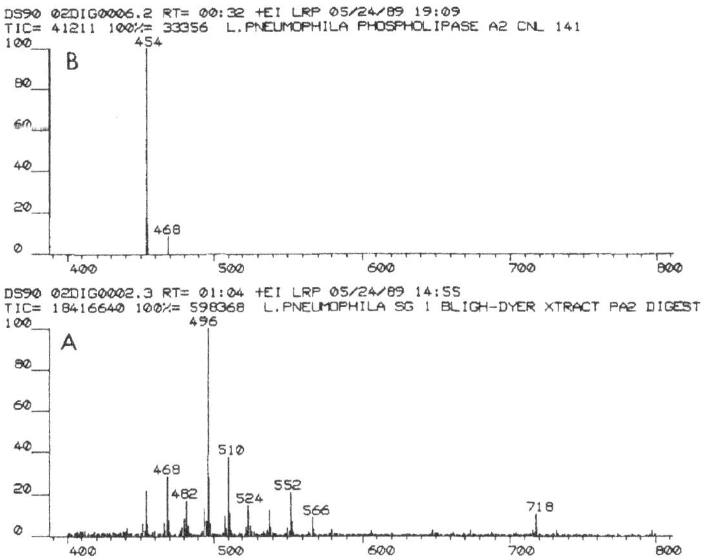


Figure 6. (A) Positive FAB spectrum of *L. Pneumophila* phospholipids after digestion with phospholipase $A_2$, and (B) constant neutral loss of 141 u from the same sample.

showing that a high proportion of their fatty acid substituents are unsaturated. An ion of *m/z* 184 indicates that some of these correspond to 3-*sn*-phosphatidylcholines. A series of glycerol backbone cleavage ions of *m/z* values between 523 and 577 suggest the presence of 3-*sn*-phosphatidylglycerol and 3-*sn*-phosphatidyl ethanolamine species, differing in their fatty acid composition and degree of methylation.

The first stage in the interpretation of such complex data is to assign the various signals to their respective phospholipid types by means of collisional activation and linked scanning. Thus Figure 5b shows the result of a CNL scan for 141 u, which selectively records 3-*sn*-phosphatidylethanolamines. We can see immediately that the ions of *m/z* 664, 678 690/692 and 704 are the $(M + H)^+$ of 3-*sn*-phosphatidylethanolamines differing in their substitution pattern. Similarly a CNL scan for 155 u transmits primarily the ions of *m/z* 704 and 718. This is a nice illustration of the value of the constant neutral loss experiment over the B/E scan, since *m/z* 704 proves to be the $(M + H)^+$ of an isobaric mixture of 3-*sn*-phosphatidylethanolamines and 3-*sn*-phosphatidyl-*N*-methylethanolamines. Most of the remaining signals can be similarly assigned to phosphatidyl-di-*N*-methylethanolamines and phosphatidylglycerols by appropriate constant neutral loss scans. The major ions in the region *m/z* 718- 760 were shown by B/E scanning to be the $(M + H)^+$ ions of 3-*sn*-phosphatidylcholines.

The next stage of interpretation is to try to determine the nature and relative positions of the fatty acid substituents in the various phospholipid families that have been identified in the mixture. Digestion of a portion of the Bligh-Dyer extract with phospholipase $A_2$ and re-analysis by FAB MS produced the spectrum shown in Figure 6a. It is obvious that the unsaturated substituents are primarily located on position 2, because the characteristic pattern of doublets has been lost, and the majority of the signals between *m/z* 454 and m/z 552 are the $(M + H)^+$ ions of saturated lysophospholipids. We can again use collisional activation and linked scanning to assign the signals. Thus a CNL scan for 141 u transmits ions of *m/z* 454 and 468) which are the $(M + H)^+$ ions of lysophosphatidylethanolamines, substituted in position 1 with hexadecanoic and heptadecanoic acids respectively (Figure 6b). Because there is a rather surprisingly homogenous substitution pattern on position 1 we can unambiguously assign the fatty acid composition of the major 3-*sn*-phosphatidylethanolamines (Table 3).

TABLE 3. Fatty acid composition of *L. pneumophila* phosphatidylethanolamines

| $(M + H)^+$ | Position 1 | Position 2 |
|---|---|---|
| 664 | 16:0 | 14:0 |
| 676 | 16:0 | 15:1 |
| 678 | 16:0 | 15:0 |
| 690 | 16:0 | 16:1 |
| 692 | 16:0 | 16:0 |
| 704 | 16:0 | cyc 17 |

Fatty acid compositions can be similarly derived for the 3-*sn*-phosphatidyl-*N*-methylethanolamines and the 3-*sn*-phosphatidylglycerols. The lyso-phosphatidylcholines (signals at *m/z* 496 and beyond), however, display a greater degree of heterogeneity in position one than do the other phospholipids. There is a further structural ambiguity that cannot be easily resolved by using the techniques described here; the hexadecanoic acid in *L.pneumophila* is actually a mixture of iso-branched (approximately 70%) and straight-chain (30%) isomers. Hence, the hexadecanoate substituted phospholipids form an isobaric mixture that are not readily distinguished by linked scanning methods. Currently the best method of characterization appears to be stereospecific liberation of the fatty acids with phospholipases, followed by their identification by GC/MS.

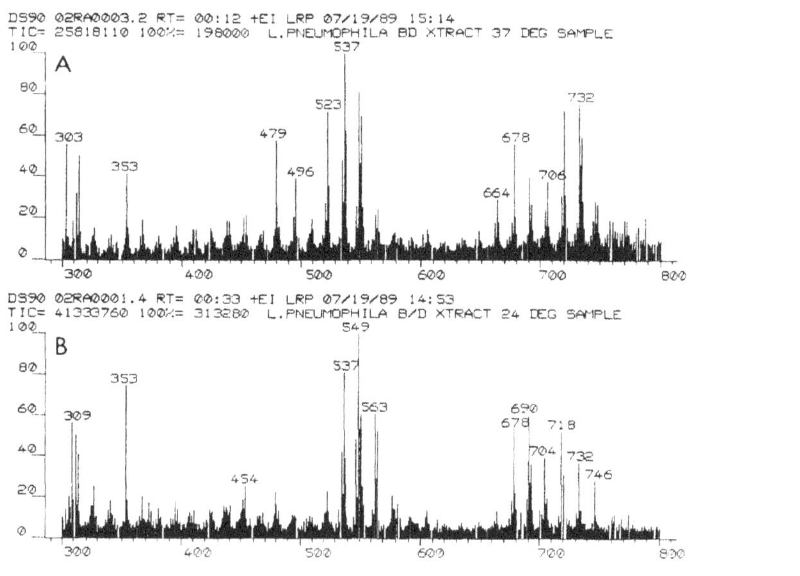


Figure 7. Comparison of the positive FAB spectra of phospholipids from *L. pneumophila* grown at 37 °C (A) and 24°C (B).

### 3.2. STUDIES OF BACTERIAL PHYSIOLOGY BY FAB-MS.

Bacteria can grow only if their membranes are in a largely fluid state; that is, they must be above the gel to liquid-crystalline transition temperature of their cellular lipids. Thermotolerant organisms, such as *Legionella*, that can survive across a wide temperature range require adaptive mechanisms to ensure that their membranes are fluid at a given growth temperature [20]. The strategies that micro-organisms have evolved for growth at reduced temperature generally involve the synthesis of membrane constituents of lower melting point. Such adaptive mechanisms include increased production of branched and unsaturated fatty acids, the reduction of the acyl chain length or alteration of the phospholipid polar head group.

FAB MS offers a rapid means of qualitatively comparing phospholipid profiles at different growth temperatures. Figure 7 shows spectra of phospholipid extracts of *L. pneumophila* grown in continuous culture at 37 °C and 24 °C. As the temperature is reduced, the proportion of phospholipids containing an unsaturated substituent is increased; thus the signals for ions of *m/z* 690, 704 and 718 are significantly enhanced at low temperature. This tendency is particularly pronounced in the 3-*sn*-phosphatidylcholines as the $(M + H)^+$ ions of the fully saturated species at *m/z* 734 and 748 are virtually absent at 24 °C. In general the unsaturated phospholipid species include only one double bond containing substituent. Digestion with phospholipase $A_2$ confirmed that these were preferentially located on position 2. The low abundance of the *m/z* 664 ion at 24 °C reflects the reduced synthesis of tetradecanoic acid at low temperature while the enhancement of *m/z* 678 indicates increased production of 14-methylhexadecanoic acid. Finally, there appears to be an overall reduction of the relative proportion of 3-*sn*-phosphatidylcholines as the temperature is lowered.

These data suggest that in *L. pneumophila* the most significant mechanism at work is increased production of unsaturated fatty acyl chains at low growth temperature. There is no evidence of any tendency towards acyl chain shortening, but it is significant that the reduction in saturated acid content is achieved almost entirely at the expense of straight chain and iso branched acids, which have a significantly higher melting temperature than their anteiso branched isomers [21]. The reduced content of 3-*sn*-phosphatidylcholines at low temperature seems surprising, because 3-*sn*-phosphatidylcholines have a 20 °C lower melting temperature than 3-*sn*-phosphatidylethanolamines with identical acyl substituents. This effect is however likely to be much less pronounced in mixed phospholipids, because it is due to packing factors and the greater ability of the ethanolamine head group to form intermolecular hydrogen bonds. It would seem, therefore, that increasing the proportion of unsaturated substituents in phospholipids of all types has a greater effect on phase transition temperatures than alterations to the polar head group.

### 3.3. NEGATIVE ION FAB-MS OF ISOPRENOID QUINONES

Isoprenoid quinones are constituents of bacterial cell membranes that play a vital role in respiratory electron transport. Two common types, menaquinones and ubiquinones are illustrated in Figure 8. There is considerable variation in the length and degree of unsaturation of the side chain, and in the structure of the quinone nucleus, between and within bacterial genera. As a result quinone determination is useful in chemotaxonomic studies [22]. In the *Legionellaceae* the only quinones present belong to the ubiquinone series [3], but these materials are unusual in that their side-chains contain more than 10 isoprene units.

Isoprenoid quinones are easily separated by high performance liquid chromatography, using ultraviolet or fluorescence detection, but their definitive identification requires the use of mass spectrometry. Informative spectra may be recorded using electron ionization, but the thermal lability of isprenoid quinones means that this is technically demanding, and the sensitivity is generally poor. This is particularly true of the high molecular weight ubiquinones found in the *Legionellaceae*. Given these difficulties, we have investigated the potential of FAB-MS for their characterization. The data obtained in the positive ion mode are not especially encouraging; ubiquinone 10 (Q10) for example produced an abundant ion *m/z* 864, which being two mass units heavier than the calculated molecular mass, must correspond to a protonated hydroquinone. Fragment ions are comparable to those observed in EI mass spectra and are indicative of the type of quinone; thus ions at *m/z* 197 and 235 (derived from the benzoquinone portion of the molecule) are diagnostic of ubiquinones [23]. The sensitivity however is no better than that routinely obtainable by EI MS, and the observation of protonated hydroquinone molecular ions makes the mass- measurement of unknowns ambiguous.

We therefore investigated the negative ion spectra of ubiquinones, reasoning that the

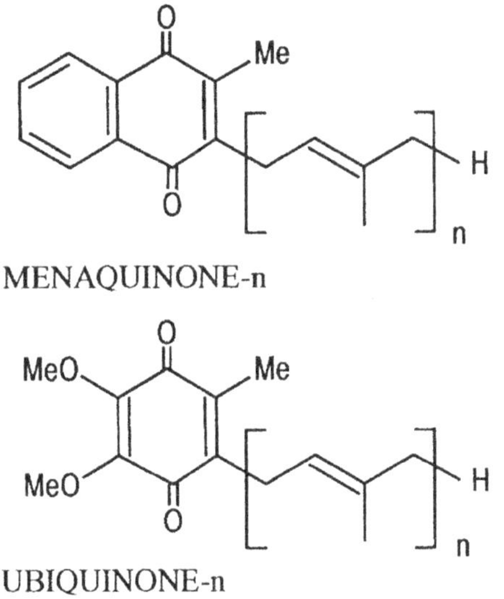


Figure 8. Structures of isoprenoid quinones belonging to the ubiquinone and menaquinone series.

biochemical role of quinones as electron acceptors implied high electron affinities. The spectra are dominated by abundant molecular radical anions rather than deprotonated species (eg Q6 produced an ion of *m/z* 590 not m/z 589). This we interpret as the formation of semiquinone anion radicals by one electron reduction. Other workers have observed similar behaviour in compounds of high electron affinity containing the napthoquinone nucleus, both under conditions of electrochemically assisted negative ion FAB [24], and in conventional negative FAB when sulfolane is used as the matrix [25].

The abundant radical anions produced by quinones in the negative ion mode suggested that direct analysis of mixtures might be feasible. The spectra of quinone extracts from *Legionella* species (Figure 9) exhibited signals with *m/z* values between 862 and 1132, corresponding to the molecular anions of Q10 to Q14 (Table 4).

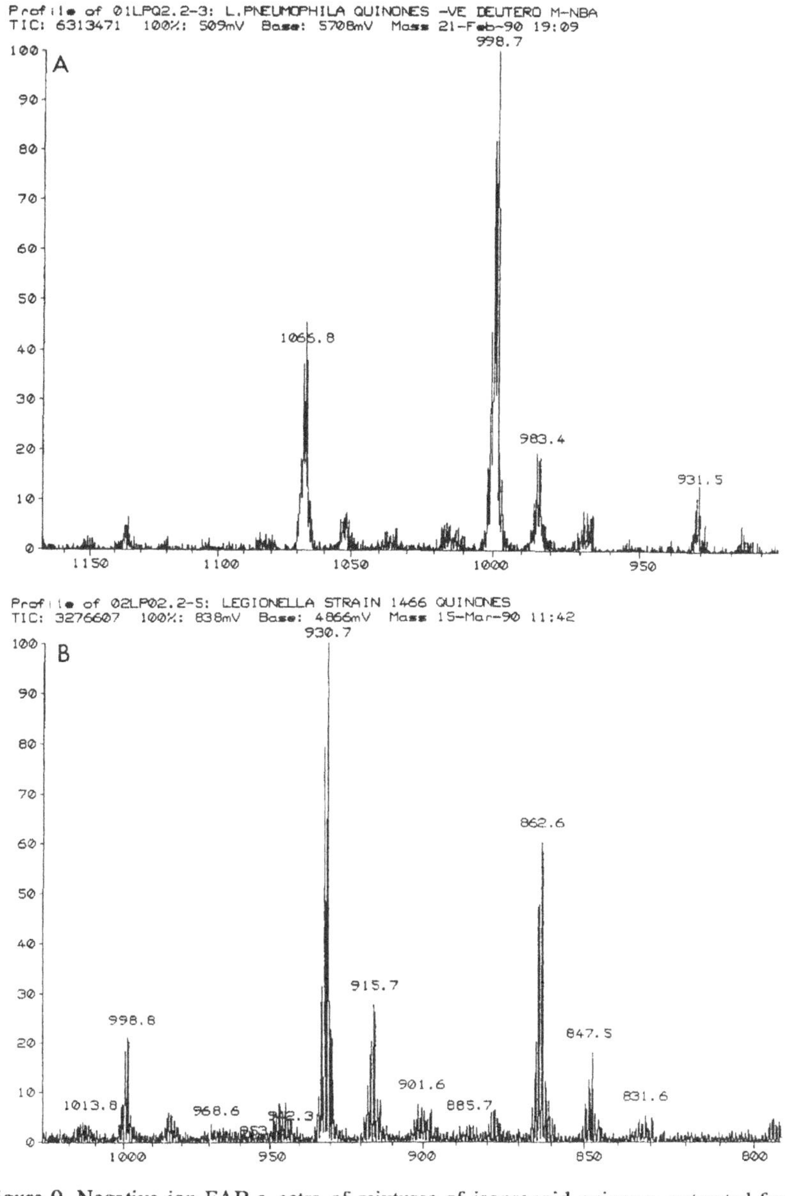


Figure 9. Negative ion FAB spectra of mixtures of isoprenoid quinones extracted from *L. pneumophila* (A) and "*L. londiniensis*" (B).

Table 4. Molecular anions of *Legionella* Ubiquinones.

| Ubiquinone | *M/Z* of Molecular anion |
|---|---|
| Q10 | 862 |
| Q11 | 930 |
| Q12 | 998 |
| Q13 | 1066 |
| Q14 | 1134 |

For example, the spectrum of *L. pneumophila* (Figure 9) contained strong signals at *m/z* 998 (Q12) and 1066 (Q13), and weaker species at *m/z* 930 (Q11) and 1132 (Q14). By contrast the spectrum of strain 1466 (*"L. londiniensis"*) was dominated by species of lower *m/z* particularly Q10 (*m/z* 862), Q11 (*m/z* 930) and Q12 (*m/z* 998). The quinone compositions determined by FAB-MS are in reasonable agreement with reversed phase HPLC derived data, which implies that supression effects are relatively insignificant. This is unsurprising in that quinones belonging to the same homologous series would not be expected to differ much in surface activity. It may also be relevent that the electron affinities of ubiquinones are independent of the length of the isoprene side-chain; radical anions are thus likely to be formed with equal facility by all members of the series.

## 4. Conclusions

GC/MS of cellular lipids has become an almost routine identification technique in some microbiology laboratories. FAB MS by contrast is much less well established, even though it allows the analysis of underivatized materials and enables the use of simpler and more rapid methods of sample preparation.

The present chapter has described the role of FAB MS as a tool for the identification and classification of micro-organisms, and has also indicated its uses for the study of microbial physiology. As one of the most powerful available methods for probing the structure of the bacterial cell surface, FAB MS has enormous potential for the investigation of such challenging poblems as the molecular mechanisms of microbial pathogenesis.

## 5. References

1. Minnikin, D. E, and Goodfellow, M. (1985) Chemical Methods in Bacterial Systematics, Academic Press, London.
2. Lechavalier, H. and Lechavalier, M. P. (1988) 'Chemotaxonomic use of lipids-an overview' in C. Ratledge, and S. G Wilkinson (eds.), Microbial Lipids (volume 1), Academic press, London, pp 869-902.
3. Wait, R. (1988) 'Confirmation of the identity of *legionellae* by whole cell fatty acids and isoprenoid quinone profiles', in A Laboratory manual for *Legionella*, T. G. Harrison, and A. G. Taylor, (eds.) J. Wiley, London pp 69-101.
4.Heller, D. N., Murphy, C. M., Cotter, R., J., Fenselau, C., and Uy, O., M. (1988) 'Constant neutral loss scanning for the characterization of bacterial phospholipids desorbed by fast atom bombardment', *Anal. Chem.* **60**, pp 2787-2791.
5. Wait, R., Hudson, M. J., and Thompson, M. H. (1989) 'Studies on the polar lipids of the Legionnaires' disease bacterium using fast atom bombardment, linked scanning and enzymatic digestion', Proceedings of the seventeenth meeting of the British Mass Spectrometry Society,

BMSS, London, 13-16.
6. Ratledge, C. and Wilkinson, S. G. (1988) Microbial Lipids (volume 1) Academic press, London.
7. Heller, D. N., Cotter, R. J., Fenselau, C., and Uy, O. M. (1987) 'Profiling of bacteria by fast atom bombardment mass spectrometry', *Anal. Chem.* **59**, 2806-2809.
8. Bligh, E. G., and Dyer W. J. (1959) 'A rapid method of total lipid extraction and Purification', *Can. J. Biochem. Physiol.* **37**, 911-917.
9. Fenwick, G. R, Eagles, J. and Self, R. (1983) 'Fast atom bombardment mass spectrometry of intact phospholipids and related compounds', *Biomed. Mass. Spectrom.* 10, 382-386.
10. Ohashi, Y. (1984) 'Structure determination of phospholipids by secondary ion mass spectrometric techniques: differentiation of isomeric esters', *Biomed. Mass. Spectrom.* **11**, 383-385.
11. Jensen, N. J, Tomer, K. B, and Gross, M. L. (1986), 'Fast atom bombardment and tandem mass spectrometry of phosphatidylserine and phosphatidylcholine', *Lipids,* **21**, 580-587.
12. Munster, H, Stein, J. and Budzikiewicz, H. (1986) 'Structural analysis of underivatized phospholipids by negative ion fast atom bombardment mass spectrometry' *Biomed. Environ. Mass. Spectrom,* **13**, 423-427.
13. Munster, H., and Budzikiewicz, H. (1987) 'Structural analysis of phospholipids by fast atom bombardment/collisional activation with a tandem mass spectrometer', *Rapid Commun. Mass. Spectrom.* **1**, 126-128.
14. Jensen, N. J, Tomer, K. B, and Gross, M. L. (1987) 'FAB MS/MS for Phosphatidylinositol, -glycerol, -ethanolamine and other complex Phospholipids', *Lipids,* **22**, 480-489.
15. Jensen, N. J, and Gross, M. L. (1988) 'A comparison of mass spectrometry methods for structural determination and analysis of phospholipids', *Mass Spectrom. Rev.* 7, 41-69.
16. Chen, S., Kirschner, G., and Traldi, P (1990) 'Positive ion fast atom bombardment mass-spectrometric analysis of the molecular species of glycerophosphatidyl serine', *Anal. Biochem.* **191**, 100-105.
17. Tomer, K. B, Jensen, N. J, and Gross, M. L. (1986) 'Fast atom bombardment and tandem mass spectrometry for determining structural modifications of fatty acids', *Anal. Chem.* **58** 2429-2433.
18. Adams, J. (1990) 'Charge remote fragmentations; analytical applications and fundamental studies'. *Mass Spectrom. Rev.* **9**, 141-186.
19. Collins, M. D. and Cummins, C. S. (1986) 'Genus *Corynebacterium*' in P. H. A. Sneath, N. S. Mair, and M. E. Sharpe (eds.), Bergey's Manual of systematic Bacteriology (volume 2), Williams and Wilkins, Baltimore.
20. Melchior, D. L. (1982) 'Lipid phase transitions and regulation of membrane fluidity in prokaryotes' *Current Topics in Membranes and Transport,* **17**, 263-317.
21. Rilfors, L, Wieslander, A, and Stahl, S. (1978) 'Lipid and protein composition of membranes of *Bacillus megaterium* variants in the temperature range 5 to 70 $^0$C' *J. Bact.*, **135**, 1043-152.
22. Collins, M. D. (1985) 'Isoprenoid quinone analysis in bacterial classification and identification' in D. E. Minnikin and M. Goodfellow (eds), Chemical Methods in Bacterial Systematics, Academic Press, London, 267-287.
23. Collins, M. D. (1985) 'Analysis of isoprenoid quinones' *Methods in Microbioloogy,* **18,** 329-366.
24. Phillips, L. R. and Bartmess, J. E. (1989) 'Electrochemically assisted fast atom bombardment negative ion mass spectrometry of quinones' *Biomed. Environ. Mass. Spectrom,* **18**, 878-883.
25. Lloyd, J. R., and Cotter, M. L. (1986) 'Sulfolane as a matrix for negative ion fast atom bombardment mass spectrometry' *Biomed. Environ. Mass. Spectrom. 13, 447.*

# STRUCTURAL ANALYSIS OF PHOSPHATIDYLINOSITOL FROM CARROT CELL MEMBRANES BY FAST ATOM BOMBARDMENT AND TANDEM MASS SPECTROMETRY

Richard B. van Breemen
*Department of Chemistry*
*North Carolina State University*
*Raleigh, North Carolina 27695-8204*

**ABSTRACT.** Inositol phospholipids from carrot cell membranes grown in suspension culture were partially purified by silicic acid column chromatography and then determined by negative ion fast atom bombardment mass spectrometry (FAB). Deprotonated molecules, $[M - H]^-$, were detected of the phosphatidylinositol (PI) and other phospholipids in the column effluent. Following FAB desorption, the structure of carrot PI was determined by using tandem mass spectrometry (MS/MS) with collisional activation. First, inositol phospholipids were differentiated from other phospholipids in the column effluent by using constant neutral loss MS/MS to identify which $[M - H]^-$ ions fragmented to eliminate neutral inositol (i.e., $[M - H - 162]^-$ ions). Next, collisional activation of the $[M - H]^-$ ion of PI at *m/z* 833 was combined with B/E linked scanning to obtain additional structural information such as the presence of the glycerol and fatty acid moieties, and additional confirmation of the inositol phosphate. Charge-remote fragmentation was used to determine the structure of each fatty acid group bound to PI. Finally, B/E linked scanning of $[M - H - \text{inositol}]^-$ ion at *m/z* 671 was used to determine the location of each fatty acid group on the glycerol chain of PI.

## 1. Introduction

Signal transduction in animal cells involves phosphatidylinositol (PI), phosphatidylinositol monophosphate (PIP) and phosphatidylinositol bisphosphate ($PIP_2$) as sources of second messengers [1, 2]. When an extracellular stimulus binds to a specific receptor in the plasma membrane of the cell, phospholipase C is activated to hydrolyze $PIP_2$ to form diacylglycerol and inositol-1,4,5-trisphosphate ($IP_3$). These hydrolysis products serve as second messengers of the original stimulus and initiate a cascade of metabolic events within the cell by increasing the concentration of cytosolic calcium. After the calcium concentration has been reduced to resting levels, further response to external stimuli depends upon regeneration of $PIP_2$ from PIP, which is formed from PI. Regulation of the formation of PI, PIP, and $PIP_2$ provides feedback control over the cellular responses to external stimuli. Although the presence and function of these and related inositol phospholipids in animal cells are established, studies are in progress to identify what phospholipids are present in plant membranes and to determine their functions in signal transduction [3-5].



*M. L. Gross (ed.), Mass Spectrometry in the Biological Sciences: A Tutorial*, 443–451.

The structures of plant phospholipids such as phosphatidylinositol (PI) can be determined by mass spectrometry [6, 7] by using only microgram quantities or less [8]. Therefore, mass spectrometry is particularly useful for lipid analysis when sample quantities are limited. Because tandem mass spectrometry (MS/MS) can serve as a sample purification step, MS/MS is also a valuable tool when impurities exist in the analyte. Because of the polarity of phospholipids, these compounds have required derivatization prior to analysis by gas chromatography-mass spectrometry [9]. Underivatized phospholipids have been analyzed by several desorption mass spectrometric techniques including $^{252}$Californium plasma desorption mass spectrometry [10], desorption chemical ionization [11], laser desorption [12], field desorption [13], and fast atom bombardment (FAB) mass spectrometry [6-8, 14, 15]. FAB mass spectrometry has been particularly useful for structural analysis of phospholipids because the sustained production of sample ions during FAB ionization has facilitated their subsequent collisional activation and MS/MS analysis [6, 8].

The use of a FAB matrix that is a good proton acceptor, such as triethanolamine, promotes the formation of negatively charged phosphate groups on phospholipids in the matrix solution. These deprotonated molecules are then desorbed during negative ion FAB and analyzed by the mass spectrometer [6, 8]. In the experiments described below, several MS/MS techniques will be used to determine the structure of a partially purified PI extracted from carrot cell membranes.

## 2. Materials And Methods

Carrot cells were grown in suspension culture according to previously published procedures [16], and 5 g were extracted according to the method of Boss and Massel [17]. A small amount of [$^3$H]inositol-labelled lipids were mixed into the lipid extract prior to loading onto a silicic acid column to serve as a tracer during chromatographic separation as described by van Breemen and coworkers [8]. During elution of the lipids from the silicic acid column, fractions were collected and assayed for radioactivity by scintillation counting. Fractions containing tritium were dried *in vacuo* and analyzed by FAB mass spectrometry. Soybean PI, bovine brain PI, and soybean lysophosphatidylinositols (LPI) were purchased from Sigma Chemical Company (St. Louis, MO) and used as reference standards for mass spectrometry.

Negative ion FAB mass spectra were obtained by using a JEOL (Tokyo, Japan) JMS-HX110HF double focusing mass spectrometer, equipped with a JMA-DA5000 data system, fast atom bombardment, collision cell in the first field-free region, and constant neutral loss and B/E linked scanning. In constant neutral loss MS/MS, fragment ions were detected that were formed by elimination of a preselected neutral molecule. B/E linked scanning is another type of MS/MS measurement in which fragment ions of a selected precursor ion are detected. In both types of MS/MS scans, the fragment ions that were detected were formed in a collision cell between the ionization source and the electrostatic sector (E) and magnetic sector (B).

For analysis by FAB mass spectrometry, a 1 μg/μL solution of each sample was prepared in chloroform, methanol, ammonia (15 M aq.), and water (90:90:7:22, v/v/v/v), and 1 μL was mixed with 1 μL of triethanolamine on the FAB probe tip. Xenon gas at 6 keV was used for FAB ionization. Calibration was carried out using sodium iodide in glycerol. The accelerating voltage was 10 kV, and the resolving power was approximately 1,000. During collisional activation, the helium gas pressure was adjusted so that the precursor ion abundance was attenuated 70%.

## 3. Results And Discussion

On the basis of co-elution with a radiolabelled standard during silicic acid column chromatography, the most abundant inositol phospholipid isolated from carrot cell membranes was phosphatidylinositol (PI). The negative ion FAB mass spectrum of the column fraction corresponding to carrot PI is shown in Figure 1. Because the PI fraction was contaminated by several other lipids, numerous ions were detected in the high mass region of the negative ion FAB mass spectrum (i.e., *m/z* 833, 766, 742, 720, 695, and 671 in Figure 1). To identify the $[M - H]^-$ ion of PI, constant neutral loss scanning was carried out on the mixture to detect those ions that fragmented to eliminate a neutral molecule of inositol, $[M - H - 162]^-$ according to the method of Heller and coworkers [18] (Figure 2). The highest mass ions that were detected in this constant neutral loss mass spectrum were observed at *m/z* 671, 433, and 409. The most abundant of these ions, *m/z* 671, was formed by the loss of inositol from the $[M - H]^-$ ion of PI at *m/z* 833 (Figure 2). The less abundant ions at *m/z* 433 and 409 in Figure 2 were formed by loss of inositol from the two lysophosphatidylinositols (LPI's) at *m/z* 595 and 571, respectively. LPI is identical to PI except that it contains one instead of two fatty acid groups. Therefore, the ion of *m/z* 833 in Figure 1 is the deprotonated molecule of carrot PI, and the ions of *m/z* 571 and 595 are the two related LPI molecules. The two LPI ions were either present in the PI sample following chromatography or were formed by fragmentation of PI in the ion source of the mass spectrometer.

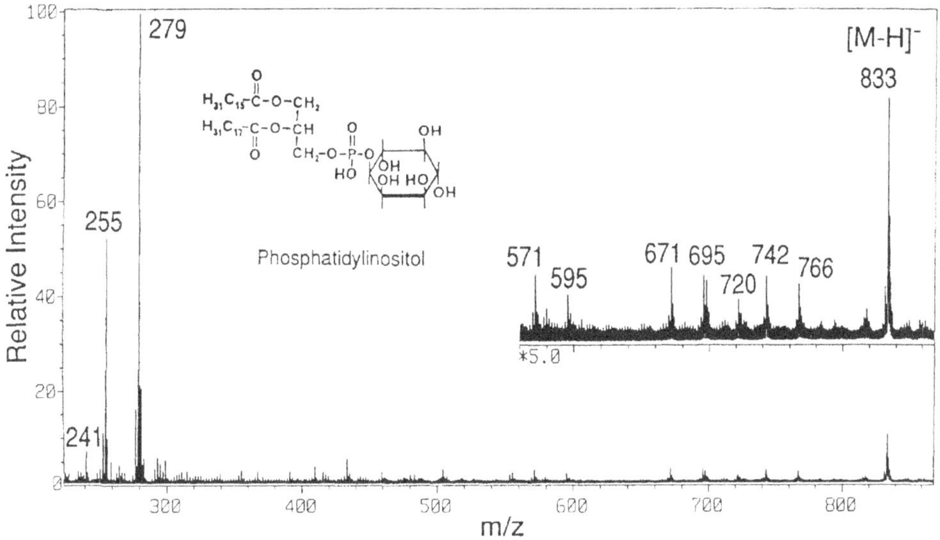


Figure 1. Negative ion FAB mass spectrum of palmitoyl linoleoyl phosphatidylinositol (PI) contained in an extract of carrot cell membranes that was partially purified by silicic acid column chromatography. The FAB matrix was triethanolamine. Republished from ref. 8 with permission.

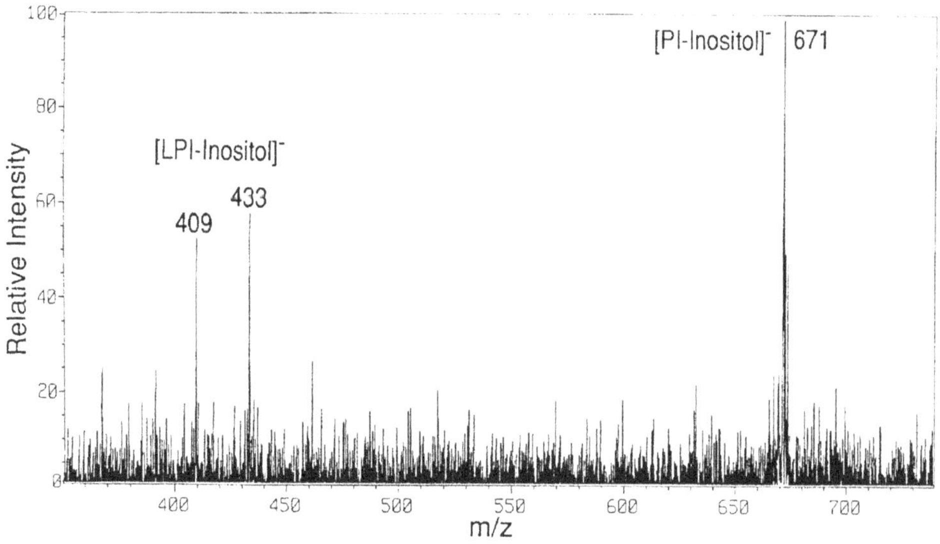


Figure 2. Negative ion constant neutral loss mass spectrum of the PI mixture analyzed in Figure 1 above. FAB ionization and collisional activation were used. In this analysis, ions were detected that had fragmented to eliminate neutral inositol, $[M - H - 162]^-$. Therefore, the precursors of the fragment ions at *m/z* 409, 433, and 671 could be assigned as the inositol phospholipids corresponding to deprotonated PI at *m/z* 833 and two deprotonated LPI molecules at *m/z* 595 and 571, respectively (see Figure 1).

Next, B/E linked scanning of the deprotonated molecule of phosphatidylinositol at *m/z* 833 was carried out to identify fragment ions that are specific to this compound without interference from matrix or other contaminating ions (Figure 3). In the B/E linked scan, abundant fragment ions that are the carboxylate anions palmitate and linoleate were detected at *m/z* 255 and *m/z* 279, and fragment ions formed by the elimination of linoleic acid and palmitic acid were observed at *m/z* 553 and *m/z* 577, respectively. (The assignments of these fatty acid groups as palmitic and linoleic acid were confirmed by charge-remote fragmentation as described below.) As observed in the constant neutral loss mass spectrum (Figure 2), loss of inositol from the $[M - H]^-$ ion of PI also formed phosphatidic acid at *m/z* 671 in the B/E linked scan shown in Figure 3. The presence of inositol was further confirmed by detection of dehydrated inositol phosphate at *m/z* 241. Dehydrated glycerophosphate, $[C_3H_6PO_5]^-$ was observed at *m/z* 153. Detection of dehydrated inositol phosphate and dehydrated glycerophosphate confirmed that this compound is a phospholipid.

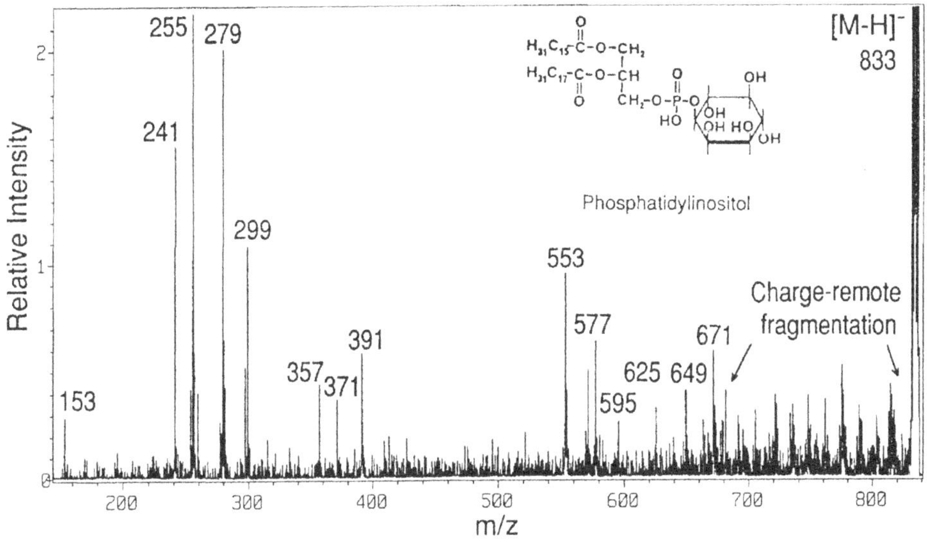


Figure 3. B/E linked scan of the deprotonated molecule of palmitoyl linoleoyl phosphatidylinositol at *m/z* 833 obtained using negative ion FAB and collisional activation. Republished from ref. 8 with permission.

In Figure 1, abundant fragment ions of PI were detected at *m/z* 255 and 279. Based on their molecular weights, these ions were tenatively identified as palmitate and linoleate, respectively. Because the ions of *m/z* 255 and 279 are also the most abundant fragment ions observed in the B/E linked scans of deprotonated PI, these ions probably are the palmitoyl and linoleoyl groups of PI instead of contaminants of the PI sample. The structures of these two fatty acids were determined by charge-remote fragmentation as described by Jensen and Gross for palmitate [7] and Jensen and coworkers for linoleate [6]. The palmitate ion at *m/z* 255 and then the linoleate ion at *m/z* 279 in the negative ion FAB mass spectrum of PI were collisionally activated, and the fragment ions were recorded by B/E linked scanning at unit resolution (Figure 4). The B/E linked scan of palmitate (Figure 4A) contained a series of fragment ions of the alkyl chain differing by 14 u (the mass of a methylene group). The B/E linked scan of linoleate (Figure 4B) contained a series of fragment ions that are interrupted by the presence of carbon-carbon double bonds. The location of these bonds can be confirmed by the change in pattern with respect to that of a saturated fatty acid and by the shift in molecular weights of fragment ions owing to the reduced number of hydrogens in the alkyl chain.

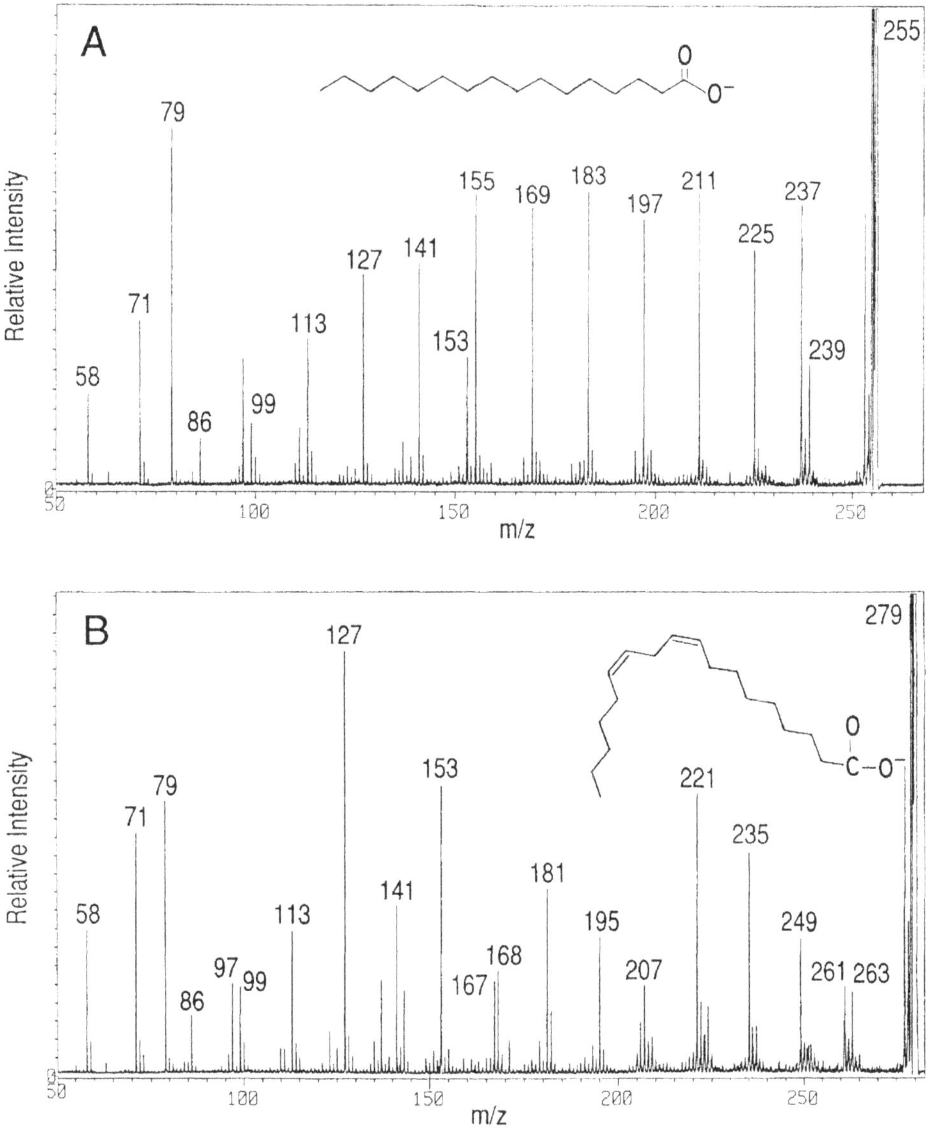


Figure 4. B/E linked scans of (A) palmitate at *m/z* 255 and (B) linoleate at *m/z* 279 by using collisional activation. The precursor ions were formed from PI in the ion source of the mass spectrometer during negative ion FAB.

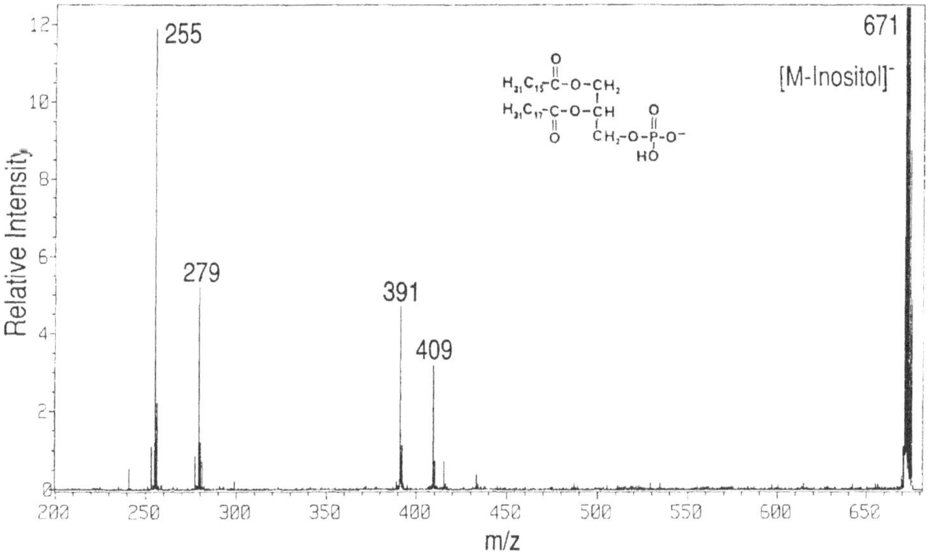


Figure 5. B/E linked scan of [M - $C_6H_{11}O_5$]$^-$ at *m/z* 671 showing a 2:1 ratio of *m/z* 255 (palmitate) to *m/z* 279 (linoleate). Based on comparison with standards, this ratio indicates that the palmitoyl group is in the *sn*-1 position and the linoleoyl group is located at the *sn*-2 position of PI.

B/E linked scanning was also used to determine the positions, *sn*-1 and *sn*-2, of each of the two fatty acids on the glycerol moiety of the phospholipid. The structure of PI shown in Figures 1 and 3 shows the palmitoyl group in the *sn*-1 position and linoleoyl group in the *sn*-2 position. In both plant and animal phospholipids, saturated fatty acids are found primarily in the *sn*-1 position whereas unsaturated fatty acids are found in the *sn*-2 position [19]. Jenson and Gross [7] assigned the positions of the two fatty acids in phosphatidylcholine by using B/E linked scanning, collisional activation, and positive ion FAB. Although our initial attempts to determine the site of attachment of each fatty acid in PI were unsuccessful using negative ion FAB with B/E linked scanning and collisional activation of [M - H]$^-$ ions, this approach was successful when phosphatidic acid ions, formed by loss of inositol in the ion source, were selected as the precursor ions (Figure 5) [8].

In the negative ion FAB mass spectrum of carrot PI (Figure 1), the fragment ion having the structure of palmitoyl linoleoyl phosphatidic acid (loss of inositol, [M - $C_6H_{11}O_5$]$^-$) was detected at *m/z* 671. B/E linked scanning of the *m/z* 671 ion following collisional activation produced two carboxylate fragments at *m/z* 255 and *m/z* 279, which are palmitate and linoleate (Figure 5). The relative abundance of the *m/z* 255 ion compared to *m/z* 279 was a ratio of approximately 2:1 (Figure 5). For comparison, collisional activation and B/E linked scanning of the *m/z* 671 ion from soybean palmitoyl linoleoyl phosphatidylinositol was analyzed. The B/E scan of the soybean standard was identical to that of the carrot PI. In both cases, the palmitate ion from the *sn*-1 position was approximately twice as abundant as the linoleate ion arising from the *sn*-2 position. The [M - $C_6H_{11}O_5$]$^-$ ion formed during FAB ionization of another standard, stearoyl arachidonoyl phosphatidylinositol from bovine brain, was also analyzed by this approach. Bovine brain PI contains stearic acid at the *sn*-1 postion and arachidonic acid at the *sn*-2 positon. As expected, the B/E linked scan of $[M - C_6H_{11}O_5]^-$ from bovine brain PI showed that carboxylate fragments arising from the *sn*-1 position were over two-fold more abundant than the carboxylate ions from the *sn*-2 position. On the basis of the data discussed above, MS/MS analysis of phosphatidic acid from inositol phospholipids provides carboxylate fragment ions corresponding to each fatty acid group in a ratio indicative of their positions, *sn*-1 or *sn*-2, on the phospholipid. Further examples of this relationship in B/E linked scans of phosphatidic acids from inositol phospholipids and other phospholipids are discussed in the chapters by Caldwell and Gross and by Gage and coworkers.

## 4. Conclusions

Because lipid contaminants are present in samples of inositol phospholipids obtained from carrot cell membranes, negative ion FAB mass spectra of these samples contained mixtures of ions of the expected inositol phospholipids as well as numerous contaminants. Therefore, the MS/MS technique of constant neutral loss scanning with collisional activation was used to identify those components of the mixtures that contained inositol groups by the appearance of ions formed by loss of inositol. Next, the MS/MS method of B/E linked scanning with collisional activation facilitated the identification of structurally significant fragment ions of the deprotonated molecule of phosphatidylinositol at *m/z* 833. For example, palmitoyl and linoleoyl groups were identified by their carboxylate ions at *m/z* 255 (palmitate) and *m/z* 279 (linoleate), and by the loss of palmitic acid and loss of linoleic acid to give the ions of *m/z* 577 and 553, respectively. Elimination of neutral inositol, $[M - C_6H_{11}O_5]^-$, was observed at *m/z* 671, and the complementary fragment ion corresponding to dehydrated inositol phosphate was detected at *m/z* 241. Dehydrated glycerophosphate was detected at *m/z* 153.

The identities of the two fatty acids present in carrot phosphatidylinositol were confirmed by charge-remote fragmentation. Specifically, B/E linked scanning following collisional activation of the fatty acid carboxylate ions at *m/z* 255 and 279 was used to determine the sites of unsaturation in linoleate and to confirm the straight chain alkyl group of the palmitate. B/E linked scanning of the $[M - C_6H_{11}O_5]^-$ ion at *m/z* 671 was used to assign palmitate to the *sn*-1 position and linoleate to the *sn*-2 position of PI from carrot cells.

Taken together, these MS/MS measurements provide structural confirmation of the presence of the inositol group, the phosphate group, the glycerol chain, and the identities and locations of the fatty acids within PI. Compared to standard FAB mass spectrometry, B/E linked scanning also helped to eliminate matrix ions and other chemical noise that can obscure sample ions and complicate the interpretation of fragmentation patterns. Chemical noise can be reduced further by carrying out these MS/MS measurements using a four sector mass spectrometer instead of a two sector instrument. Although no abundant interfering ions were detected in the MS/MS spectra reported here, contaminant ions arising from the matrix have been observed in other B/E linked scans using a two sector mass spectrometer [20]. The use of a two sector mass spectrometer also resulted in the selection of precursor ions at less than unit resolution. Therefore, contributions from ions at slightly higher and lower *m/z* values than that of the selected precursor ion were transmitted during B/E linked scanning. However, a two sector mass spectrometer costs considerably less to purchase and operate than four sector versions and therefore is more widely available.

## 5. Acknowledgements

This research was supported by the North Carolina Biotechnology Center. The carrot cell extracts were provided by Jeffery J. Wheeler and Wendy F. Boss of the Department of Botany, North Carolina State University.

## 6. References


1. Berridge, M.J., and Irvine, R.F. (1984) *Nature* **312**, 315.
2. Putney, Jr., J.W. (1986) in *Receptor Biochemistry and Methodology*, (Venter, J.C., and Harrison, L.C., eds.) Vol. 7, Alan R. Liss, New York.
3. Boss, W.F. (1989) in *Second Messengers in Plant Growth and Development* pp. 29-56, Alan R. Liss, New York.
4. Abdel-Latif, A.A., Smith, J.P., Akhtar, R.A. (1985) in *Inositol and Phosphoinositides: Metabolism and Regulation* (Bleasdale, J.E., Eichberg, J., and Hauser, G., eds.) pp. 275-298, Humana Press, Clifton, NJ.
5. Sommarin, M., and Sandelius, A.S. (1988) *Biochim. Biophys. Acta* **958**, 268-278.
6. Jensen, N.J., Tomer, K.B., and Gross, M.L. (1987) *Lipids* **22**, 480-489.
7. Jensen, N.J., and Gross, M.L. (1988) *Mass Spectrom. Rev.* **7**, 41-69.
8. van Breemen, R.B., Wheeler, J.J., and Boss, W.F. (1990) *Lipids* **25**, 328-334.
9. Sherman, W.R., Ackermann, K.E., Berger, R.A., Gish, B.G., and Zinbo, M. (1986) *Biomed. Environ. Mass Spectrom.* **13**, 333-341.
10. Demirev, P.A. (1987) *Biomed. Environ. Mass Spectrom.* **14**, 241-246.
11. Aberth, W., Straub, K., and Burlingame, A. (1982) *Anal. Chem.* **54**, 2029.
12. Cotter, R., and Tabet, J.-C. (1983) *Int. J. Mass Spec. Ion Proc.* **53**, 151.
13. Lehman, W.D., and Kessler, M. (1983) *Chem. and Phys. of Lipids* **32**, 123.
14. Munster, H., and Budzikiewicz, H. (1987) *Rap. Commun. Mass Spectrom.* **1**, 126-128.
15. Heller, D.N., Cotter, R.J., Fenselau, C., and Uy, O.M. (1987) *Anal. Chem.* **59**, 2806-2809.
16. Boss W.F., Brightman A.O., and Grimes, H.D. (1984) *Protoplasma* **120**, 125.
17. Boss, W.F., and Massel, M.O. (1985) *Biochem. Biophys. Res. Commun.* **132**, 1018-1023.
18. Heller, D.N., Murphy, C.M., Cotter, R.J., Fenselau, C., and Uy, O.M. (1988) *Anal. Chem.* **60**, 2787-2791.
19. Sumida, S., and Mudd, J.B. (1970) *Plant Physiol.* **45**, 712-718.
20. Martin, L.B., Schreiner, A.F., and van Breemen, R.B. (1991) *Anal. Biochem.* **193**, 6-15.

# INDEX